Design Automation for Quantum Computing

Prabhat Mishra
Editor

Design Automation for Quantum Computing

Springer

Editor
Prabhat Mishra
University of Florida
Gainesville, FL, USA

ISBN 978-3-032-09302-8 ISBN 978-3-032-09303-5 (eBook)
https://doi.org/10.1007/978-3-032-09303-5

This Springer imprint is published by the registered company Springer Nature Switzerland AG
The registered company address is: Gewerbestrasse 11, 6330 Cham, Switzerland

Preface

Despite technological advances, classical computers cannot efficiently solve several classes of important problems, such as optimization and logistics, quantum simulation, and data sampling. Quantum computers provide a promising alternative to solve various complex problems significantly faster than the classical computers. Quantum computing uses qubits, which can exist in a superposition of zero and one states, and can utilize entanglement to perform tasks efficiently. Quantum algorithms make use of these quantum phenomena to offer advantages over classical computing. For example, the performance of classical computers increases linearly with the number of bits (transistors). In contrast, the performance of quantum computers increases exponentially with the number of qubits. In order to utilize the potential of quantum computing, scientists and engineers need to solve various design automation challenges starting from specification to implementation and analysis.

This book provides a comprehensive overview of quantum computing from the perspective of computer scientists and engineers. The readers will get a clear understanding of various design automation steps, including specification of quantum algorithms, compiling algorithms to produce quantum circuits, synthesizing quantum circuits to map them to physical quantum computers, verification and debug, and error correction for noise-resilient quantum computing. This will enable computer scientists and engineers to design a reliable quantum computing framework to solve the challenges in the 21st century. The presentation of topics has been divided into six categories with each category focusing on a specific aspect of the big picture. A brief outline of the book is provided below.

1. **Introduction:** The first part of the book includes one introductory chapter on design automation and quantum computing.
    - The first chapter provides an overview of quantum computing and outlines various design automation challenges.
2. **Quantum Algorithms and Architectures:** The second part of the book focuses on quantum architectures and algorithms.

- The second chapter provides an overview of quantum computing architectures.
- The third chapter describes state preparation for variational quantum algorithms.
- The fourth chapter outlines various quantum machine learning models.

3. **Quantum Compilation and Error Correction:** The third part of the book deals with quantum compilation and error correction methods.

   - The fifth chapter describes quantum compilation methods to transform quantum algorithms into quantum circuits as well as mapping them to quantum computers.
   - The sixth chapter provides an overview of quantum error correction techniques.

4. **Quantum Verification and Debug:** The fourth part of the book looks at quantum verification and debug methods.

   - The seventh chapter provides an overview of quantum verification paradigms that ensure fidelity and functional correctness in quantum systems.
   - The eighth chapter presents various techniques for quantum circuit simulation.
   - The ninth chapter describes quantum testing and validation methods.
   - The tenth chapter explores automated debugging of quantum systems.

5. **Quantum Security and Post-quantum Cryptography:** The fifth part of the book covers quantum security as well as post-quantum cryptographic solutions.

   - The eleventh chapter explores security attacks abusing pulse-level quantum circuits.
   - The twelfth chapter presents antivirus for quantum computers.
   - The thirteenth chapter describes trusted execution environments for quantum computers.
   - The fourteenth chapter explores post-quantum cryptographic solutions.

6. **Conclusions and Future Directions:** The last part concludes the book with a summary and discussion on future directions.

   - The last chapter concludes the book with an executive summary as well as discussion on designing fault-tolerant quantum systems.

We hope you enjoy reading this book and find the information useful for developing automated tools and techniques for design automation of quantum systems.

Gainesville, FL, USA
February 2026

Prabhat Mishra

Acknowledgments

This book would not be possible without the contributions of many researchers and experts in the field of design automation and quantum computing. I would like to gratefully acknowledge the contributions from the following authors who have contributed book chapters:

- Abhoy Kole, University of Bremen, Germany
- Ameya Bhave, University of Texas at Dallas, USA
- Anton Maidl, German Aerospace Center, Germany
- Archisman Ghosh, Pennsylvania State University, USA
- Aruna Jayasena, University of Tennessee, USA
- Avimitra Chatterjee, Pennsylvania State University, USA
- Bardia Taghavi, Florida Atlantic University, USA
- Chuanqi Xu, Yale University, USA
- Daniel Schönberger, Technical University of Munich, Germany
- Emma Andrews, University of Florida, USA
- Ferhat Erata, Amazon, USA
- Hanrui Wang, Eigen AI, USA
- Hansika Weerasena, University of Florida, USA
- Jakub Szefer, Northwestern University, USA
- Kanad Basu, Rensselaer Polytechnic Institute, USA
- Ludwig Schmid, Technical University of Munich, Germany
- Lukas Burgholzer, Technical University of Munich, Germany
- Masahiro Fujita, University of Tokyo, Japan
- Mehran Mozaffari Kermani, University of South Florida, USA
- Merve Karabulut, Florida Atlantic University, USA
- Navnil Choudhury, Rensselaer Polytechnic Institute, USA
- Reza Azarderakhsh, Florida Atlantic University, USA
- Robert Wille, Technical University of Munich, Germany
- Sahan Sanjaya, University of Florida, USA
- Sanjay Deshpande, Northwestern University, USA
- Song Han, Massachusetts Institute of Technology, USA

- Swaroop Ghosh, Pennsylvania State University, USA
- Theodoros Trochatos, Yale University, USA
- Yannick Stade, Technical University of Munich, Germany
- Yao Lu, Yale University, USA
- Yongshan Ding, Yale University, USA
- Yusuke Kimura, Fujitsu, Japan

Any opinions, findings, or conclusions presented in this book are those of the authors.

Contents

Introduction

Design Automation Challenges in Quantum Computing

Prabhat Mishra

Abstract Quantum computers offer the ability to solve specific simulation and optimization problems significantly faster than the classical computers. In order to utilize the potential of quantum computing, scientists and engineers need to solve various design automation challenges starting from specification to implementation and analysis. In this chapter, we first look at the basics of quantum computing. Next, we discuss design automation challenges in developing quantum systems.

1 Introduction

Classical computers utilize hardware and software to solve a wide variety of problems. Despite technological advances in modern electronic systems, there are many real-world problems that are hard even for classical computers. For example, a classical computer may take too long (computationally intensive) or produce low-accuracy results for factoring large numbers, simulating the atomic-level intricacies of advanced materials, modeling molecular dynamics and chemical reactions, or solving complex optimization problems. Interestingly, quantum computers can efficiently solve these complex problems.

Quantum computing is advancing rapidly over the years with the potential to significantly outperform classical computing in solving specific classes of simulation and optimization problems. Figure 1 illustrates four fundamental differences between classical and quantum computing, and is discussed below:

1. While classical computers perform calculations using Boolean values (bits), which can be either "0" (false) or "1" (true), quantum computers utilize qubits, which can exist in a superposition of $|0\rangle$ and $|1\rangle$ at the same time.
2. As discussed in Sect. 2.4, a quantum computer can potentially solve any problem that a classical computer can solve. However, quantum and classical computers

P. Mishra (✉)
University of Florida, Gainesville, Florida, USA
e-mail: prabhat@ufl.edu

P. Mishra (ed.), *Design Automation for Quantum Computing*,
https://doi.org/10.1007/978-3-032-09303-5_1

	Classical Computing	Quantum Computing
Units	Calculates with transistors (**bits**), which can be on (0) or off (1)	Calculates with **qubits**, which can represent 0 and 1 at the same time
Tasks	Suitable for solving everyday activities and applications	Suitable for atomic simulations and complex optimization problems.
Ability	Computation ability increases linearly with the no. of transistors	Performance increase is exponential with the number of qubits
Noise	Classical computers have low error rates and can operate in room temp.	Quantum computers have high error rates and need to be kept ultracold

Fig. 1 Classical versus quantum computing

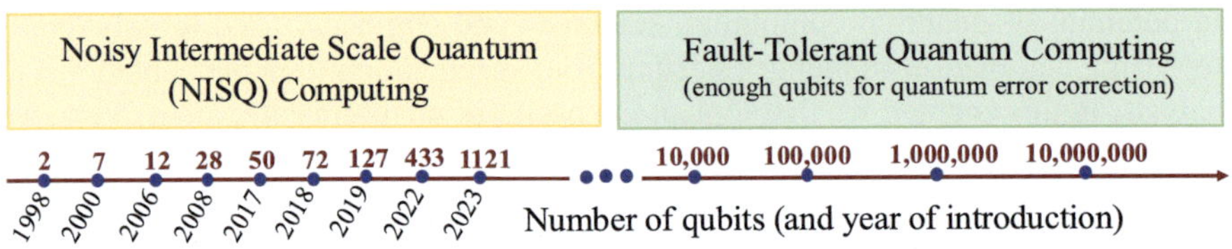

Fig. 2 Landscape of quantum computing. The quantum computers today with a small number of qubits (NISQ) can solve problems while maintaining low error probability. The future quantum computers with a large number of qubits can perform fault-tolerant computing utilizing quantum error correction

are suitable for solving different classes of problems. For example, classical computers can efficiently solve everyday tasks, while quantum computers are ideal for solving bounded-error quantum polynomial time algorithms.

3. The performance of quantum computers scales significantly faster than their classical counterpart. For example, the performance of classical computers increases linearly with the number of bits (transistors). In contrast, the performance of quantum computers increases exponentially with the number of qubits. Specifically, quantum computers have the potential to transform all 2^n possible states by using n entangled qubits.
4. Noise (error rate) is a major challenge in designing reliable quantum computers. While classical computers have low error rates and can operate in room temperature, quantum computers have very high error rates and need to be kept ultracold. In fact, noise analysis and mitigation strategies are vital in designing reliable quantum computers.

Figure 2 shows the landscape of general-purpose quantum computing in terms of qubits. It all started in 1998 with a two-qubit quantum computer. Twenty-five years later, IBM Condor features 1121 qubits. While there are application-specific quantum computers featuring more qubits, we do not show them in the figure. For example, D-Wave quantum computer has more than 5000 qubits that is specifically designed as quantum annealer. We are today in the NISQ (noisy intermediate-scale quantum)

Fig. 3 Google's Sycamore quantum computer [4]. The quantum computer is connected to classical computers as well as other hardware devices to enable classical-quantum computing. While the quantum computer handles the quantum computing tasks, the classical devices perform various activities, including quantum state preparation, pulse generation for mapping of quantum circuits, and analysis of results

era where quantum noise is a major concern. There are classes of algorithms in quantum machine learning and quantum communication that can be performed in NISQ systems, while maintaining low error probability and superior performance than classical computation. For example, the Quantum Approximate Optimization Algorithm (QAOA) [1] and Variational Quantum Eigensolver (VQE) [2, 3] leverage both classical and quantum hardware. Due to the limited number of qubits, error-corrected quantum computation at scale is not feasible using NISQ systems. In the future, we will be able to perform fault-tolerant quantum computing with thousands to millions of qubits.

The construction of quantum devices necessitates intricate engineering to harness and manipulate quantum mechanical effects. In other words, while the quantum computers deal with the quantum computing task, they utilize support from classical computing devices, such as Central Processing Unit (CPU), Graphics Processing Unit (GPU), and Field Programmable Gate Array (FPGA). Figure 3 shows that the Google Sycamore quantum computer is connected to classical computing devices to provide various support required for quantum computing, such as preparation of the initial state of the quantum computer, mapping of the required functionality, measurement of the result, and providing feedback, if needed.

The remainder of this chapter is organized as follows. We first introduce the readers to the basics of quantum computing. Next, we discuss various design automation stages and associated challenges in designing quantum computing systems.

2 Introduction to Quantum Computing

This section describes the basic building blocks for designing quantum systems. We first describe the basic unit of quantum computing (qubit). Next, we discuss various attributes, including superposition and entanglement. We then compare how quantum gates and circuits differ from their classical counterparts. Finally, we briefly describe quantum algorithms and architectures.

2.1 *Qubits and Superposition*

Any computation deals with data processing that accepts some input data, performs the computations, and produces the result data. In classical computation, the data is represented using bits, which represent Boolean values (0 or 1). In case of quantum computation, the basic unit is "quantum bit," referred to as qubit. While classical bits can be represented by an electrical switch (on or off), qubits are generally represented by the spin of electrons or the polarization of photons. In quantum computing, the basis states are represented as $|0\rangle$ and $|1\rangle$. Unlike classical bits, a qubit represents both $|0\rangle$ and $|1\rangle$ at the same time. Note that qubit supports $|0\rangle$, $|1\rangle$, as well as a superposition of them. Specifically, a qubit is a two-level quantum system described by a linear combination of the computational basis states:

$$|\psi\rangle = \alpha\,|0\rangle + \beta\,|1\rangle\,, \quad \text{where } \alpha, \beta \in \mathbb{C}, \; |\alpha|^2 + |\beta|^2 = 1.$$

Unlike a classical bit, which can be either 0 or 1, a qubit can exist in a superposition of both states simultaneously. The complex amplitudes α and β determine the probability of observing each basis state upon measurement: the qubit collapses to $|0\rangle$ with probability $|\alpha|^2$, and to $|1\rangle$ with probability $|\beta|^2$. In vector notation, a qubit state can be expressed as

$$|\psi\rangle = \begin{bmatrix} \alpha \\ \beta \end{bmatrix},$$

where the standard basis states are represented as

$$|0\rangle = \begin{bmatrix} 1 \\ 0 \end{bmatrix}, \quad |1\rangle = \begin{bmatrix} 0 \\ 1 \end{bmatrix}.$$

This representation is widely used in quantum computing to perform mathematical operations on qubit states using linear algebra.

2.2 *Quantum Entanglement*

Quantum entanglement is a phenomenon where the states of two or more qubits become correlated in such a way that if we measure one entangled qubit, we know the state of the other qubit, even if they are physically separated. For example in case of the Bell circuit shown in Fig. 10b, if the first qubit is measured and found to be in the state $|0\rangle$, the second qubit will also collapse to $|0\rangle$. Similarly, if the first qubit is measured and found to be in the state $|1\rangle$, the second qubit will also collapse to $|1\rangle$. This form of non-local correlation defies classical intuition and has implications for quantum communication, teleportation, and cryptographic protocols.

An important consideration in quantum computing is how to entangle qubits. We can perform entanglement based on photons as well as electrons. When a laser beam sends photons through a special crystal, most of the photons goes through the crystal. However, a few photons gets split into two photons. Since the energy and momentum must be conserved, the state describing the polarization of the two photons is entangled. Similarly, entangled electrons can occur in semiconductors and these electrons have been used in many experiments. Note that it is difficult to move entangled electrons far apart since they have a tendency to interact with the environment. Although entangled photons are easier to separate, they are difficult to measure to be useful in experiments. Loophole-free Bell test combines the advantages of these approaches [5].

From engineering perspective, we can simply use a CNOT gate that accepts normal (unentangled) qubits as inputs to produce entangled qubits. Although CNOT and other quantum gates will be described in Sect. 2.3, a CNOT gate can be described using the following orthogonal matrix. The 4×4 matrix can be constructed by taking standard basis for $\mathbb{R}^4$ and interchanging the order of the last two vectors.

$$\begin{bmatrix} 1 & 0 & 0 & 0 \\ 0 & 1 & 0 & 0 \\ 0 & 0 & 0 & 1 \\ 0 & 0 & 1 & 0 \end{bmatrix}.$$

The CNOT gate accepts two unentangled qubits. It produces two entangled qubits that can be obtained by multiplying the matrix.

$$\begin{bmatrix} 1 & 0 & 0 & 0 \\ 0 & 1 & 0 & 0 \\ 0 & 0 & 0 & 1 \\ 0 & 0 & 1 & 0 \end{bmatrix} \begin{bmatrix} \frac{1}{\sqrt{2}} \\ 0 \\ \frac{1}{\sqrt{2}} \\ 0 \end{bmatrix} = \begin{bmatrix} \frac{1}{\sqrt{2}} \\ 0 \\ 0 \\ \frac{1}{\sqrt{2}} \end{bmatrix} = \frac{1}{\sqrt{2}} \begin{bmatrix} 1 \\ 0 \\ 0 \\ 1 \end{bmatrix} \Rightarrow \frac{1}{\sqrt{2}} \begin{bmatrix} 1 \\ 0 \end{bmatrix} \otimes \begin{bmatrix} 1 \\ 0 \end{bmatrix} + \frac{1}{\sqrt{2}} \begin{bmatrix} 1 \\ 0 \end{bmatrix} \otimes \begin{bmatrix} 1 \\ 0 \end{bmatrix}.$$

2.3 Quantum Gates and Circuits

Before we delve into quantum gates, let us briefly review three classical gates (AND, OR, NOT) and their functionality. Figure 4 shows the pictorial representation as well as truth tables of these gates. For example, an AND gate accepts two inputs and produces one output. Both input and output carry Boolean values, either true (1) or false (0). Based on the Boolean logic, the output of the AND gate is true (1) if both inputs are true (1), false (0) otherwise. Similarly, an OR gate output is false (0) if both inputs are false (0), true (1) otherwise. A NOT gate produces true (1), if the input is false (0), false (0) otherwise.

Let us discuss three reversible gates (CNOT, Toffoli, and Fredkin) from classical viewpoint using Boolean values. The CNOT gate is a controlled NOT gate that accepts two inputs and produces two outputs. The first input is used as a control bit and becomes the first output. If the first input is true (1), the second input gets inverted (acts as a NOT operation) and goes to the second output. If the first input is false (0), the second input passes through the gate to the second output. Therefore, the functionality of a CNOT gate with input values x and y can be viewed as $f_{CNOT}(x, y) = (x, x \oplus y)$. Figure 5 shows the pictorial representation as well as the truth tables of the CNOT gate.

The Toffoli gate accepts three inputs and produces three outputs. The first two inputs are used as a control bits. If the first two inputs are true (1), the third input gets inverted (acts as a NOT operation), otherwise it remains the same. This gate is also referred to as controlled CNOT (CCNOT) gate. Therefore, the functionality of a Toffoli gate with input values x, y, and z can be viewed as $f_{CCNOT}(x, y, z) =$

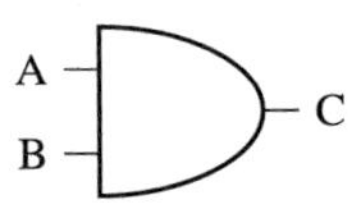

(a) Symbol of the AND gate

A	B	C
0	0	0
0	1	0
1	0	0
1	1	1

(b) Truth table of the AND gate

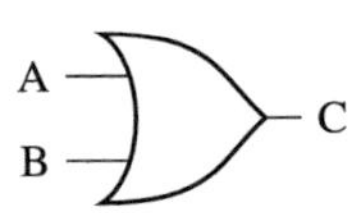

(c) Symbol of the OR gate

A	B	C
0	0	0
0	1	1
1	0	1
1	1	1

(d) Truth table of the OR gate

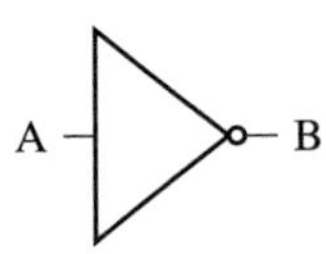

(e) Symbol of the NOT gate

A	B
0	1
1	0

(f) Truth table of the NOT gate

Fig. 4 Three classical gates and their functionality

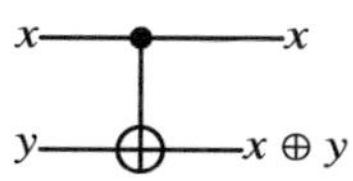

(a) Representation of the CNOT gate

x	y	x	x⊕y
0	0	0	0
0	1	0	1
1	0	1	1
1	1	1	0

(b) Truth table of the CNOT gate

Fig. 5 Representation and functionality of the CNOT gate using Boolean values

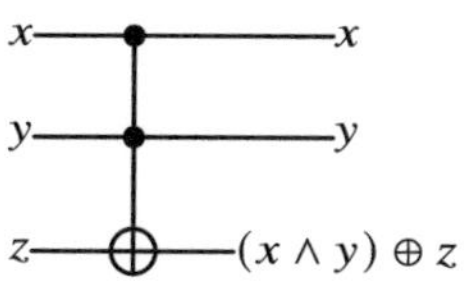

(a) Representation of the Toffoli gate

x	y	z	x	y	(x∧y)⊕z
0	0	0	0	0	0
0	0	1	0	0	1
0	1	0	0	1	0
0	1	1	0	1	1
1	0	0	1	0	0
1	0	1	1	0	1
1	1	0	1	1	1
1	1	1	1	1	0

(b) Truth table of the Toffoli gate

Fig. 6 Representation and functionality of the Tofolli gate using Boolean values

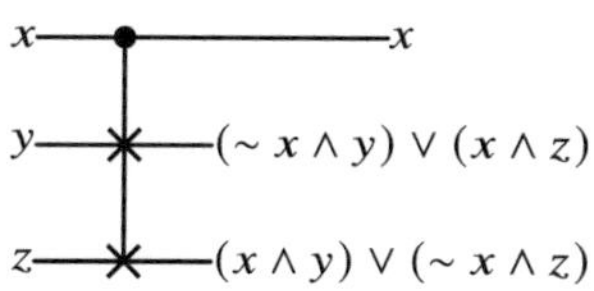

(a) Representation of the Fredkin gate

x	y	z	x	(∼x∧y)∨(x∧z)	(x∧y)∨(∼x∧z)
0	0	0	0	0	0
0	0	1	0	0	1
0	1	0	0	1	0
0	1	1	0	1	1
1	0	0	1	0	0
1	0	1	1	1	0
1	1	0	1	0	1
1	1	1	1	1	1

(b) Truth table of the Fredkin gate

Fig. 7 Representation and functionality of the Fredkin gate using Boolean values

$(x, y, (x \wedge y) \oplus z)$. Figure 6 shows the pictorial representation as well as the truth tables of the Toffoli gate.

The Fredkin gate accepts three inputs and produces three outputs. The first input is used as a control bit and becomes the first output. If the first input is false (0), the second and third inputs remains the same. If the first input is true (1), it swaps the second and third inputs. This gate is also referred to as controlled swap (CSWAP) gate. Therefore, the functionality of a Fredkin gate with input values x, y, and z can be viewed as: when x is 0 $f_{CSWAP}(0, y, z) = (0, y, z)$ and when x is 1, $f_{CSWAP}(1, y, z) = (1, z, y)$. Figure 7 shows the pictorial representation as well as the truth tables of the Fredkin gate.

Let us look at the quantum CNOT gate that can produce entangled outputs. Figure 8 shows the diagram and functionality of the quantum CNOT gate. Although

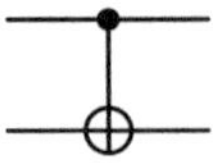

(a) Symbol

x	y	x	x⊕y
$\|0\rangle$	$\|0\rangle$	$\|0\rangle$	$\|0\rangle$
$\|0\rangle$	$\|1\rangle$	$\|0\rangle$	$\|1\rangle$
$\|1\rangle$	$\|0\rangle$	$\|1\rangle$	$\|1\rangle$
$\|1\rangle$	$\|1\rangle$	$\|1\rangle$	$\|0\rangle$

(b) Functionality

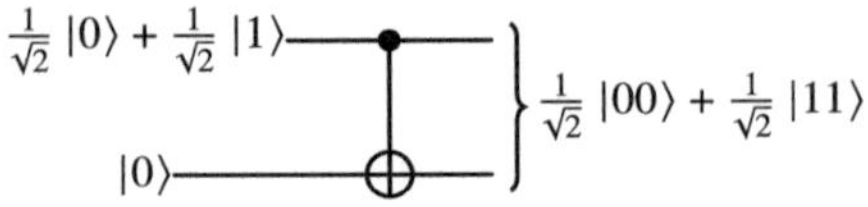

(c) Entangled qubits

Fig. 8 Representation and functionality of the quantum CNOT gate

Fig. 5a (classical/reversible CNOT) looks the same as Fig. 8a (quantum CNOT), there are two major differences. Unlike classical CNOT that operates on Boolean values (0 and 1), x and y in quantum CNOT can have any qubit values. Figure 8b shows the truth table using $|0\rangle$ and $|1\rangle$ instead of Boolean values. The functionality of the quantum CNOT gate is different from classical CNOT gate. For example, in case of classical CNOT gate, the first input becomes the first output (top wire carries the input to the output). This is true for quantum CNOT only if the first input is either $|0\rangle$ or $|1\rangle$; however, it is not true for other qubit values. For example, if we provide $\frac{1}{\sqrt{2}}|0\rangle + \frac{1}{\sqrt{2}}|1\rangle$ as the first input and $|0\rangle$ as the second input, it produces the entangled output $\frac{1}{\sqrt{2}}|00\rangle + \frac{1}{\sqrt{2}}|11\rangle$, as shown in Fig. 8c. We cannot assign different values to the top and bottom qubits. These two qubits are entangled, and a measurement on one will affect the state of the other.

Hadamard gate is another important quantum gate that is often used to put the standard basis vectors into superpositions. For example, if we apply $|0\rangle$ to the input of the Hadamard gate, it will produce $\frac{1}{\sqrt{2}}|0\rangle + \frac{1}{\sqrt{2}}|1\rangle$ (based on Fig. 9b).

We can connect a set of quantum gates to construct quantum circuits. For example, let us connect a Hadamard gate and a CNOT gate to construct the Bell circuit (shown in Fig. 10a), which is used to create Bell states. Since it has two input qubits, we can apply four possible qubit pairs that form the standard basis: $(|0\rangle, |0\rangle)$, $(|0\rangle, |1\rangle)$, $(|1\rangle, |0\rangle)$, and $(|1\rangle, |1\rangle)$. Figure 10b shows the scenario when we apply $|0\rangle$ to both inputs. As discussed earlier, the Hadamard gate with input $|0\rangle$ will produce $\frac{1}{\sqrt{2}}|0\rangle + \frac{1}{\sqrt{2}}|1\rangle$. Since the other qubit is $|0\rangle$, the state of the system of two qubits is $(\frac{1}{\sqrt{2}}|0\rangle + \frac{1}{\sqrt{2}}|1\rangle)\,|0\rangle = \frac{1}{\sqrt{2}}|00\rangle + \frac{1}{\sqrt{2}}|10\rangle$. We need to apply CNOT gate to this state, which will flip $|10\rangle$ to $|11\rangle$ (based on the truth table in Fig. 8b). This will lead to the final state $\frac{1}{\sqrt{2}}|00\rangle + \frac{1}{\sqrt{2}}|11\rangle$, as shown in Fig. 10b. Similarly, we can compute the output of the Bell circuit for the other input pairs, as shown in Fig. 10.

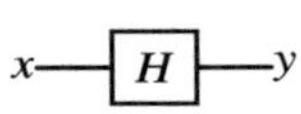

(a) Symbol

x	y
$\|0\rangle$	$\frac{1}{\sqrt{2}}\|0\rangle + \frac{1}{\sqrt{2}}\|1\rangle$
$\|1\rangle$	$\frac{1}{\sqrt{2}}\|0\rangle - \frac{1}{\sqrt{2}}\|1\rangle$

(b) Truth table

$$H = \begin{bmatrix} \frac{1}{\sqrt{2}} & \frac{1}{\sqrt{2}} \\ \frac{1}{\sqrt{2}} & -\frac{1}{\sqrt{2}} \end{bmatrix} = \frac{1}{\sqrt{2}}\begin{bmatrix} 1 & 1 \\ 1 & -1 \end{bmatrix}$$

(c) Superposition

Fig. 9 Representation and functionality of the quantum Hadamard gate

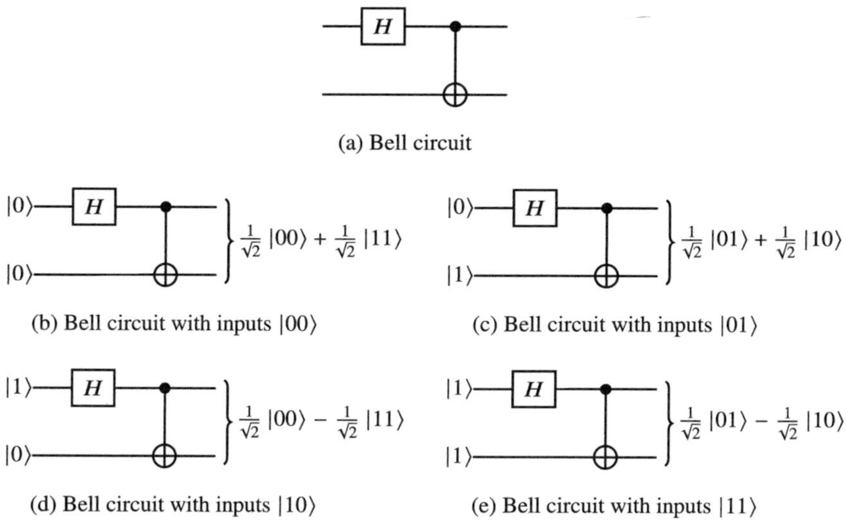

(a) Bell circuit

(b) Bell circuit with inputs $|00\rangle$

(c) Bell circuit with inputs $|01\rangle$

(d) Bell circuit with inputs $|10\rangle$

(e) Bell circuit with inputs $|11\rangle$

Fig. 10 Representation and functionality of the Bell circuit

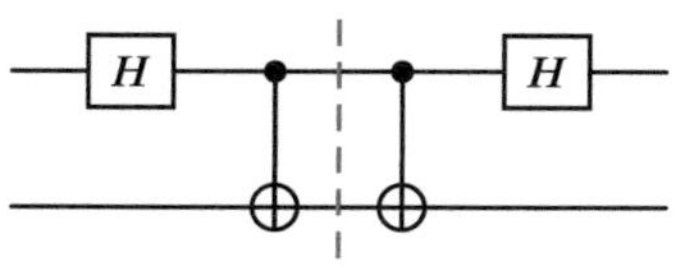

Fig. 11 This quantum circuit maintains the original state. The second half (inverse Bell circuit) undoes the actions of the first half (Bell circuit)

It is clear to see that the Bell circuit produces entangled outputs in all four scenarios. Both CNOT and Hadamard gates have an interesting property: if we apply the same gate twice, the second gate will undo the computation by the first gate. Therefore, if we construct the circuit in Fig. 11, it does not change anything. The inverse Bell circuit (second half) undoes the action of the Bell circuit (first half). In other words, if we provide inputs to the inverse Bell circuit from the Bell basis, it will produce the vectors in the standard basis. These interesting properties of the Bell circuit has applications in various domains, including superdense cooling and quantum teleportation.

2.4 *Quantum Algorithms*

Let us revisit our discussion about quantum computation versus classical computation. If we use only $|0\rangle$ and $|1\rangle$ for quantum computation (no superpositions) and if it is noise-free, performing quantum computation using quantum CNOT gate is identical to running classical computation using classical CNOT gate. Similarly, performing quantum computation using quantum Fredkin gate is identical to running classical computation using classical Fredkin gate if we use only $|0\rangle$ and $|1\rangle$ for quantum computation. Since the classical Fredkin gate is a universal gate and quantum Fredkin gate is more powerful than its classical counterpart, we can construct a quantum circuit to perform any computation that is possible by classical computation. In other words, a quantum computer can potentially solve any problem that can be solved using classical computers. In reality, classical computers can solve everyday applications more efficiently than quantum computers. However, there are various optimization and simulation scenarios where quantum computers can significantly outperform their classical counterpart. Let us discuss the specific application scenarios and the underlying quantum algorithms where quantum computers can outperform classical computers.

Quantum algorithms have the inherent speedup advantage over its classical counterpart since classical algorithms can operate only on two values (0 and 1) while quantum algorithms can operate on the superposition of all possible inputs. As a result, unlike classical systems that run on one input at a time, quantum systems can run on all possible inputs at the same time. In reality, this can lead to a lot of results, and most of them can be unexpected answers. Therefore, the development of quantum algorithms cannot simply rely on the superposition of states. Moreover, quantum hardware can be very noisy, requiring multiple iterations of the same computation or utilization of quantum error corrections methods, which can further reduce the quantum advantage. In this section, we focus on several quantum algorithms. We will consider the effect of noise and error correction in later sections.

Exact simulation of quantum systems, such as modeling and simulation of interactions of atoms and configuration of electrons in chemistry, is hard for classical computers. In practice, scientists use approximation methods or heuristics to find beneficial results. This may be good enough in some applications; however, the exact solution may be needed in many safety-critical scenarios, such as drug discovery or treatment. Quantum computers are suitable for exact simulation of quantum systems. Similarly, quantum annealing is very efficient for solving certain optimization problems compared to running simulated annealing on classical computes. It can find the best solution (e.g., global minimum) significantly faster utilizing inherent quantum features, such as quantum tunneling. Quantum Fourier Transform (QFT) is another popular quantum algorithm, and, in fact, it is a building block for many quantum algorithms in the literature. QFT algorithm transforms quantum states to encode periodic patterns or phases. Quantum spectral clustering can produce beneficial clusters in linear time [6]. Grover's search algorithm provides quadratic improvement in searching in unstructured database compared to classical search algorithms [7].

Quantum machine learning algorithms leverage the principles of quantum mechanics to accelerate machine learning tasks, often implementing classical machine learning architectures as quantum algorithms. These quantum machine learning algorithms can be broadly divided into two categories: (i) hybrid approaches with classical and quantum components and (ii) pure quantum approaches. The algorithms in the first category combine classical and quantum computing resources, where certain parts of the computation are handled by a quantum computer while others are handled by a classical computer. There are a wide variety of classical-quantum hybrid algorithms to utilize both classical and quantum neural networks and to solve combinatorial optimization problems, such as Quantum Approximate Optimization Algorithm (QAOA) and Variational Quantum Eigensolver (VQE). A common feature across these variational quantum algorithms is that the quantum circuit prepares a state in the quantum computer, and a classical optimizer then updates the parameters of the quantum circuit [8]. The algorithms in the second category are designed specifically to run entirely on quantum hardware, such as quantum reinforcement learning.

Shor's algorithm is a well-studied quantum algorithm since it demonstrated the ability of a quantum computer to perform factoring of large numbers in polynomial time [9]. This is significant because many cryptographic algorithms (e.g., RSA encryption) are considered secure with the assumption that it is easy to multiply two large prime numbers to produce the result, but it is infeasible for classical computer to find the prime factors from the result in a reasonable time. In reality, it is not possible to perform factoring using today's quantum computers when large prime numbers are used due to limited number of reliable qubits. However, the fear that quantum computers in the future would be able to break encryption methods has led to significant investment from government agencies to develop post-quantum cryptographic algorithms. The basic idea is that even if we have a quantum computer with enough reliable qubits in the future, they will not be able to break these post-quantum cryptographic algorithms which rely on innovative solutions (e.g., lattice-based cryptography instead of multiplication of prime numbers).

2.5 *Quantum Hardware Architectures*

To prove that the quantum algorithms can outperform their classical counterpart on specific applications, we need to perform two activities: (1) transform these quantum algorithms into quantum circuits and (2) execute these quantum circuits using either quantum simulators or quantum computers. There are popular quantum simulators, such as IBM Qiskit, Google Cirq, and Microsoft Quantum Development Kit. In this section, we focus on hardware architectures for quantum computers.

There are various obstacles in building quantum computers. For example, decoherence is a major problem that results from the fact that a qubit interacts with the environment that is not part of the computation [10, 11]. To ensure reliable quantum computing, we should be able to initialize a qubit to a specific state and expect that the qubit will maintain that state until we need to use the qubit to form gates or circuits and execute them. Photons are more suitable for quantum communication than quantum computation. This is due to the fact that photons are easy to initialize and they are not prone to interact with the environment. However, photons are difficult to store so that we can use them when needed during the computation. On the other hand, scalability is a major issue in implementing qubits using electron

Quantum Technology	Providers	Applicability
Superconducting Qubits	IBM, Google, Rigetti, etc.	General-Purpose
Topological Qubits	Microsoft	General-Purpose
Trapped Ions	IonQ, Quantinuum, etc.	General-Purpose
Photonic Devices	Xanadu, PsiQuantum, etc.	General-Purpose
Quantum Annealing	D-Wave	Application-Specific

Fig. 12 Quantum computers based on different technologies

spins. Trapped ions utilize energy levels of ions to implement qubits and they can be manipulated with lasers.

One way to minimize the interactions with the environment is to shield quantum computers from external light and heat. In extreme cold temperatures, superconductors have quantum properties that are exploited to design superconducting qubits. For example, IBM has developed quantum computers with increasing number of superconducting qubits over the years [12]. Topological qubits offer a promising alternative to avoid environmental interactions. Topological qubits rely on the topological properties of matter for their stability and error correction. Unlike traditional qubits that store information in the properties of individual particles, topological qubits encode information in the arrangement and interaction of particles. This approach offers inherent protection against environmental interactions and decoherence, making them suitable for designing robust and scalable quantum computers. For example, Microsoft developed a quantum chip that can observe and control Majorana particles to produce more reliable and scalable qubits [13].

Figure 12 shows that the modern quantum computers can be divided into two broad categories: application-specific and general-purpose. For example, the quantum computers developed by D-Wave are suitable for solving optimization problems using quantum annealing. On the other hand, the general-purpose quantum computers support universal gate-based quantum circuits, and therefore they can be used for solving any quantum algorithms. The general-purpose quantum computers can be further categorized based on the underlying technology: superconducting qubits, trapped ions, topological qubits, and photonic devices.

3 Design Automation Challenges

Increasing complexity of quantum algorithms coupled with the diversity of quantum technologies makes a critical need to develop automated methods to execute quantum algorithms on diverse quantum computers. Figure 13 shows the major components in design automation for quantum computing. Development of quantum algorithms is a crucial first step to describe how we plan to solve a specific application using a

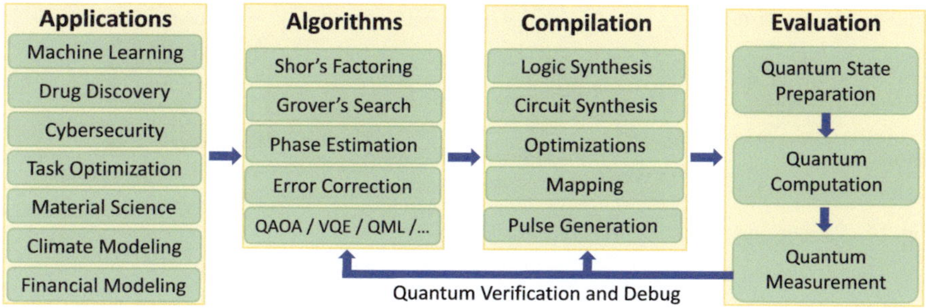

Fig. 13 Design automation steps in quantum computing. The first step is to develop algorithms to solve real-world applications. The next step compiles these algorithms to quantum circuits and maps them to the evaluation platform (quantum computer or quantum simulator). The evaluation consists of state preparation followed by quantum computation and measurement. The results need to be verified to ensure reliability and accuracy. In case of mismatch, we need to modify the implementation

quantum computer. The next step is to compile the algorithm to produce an equivalent quantum circuit. In order to map this circuit to the quantum computer, we need to perform various steps depending on the target quantum computer. For example, we need to generate electrical pulses to construct gates using the physical qubits in the quantum computers. Various optimizations, such as error correction, are required for reliable quantum computing. The quantum computation can be performed using physical quantum computer or quantum simulator. Once the initial state is prepared, actual quantum computation is performed. Finally, we need to measure and analyze the results. We need to perform quantum verification to ensure the functional accuracy and reliability of the underlying quantum computation. Similarly, we need to identify any security vulnerabilities and develop suitable countermeasures. The remainder of this section describes these steps in detail.

3.1 Quantum State Preparation

State preparation is a vital component in quantum information processing. Specifically, quantum computers must start in a known state in order to correctly apply gates and perform a computation. Similarly, quantum communication protocols require the preparation of an initial entangled state, and quantum memory relies on state preparation to read and write quantum states. A major hurdle in preparing an arbitrary state is the creation of highly entangled states. Note that entanglement also occurs between qubits and external degrees of freedom, leading to undesired effects such as decoherence and noise. The objective of state preparation algorithms is to develop strategies to strengthen the wanted entanglement while mitigating the unwanted effects.

One of the major components for realizing useful quantum computers is the ability to prepare the computer to an arbitrary entangled state. Presently, the primary

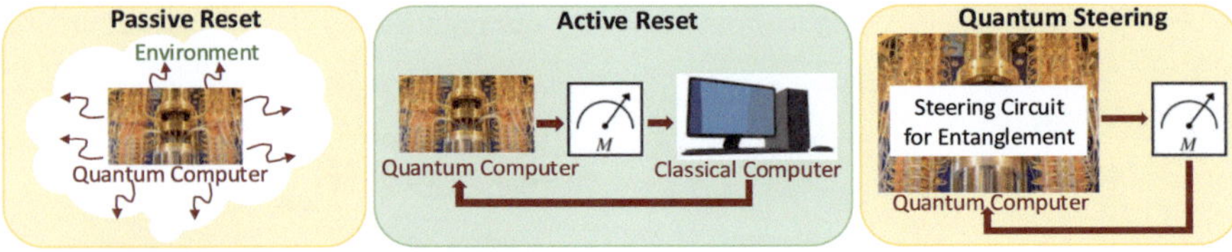

Fig. 14 Various quantum state preparation methods

state preparation method is to prepare the computer in a fiducial state, followed by a sequence of calibrated gates. Fiducial state preparation can be broadly divided into two categories: *passive reset* [14–16] and *active reset* [17, 18], as shown in Fig. 14. The passive reset is achieved by waiting for the quantum computer to thermalize to the ground state [14]—by putting the system in an environment that is at a sufficiently low temperature. In case of active reset, the state of the qubits is collapsed via measurement and single-qubit rotations, conditioned on the outcome of measurements, are performed to correct the state. In other words, it makes projective measurements of a set of observables, and then conditioned-rotations are applied to correct the results of probabilistic outcomes. After this fiducial state is prepared, a sequence of single- or two-qubit gates is applied to evolve the state to the desired entangled state.

Both passive and active resets (shown in Fig. 14) have practical limitations. Passive reset suffers from prolonged waiting times that may become infeasible for future quantum computers, especially those exhibiting extended coherence periods. Thermal dynamics also pose a problem, with the necessity of near-zero temperature environments to maintain a system near its ground state. Although active reset is designed to address some of the limitations of passive reset, it also faces practical limitations. The fidelity of the measurement process can be imperfect, leading to erroneous state preparation. Moreover, imperfections in quantum gates due to calibration errors, external interference, or intrinsic limitations can be a bottleneck in evolving the initial pure fiducial state to the desired entangled state. There are many promising avenues for addressing these challenges, including state preparation via dissipation [19–21] and state preparation using measurement-induced steering [8, 22–25].

3.2 *Quantum Compilation*

Figure 15 shows an overview of a typical quantum compilation framework that consists of three major tasks. The first task enables specification of quantum algorithms and associated constraints using a high-level language (e.g., Rust or Python). The second task transforms a given specification to a quantum circuit using various methods, such as cosine-sine decomposition [26] or Solovay–Kitaev algorithm [27]. The

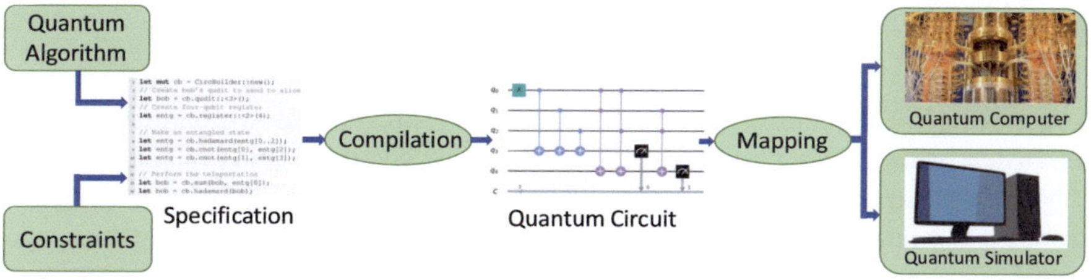

Fig. 15 An overview of a typical quantum compilation framework. The first task captures a quantum algorithm as well as constraints using a language-based specification. The second task performs quantum compilation and associated optimizations. The third task performs mapping of the quantum circuit to enable evaluation using either a physical quantum computer or simulation of quantum systems

third task performs mapping to enable evaluation using a physical quantum computer or simulation of quantum systems using a classical computer.

This transition from a high-level quantum algorithm to a quantum circuit consists of a series of steps which may depend on the particular quantum computer architecture [28, 29]. In general, the high-level algorithm and corresponding quantum circuit are invariant to the underlying assumptions of an architecture or technology. Specifically, after a quantum compiler produces the target operations, a real-time controller device defines the actual interaction with a quantum computer, such as by coordinating many control devices to send microwave pulses. Compiler optimizations can reduce noise and increase computation speed. While long computations on near-term quantum computers may be prohibitive due to constraints in circuit depth, such computations can be performed by designing noise-tolerant high-level algorithms and optimizing circuit translation with respect to noise.

Quantum compiler takes an arbitrary unitary matrix as input and returns the decomposition as a sequence of elementary operations. Due to the common assumption that a quantum computer operates on qubits, the input unitary matrix is often restricted to some power of two. Qubiter [26] is one of the first quantum compilers to be proposed, it employs Quantum Bayesian Nets [30] to model quantum systems graphically and utilizes the cosine-sine decomposition for compilation. However, when handling large qubit systems, or when required to work with a particular quantum technology with noise models, general-purpose quantum compilers introduce unacceptable time overhead. For this reason, many state-of-the-art quantum compilers restrict inputs to single-qubit and two-qubit gates. Most present-day quantum compilers such as the ones employed in Qiskit [31] use the KAK decomposition [32] to exactly decompose a two-qubit operation.

3.3 Quantum Security

A vast majority of quantum researchers do not have direct access to quantum computers. Instead, they utilize cloud-based services where we need to rely on the reliability of both classical and quantum computers. For example, Fig. 16 shows an overview of quantum computing using cloud-based quantum computers. In this scenario, designers develop their quantum circuits and submit computing jobs to the cloud through encrypted communication. The cloud job scheduling servers then send pulse-level circuits to microwave electronics to generate corresponding control pulses. Qubits in quantum computers must be regulated and controlled by control pulses to implement quantum algorithms and get results back to users. Therefore, we need to consider both quantum and classical security vulnerabilities to enable trustworthy quantum computing [33–42].

The delicate nature of quantum phenomena makes the devices susceptible to errors, sensitivity to environmental factors, and various types of vulnerabilities, including cross-talk among qubits. Cross-talk, for example, poses a critical concern in the context of shared quantum systems, as multiple users accessing the same quantum computer may inadvertently or intentionally obtain information from one another, compromising the security and confidentiality of the data being processed. To address the security vulnerabilities inherent in quantum devices, researchers must develop new techniques and protocols that safeguard against cross-talk and other forms of information leakage. For instance, employing isolation and partitioning techniques to separate users' quantum operations can ensure that each user's computation is confined to a designated portion of the quantum system. This approach would minimize the potential for cross-talk and information leakage among users, although it may entail additional resource overhead and reduced computational efficiency. Moreover, developing quantum error correction techniques that can detect and mitigate the effects of cross-talk and other sources of noise is critical for enhancing the security of quantum devices [23]. By implementing robust error correction methods, it is possible to minimize the impact of cross-talk on the accuracy and reliability of quantum computations, reducing the likelihood of unauthorized information access. However, it is a major challenge to develop an effective error correction strategy since it must account for the unique properties and behaviors of quantum systems.

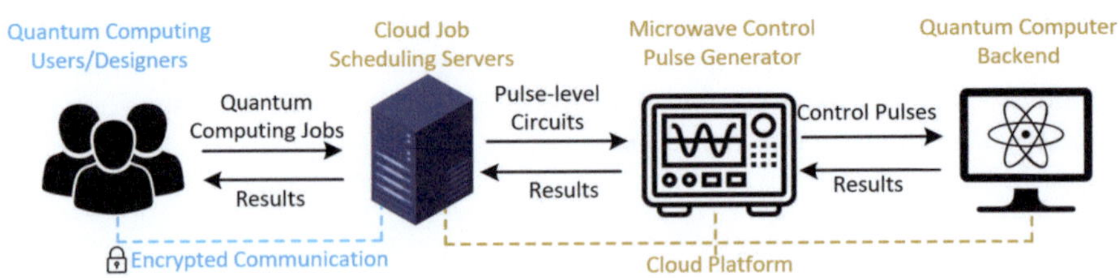

Fig. 16 Overview of quantum computing using cloud-based quantum computers [42]

As shown in Fig. 3, classical hardware, such as CPUs, FPGAs, and ASICs, plays a crucial role in the operation of quantum computing systems, as they are responsible for interfacing with quantum devices and orchestrating the necessary control and data processing tasks. As the availability of quantum computers in the cloud becomes more widespread, classical computing components remain a vital puzzle piece in the successful implementation and utilization of these powerful systems. The cloud-based nature of quantum computing services necessitates the secure transmission of jobs and data between classical and quantum systems, further emphasizing the importance of classical computing components in the overall operation of quantum systems. To ensure the confidentiality, integrity, and availability of data exchanged between classical computers and quantum devices, it is essential to employ post-quantum cryptography (PQC) techniques that can withstand the computational power of quantum machines. In this context, classical computing components not only serve as indispensable tools for interfacing with quantum devices but also a potential point of failure in the rapidly evolving security landscape of quantum computing.

3.4 Quantum Error Correction

The enhanced computational ability of quantum systems is driven by the key quantum properties, such as superposition and entanglement. In reality, qubits also entangle with unwanted degrees of freedom (the environment), leading to decoherence and a loss of information [10, 43]. In other words, the quantum computer must satisfy two conflicting requirements. Qubits need to be externally controlled, measured, and entangled. On the other hand, qubits must be isolated from their environment to avoid unwanted entanglement. As a result of these conflicts, quantum computers are expected to be noisy, where errors propagate and grow during the execution.

Quantum error-correcting codes (ECC) provide a realistic path toward fault-tolerant quantum computing. ECC have three main requirements: (1) provide an encoding of physical qubits to logical qubits, (2) the ability to detect when an error has occurred, and (3) a mechanism to correct the logical qubit. In other words, by creatively entangling several physical qubits as one logical qubit, errors can be detected and corrected via quantum error-correcting codes [44–46]. However, realizing error-correcting codes requires real-time (mid-circuit) qubit measurement and qubit reset for syndrome extraction and correction. For example, Fig. 17 shows three major stages in quantum computing: state preparation, quantum computation, and quantum measurement. Qubits are first initialized to a known fiducial state during state preparation. To enable error correction, we need the ability to reset, reinitialize, and reuse of qubits during the quantum computation. The final state is then measured during quantum measurement.

Two key facts make traditional classical codes inapplicable for quantum computers: (a) measurement collapses the state of a qubit and (b) by the no-cloning theorem, information cannot be copied. Instead, quantum error-correcting codes spread the information of one logical qubit across many high-entangled physical qubits. Errors

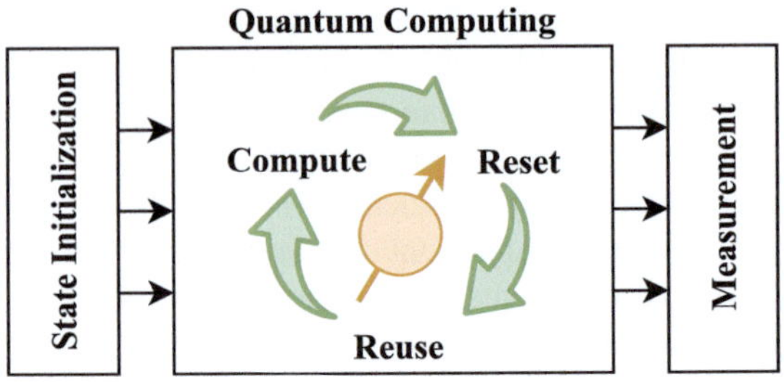

Fig. 17 Overview of quantum error correction during quantum computation

are detected using syndrome measurements, where many-qubit measurements are performed such that it does not destroy the information encoded in the logical qubit. If an error is detected, operations corresponding to the type of error are used to revert the error. In other words, an error-correcting code, E, will ideally undo the action of a noisy process $\mathcal{E}$ and return the original density matrix: $E(\rho\prime) = E \circ \mathcal{E}(\rho) = \rho$, where ρ and $\rho\prime$ are original (noise-free) and modified (noise-infected) density matrices. Quantum error-correcting codes depend on a number of key parameters, such as the number of physical qubits necessary, the number of encoded logical qubits, and the maximum physical error rates threshold [47, 48].

Surface codes are widely used for quantum error correction [49]. Surface codes build logical qubits by assuming topological features of physical qubits [50, 51]. Specifically, surface codes assume the physical layout of qubits is given by a lattice. Recent experiments have successfully demonstrated surface code implementation by simulating the code on programmable quantum processors, such as on Rydberg atoms [52] and superconducting qubits [50]. Two key steps are performed to prepare the computer for surface code simulation. The first step is to reset the computer to a fiducial, all-zeros, state—either by waiting for the qubits to naturally decay, or by measuring, classically reading, and correcting the qubits. The second key step is to apply highly calibrated (single and entangling) gates to the all-zeros state to prepare the simulated state of a surface code. Quantum error correction is a promising pathway to design noise-resilient quantum systems [23].

3.5 Quantum Simulation and Benchmarking

Quantum simulators can be divided into three broad categories: analog simulators, digital simulators, and hybrid simulators. Analog quantum simulators are ideal for simulating specific types of quantum systems. For example, researchers can use trapped ions to simulate spin systems or cold atoms in optical lattices. Digital quantum simulators are more flexible than their analog counterpart since they use quantum gates and algorithms to simulate quantum systems. Digital simulators rely on classical computers to mimic quantum behavior, which inherently limits their performance compared to physical quantum computers. Hybrid quantum simulators combine classical computing and quantum systems to solve complex problems.

Digital quantum simulators are widely used for design and evaluation of quantum algorithms. There are important considerations for using quantum simulators. Most quantum researchers cannot access physical quantum computers due to cost or other considerations. Therefore, they use quantum simulators to execute the quantum circuits. While some quantum computers are available for research purposes through cloud service, they either have limited number of qubits or it is free for a limited time. Moreover, it is beneficial to use quantum simulators in early stages of the design cycle when many iterations are required to finalize the implementation while satisfying the constraints. There are popular quantum simulators, such as IBM Qiskit, Google Cirq, and Microsoft Quantum Development Kit. IBM Qiskit is an open-source software simulator for quantum computing that allows users to write quantum circuits and optimize them for the target quantum hardware or simulators. Google Cirq is a Python software library for writing, manipulating, and optimizing quantum circuits, and then running them on quantum computers and quantum simulators. Similarly, there are software simulators from IonQ, Rigetti, and Quantinuum.

Quantum simulators are useful for quantum benchmarking efforts. Quantum benchmarks allow us to measure the performance of quantum computers, such as gate fidelity, entanglement generation, and noise levels, across different algorithms or optimizations. Quantum benchmarks also allow us to compare between different quantum computers. By analyzing the results of these benchmarks, designers can identify areas for improvement and track progress toward building robust quantum computers. While randomized benchmarking [53, 54] is scalable and can provide average error rates of gates by conducting long sequences of random gates, it may not capture system-wide properties. Metrics such as quantum volume [55] are linked with system error rates and provide a quantification to the effective number of qubits of a quantum computer. Alternatively, one may observe the performance of many small quantum workloads via an application benchmark, such as quantum Fourier transform and phase estimation, and obtain performance estimates [56]. However, these application benchmarks may be biased to the choice of workloads. There are recent initiatives to develop quantum benchmarks that are applicable across quantum computers [57].

3.6 Quantum Measurement

Measurement in quantum systems involves coupling a qubit with a measurement device (apparatus), like a resonator in superconducting systems. The quantum state influences the state of the measurement device, which is then read out. This readout is processed to yield measurement signals. However, these signals do not directly reveal the quantum state—instead, they represent certain physical properties (like frequency shifts in a resonator) that correlate with the quantum state. Hence, a classifier is needed

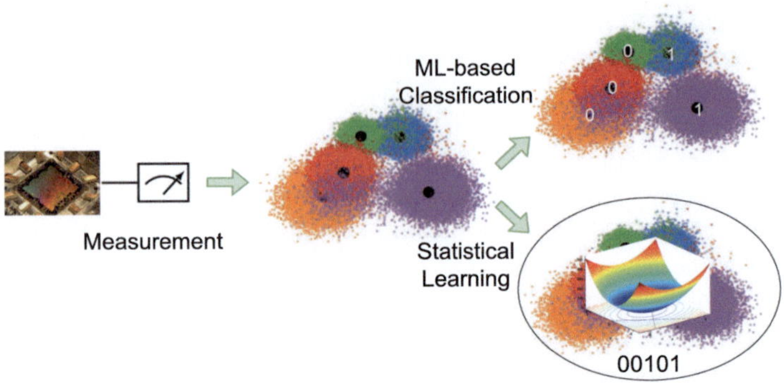

Fig. 18 An overview of quantum measurement. Machine learning (ML)-based methods perform multiple (independent) classifications whereas statistical methods perform one classification using cumulative distribution functions

to map these measurement signals back to the quantum states. In other words, the classifier decodes the signals from the measurement device, determining whether it represents a 0 or a 1 in the case of a qubit. This step is known as quantum state discrimination (classification) and is a crucial component in unveiling the results of a quantum computation. Note that the measurement mechanism itself is prone to various noise sources which are both classical and quantum in nature. For example, classical noise may be introduced as analog signals must travel through various stages of a dilution refrigerator to an analog-to-digital converter. Quantum noise may enter as the measurement device itself is a quantum system that may entangle with unwanted degrees of freedom [58].

Figure 18 shows an overview of the measurement procedure in quantum computers. Common physical representations of quantum computers, such as superconducting transmons, characterize a collapsed qubit as a two-component vector—the in-phase and quadrature components (IQ) of a wave observed as it passes through the resonator [59, 60]. The quantum measurement device yields sets of results, including IQ points. Subsequently, a classifier labels each point as either $|0\rangle$ or $|1\rangle$. A vast majority of existing approaches utilize machine learning for classification [61–63], while recent approaches utilize various statistical learning methods, such as cumulative distribution functions [64].

Machine learning techniques are widely used for qubit state discrimination [61–63, 65–69]. For example, in case of linear discriminant analysis, a hyperplane is selected to partition the IQ vector space into regions of $|0\rangle$ and $|1\rangle$ based on the measured IQ outputs from the training data, which the model assumes follows a Gaussian distribution. An overall qubit state is obtained by analyzing the frequencies associated with each prediction class. Due to noise in quantum systems (imperfect measurement devices, environmental contamination, and qubit cross-talk), sampled data contains high variance and may appear in the partition, leading to misclassifications. There are recent approaches that perform quantum measurement classification using cumulative distributions functions to reduce the qubit readout error compared to machine learning-based solutions [64, 70].

3.7 *Quantum Verification and Debug*

Quantum verification is necessary to ensure that a quantum computer is producing the expected results. While there are various avenues for quantum verification, we briefly describe three approaches in this section: simulation-based validation, formal verification, and quantum characterization. While quantum characterization is unique to quantum systems [71], the underlying principles of the other two approaches (simulation and formal verification) are similar to verifying classical systems [72–78].

Formal verification is used to guarantee the correctness of quantum algorithms. Although there are various use cases of formal verification, we briefly describe property (model) checking and equivalence checking. Model checking verifies if a quantum computation adheres to specific properties, ensuring its correctness and reliability. It involves defining properties using temporal logic and verifying their satisfaction by the system's evolution. This technique extends the classical model checking approach to quantum systems. For example, it extends classical binary decision diagrams to quantum decision diagrams. Similarly, equivalence checking is useful to check if two implementations are functionally equivalent. For example, we can check equivalence between a quantum algorithm and respective quantum circuit (after compilation) to ensure that the compilation did not introduce any errors. Similarly, we can perform equivalence checking between two versions of quantum circuits that are expected to be functionally equivalent but one of them have gone through specific optimizations. While formal verification can provide verification guarantees, they cannot handle large designs (e.g., quantum circuits with large number of gates) due to the fact that the complexity of the backend quantum satisfiability solvers or quantum decision diagrams grows exponentially with the increase in the number of qubits.

Simulation-based validation can handle large designs but cannot provide verification guarantees. For example, when we apply specific input patterns (tests) and compare with the expected output, it only verifies the accuracy of the implementation for that specific test pattern. Similarly, when we introduce assertions in the quantum circuit, it produces assertion failure only if the execution sequence violates the assertion. In other words, if the assertion does not fail during execution (simulation), it does not guarantee that the implementation is correct since the scope of the simulation is limited by the applied test patterns. Unless we try all possible input patterns, which can be exponential, simulation-based validation cannot provide verification guarantees. In reality, designers use an effective combination of simulation-based validation and formal verification.

When we identify a mismatch between the actual output and the expected output during verification, we need to debug to detect the root-cause and fix it. When property checking fails, it provides a counterexample which can be used to fix the problem. Similarly, an assertion failure during simulation is easier for root-cause analysis. However, it would be hard to debug in the context of black-box verification, such as execution inside a physical quantum computer. There are different quantum characterization methods that can retrieve the underlying quantum process based on a given set of measurement data, such as state [79] and process tomography [80, 81], gateset tomography [82], measurement tomography [83], and fast quantum characterization via Riemannian optimization [71].

4 Summary

Quantum computing has the potential to significantly outperform their classical counterpart in solving several classes of optimization problems. In this chapter, we introduced the readers to the world of quantum computing. We have also outlined similarities and differences of various concepts between classical and quantum computing domains. Next, we described various design automation stages in quantum computing and associated challenges. The subsequent chapters describe practical solutions to address these challenges.

References

1. Edward Farhi, Jeffrey Goldstone, and Sam Gutmann. A quantum approximate optimization algorithm. arXiv preprint arXiv:1411.4028, 2014.
2. McClure, Doug T., Hanhee Paik, Lev S. Bishop, Matthias Steffen, Jerry M. Chow, and Jay M. Gambetta. Rapid Driven Reset of a Qubit Readout Resonator. *Phys. Rev. Appl.*, 5(1), 2016.
3. Peruzzo, Alberto, Jarrod McClean, Peter Shadbolt, Man-Hong. Yung, Xiao-Qi. Zhou, Peter J. Love, Alán. Aspuru-Guzik, and Jeremy L. O'Brien. 2014. A variational eigenvalue solver on a photonic quantum processor. *Nature Communications* 5 (1): 4213.
4. Matthew Sparkes. Google demonstrates vital step towards large-scale quantum computers. *New Scientist*, 2021.
5. Bas Hensen, Hannes Bernien, Anaïs E Dréau, Andreas Reiserer, Norbert Kalb, Machiel S Blok, Just Ruitenberg, Raymond FL Vermeulen, Raymond N Schouten, Carlos Abellán, et al. Loophole-free bell inequality violation using electron spins separated by 1.3 kilometres. *Nature*, 526(7575):682–686, 2015.
6. Daniel Volya and Prabhat Mishra. Quantum spectral clustering of mixed graphs. In *2021 58th ACM/IEEE design automation conference (DAC)*, pages 463–468. IEEE, 2021.
7. Grover, Lov K. A fast quantum mechanical algorithm for database search. In *Proceedings of the twenty-eighth annual ACM symposium on Theory of computing*, pages 212–219, 1996.
8. Sahan Sanjaya, Daniel Volya, and Prabhat Mishra. Variational quantum algorithms via measurement-induced passive steering. In *2024 IEEE International Conference on Quantum Computing and Engineering (QCE)*, volume 1, pages 481–487. IEEE, 2024.
9. Peter W Shor. Polynomial-time algorithms for prime factorization and discrete logarithms on a quantum computer. *SIAM review*, 41(2):303–332, 1999.

10. Daniel Volya and Prabhat Mishra. Impact of noise on quantum algorithms in noisy intermediate-scale quantum systems. In *IEEE International Conference on Computer Design (ICCD)*, pages 1–4. IEEE, 2020.
11. Daniel Volya and Prabhat Mishra. Modeling of noisy quantum circuits using random matrix theory. In *2022 IEEE 40th International Conference on Computer Design (ICCD)*, pages 132–138. IEEE, 2022.
12. IBM Quantum. https://www.ibm.com/quantum, 2025.
13. Microsoft Azure Quantum, Morteza Aghaee, Alejandro Alcaraz Ramirez, Zulfi Alam, Rizwan Ali, Mariusz Andrzejczuk, Andrey Antipov, Mikhail Astafev, Amin Barzegar, Bela Bauer, et al. Interferometric single-shot parity measurement in inas–al hybrid devices. *Nature*, 638(8051):651–655, 2025.
14. Paul Adrien Maurice Dirac. The quantum theory of the emission and absorption of radiation. *Proceedings of the Royal Society of London. Series A, Containing Papers of a Mathematical and Physical Character*, 114(767):243–265, 1927.
15. Harty, T. P., D. T. C. Allcock, C. Ballance, L. Guidoni, H. A. Janacek, N. M. Linke, D. N. Stacey, and D. M. Lucas. 2014. High-fidelity preparation, gates, memory, and readout of a trapped-ion quantum bit. *Physical review letters* 113 (22): 220501.
16. Chad Rigetti, Jay M Gambetta, Stefano Poletto, Britton LT Plourde, Jerry M Chow, Antonio D Córcoles, John A Smolin, Seth T Merkel, Jim R Rozen, George A Keefe, et al. Superconducting qubit in a waveguide cavity with a coherence time approaching 0.1 ms. *Physical Review B–Condensed Matter and Materials Physics*, 86(10):100506, 2012.
17. Daniel Basilewitsch, Jonas Fischer, Daniel M Reich, Dominique Sugny, and Christiane P Koch. Fundamental bounds on qubit reset. *Physical Review Research*, 3(1):013110, 2021.
18. Caroline Tornow, Naoki Kanazawa, William E Shanks, and Daniel J Egger. Minimum quantum run-time characterization and calibration via restless measurements with dynamic repetition rates. *Physical Review Applied*, 17(6):064061, 2022.
19. Barbara Kraus, Hans P Büchler, Sebastian Diehl, Adrian Kantian, Andrea Micheli, and Peter Zoller. Preparation of entangled states by quantum markov processes. *Physical Review A–Atomic, Molecular, and Optical Physics*, 78(4):042307, 2008.
20. Frank Verstraete, Michael M Wolf, and J Ignacio Cirac. Quantum computation and quantum-state engineering driven by dissipation. *Nature physics*, 5(9):633–636, 2009.
21. Hendrik Weimer, Markus Müller, Igor Lesanovsky, Peter Zoller, and Hans Peter Büchler. A rydberg quantum simulator. *Nature Physics*, 6(5):382–388, 2010.
22. Sthitadhi Roy, JT Chalker, IV Gornyi, and Yuval Gefen. Measurement-induced steering of quantum systems. *Physical review research*, 2(3):033347, 2020.
23. Daniel Volya and Prabhat Mishra. Quantum steering of surface error correcting codes. In *2023 IEEE International Conference on Quantum Computing and Engineering (QCE)*, volume 1, pages 1394–1400. IEEE, 2023.
24. Volya, Daniel, and Prabhat Mishra. 2024. State preparation on quantum computers via quantum steering. *IEEE Transactions on Quantum Engineering (TQE)* 5: 1–14.
25. Daniel Volya, Zhixin Pan, and Prabhat Mishra. Feedback-based steering for quantum state preparation. In *2023 IEEE International Conference on Quantum Computing and Engineering (QCE)*, volume 1, pages 1308–1318. IEEE, 2023.
26. Robert R Tucci. A rudimentary quantum compiler (2cnd ed.). arXiv preprint quant-ph/9902062, 1999.
27. Christopher M Dawson and Michael A Nielsen. The solovay-kitaev algorithm. *arXiv preprint quant-ph/0505030*, 2005.
28. Daniel Volya and Prabhat Mishra. Quantum data compression for efficient generation of control pulses. In *Asia and South Pacific Design Automation Conference (ASPDAC)*, pages 216–221, 2023.
29. Daniel Volya and Prabhat Mishra. Qudcom: Towards quantum compilation for qudit systems. arXiv preprint arXiv:2311.07015, 2023.
30. Robert R Tucci. Quantum bayesian nets. arXiv preprint quant-ph/9706039, 1997.
31. Qiskit. https://www.ibm.com/quantum/qiskit, 2025.

32. Robert R Tucci. An introduction to cartan's kak decomposition for qc programmers. arXiv preprint quant-ph/0507171, 2005.
33. Farahmandi, Farimah, and Yuanwen Huang. 2020. *and Prabhat Mishra*. System-on-Chip Security: Validation and Verification. Springer.
34. Jayasena, Aruna, Richard Bachmann, and Prabhat Mishra. 2025. Ciseleaks: Information leakage assessment of cryptographic instruction set extension prototypes. *IEEE Transactions on Information Forensics and Security (TIFS)* 20: 1551–1565.
35. Aruna Jayasena and Prabhat Mishra. Firmwall: Directed symbolic execution of firmware binaries for defending against unauthorized system calls. *IEEE Transactions on Information Forensics and Security (TIFS)*, 2025.
36. Aruna Jayasena, Sai Suprabhanu Nallapaneni, and Prabhat Mishra. Fuss: Coverage-directed hardware fuzzing with selective symbolic execution. *ACM Transactions on Embedded Computing Systems (TECS)*, 2025.
37. Prabhat Mishra, Swarup Bhunia, and Mark Tehranipoor. *Hardware IP security and trust*. Springer, 2017.
38. Pan, Zhixin, and Prabhat Mishra. 2022. A survey on hardware vulnerability analysis using machine learning. *IEEE Access* 10: 49508–49527.
39. Zhixin Pan and Prabhat Mishra. *Explainable AI for Cybersecurity*. Springer, 2023.
40. Zhixin Pan, Jennifer Sheldon, and Prabhat Mishra. Hardware-assisted malware detection using explainable machine learning. In *2020 IEEE 38th International Conference on Computer Design (ICCD)*, pages 663–666. IEEE, 2020.
41. Sahan Sanjaya, Aruna Jayasena, and Prabhat Mishra. Information leakage through physical layer supply voltage coupling vulnerability. *IEEE Transactions on Very Large Scale Integration (VLSI) Systems*, 33(6):1715–1728, 2025.
42. Daniel Volya, Tao Zhang, Nashmin Alam, Mark Tehranipoor, and Prabhat Mishra. Towards secure classical-quantum systems. In *2023 IEEE International Symposium on Hardware Oriented Security and Trust (HOST)*, pages 283–292. IEEE, 2023.
43. Gil Kalai. The argument against quantum computers. In *Quantum, probability, logic: The work and influence of Itamar Pitowsky*, pages 399–422. Springer, 2020.
44. Google Quantum, A. I. 2023. Suppressing quantum errors by scaling a surface code logical qubit. *Nature* 614 (7949): 676–681.
45. Sergey Bravyi, Andrew W Cross, Jay M Gambetta, Dmitri Maslov, Patrick Rall, and Theodore J Yoder. High-threshold and low-overhead fault-tolerant quantum memory. *Nature*, 627(8005):778–782, 2024.
46. Paetznick, A., M. P. da Silva, C. Ryan-Anderson, J. M. Bello-Rivas, J. P. Campora III, A. Chernoguzov, J. M. Dreiling, et al. Demonstration of logical qubits and repeated error correction with better-than-physical error rates. arXiv preprint arXiv:2404.02280, 2024.
47. Earl T Campbell, Barbara M Terhal, and Christophe Vuillot. Roads towards fault-tolerant universal quantum computation. *Nature*, 549(7671):172–179, 2017.
48. Austin G Fowler, Ashley M Stephens, and Peter Groszkowski. High-threshold universal quantum computation on the surface code. *Physical Review A–Atomic, Molecular, and Optical Physics*, 80(5):052312, 2009.
49. Austin G Fowler, Matteo Mariantoni, John M Martinis, and Andrew N Cleland. Surface codes: Towards practical large-scale quantum computation. *Physical Review A–Atomic, Molecular, and Optical Physics*, 86(3):032324, 2012.
50. Satzinger, K. J., Y-J. Liu, A. Smith, C. Knapp, M. Newman, C. Jones, Zhaoshi Chen, et al. Realizing topologically ordered states on a quantum processor. *Science*, 374(6572):1237–1241, 2021.
51. Roope Uola, Ana CS Costa, H Chau Nguyen, and Otfried Gühne. Quantum steering. *Reviews of Modern Physics*, 92(1), 2020.
52. Giulia Semeghini, Harry Levine, Alexander Keesling, Sepehr Ebadi, Tout T Wang, Dolev Bluvstein, Ruben Verresen, Hannes Pichler, Marcin Kalinowski, Rhine Samajdar, et al. Probing topological spin liquids on a programmable quantum simulator. *Science*, 374(6572):1242–1247, 2021.

53. Dankert, Christoph, Richard Cleve, Joseph Emerson, and Etera Livine. 2009. Exact and approximate unitary 2-designs and their application to fidelity estimation. *Physical Review A-Atomic, Molecular, and Optical Physics* 80 (1): 012304.
54. Jay M Gambetta, Antonio D Córcoles, Seth T Merkel, Blake R Johnson, John A Smolin, Jerry M Chow, Colm A Ryan, Chad Rigetti, Stefano Poletto, Thomas A Ohki, et al. Characterization of addressability by simultaneous randomized benchmarking. *Physical review letters*, 109(24):240504, 2012.
55. Andrew W Cross, Lev S Bishop, Sarah Sheldon, Paul D Nation, and Jay M Gambetta. Validating quantum computers using randomized model circuits. *Physical Review A*, 100(3):032328, 2019.
56. Thomas Lubinski, Joshua J Goings, Karl Mayer, Sonika Johri, Nithin Reddy, Aman Mehta, Niranjan Bhatia, Sonny Rappaport, Daniel Mills, Charles H Baldwin, et al. Quantum algorithm exploration using application-oriented performance benchmarks. arXiv preprint arXiv:2402.08985, 2024.
57. Daniel Volya and Prabhat Mishra. Quantum benchmarking via random dynamical quantum maps. In *2024 IEEE International Conference on Quantum Computing and Engineering (QCE)*, volume 1, pages 1277–1283. IEEE, 2024.
58. Onur Hosten, Nils J Engelsen, Rajiv Krishnakumar, and Mark A Kasevich. Measurement noise 100 times lower than the quantum-projection limit using entangled atoms. *Nature*, 529(7587):505–508, 2016.
59. Romeo Bianchetti, Stefan Filipp, Matthias Baur, Johannes M Fink, Christian Lang, Lars Steffen, Maxime Boissonneault, Alexandre Blais, and Andreas Wallraff. Control and tomography of a three level superconducting artificial atom. *Physical review letters*, 105(22):223601, 2010.
60. Philip Krantz, Morten Kjaergaard, Fei Yan, Terry P Orlando, Simon Gustavsson, and William D Oliver. A quantum engineer's guide to superconducting qubits. *Applied physics reviews*, 6(2), 2019.
61. Jihye Kim, Byungdu Oh, Yonuk Chong, Euyheon Hwang, and Daniel K Park. Quantum readout error mitigation via deep learning. *New Journal of Physics*, 24(7):073009, 2022.
62. Easwar Magesan, Jay M Gambetta, Antonio D Córcoles, and Jerry M Chow. Machine learning for discriminating quantum measurement trajectories and improving readout. *Physical review letters*, 114(20), 2015.
63. Luis A Martinez, Yaniv J Rosen, and Jonathan L DuBois. Improving qubit readout with hidden markov models. *Physical Review A*, 102(6):062426, 2020.
64. Utt, Zachery, Daniel Volya, and Prabhat Mishra. 2024. Quantum measurement classification using statistical learning. *ACM Transactions on Quantum Computing (TQC)* 5 (2): 1–16.
65. Utkarsh Azad and Helena Zhang. Machine learning based discrimination for excited state promoted readout. In *IEEE/ACM Symposium on Edge Computing (SEC)*, pages 362–367. IEEE, 2022.
66. Duan, Peng, Zi-Feng. Chen, Qi. Zhou, Wei-Cheng. Kong, Hai-Feng. Zhang, and Guo-Ping. Guo. 2021. Mitigating crosstalk-induced qubit readout error with shallow-neural-network discrimination. *Physical Review Applied* 16 (2): 024063.
67. Dmitry Gavinsky, Julia Kempe, Oded Regev, and Ronald De Wolf. Bounded-error quantum state identification and exponential separations in communication complexity. In *Proceedings of the thirty-eighth annual ACM symposium on Theory of Computing*, pages 594–603, 2006.
68. Benjamin Lienhard, Antti Vepsäläinen, Luke CG Govia, Cole R Hoffer, Jack Y Qiu, Diego Ristè, Matthew Ware, David Kim, Roni Winik, Alexander Melville, et al. Deep-neural-network discrimination of multiplexed superconducting-qubit states. *Physical Review Applied*, 17(1), 2022.
69. David Quiroga, Prasanna Date, and Raphael Pooser. Discriminating quantum states with quantum machine learning. In *2021 International Conference on Rebooting Computing (ICRC)*, pages 56–63. IEEE, 2021.
70. Zachery Utt, Daniel Volya, and Prabhat Mishra. Quantum measurement discrimination using cumulative distribution functions. In *2023 Design, Automation & Test in Europe Conference & Exhibition (DATE)*, pages 1–6. IEEE, 2023.

71. Daniel Volya, Andrey Nikitin, and Prabhat Mishra. Fast quantum characterization via riemannian optimization. In *2024 IEEE International Conference on Quantum Computing and Engineering (QCE)*, volume 1, pages 1270–1276. IEEE, 2024.
72. Farahmandi, Farimah, and Prabhat Mishra. 2018. Automated test generation for debugging multiple bugs in arithmetic circuits. *IEEE Transactions on Computers* 68 (2): 182–197.
73. Jayasena, Aruna, and Prabhat Mishra. 2024. Directed test generation for hardware validation: A survey. *ACM Computing Surveys* 56 (5): 1–36.
74. Jayasena, Aruna, and Prabhat Mishra. 2024. Hive: Scalable hardware-firmware co-verification using scenario-based decomposition and automated hint extraction. *IEEE Transactions on Computer-Aided Design of Integrated Circuits and Systems* 43 (10): 3278–3291.
75. Lyu, Yangdi, and Prabhat Mishra. 2020. Scalable concolic testing of rtl models. *IEEE Transactions on Computers* 70 (7): 979–991.
76. Lyu, Yangdi, Xiaoke Qin, Mingsong Chen, and Prabhat Mishra. 2018. Directed test generation for validation of cache coherence protocols. *IEEE Transactions on Computer-Aided Design of Integrated Circuits and Systems* 38 (1): 163–176.
77. Prabhat Mishra and Farimah Farahmandi. *Post-Silicon Validation and Debug*. Springer, 2019.
78. Witharana, Hasini, Hansika Weerasena, and Prabhat Mishra. 2024. Formal verification of virtualization-based trusted execution environments. *IEEE Transactions on Computer-Aided Design of Integrated Circuits and Systems* 43 (11): 4262–4273.
79. John A Smolin, Jay M Gambetta, and Graeme Smith. Efficient method for computing the maximum-likelihood quantum state from measurements with additive gaussian noise. *Physical review letters*, 108(7):070502, 2012.
80. George C Knee, Eliot Bolduc, Jonathan Leach, and Erik M Gauger. Quantum process tomography via completely positive and trace-preserving projection. *Physical Review A*, 98(6), 2018.
81. JF Poyatos, J Ignacio Cirac, and Peter Zoller. Complete characterization of a quantum process: the two-bit quantum gate. *Physical Review Letters*, 78(2):390, 1997.
82. Nielsen, Erik. 2021. John King Gamble, Kenneth Rudinger, Travis Scholten, Kevin Young, and Robin Blume-Kohout. *Gate set tomography. Quantum* 5: 557.
83. Yanzhu Chen, Maziar Farahzad, Shinjae Yoo, and Tzu-Chieh Wei. Detector tomography on ibm quantum computers and mitigation of an imperfect measurement. *Physical Review A*, 100(5), 2019.

Quantum Algorithms and Architectures

Quantum Computing Architectures

Hansika Weerasena and Prabhat Mishra

Abstract Quantum computing is promising to solve certain classes of problems more efficiently than classical computers by leveraging the principles of quantum mechanics. As quantum hardware continues to evolve, it is important to view these systems from a computer architecture perspective. For example, it is necessary to examine how qubits are arranged, controlled, and interconnected to form gates and circuits. Similarly, we need to explore how data and entanglement are routed within and between quantum processors. This chapter provides an overview of the architectural structure of quantum processing units, including both single-core and modular multi-core configurations. It also explores how communication is managed in quantum computing systems, introducing the concept of quantum on-chip networks. Finally, we discuss quantum computing simulators that are essential tools for exploring quantum computer architectures.

1 Introduction

Quantum computing represents a fundamentally different paradigm from traditional computing. It is not the next generation of high-performance computing, it is entirely new computing model rooted in the principles of quantum mechanics. While classical computers are ideal for a wide range of tasks, quantum computers are designed to solve a specific class of simulation and optimization problems significantly faster than classical computers. The superiority of quantum computers comes from its inherent ability to perform parallel processing using superposition and entanglement. For example, the performance of classical computers increases linearly with the number of bits (transistors). In contrast, the performance of quantum computers increases exponentially with the number of qubits. Specifically, quantum computers have the potential to transform all 2^n possible states by using n entangled qubits.

H. Weerasena · P. Mishra (✉)
University of Florida, Gainesville, USA
e-mail: prabhat@ufl.edu

P. Mishra (ed.), *Design Automation for Quantum Computing*,
https://doi.org/10.1007/978-3-032-09303-5_2

(a) Classical computer

(b) Google Sycamore quantum computer [23]

Fig. 1 Classical versus quantum computers

Figure 1 shows examples of classical and quantum computers. Most computer scientists and engineers understand the components of classical computers. For example, a typical desktop or laptop will include various components, including central processing unit (CPU), graphical processing unit (GPU), field programmable gate array (FPGA), memory modules, storage devices, power supplies, buses for internal communication, network interfaces, and peripherals, such as mouse, keyboard, and monitor. These components are tightly packed in a typical mobile phone using system-on-chip (SoC) architecture. In contrast, quantum computers employ a radically different architecture. If we look inside a quantum computer, qubits are arranged in a specific pattern that can be programmed to form quantum gates and circuits. Since quantum computers today collaborates with classical computers to efficiently solve problems, a quantum computer is typically connected with a wide variety of classical components, including CPU, GPU, and FPGA. This chapter explores the components, structure, and organization of quantum computers from the system and architectural perspectives.

To get an idea of where quantum computing architectures are today, let us discuss a little bit about history. The concept of quantum computing was first articulated by Nobel prize-winning theoretical physicist Richard Feynman in his 1981 lecture "Simulating Physics with Computers" [10]. Feynman observed that classical computers struggle to efficiently simulate quantum phenomena due to the exponential growth of quantum state space. He proposed a fundamentally new model—a quantum computer—that would operate under the rules of quantum mechanics and natively simulate physical systems. This sparked the theoretical foundation for quantum computing, further formalized in 1985 by David Deutsch's model of a universal quantum computer. By 1994, Peter Shor's quantum algorithm for integer factorization demonstrated the potential to outperform classical systems exponentially for certain problems, marking a major inflection point in quantum algorithm research.

Despite early theoretical breakthroughs, practical quantum computing remained elusive throughout the 1980s and 1990s due to severe hardware limitations. The fundamental unit of information in quantum computing is called a quantum bit (qubits). Implementing and maintaining qubits required technological capabilities that did not yet exist. Early demonstrations, such as small-scale nuclear magnetic resonance (NMR)-based systems in the late 1990s and early 2000s, were important milestones but lacked scalability and universality. It wasn't until the late 2010s that hardware matured enough to support viable quantum processors. Google's 53-qubit Sycamore processor [23] achieved a form of "quantum supremacy" in 2019, and companies like IBM, IonQ, and Rigetti began deploying cloud-accessible, programmable quantum systems with growing number of qubits and improved architectural control. Quantum computing research is growing rapidly in various fronts, including advancement in qubit technologies, exploration of robust and fault-tolerant materials, improvement in control mechanisms, and leveraging deep learning for design space exploration. These converging innovations are accelerating progress, enabling the development of robust and scalable quantum systems.

In classical domain, computer architecture refers to the high-level design and functional behavior of a system as visible to the programmer, including the instruction set, data formats, and memory addressing modes. In contrast, computer organization focuses on how these abstract concepts are physically implemented, covering components like control units, datapaths, pipelines, and memory hierarchies. While computer architecture defines what the system should do, computer organization explains how it is performed. This chapter aims to explore quantum computing through the lens of computer architecture, offering a system-level understanding of how quantum machines are built and operated. Unlike most physics-centric discussions, we focus on architectural layers, from basic building blocks to complex system integration. This chapter adopts a bottom-up approach, starting with how information is represented using qubits versus classical bits, and gradually building toward larger systems that span quantum processing units (QPUs), multi-core quantum architectures, and quantum communication networks.

This chapter assumes that the readers have the basic knowledge of classical computer architecture. Whether you are curious about what lies inside a quantum computer or you are looking to explore new architectural paradigms, this chapter aims to guide you through the rapidly evolving landscape of quantum computing. Specifically, we begin by comparing traditional and quantum representations of information, followed by an overview of various physical technologies for realizing qubits, such as superconducting circuits and trapped ions. We then introduce the concept of quantum logic gates and explain how they relate to classical gate operations. Next, we examine the organization of single-core QPUs, including control circuitry and topology layouts used by companies like IBM and Google. This naturally leads into a discussion of system-level architecture, including cooling mechanisms, classical-quantum interfaces, and error correction subsystems. The subsequent sections explore modular quantum systems, the emerging field of quantum network-on-chip, and recent developments in quantum communication protocols.

The remainder of this chapter is organized as follows. Section 2 introduces quantum bits (qubits) and compares it with classical bits. Section 3 describes quantum gates and quantum circuits. Section 4 presents quantum processing units of different companies, including IBM, Google, and Rigetti. Section 5 describes modular quantum computing architectures. Section 6 outlines on-chip communication in quantum computing systems. Finally, Sect. 7 describes quantum computing simulators that are essential tools for exploring quantum computer architectures.

2 From Classical Bits to Quantum Bits

The difference between traditional and quantum computing extends all the way down to the definition and physical implementation of the fundamental unit of digital computation and information. In classical computing, the fundamental unit of information is called a bit, whereas in quantum computing, it is a qubit. In this section, we discuss both classical bit and quantum bit (qubit) in terms of their logical behavior, physical realization, and mathematical representation. Understanding these basic units is essential for analyzing and comparing classical and quantum computer architectures from the ground up.

2.1 *Classical Bit*

A classical bit is a binary unit that can have one of two values: "0" or "1." A useful analogy is a electrical switch that can be either on or off, representing 1 or 0, respectively. Classical computing involves data and instructions, both of which can be represented in binary. For example, to add the numbers 1 and 2, a machine-level representation might be 00 01 10, where the first two bits (00) indicate the addition operation and the remaining bits (01 and 10) represent the operands (1 and 2) in binary. Every classical computing device, such as supercomputers, servers, and laptops, relies on storing and manipulating information using bits.

2.2 *Quantum Bit (Qubit)*

A qubit is also capable of representing binary states like $|0\rangle$ and $|1\rangle$. Unlike classical bits, a qubit can exist in both states simultaneously due to a quantum phenomenon called superposition. At a single moment, there is a probability of it being $|0\rangle$ and $|1\rangle$, and this can be mathematically expressed as a linear combination of both states: $\alpha\,|0\rangle + \beta\,|1\rangle$, where α and β are complex probability amplitudes that satisfy $|\alpha|^2 + |\beta|^2 = 1$. However, when a qubit is measured, this superposition collapses to either $|0\rangle$ or $|1\rangle$, with probabilities $|\alpha|^2$ and $|\beta|^2$, respectively. A commonly used analogy is Schrödinger's cat thought experiment, where a cat inside a box is both alive and dead until observed. This illustrates the nature of quantum superposition and measurement. Furthermore, if we have n qubits, they can represent 2^n possible states simultaneously.

In contrast, representing those same 2^n states with classical bits would require 2^n individual classical bits. The ability to encode and process exponentially many states is one of the key advantages of quantum computing.

2.3 *Physical Implementation of Qubits*

Both bits and qubits are logical concepts that require physical implementation in order to build a working computer. In the early days of classical computing, there were various attempts to represent bits using technologies such as vacuum tubes and relays. Eventually, researchers realized that transistors, acting as switches, provided the most efficient and scalable way to represent and manipulate bits. Today, modern computers use billions of transistors to physically encode bits. On the other hand, quantum computing is still in its early stages, similar to the vacuum tube era of classical computing. Currently, there are multiple competing technologies used to implement qubits physically, including superconducting circuits, trapped ions, photonic qubits, and spin qubits. Figure 2 summarizes several of the prominent qubit technologies, the physical mechanisms they use to represent quantum states, and the companies actively pursuing each technology. Among these, superconducting qubits are currently the most widely adopted, but other approaches show promising potential in terms of scalability, coherence times, and ease of integration.

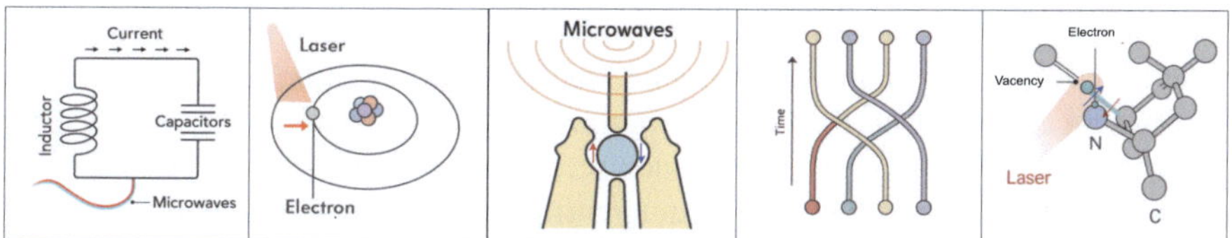

Superconducting	Trapped Ions	Trapped Ions	Topological Qubits	Diamond Vacancies
In a superconducting circuit, current flows without resistance, oscillating back and forth through the loop. When a microwave signal is applied, it excites the current into quantum superposition states.	Ions are atoms with an electrical charge. They have quantum energy levels determined by the positions of their electrons. Carefully tuned lasers are used to cool and trap these ions, allowing them to be placed into superposition states.	These tiny "artificial atoms" are formed by trapping an extra electron inside a small region of pure silicon. By applying microwaves, the electron's quantum state can be manipulated and controlled.	In certain semiconductor structures, electrons can form quasiparticles whose motion follows braided paths over time. These braids are highly stable and can be used to encode quantum information in a way that resists errors.	A nitrogen atom and a missing carbon site in the diamond lattice capture an extra electron. The electron's quantum spin, along with spins of nearby carbon nuclei, can be tuned and controlled with light.
Google, D-Wave, Raytheon, Intel, Rigetti, IBM	AQT , QCI , Honeywell Quantum Factory, IonQ	Leti (CEA-Leti), Silicon Quantum Computing, Intel	Microsoft	Quantum Diamond Technologies

Fig. 2 Popular quantum computing technologies to implement qubits

2.4 Mathematical Representation and Properties of Qubits

The classical bit can take on one of two discrete values, 0 or 1. These binary values can be interpreted as voltage levels in a circuit, low for 0 and high for 1. At the logical level, it can be viewed as false (0) and true (1). In contrast, the mathematical representation of a qubit is more complex. Qubits are described using the Dirac notation, where the basis states are denoted as $|0\rangle$ and $|1\rangle$. A general qubit state $|\psi\rangle$ is represented as a linear combination (or superposition) of these basis states:

$$|\psi\rangle = \alpha\,|0\rangle + \beta\,|1\rangle$$

Here, α and β are complex numbers that represent probability amplitudes that satisfy $|\alpha|^2 + |\beta|^2 = 1$. Understanding the core principles of quantum mechanics is essential for understanding quantum computing architecture. There are four important properties that define a qubit behavior.

- **Superposition:** Unlike classical bits, which can be in either state 0 or 1, a qubit can exist in a linear combination of both basis states: $|0\rangle$ and $|1\rangle$. This means a qubit can represent multiple states simultaneously, enabling quantum parallelism. A qubit in state $\alpha\,|0\rangle + \beta\,|1\rangle$ carries information in its probability amplitudes α and β, where $|\alpha|^2 + |\beta|^2 = 1$.
- **Entanglement:** This is the quantum property Einstein called "spooky action in a distance." Entanglement is a quantum phenomenon where two or more qubits become linked such that the state of one qubit directly influences the state of another, no matter the distance between them. This unique correlation is fundamental for quantum communication and many quantum algorithms.
- **Coherence:** To preserve quantum information during computation, qubits must remain in a coherent state. Coherence refers to the ability of a qubit to maintain phase relationship between its quantum amplitudes over time. Environmental noise, thermal fluctuations, and imperfections can cause decoherence, collapsing the qubit's state and limiting computational fidelity.
- **Measurement:** Measuring a qubit projects its superposed state onto one of the classical basis states, $|0\rangle$ or $|1\rangle$, with probabilities defined by the amplitudes. Measurement destroys the quantum state, making it a critical and final step in most quantum algorithms.

3 Quantum Gates and Circuits

Just as bits and qubits form the foundation of information representation, logic gates define how computations are performed. This section explores classical and quantum gates from an architectural perspective, examining their logical behavior, mathematical representation, physical realization, and role in processor design.

3.1 Classical Gates

Classical logic gates are the fundamental building blocks of digital circuits in classical computing. They take one or more binary inputs and produce a binary output based on well-defined logical operations. Common types of logic gates include AND, OR, NOT, etc. These gates are combined to implement more complex computational behavior. Table 1 shows the truth table of a two input AND gate, demonstrating how bits are manipulated through logical operations (C = A & B). The output is true (1) only when both inputs are true (1). These gates have a physical realization through arrangements of transistors, typically implemented using Complementary Metal-Oxide-Semiconductor (CMOS) technology. For instance, an AND gate can be constructed using several transistors configured to achieve the required logic function.

3.2 Quantum Gates

Similar to classical logic gates, quantum gates are the fundamental building blocks of quantum circuits. While classical gates change the state of a bit between 0 and 1, quantum gates are operations that evolve the quantum state of qubits through linear algebraic transformations. These operations manipulate the qubit's wave function by altering the probability amplitudes associated with its basis states. Common types of quantum gates include single-qubit gates such as Pauli-X, Y, Z, Hadamard (H), Phase (S), and T-gates; two-qubit gates like Controlled-NOT (CNOT), Controlled-Z (CZ), and SWAP; and multi-qubit gates such as the Toffoli (CCNOT) gate. Table 2 provides examples of single-qubit quantum gates, showing their matrix forms and the resulting states for each input basis state. For instance, the Pauli-X gate (first row) acts as a quantum version of the classical NOT gate: It flips a qubit's state, transforming an input of $|0\rangle$ into $|1\rangle$ and an input of $|1\rangle$ into $|0\rangle$.

Quantum gates are fundamentally different from classical gates in that they do not have a fixed physical implementation. Instead, they are realized dynamically using control signals. These precisely shaped control signals cause the qubit's quantum state to rotate on the Bloch sphere according to the unitary transformation defined by the gate. From architectural perspective, quantum gates should be viewed as instructions or micro-operations rather than static physical circuits. When a program specifies a Hadamard gate on a particular qubit, the control hardware generates

Table 1 Truth table for a 2-input AND gate

Input A	Input B	Output C
0	0	0
0	1	0
1	0	0
1	1	1

Table 2 Examples of single-qubit quantum gates with their matrix forms and resulting states for each input basis state

Gate	Matrix	Output when input is $\|0\rangle$	Output when input is $\|1\rangle$
X (NOT)	$\begin{bmatrix} 0 & 1 \\ 1 & 0 \end{bmatrix}$	$\|1\rangle$	$\|0\rangle$
Z	$\begin{bmatrix} 1 & 0 \\ 0 & -1 \end{bmatrix}$	$\|0\rangle$	$-\|1\rangle$
H (Hadamard)	$\frac{1}{\sqrt{2}}\begin{bmatrix} 1 & 1 \\ 1 & -1 \end{bmatrix}$	$\frac{\|0\rangle+\|1\rangle}{\sqrt{2}}$	$\frac{\|0\rangle-\|1\rangle}{\sqrt{2}}$

the appropriate pulse sequence to enact the desired transformation. Any quantum operation can be decomposed into a sequence of these basic gates. In other words, a finite set of fundamental quantum gates is sufficient to approximate any arbitrary quantum computation to a desired level of accuracy.

3.3 *Physical Realization of Quantum Gates*

In classical computing, logic gates are implemented as static circuits built from transistors that permanently define their behavior. Quantum gates, however, lack fixed physical structures; instead, they are dynamically reconfigurable operations applied to qubits at runtime, similar to how microcode in a CPU activates different functional units based on the executed instruction. In practice, quantum gates function as instructions that are enacted by the control subsystem of a Quantum Processing Unit (QPU). This execution engine applies the required transformation to a qubit by sending precisely timed and shaped control signals, which vary depending on the qubit technology. For example, in superconducting qubits, microwave pulses are used to manipulate the quantum state, while in trapped ion systems, laser beams serve the same purpose.

Two-qubit entangling gates such as CNOT and CZ are realized through physical couplers that enable controlled interactions between qubits, as shown in Fig. 3. Architecturally, these couplers are analogous to hardware interconnects that allow data dependencies and multi-bit operations in classical processors. However, unlike classical interconnects that physically transfer bits along wires, quantum couplers do not move qubits between locations. Instead, the QPU's control system sends activation signals to the coupler, allowing the stationary qubits to temporarily interact and modify their joint quantum state to perform the desired entangling operation. The fidelity and speed of these gate operations directly influence the cycle time and

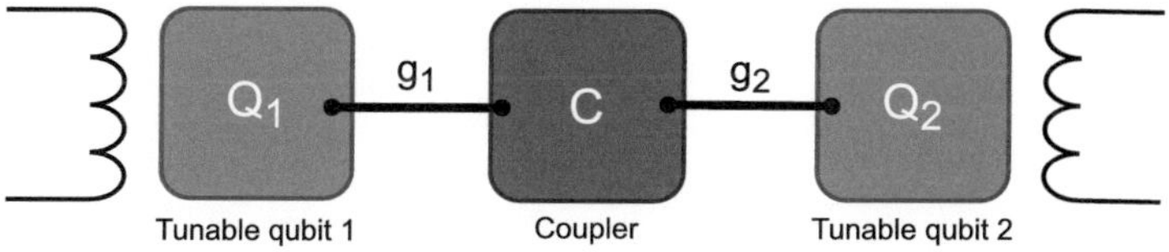

Fig. 3 Tunable coupler architecture for implementing two-qubit gates in superconducting quantum processors. The coupler (C) mediates controllable interactions between two adjacent tunable qubits (Q_1, Q_2) through adjustable coupling strengths g_1 and g_2, enabling high-fidelity entangling operations such as CNOT and CZ gates

overall performance of quantum processors, much like how gate delays and timing characteristics define the performance of CMOS-based digital circuits.

3.4 Entanglement in Quantum Processors

Entanglement is a fundamental property of quantum systems and a key enabler of quantum algorithms. Architecturally, achieving entanglement requires creating a shared quantum interaction between two or more qubits so that their states become correlated in a way that cannot be described independently. In superconducting QPUs, this is implemented using tunable couplers or shared resonators that mediate interactions between adjacent qubits. By applying precisely shaped microwave pulses, the control system adjusts the energy levels of the qubits and the coupler, enabling controlled exchange of quantum information that generates entanglement. In trapped-ion systems, entanglement is achieved by applying laser pulses that couple the ions' internal states through their collective vibrational modes. From a system design perspective, these mechanisms are analogous to configurable interconnect fabrics in classical multiprocessors: they allow local modules (qubits) to establish temporary high-bandwidth interactions without physically moving data. The efficiency and fidelity of entanglement generation directly affect the performance and scalability of quantum processors, much like interconnect bandwidth and latency define performance limits in classical architectures. Figure 3 shows an intermediate tunable coupler for implementing two-qubit gates. This is the dominant approach in superconducting quantum processors developed by different companies, such as IBM, Google, and Rigetti.

3.5 Classical Versus Quantum Circuits

In classical computing, digital circuits are constructed by interconnecting multiple logic gates to perform complex computations. These circuits define fixed hardware pathways where bits flow through gates like AND, OR, and NOT to form

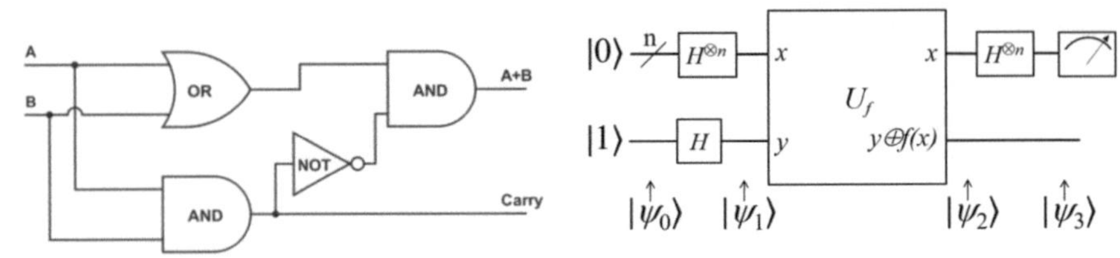

Fig. 4 Comparison between classical and quantum circuits. (Left) A classical combinational circuit implementing binary addition using interconnected logic gates (AND, OR, NOT). (Right) A quantum circuit for Deutsch-Jozsa algorithm [7], where a sequence of quantum gates, and measurement dynamically manipulates qubits to compute the function's parity

combinational logic blocks, such as adders and multiplexers, and sequential elements like registers. The arrangement of these gates determines the computation performed by the processor's datapath. For example, a full adder as shown in Fig. 4 (left) can be constructed using AND, OR, and XOR gates, and it is used to perform binary addition of numbers. Hardware designers build fixed, reusable components such as Arithmetic Logic Unit (ALU), multiplexers, and adders from these gates, and algorithms expressed as instruction sets dynamically reuse these components during execution. For example, an ADD instruction does not build a new adder, it activates the existing ALU composed of AND, OR, and NOT gates.

In quantum computing, the term "circuit" refers to a temporal sequence of quantum gate operations applied to qubits rather than a permanent wiring of physical components. Unlike classical processors, quantum hardware does not contain fixed functional blocks such as ALUs. Instead, it consists of qubits and their couplings, with computation carried out through a set of primitive quantum operations, single- and multi-qubit gates, that serve as directly executable instructions. Any quantum algorithm is expressed as a sequence of these gates, which dynamically manipulate qubits to achieve the desired computation. For instance, the DeutschÃ¢â,¬âŁœJozsa algorithm can be expressed as a quantum circuit, as shown in Fig. 4 (right). It is one of the earliest quantum algorithms that demonstrated quantum speedup [7]. In this circuit, Hadamard and CNOT gates are applied sequentially to prepare superposition states, create entanglement, and finally measure the qubits to determine whether the function is constant or balanced. The same quantum hardware can execute different algorithms simply by changing the sequence and timing of quantum gates constructed by the QPU's control electronics, without requiring any physical reconfiguration of the underlying quantum chip. The quantum circuit [17] is the most popular model of quantum computation, where quantum operators are described as quantum gates.

4 Quantum Processing Unit

A Quantum Processing Unit (QPU) is the state-of-the-art processing hardware at the core of modern quantum computers, analogous to the Central Processing Unit (CPU) in classical systems. Much like a CPU serves as the "brain" of a traditional computer, a QPU functions as the execution engine of a quantum system, performing quantum gates on qubits to carry out computations. Architecturally, a QPU is more than just a quantum chip, it integrates physical qubits along with supporting control electronics and classical computer hardware. These additional components manage essential functions such as storing instructions in memory, generating and amplifying input signals, processing outputs, and isolating useful signals from noise. Together, they enable the precise manipulation of qubits and reliable execution of quantum algorithms, forming the foundation of quantum computing architectures.

The QPU architecture cannot be discussed without considering the surrounding system-level infrastructure. Figure 5 provides an overview of a QPU and its supporting subsystems. While the core of the QPU operates under strict quantum mechanical constraints, it relies heavily on classical electronics and infrastructure for control, measurement, and error correction. The components are categorized into core QPU elements and system-level infrastructure. Note that these components conceptually apply to all QPU technologies, but their physical realizations may differ. For example, couplers may be microwave resonators for superconducting qubits, laser-mediated interactions for trapped ions, or optical components for photonic qubits.

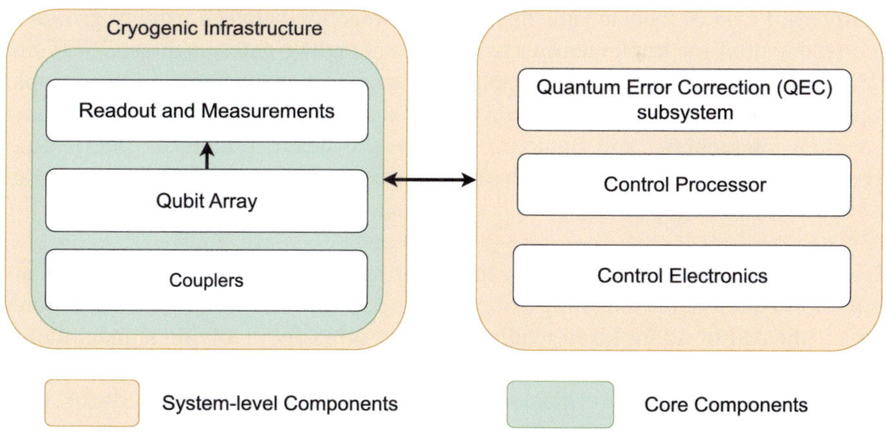

Fig. 5 System-level view of a Quantum Processing Unit (QPU) architecture. The core QPU components, including the qubit array, couplers, and readout interfaces, are housed within a cryogenic environment to preserve quantum coherence. The system-level components, such as control electronics, classical processors, and error correction logic, are external units for managing gate execution, generating control signals (e.g., microwave or laser pulses), and processing measurement results

We first discuss both core and supporting components of quantum processing units. Next, we explore several state-of-the-art quantum processors developed by leading technology vendors, including IBM, Google, and Rigetti. Each quantum processor highlights different architectural choices, qubit technologies, and system-level trade-offs in building scalable quantum computing platforms.

4.1 Core Components in Quantum Processing Units

In this section, we discuss three core components in quantum processing units (QPU): qubit array, couplers, and readout and measurement units. These are the essential quantum elements that reside within the QPU boundary and directly participate in quantum computation.

Qubit Array: The physical medium where quantum information is stored and manipulated. It consists of an arrangement of qubits that serve as the core computational units of the QPU. Depending on the underlying technology (e.g., superconducting circuits, trapped ions, photonic waveguides), qubits may require specialized isolation, cryogenic cooling, and coherence-preserving environments to minimize noise and decoherence. Architecturally, the layout of the qubit array (e.g., linear, 2D grid, etc.) directly impacts gate connectivity, fidelity, and the feasibility of implementing complex quantum circuits.

Couplers: Physical mechanisms that enable interaction between pairs or groups of qubits, essential for implementing two-qubit entangling gates such as CNOT and CZ. The physical realization of couplers varies across technologies. For example, superconducting qubits use microwave resonators, trapped ions use laser-induced Coulomb interactions, and photonic systems use beam-splitters or interferometers. Architecturally, couplers define the qubit connectivity graph, influencing gate scheduling, circuit depth, and overall system scalability.

Readout and Measurement Units: Interfaces responsible for collapsing the quantum state of qubits and converting the result into classical bits. Each qubit is measured via a dedicated or multiplexed readout channel, typically involving signal amplification (e.g., using Josephson parametric amplifiers in superconducting systems) and digitization. These units are critical for executing measurement-based feedback, error correction, and final result extraction in quantum algorithms.

4.2 System-Level Components

There are various components that reside outside the quantum core and play a critical role in orchestrating, controlling, and interpreting quantum operations. In this section, we briefly describe four components: cryogenic infrastructure, control electronics, classical control processor, and error correction subsystem.

Cryogenic Infrastructure: This infrastructure is required in quantum technologies such as superconducting qubits, where core QPU components must operate at millikelvin temperatures to preserve quantum coherence and reduce thermal noise. This infrastructure provides thermal isolation, vibration damping, and stable environmental control. Figure 1b depicts a portion of the multi-stage dilution refrigerator, consisting of tiered cooling platforms that reach progressively lower temperatures as it moves downward. At the base, its coldest stage, the quantum processor (qubit chip) is mounted and operated. While the dilution refrigerator is technically part of the system-level infrastructure, it tightly encloses and enables the QPU, making it functionally inseparable in many implementations.

Control Electronics: Control electronics is responsible for generating the precisely shaped electrical or optical signals, such as microwave pulses for superconducting qubits or laser beams for trapped ions, that implement quantum gate operations. Architecturally, they serve as the interface between classical instruction dispatch and low-level qubit manipulation. These electronics often include waveform generators, mixers, and digital to analog converters (DACs), and are connected to the qubits via filtered and attenuated transmission lines to preserve signal integrity at cryogenic temperatures. The specific realization of control hardware varies significantly with qubit technology, ranging from room-temperature radio frequency (RF) sources for trapped ions to multi-stage attenuated microwave lines for superconducting systems.

Classical Control Processor: Classical processor functions as the execution engine for the quantum instruction stream. It schedules gate operations, coordinates timing across qubit control lines, triggers measurement, and handles classical post-processing tasks. Architecturally, it plays a role similar to a CPU's control unit in classical computing but with additional responsibilities related to the hybrid nature of quantum-classical execution. In many platforms, this processor must operate in real time to support tight feedback loops, especially for mid-circuit measurements or quantum error correction. Its implementation may vary from general-purpose FPGAs and CPUs in superconducting systems to more specialized digital logic in photonic or trapped-ion architectures.

Error Correction Subsystem: This subsystem implements quantum error correction (QEC) codes by continuously analyzing the measurement results and applying feedback operations to mitigate decoherence-induced errors. QEC spans both quantum and classical domains: ancilla qubits and gate operations reside within the QPU, while the measurement $\rightarrow$ decoding $\rightarrow$ feedback loop is managed by classical control processors. The specific realization and integration of QEC mechanisms vary depending on the underlying qubit technology.

Although these components are physically separate and operate under different constraints, they must be tightly synchronized to achieve reliable quantum computation. This division reflects a layered design: quantum logic at the cryogenic core and classical orchestration in the surrounding system. The modular nature of these components allows the same control architecture to support different qubit technologies, with variations in how qubits and couplers are physically realized.

4.3 IBM Quantum Processors

IBM QPU architectures are built around superconducting qubits, arranged on silicon chips and controlled via precisely shaped microwave pulses. They support both single-qubit gates and two-qubit gates, enabling universal quantum computation. The IBM QPUs implement entanglement via couplers, which allow controlled interactions between specific qubit pairs. IBM QPUs also have cryogenic infrastructure to maintain coherence.

Early designs, such as `ibmq_toronto` (27 qubits) and `ibmq_melbourne` (15 qubits), used relatively simple mesh and ladder topologies, respectively (see Fig. 6). In these diagrams, each node represents a physical qubit and each link indicates a two-qubit coupling, i.e., where entangling gates like CNOT can be applied. The limited connectivity in these early architectures constrained parallelism and required additional SWAP operations to route quantum information. Over the span of a few years, IBM advanced QPU designs across four main generations:

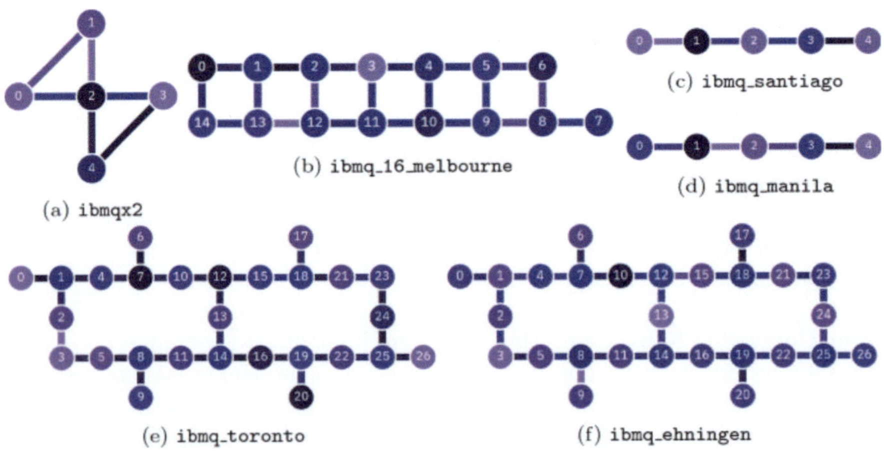

Fig. 6 Early generation IBM architectures with different number of qubits and topologies [11]. **a** ibmqx2: A 5-qubit device with a sparse, triangular connection pattern. **b** ibmq-16-melbourne: A 15-qubit device with a linear+ladder topology. **c, d** ibmq-santiago and ibmq-manila: Small 5-qubit devices with linear connections. **e, f** ibmq-toronto and ibmq-ehningen: 27-qubit devices with IBM's heavy-hex topology designed to reduce crosstalk and optimized for error correction

- *Eagle* (127 qubits): It marked the first 100+ qubit system in IBM's lineup, using fixed-frequency transmons and a dense heavy-hex lattice topology. The heavy-hex lattice reduces crosstalk and enables more efficient quantum error correction by minimizing the number of directly connected qubits while still allowing sufficient connectivity for algorithms. Eagle processors are publicly accessible in IBM cloud, for example, ibm_brisbane QPU is running Eagle r3 with 127 qubits.
- *Osprey* (433 qubits): It demonstrated at-scale chip density and wiring capabilities, focusing on exceeding qubit counts. It is used for research and scaling experiments rather than public use.
- *Condor* (1,121 qubits): Currently the largest qubit count achieved, employing cross-resonance gates. Architecturally, Condor pushes integration and packaging limits; its performance is comparable to Osprey due to similar error profiles.
- *Heron* (133/156 qubits): Heron demonstrates superior gate performance with 3x less error rates compared to Eagle due to its tunable coupler technology. Heron r2/r3 also uses heavy-hex lattice topology with each qubit connected with up to three neighbors. Heron is accessible to users via the cloud. Figure 7 shows the topology of Heron r3 from the ibm_pittsburgh QPU. In both early and modern IBM QPU diagrams, each node represents a single physical qubit, and each link corresponds to a tunable coupler enabling entangling operations.

While processors like Condor and Osprey demonstrate IBM's ability to scale qubit counts in a single processor which can test the limits of qubit counts in a QPU. Heron represents a strategic shift toward practical performance, with its refined architecture and significantly lower error rates making it IBM's most performant chip to date. According to IBM quantum roadmap [12], their focus is on two major areas increasing scalability and fault tolerance in QPUs, eventually leading to the first quantum supercomputer. In terms of fault tolerance, their next commercially available

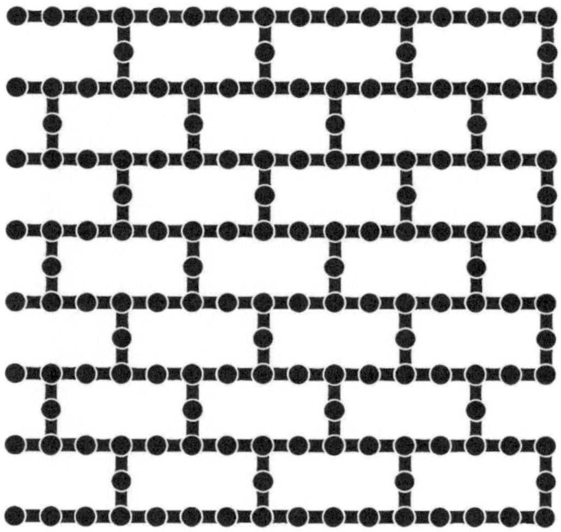

Fig. 7 IBM Heron r3 topology (ibm_pittsburgh) with 156 qubits [20]. The "heavy-hex" lattice topology allows up to three neighbors for a qubit. The couplers in between qubits are tunabl

processor series of Nighthawk from 2025-2028 plan to use square lattice, connecting each qubit to up to four neighbors. Nighthawk will have 120 qubits and supposed to deliver upto 16x effective circuit depth than heavy-hex lattice based QPUs of its predecessors. Nighthawk's square lattice design aims to increase parallelism in two-qubit operations, thereby reducing circuit depth and improving the feasibility of implementing error correction codes at scale. In terms of scalability, their focus is on modular quantum computing such as the proposed Quantum System Two™ which will be discussed in Sect. 5. IBM has already started working on a Quantum data center in Poughkeepsie, NY that will host future modular fault-tolerant quantum supercomputer by early 2030s.

4.4 Google Quantum Processors

Google's quantum processors are based on superconducting transmon qubits fabricated on silicon or sapphire substrates, controlled using microwave pulses. Two-qubit entangling gates are implemented via tunable couplers that can be configured to realize iSWAP or CZ gates depending on pulse parameters. The qubits are arranged in a predominantly two-dimensional (2D) rectangular grid, where each qubit couples to up to four nearest neighbors, balancing high connectivity for surface code layouts with manageable wiring complexity. Single-qubit rotations are driven by resonant microwave pulses, while two-qubit gates are activated by tuning the coupler to generate the desired exchange and phase interactions.

Bristlecone processor (2018) featured 72 qubits in a large 2D grid to study scalability, calibration challenges, and error behavior. In 2019, Google introduced *Sycamore*, a 54-qubit (53 active) chip with tunable couplers, which achieved the "quantum supremacy" milestone via random-circuit sampling. Figure 8 shows the topology of a Sycamore QPU [1]. The node and link abstractions are conceptually similar to

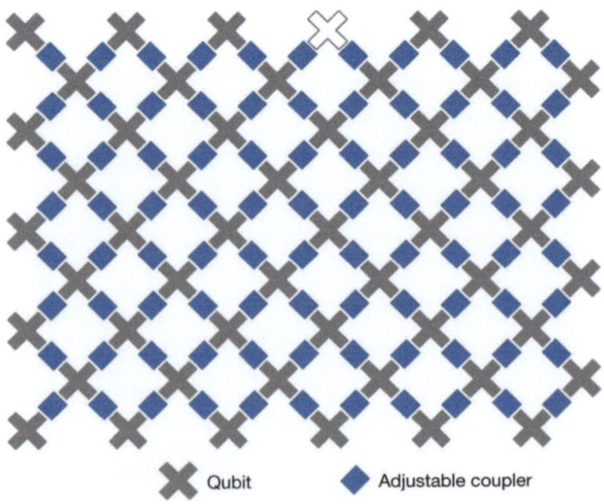

Fig. 8 Layout of Google Sycamore processor, showing a rectangular array of 54 qubits (grey), each connected to its four nearest neighbors with couplers (blue) [1]

IBM's architecture, with each node representing a physical qubit and each link representing a tunable coupler between neighboring qubits. Sycamore also follows a 2D grid topology as its predecessor. The 2D nearest-neighbor layout directly supports the surface code's stabilizer structure, minimizing the number of SWAP operations for syndrome extraction while keeping control wiring feasible in cryogenic environments. This geometry also enables modular scaling by tiling larger grids without sacrificing locality.

Sycamore also established cross-entropy benchmarking (XEB) as a standard performance metric and became the baseline design ("Sycamore-class") for subsequent processors used in quantum error correction studies. Google's research has since focused on improving entangling gate fidelities through refined coupler designs and advanced pulse shaping; scaling surface-code experiments from distance-3 to distance-5 logical patches, achieving clear error suppression beyond the practical threshold; advancing cryogenic integration with higher wiring density, multiplexed readout, and uniform coupler performance; and conducting detailed noise studies to co-optimize device physics, calibration schedules, and error models.

Google's *Willow* processor increased quantum coherence times from $\sim$20 μs to over 100 μs without sacrificing architectural advantages, benefiting from fabrication advances at their new Santa Barbara facility. Willow's logical qubits operate well below the surface-code threshold, demonstrating an exponential error-rate reduction when scaling encoded grids from 3×3 to 5×5 to 7×7 logical qubits. Using the random-circuit sampling benchmark [14], Willow performed a computation in 25 minutes that would take the best supercomputer $\sim 10^{25}$ years, reinforcing quantum supremacy claims. Willow comprises 105 qubits with an average connectivity of 3.47, as shown in Fig. 9.

Guided by a six-stage plan, Google aims to progress from the current hundred-qubit devices toward large-scale, error-corrected systems. Milestones include improving quantum error correction (2023), building long-lived logical qubits (10^{-6} error rate), implementing high-fidelity logical gates, engineering large-scale 2D grids of 10^5 qubits, and ultimately constructing a large-scale fault-tolerant quantum computer with logical error rates below 10^{-13}.

4.5 *Rigetti Quantum Processors*

Rigetti's quantum processors use superconducting qubits arranged in planar arrays with tunable couplers and fast gate operations. Qubit interactions, both single- and two-qubit gates, are controlled via microwave or DC pulses, and the control stack is optimized for low-latency execution. Rigetti also operates its own dedicated fabrication facility for superconducting qubits. To address scalability, they emphasize multi-chip architectures, linking smaller dies with coherent interconnects to enable entanglement across chips.

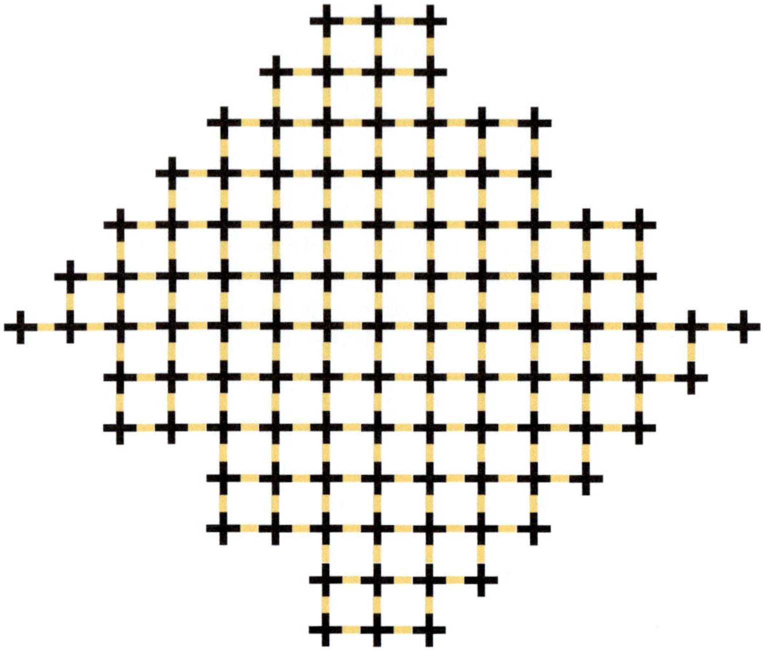

Fig. 9 Layout of Google Willow processor [16], showing a rectangular array of 105 qubits, each connected to its four nearest neighbors with couplers

Early systems such as Aspen-1 through Aspen-4 featured modest qubit counts (13Ã¢â‚¬â€œ16 qubits) in sparse mesh layouts. These initial designs supported benchmarking efforts and software stack development, including the Forest platform and Quil compiler. The Aspen-8 chip (32 qubits) advanced to a regular planar layout with three-fold connectivity, where each qubit couples to three neighbors. It delivered fast gate times (60 ns for single-qubit; 160 ns for two-qubit) and relatively long coherence ($\sim$ 20 μs), while supporting dual-mode entangling gates (CZ and XY), which reduce circuit depth. Aspen-9 and Aspen-10 further extended this architecture with similar layouts and incremental improvements in gate fidelity. Aspen-11 introduced a larger 40-qubit device featuring an octagonal lattice, designed to enhance connectivity and benchmark scaling trends.

In 2022, Rigetti developed first ever multi-chip system by combining two 40 qubit chips marking significant steps toward industrial-scale quantum processors. The two dies are linked together with modular couplers (Rigetti's proprietary multi-chip technology). Figure 10 shows logical connectivity map of Rigetti Aspen-M where the chip utilized the octagonal lattice layout from its predecessors. Neighboring superconducting qubits naturally interact strongly, so the main challenge is suppressing this when we don't need it. In Aspen-class systems, fixed couplings are used, with qubit frequencies controlling interaction strength. Every other qubit is "parked" at a higher frequency to keep neighbors detuned and weakly interacting at rest. During a

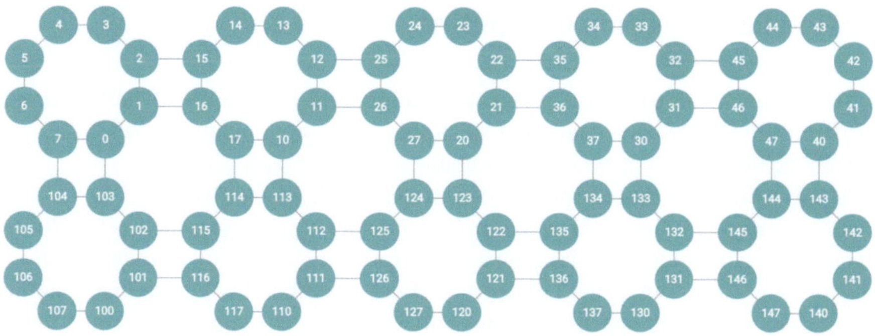

Fig. 10 The layout of 80 qubits in Rigetti's Aspen-M chip series that features an octagon lattice [44]

two-qubit gate, their frequencies are rapidly brought closer to enable entanglement. This approach works but constrains how strong and fast couplings can be, since too much residual "always-on" interaction would degrade performance. Tunable couplers remove this tradeoff altogether. They allow the qubits' frequencies to be close to each other while turning their interactions off.

The fourth generation of Rigetti QPUs, the Ankaa series, introduced tunable couplers. These couplers not only enhance gate performance but also allow each qubit to connect to more neighbors (Fig. 11). As a result, Ankaa chips adopt square-lattice architectures, moving away from the earlier octagonal layouts. The tunable couplers support ISWAP and CZ gates, allowing flexible gate synthesis and tile-based expansion for larger devices. Ankaa is a 84-qubit single-chip processor. Overall, Rigetti pushes toward larger, more interconnected superconducting quantum systems by using dense topologies, dual-gate primitives (CZ/ISWAP), and advanced packaging strategies to retain signal integrity and performance across expanding QPU sizes.

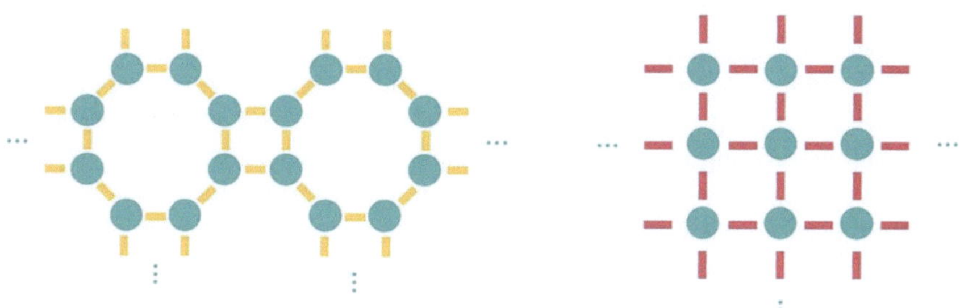

Fig. 11 Fourth generation Ankaa chips (right) moved to square-lattice compared to aspen-M's octagonal layout (left) [2]

5 Modular Quantum Computing

To fully unlock the potential of quantum computing, thousands of fault-tolerant, interconnected qubits are required [15]. However, current quantum technologies are far from achieving this goal, as there remain significant technological limitations on the number of qubits that can be embedded in a single quantum chip [27].

So far, we have discussed *monolithic quantum computing*, which refers to the execution of a quantum algorithm on a single quantum processor. In such systems, a quantum algorithm is executed either directly as described by the original quantum circuit or by replacing unavailable gates with equivalent sequences supported by the hardware. For this to be possible, the number of physical qubits within the processor must be at least equal to the width of the quantum circuit, i.e., the number of logical qubits. Each qubit in the circuit must be mapped onto a distinct physical qubit. This requirement limits scalability. As quantum algorithms grow in complexity, the need for architectures that extend beyond a single chip becomes unavoidable.

In this section, we first discuss challenges in modular quantum computing. Next, we describe how IBM and Rigetti are developing modular quantum architectures, with the broader vision of building quantum-centric supercomputers that integrate quantum processors, classical processors, and hybrid communication networks.

5.1 Challenges in Modular Quantum Computing

A promising solution adopted by both academia and industry is the *distributed quantum computing* (DQC) paradigm, which relies on quantum network infrastructures to cluster together modular and smaller quantum processors [3]. This architectural approach is directly analogous to the transition from single-core to multi-core processors in classical computing. In the DQC paradigm, individual quantum processors—each constrained in qubit count—cooperate to solve tasks that exceed the capacity of a single processor. While classical distributed systems scale linearly with the number of processors, quantum modular systems can, in principle, unlock exponential increases in computational power by clustering quantum chips. A key aspect of distributed computation is the need to perform *non-local quantum gates*—operations between qubits residing in different processors. In classical distributed computing, non-local operations are handled by copying and transmitting data between nodes. In quantum systems, such operations rely on *entanglement* shared between distant qubits, requiring both quantum and classical communication primitives.

Despite its promise, modular quantum computing faces major challenges:

1. **Interconnection of Multiple QPUs:** Multi-core quantum architectures require dedicated hardware and software resources to interconnect multiple QPUs. Unlike classical systems that exploit shared memory, quantum systems rely on shared entangled states to enable parallelism.

2. **Hardware and Software for Interconnection:** Effective modular architectures demand specialized components such as entanglement generation and distribution units, as well as classical control systems to manage the required quantum and classical communications.
3. **Entanglement Coherence:** Similar to cache coherence in classical multi-core processors, quantum systems must maintain *entanglement coherence*—the consistency of shared entangled states across processors. Since any operation on an entangled qubit impacts the global state, maintaining coherence is a fundamental challenge. Moreover, entanglement is inherently fragile and degrades over time.
4. **Scalability Limitations:** While modularity theoretically enables scaling, practical constraints exist. The physical distance between remote qubits is limited, restricting the number of cores that can be interconnected within a single system. This bounds the size of physically clustered quantum processors.
5. **Quantum Transducers for Long-Distance Links:** Scaling beyond chip-level to multi-computer or quantum data-center ("farm") architectures requires quantum transducers—devices that convert stationary qubits into photonic qubits suitable for transmission over quantum channels. Building efficient, low-loss transducers remains one of the most formidable technological challenges.

5.2 Rigetti Roadmap for Developing Modular Quantum Systems

Rigetti's development of multi-chip tunable couplers for coherent chip-to-chip interactions demonstrates progress toward overcoming these barriers and lays the groundwork for modular, chiplet-style architectures. Rigetti has demonstrated high-fidelity, low-latency quantum interconnects between chips. In 2022, they unveiled the first multi-chip quantum processor by linking two 40-qubit dies through proprietary modular couplers. This milestone marked a significant step toward industrial-scale modular quantum systems and demonstrated practical chiplet integration.

5.3 IBM Roadmap for Developing Modular Quantum Systems

IBM has made modularity central to its long-term roadmap in designing fault-tolerant quantum computers. We briefly describe three IBM modular quantum computer architectures.

- **Quantum Flamingo:** A proof-of-concept that connects two Heron R2 chips via L-couplers. These couplers enable cross-chip CNOT gates, allowing Flamingo to demonstrate scalable modular execution.

- **IBM Quantum System Two:** This modular system ties together Heron processors with advanced cryogenics, control electronics, and runtime servers. The system accommodates three Heron processors, serving as a foundation for scalable, error-corrected quantum computation.
- **Kookaburra (2025–2026):** IBM's upcoming 1386-qubit modular chip, designed with inter-chip communication support. When three Kookaburra chips are interconnected, the system will scale to 4,158 qubits. This processor is designed to integrate quantum memory with logic operations, forming a cornerstone of IBM's fault-tolerant quantum computing strategy.

6 Quantum on-Chip Communication

Both in classical and quantum computing, on-chip communication plays a pivotal role in determining the scalability and performance of processors. In a monolithic quantum processing unit (QPU), qubits must interact with each other to realize two-qubit gates, which are essential for universal computation. This requires efficient coupling mechanisms across the chip. As systems scale toward modular quantum computing, data movement and interconnectivity become critical bottlenecks. Modular architectures demand robust communication mechanisms not only within a single die but also across multiple chips to build large-scale quantum systems. In this section, we first provide an overview of classical on-chip communication. Next, we discuss quantum intra-chip connectivity. Finally, we discuss quantum inter-chip communication.

6.1 Classical On-Chip Communication

In modern multi-processor systems-on-chip (MPSoCs), different intellectual property (IP) cores must exchange information to execute complex tasks. A major use case is cache coherence, where distributed caches must remain consistent across processors. Early designs relied on shared buses, but this approach did not scale due to contention and limited bandwidth. To overcome this, the *Network-on-Chip* (NoC) paradigm was introduced, routing packets instead of relying on direct wires.

A NoC transfers two kinds of information: (i) data packets containing application payload, and (ii) control packets used for coordination and synchronization. Typically, an electrical NoC transmits bits as electrical signals over metal interconnects. Emerging approaches also explore wireless and optical links for reduced latency and higher bandwidth. Figure 12 shows a typical mesh NoC. A NoC comprises three primary components: *routers* that switch packets, *network interfaces* that connect cores to the network, and *links* that carry data. A packet traverses from source to destination via routers along the path, enabling scalable on-chip communication.

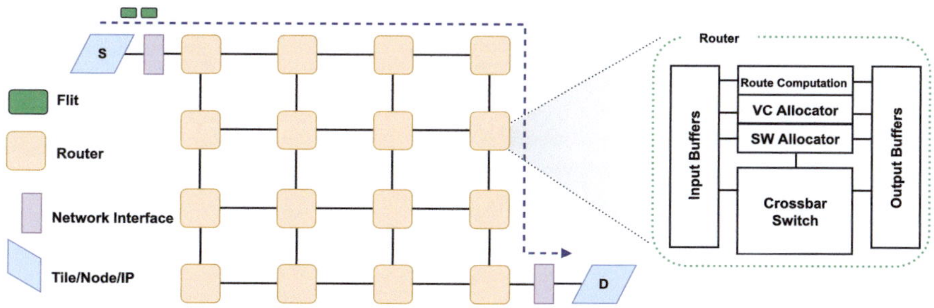

Fig. 12 NoC with 4x4 mesh topology. Each IP is connected to NoC via a network interface followed by a router. Node S (source) sends packets to D (destination) via NoC. The figure on the right shows a typical router micro-architecture [41]

6.2 *Quantum Intra-Chip Connectivity*

Within a single quantum chip, physical qubits require controlled interaction to realize multi-qubit operations such as the CNOT or CZ gate. This intra-chip connectivity is typically achieved through mediums such as superconducting resonators, coplanar waveguides, Coulomb interactions in trapped ions, or spin chains. These physical couplings allow qubits to exchange quantum information, albeit subject to noise and physical layout constraints. The necessity of intra-chip communication arises because not all qubit pairs can directly interact. Each quantum device is characterized by a *coupling graph*, where vertices represent qubits and edges represent the possibility of implementing a two-qubit gate. This restricted connectivity requires swap operations to move quantum states across the chip, directly impacting circuit depth and execution fidelity. Thus, intra-chip communication fundamentally transfers *quantum states* via controlled interactions among qubits.

6.3 *Quantum Inter-Chip Communication*

As quantum devices evolve beyond monolithic designs, modular quantum computing requires interconnecting multiple chips. The concept of a *Quantum Network-on-Chip* (QNoC) emerges as the quantum analogue of classical NoCs. In QNoC architectures, short-distance inter-chip links (e.g., within the same cryostat or optical table) enable qubits in different processors to interact. Compared to multi-computer architectures, scaling opportunities are more constrained due to the limited spatial proximity of quantum modules.

Unlike classical NoCs, QNoCs must respect fundamental quantum principles. Classical distributed systems freely duplicate and transfer information, but in quantum systems, the *no-cloning theorem* prohibits copying unknown quantum states. Furthermore, quantum measurement irreversibly destroys superposition and entan-

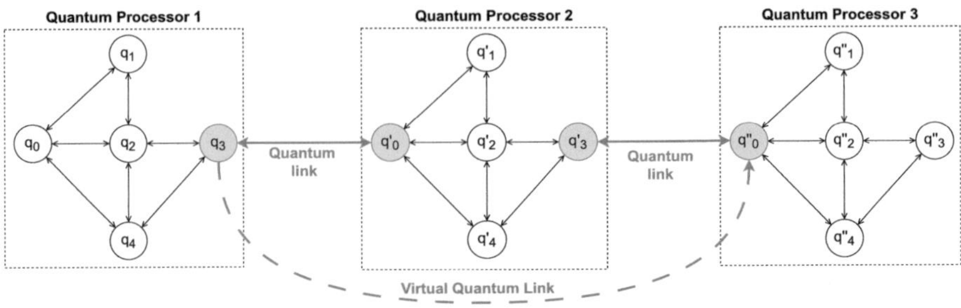

Fig. 13 Entanglement swapping enables remote processors to share quantum states, creating *virtual quantum links* for scalable QNoC architectures [3]

glement, making direct transmission infeasible. Instead, *quantum teleportation* is employed: an unknown qubit is transmitted using pre-shared entangled qubits (e.g., Bell states) and the exchange of classical information.

In the teleportation protocol, a pair of entangled qubits is distributed between source and destination. A Bell-state measurement at the sender's side, combined with the transmission of two classical bits, allows the receiver to reconstruct the unknown qubit. Thus, QNoC communication involves both quantum resources (entanglement distribution) and classical channels (transmitting measurement outcomes). Entanglement swapping further enhances connectivity by allowing intermediate nodes to extend entanglement across non-directly connected modules. For example, as shown in Fig. 13, quantum processor 2 can enable processors 1 and 3 to share a Bell pair via entanglement swapping, thereby creating a *virtual quantum link*.

While the development of quantum on-chip communication is necessary to design modular multi-chip quantum systems, the designers need to address various challenges for designing effective QNoCs:

- **Scalable Architectures:** Enabling large-scale QNoC designs that integrate thousands of qubits while minimizing communication overhead remains a fundamental challenge. Current physical implementations of qubits (e.g., superconducting, ion-trap, photonic) impose strict constraints on qubit coherence times, interconnect length, and cross-talk. Scaling up requires modular architectures with efficient inter-module quantum interconnects, yet the probability of successful entanglement distribution typically decreases with distance and network size. Additionally, resource overhead for quantum error correction grows super-linearly as the number of qubits increases, further stressing interconnect capacity. Balancing physical layout constraints, fabrication variability, and the need for reconfigurable topologies that can adapt to qubit failures makes scalable QNoC design particularly difficult.
- **Quantum Routing Protocols:** Routing must trade off between entanglement fidelity and success probability, all under tight decoherence limits. Unlike classical packets, quantum states cannot be retransmitted due to the no-cloning theorem, so adaptive protocols must handle fluctuating hardware error rates. Incorporating

purification and error correction into routing further increases resource overhead, making scalable and fault-tolerant routing a key challenge.

- **Integration with Classical NoCs:** Hybrid QNoC designs require tight synchronization between fast classical control signals and the slower, probabilistic process of entanglement distribution. Mismatches in timing or bandwidth can create bottlenecks, while ensuring low-latency feedback for error correction and qubit operations adds further complexity.
- **Security:** Security has been an important consideration in traditional Network-on-chips [13, 41]. NoC security attacks and defenses needs to consider: confidentiality, integrity, anonymity, authenticity, availability, and freshness [4–6, 39, 40]. Just as classical NoCs require secure communication to prevent data leakage, QNoCs will also need quantum-safe protocols to preserve confidentiality and integrity of quantum states.

Overall, quantum on-chip communication is a cornerstone of scalable quantum computing, bridging physical qubit interactions on a single die and interconnecting multiple quantum processors in modular architectures.

7 Quantum Architecture Simulators

Quantum computing simulators are essential tools for exploring architectural trade-offs, validating algorithms, and studying the impact of communication in both monolithic and modular QPU systems. Unlike algorithm-level simulators that only model logical gates, architecture simulators capture hardware constraints, device-level noise, and network interconnects.

IBM's Qiskit framework [19] provides one of the most widely used simulation environments. Qiskit Aer [24] offers high-performance circuit simulation with configurable noise models, while Qiskit Pulse [21] exposes low-level pulse control, bridging software abstractions with hardware physics. Its transpiler supports hardware-aware compilation and qubit routing, making it a useful tool for evaluating the architectural cost of limited qubit connectivity. Google's Cirq [9] follows a similar approach, focusing on near-term devices such as Sycamore-class processors. The Open Quantum Library (OpenQL) [18] project provides quantum computer emulator and various open source libraries for using quantum computing. OpenQL also plays a role in quantum-classical co-design by enabling both quantum circuit simulation and resource estimation.

At the network and modular computing level, the SeQUeNCe [43] simulator developed at Argonne National Laboratory is a full-system discrete-event simulator for quantum network communication. It models interconnects, entanglement distribution, quantum memory, and traffic behavior, offering insights into the architectural cost of modular quantum computing. Similarly, NetSquid (NETwork Simulator for QUantum Information using Discrete events) [8] enables end-to-end simulation of modular quantum systems and quantum networks, from physical qubits and noise processes up to application protocols. Research prototypes introduced in studies of multi-core quantum architectures, extend these ideas to model teleportation-based communication and classical feedback control in chip-scale settings.

8 Summary

This chapter presented a bottom-up exploration of quantum computing architectures, designed for readers with a background in traditional computer architecture. Starting with a comparison between classical and quantum paradigms, it explains why quantum computing is fundamentally different, not merely an evolution of classical systems, but a distinct computational model rooted in quantum mechanics. We first compared classical bits and quantum bits (qubits) from both logical and physical perspectives. Next, we outlined how qubits can be programmed to form quantum gates and circuits in different quantum technologies, including superconducting circuits, trapped ions, and photonics. Then, we described the major components in quantum processing units (QPU) developed by IBM, Google, and Riggeti. Next, we explored modular quantum computing, where multiple QPUs are linked to build scalable architectures. We also discussed both intra- and inter-chip communication using entanglement, quantum teleportation, and routing mechanisms. Finally, we discussed simulators for design space exploration of quantum architectures. To enable an effective utilization of future quantum computing architectures, there is a need for design automation methods in various phases of quantum computing, including quantum algorithms [22, 29], quantum compilation [31, 33], quantum state preparation [35, 37], quantum error correction [32], quantum security [38], quantum noise [28, 30], quantum benchmarking [34], quantum characterization [36], quantum verification [42], and quantum measurement [25, 26].

References

1. Frank Arute, Kunal Arya, Ryan Babbush, Dave Bacon, Joseph C Bardin, Rami Barends, Rupak Biswas, Sergio Boixo, Fernando GSL Brandao, David A Buell, et al. Quantum supremacy using a programmable superconducting processor. *Nature*, 574(7779):505–510, 2019.
2. Andrew Bestwick. Introducing the ankaa™-1 system – rigetti's most sophisticated chip architecture unlocks a promising path to narrow quantum advantage. Blog post, Rigetti on Medium, August 2023. Accessed: August 24, 2025.
3. Marcello Caleffi, Michele Amoretti, Davide Ferrari, Jessica Illiano, Antonio Manzalini, and Angela Sara Cacciapuoti. Distributed quantum computing: a survey. *Computer Networks*, 254:110672, 2024.
4. Charles, Subodha, Yangdi Lyu, and Prabhat Mishra. 2020. Real-time detection and localization of distributed dos attacks in noc-based socs. *IEEE Transactions on Computer-Aided Design of Integrated Circuits and Systems* 39 (12): 4510–4523.

5. Charles, Subodha, and Prabhat Mishra. 2020. Reconfigurable network-on-chip security architecture. *ACM Transactions on Design Automation of Electronic Systems (TODAES)* 25 (6): 1–25.
6. Charles, Subodha, and Prabhat Mishra. 2021. A survey of network-on-chip security attacks and countermeasures. *ACM Computing Surveys (CSUR)* 54 (5): 1–36.
7. David Collins, KW Kim, and WC Holton. Deutsch-jozsa algorithm as a test of quantum computation. *Physical Review A*, 58(3):R1633, 1998.
8. Tim Coopmans, Robert Knegjens, Axel Dahlberg, David Maier, Loek Nijsten, Julio de Oliveira Filho, Martijn Papendrecht, Julian Rabbie, Filip Rozpedek, Matthew Skrzypczyk, et al. Netsquid, a network simulator for quantum information using discrete events. *Communications Physics*, 4(1):164, 2021.
9. Cirq Developers. *Cirq*. Zenodo, August 2025.
10. Richard P Feynman. Simulating physics with computers. In *Feynman and computation*, pages 133–153. cRc Press, 2018.
11. Geng, Alexander, Ali Moghiseh, Claudia Redenbach, and Katja Schladitz. 2023. Improved frqi on superconducting processors and its restrictions in the nisq era. *Quantum Information Processing* 22 (2): 104.
12. IBM Quantum. 2025 quantum development & innovation roadmap. IBM, June 2025. IBM Quantum development and innovation roadmap PDF.
13. Prabhat Mishra and Subodha Charles. *Network-on-chip security and privacy*. Springer, 2021.
14. Alexis Morvan, B Villalonga, X Mi, S Mandrà, A Bengtsson, PV Klimov, Z Chen, S Hong, C Erickson, IK Drozdov, et al. Phase transitions in random circuit sampling. *Nature*, 634(8033):328–333, 2024.
15. Nemoto, Kae. 2023. Our future with quantum computers. *JSAP Review* 2023: 230212.
16. Hartmut Neven. Meet willow, our state-of-the-art quantum chip. Google Research Blog, December 2024. Accessed: August 24, 2025.
17. Michael A Nielsen and Isaac L Chuang. *Quantum computation and quantum information*. Cambridge university press, 2010.
18. OpenQL. The open quantum library (openql) project. Open Source Project, n.d. Accessed: August 23, 2025.
19. Qiskit. Qiskit. GitHub Organization, n.d. Accessed: August 23, 2025.
20. IBM Quantum. Ibm quantum platform. Cloud-based quantum computing platform, n.d. Accessed: August 24, 2025.
21. IBM Quantum. Pulse (qiskit api v0.27). IBM Quantum documentation, n.d. Accessed: August 23, 2025.
22. Sahan Sanjaya, Daniel Volya, and Prabhat Mishra. Variational quantum algorithms via measurement-induced passive steering. In *2024 IEEE International Conference on Quantum Computing and Engineering (QCE)*, volume 1, pages 481–487. IEEE, 2024.
23. Matthew Sparkes. Google demonstrates vital step towards large-scale quantum computers. *New Scientist*, 2021.
24. Qiskit Development Team. Qiskit aer (version 0.17.1). Online Documentation, n.d. Accessed: August 23, 2025.
25. Zachery Utt, Daniel Volya, and Prabhat Mishra. Quantum measurement discrimination using cumulative distribution functions. In *2023 Design, Automation & Test in Europe Conference & Exhibition (DATE)*, pages 1–6. IEEE, 2023.
26. Utt, Zachery, Daniel Volya, and Prabhat Mishra. 2024. Quantum measurement classification using statistical learning. *ACM Transactions on Quantum Computing (TQC)* 5 (2): 1–16.
27. Rodney Van Meter and Simon J Devitt. The path to scalable distributed quantum computing. *Computer*, 49(9):31–42, 2016.
28. Daniel Volya and Prabhat Mishra. Impact of noise on quantum algorithms in noisy intermediate-scale quantum systems. In *IEEE International Conference on Computer Design (ICCD)*, pages 1–4. IEEE, 2020.
29. Daniel Volya and Prabhat Mishra. Quantum spectral clustering of mixed graphs. In *2021 58th ACM/IEEE design automation conference (DAC)*, pages 463–468. IEEE, 2021.

30. Daniel Volya and Prabhat Mishra. Modeling of noisy quantum circuits using random matrix theory. In *2022 IEEE 40th International Conference on Computer Design (ICCD)*, pages 132–138. IEEE, 2022.
31. Daniel Volya and Prabhat Mishra. Quantum data compression for efficient generation of control pulses. In *Asia and South Pacific Design Automation Conference (ASPDAC)*, pages 216–221, 2023.
32. Daniel Volya and Prabhat Mishra. Quantum steering of surface error correcting codes. In *2023 IEEE International Conference on Quantum Computing and Engineering (QCE)*, volume 1, pages 1394–1400. IEEE, 2023.
33. Daniel Volya and Prabhat Mishra. Qudcom: Towards quantum compilation for qudit systems. arXiv preprint arXiv:2311.07015, 2023.
34. Daniel Volya and Prabhat Mishra. Quantum benchmarking via random dynamical quantum maps. In *2024 IEEE International Conference on Quantum Computing and Engineering (QCE)*, volume 1, pages 1277–1283. IEEE, 2024.
35. Volya, Daniel, and Prabhat Mishra. 2024. State preparation on quantum computers via quantum steering. *IEEE Transactions on Quantum Engineering (TQE)* 5: 1–14.
36. Daniel Volya, Andrey Nikitin, and Prabhat Mishra. Fast quantum characterization via riemannian optimization. In *2024 IEEE International Conference on Quantum Computing and Engineering (QCE)*, volume 1, pages 1270–1276. IEEE, 2024.
37. Daniel Volya, Zhixin Pan, and Prabhat Mishra. Feedback-based steering for quantum state preparation. In *2023 IEEE International Conference on Quantum Computing and Engineering (QCE)*, volume 1, pages 1308–1318. IEEE, 2023.
38. Daniel Volya, Tao Zhang, Nashmin Alam, Mark Tehranipoor, and Prabhat Mishra. Towards secure classical-quantum systems. In *2023 IEEE International Symposium on Hardware Oriented Security and Trust (HOST)*, pages 283–292. IEEE, 2023.
39. Hansika Weerasena, Subodha Charles, and Prabhat Mishra. Lightweight encryption using chaffing and winnowing with all-or-nothing transform for network-on-chip architectures. In *2021 IEEE International Symposium on Hardware Oriented Security and Trust (HOST)*, pages 170–180. IEEE, 2021.
40. Weerasena, Hansika, and Prabhat Mishra. 2024. Breaking on-chip communication anonymity using flow correlation attacks. *ACM Journal on Emerging Technologies in Computing Systems* 20 (4): 1–27.
41. Weerasena, Hansika, and Prabhat Mishra. 2024. Security of electrical, optical, and wireless on-chip interconnects: A survey. *ACM Transactions on Design Automation of Electronic Systems* 29 (2): 1–41.
42. Hasini Witharana, Daniel Volya, and Prabhat Mishra. Qcassert: Quantum device testing with concurrent assertions. In *Asia and South Pacific Design Automation Conference (ASP-DAC)*, pages 491–496. IEEE, 2024.
43. Xiaoliang Wu, Alexander Kolar, Joaquin Chung, Dong Jin, Tian Zhong, Rajkumar Kettimuthu, and Martin Suchara. Sequence: a customizable discrete-event simulator of quantum networks. *Quantum Science and Technology*, 6, sep 2021.
44. Yuan, Pei, and Shengyu Zhang. 2025. Full characterization of the depth overhead for quantum circuit compilation with arbitrary qubit connectivity constraint. *Quantum* 9: 1757.

State Preparation for Quantum Computing

Sahan Sanjaya and Prabhat Mishra

Abstract Variational quantum algorithms have emerged as a leading framework for harnessing near-term quantum processors, combining parameterized quantum circuits with classical optimization to approximate solutions of classically intractable problems. Quantum state preparation is a critical component in implementing variational quantum algorithms. Specifically, quantum state preparation sets the initial landscape explored by the variational ansatz and ultimately bounds the accuracy of the achieved solution. This chapter provides a comprehensive overview of end-to-end framework from fiducial initialization and ansatz selection, through measurement, cost-function evaluation, and classical parameter updates, emphasizing various trade-offs in noisy intermediate-scale quantum hardware. We present diverse state-preparation methods, spanning conventional techniques such as active and passive reset, dissipative dynamics, and algorithmic cooling, and highlighting their suitability for different computational goals and hardware constraints. We also discuss quantum state preparation using measurement-induced steering that can outperform conventional state preparation methods.

1 Introduction

Quantum computing represents a paradigm shift from the classical computing framework, harnessing the principles of quantum mechanics to perform information processing tasks. Unlike classical bits, which exist deterministically in one of two states (0 or 1), quantum bits (qubits) can exist in coherent superpositions of these states, enabling them to encode and process vastly richer information. As shown in Fig. 1, the quantum computing model organizes this capability into a structured workflow. It starts with preparing qubits in well-defined initial states, mostly known fiducial

S. Sanjaya (✉) · P. Mishra
University of Florida, Gainesville, USA
e-mail: ssanjaya@ufl.edu

P. Mishra
e-mail: prabhat@ufl.edu

P. Mishra (ed.), *Design Automation for Quantum Computing*,
https://doi.org/10.1007/978-3-032-09303-5_3

Fig. 1 Three important stages in quantum computing. First, qubits are initialized to a known fiducial state. Next, the qubits undergo a sequence of discrete quantum gate operations. Finally, the qubits are measured to obtain the computational result

states, manipulates these states through controlled quantum operations, and finally extracts computational results via measurements. This model serves as the foundation for a variety of quantum algorithms that promise significant speedup over their classical counterparts in specific classes of applications [1].

A fundamental aspect of this workflow is *state initialization*, where qubits are prepared in a known fiducial state, typically the computational basis state $|0\rangle$. This step is crucial, as it sets the reference point from which quantum evolution begins. In many algorithms, a uniform superposition is later created from this fiducial state using specific gates, such as *Hadamard* gate. However, the initialization to $|0\rangle$ ensures predictability and repeatability at the start of the computation. In more advanced protocols, qubits may be initialized into specially engineered states to provide algorithmic advantages or to enable specific error-correcting schemes [2–4].

Following initialization, the *quantum processing phase* involves the application of a sequence of discrete quantum gates that manipulate the states of the qubits. These gates, drawn from a universal set, enact unitary transformations that drive the evolution of the quantum state according to the designed algorithm. During computation, qubits may be reset, reinitialized, or reused, especially in scenarios involving intermediate measurements or error correction. For example, in quantum error-correcting codes, ancillary qubits are frequently recycled after syndrome extraction. This ability to dynamically manipulate and reuse qubits, combined with the non-classical features of superposition and entanglement, enables the power of quantum information processing.

Finally, the computation concludes with *quantum measurement*, where information encoded in the quantum state is extracted by collapsing it into one of the classical basis states. This probabilistic process translates quantum amplitudes into observable outcomes. In many quantum algorithms, repeated measurements are performed to estimate expectation values or probability distributions, from which the desired solution is inferred [5, 6]. Post-selection, where only specific measurement outcomes are retained for further processing, is often employed to enhance computational fidelity or to implement certain subroutines. Thus, measurement is not only the final readout but also a key part of running complex quantum protocols.

In this chapter, we focus on quantum state preparation, with an emphasis on its role within variational quantum algorithms. The remainder of this chapter is organized as follows. We first provide an overview of variational quantum algorithms. Next, we discuss different state preparation methods. Finally, we present state preparation for variational quantum algorithms using measurement-induced steering.

2 Variational Quantum Algorithms

In the journey toward achieving quantum advantage, where quantum computers outperform classical ones in specific classes of applications, we currently operate within the era of noisy intermediate-scale quantum (NISQ) devices. These systems are constrained by noise, decoherence, gate infidelities, and a limited number of qubits [7–9]. These constraints severely limit the depth and complexity of circuits that can be reliably executed. This raises an important question: how can today's NISQ devices be used to achieve quantum advantage? To address this challenge, Variational Quantum Algorithms (VQAs) have emerged as the most promising strategy. VQAs address the inherent limitations of NISQ hardware through a hybrid quantum–classical paradigm. VQAs can be viewed as the quantum analogue of classical machine learning methods, where parameterized quantum circuits evaluate problem-specific objectives while classical computers carry out parameter optimization.

The remainder of this section is organized as follows. We first provide an overview of a hybrid quantum–classical paradigm in a VQA framework. Next, we describe major components in a VQA framework, including cost function formulation, ansatz design, and classical optimization. Then, we survey the application domains where VQAs are most beneficial. Finally, we highlight the importance of state preparation in variational quantum algorithms.

2.1 *Overview of a VQA Framework*

The central idea of a VQA is to frame a computational problem, such as finding the ground state of a molecule, solving a system of linear equations, or classifying data as the task of minimizing a cost function over a quantum state space. This cost function must be efficiently computable using a quantum processor, while being expressive enough to distinguish between good and poor solutions. To achieve this, a parameterized quantum circuit is used, which prepares variational quantum states, and a classical optimizer iteratively updates the circuit parameters to minimize the cost as depicted in the Fig. 2. This interplay between quantum state preparation and classical optimization forms a feedback loop that harnesses the strengths of both quantum and classical computing domains.

VQAs are often regarded as the quantum counterpart to classical machine learning (ML) methods. Let us consider a simple example of a VQA-based classification task. Suppose we want to classify 2D data points into two categories, say, red or blue. In the VQA classifier, also known as a quantum classifier, each input data point (x,y) is encoded into a quantum state. Based on the information of the task, a cost function is defined first to quantify the difference between this prediction and the true label. This cost function plays a role analogous to the loss function in classical machine learning, mapping tunable parameters (analogous to weights in classical models) to a real-valued performance metric. A parameterized quantum circuit (ansatz) processes

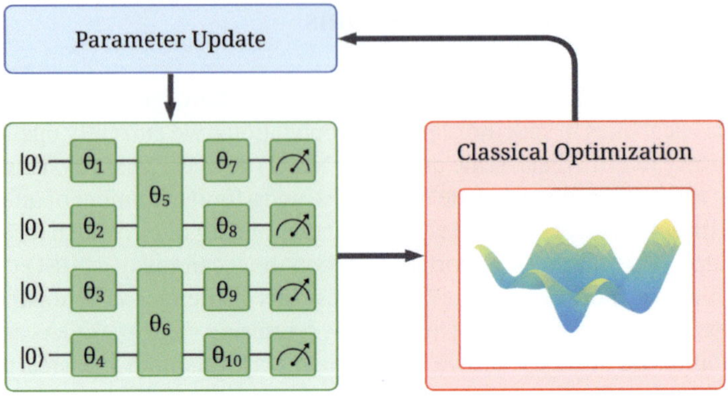

Fig. 2 Schematic of a variational quantum algorithm, a hybrid architecture that combines quantum parameterized circuits with classical optimization

the input state, and the circuit's output is measured to compute an expectation value, which serves as the ML model's prediction. Using the expectation value and the cost function, a classical optimizer then updates the parameters of the quantum circuit to minimize the cost function. This hybrid loop, quantum circuit evaluation followed by classical parameter updates, mirrors the training process of traditional machine learning models. At the same time, it leverages quantum mechanics to potentially explore better solution spaces that may be inaccessible to classical models.

2.2 *Major Components in a VQA Framework*

The success of a VQA framework depends on the development of its core components: defining the cost function, exploring various ansatz designs, and examining the choice of classical optimizers. In this section, we first outline the cost function. Next, we describe the ansatz design. Finally, we explore classical optimization algorithms.

2.2.1 Cost Function

One of the key components of a VQA is the cost function, which must be carefully designed to both reflect the objective of the task and allow for efficient evaluation on quantum hardware. The cost function defines the cost landscape (hypersurface) over which the classical optimizer adjusts the parameters of the ansatz in search of a global minima. A general form of the cost function is given by

$$C(\theta) = \sum_k f_k \left(\mathrm{Tr}\left[O_k U(\theta)\rho_k U^\dagger(\theta)\right]\right), \tag{1}$$

where $\{O_k\}$ are a set of observables, $\{\rho_k\}$ are input quantum states, and f_k are some functions that map measurement outcomes to real-valued contributions. Here, $U(\theta)$ is a parameterized unitary gate, composed of a sequence of quantum gates whose behavior depends on a vector of discrete and continuous parameters $\theta = (\theta_1, \theta_2, \ldots)$.

The choice of cost function directly influences the trainability of the algorithm, the speed of convergence, and its susceptibility to challenges such as barren plateaus. Therefore, the selected cost function should satisfy several important criteria:

- Faithfulness: Its global minimum should correspond to the correct solution to the problem (e.g., the ground state energy of a molecular system).
- Measurability: The cost should be estimable with a manageable number of quantum measurements and should not require excessive classical post-processing.
- Meaningfulness: Lower cost values should correspond to better-quality solutions with respect to the problem's objective.
- Trainability: The cost landscape should allow classical optimizers to efficiently find global minima without getting trapped in flat regions or barren plateaus.

2.2.2 Ansatz Design

Ansatz design is a critical component of the VQA framework. The ansatz defines the structure of the parameterized quantum circuit. It plays a crucial role in determining the expressibility of the quantum model, which is the ability to represent a wide range of quantum states and strongly affects the convergence behavior during optimization. A well-constructed ansatz can represent the solution space with high fidelity using a relatively small number of parameters, making training more efficient. In contrast, a poorly chosen ansatz may lack sufficient expressibility or may lead to barren plateaus, where the cost function becomes nearly flat across large regions of parameter space, making optimization extremely difficult.

The choice of ansatz determines the parameters θ, which are encoded into the unitary $U(\theta)$ appearing in the cost function in Eq. 1. This unitary transformation is typically constructed as a sequence of parameterized gates. Formally, the ansatz unitary can be expressed as a product of sequential unitaries as follows:

$$U(\theta) = \prod_{l=1}^{L} U_l(\theta_l), \tag{2}$$

where each component unitary $U_l(\theta_l)$ can be decomposed as

$$U_l(\theta_l) = \prod_{m} e^{-i\theta_m H_m} W_m, \tag{3}$$

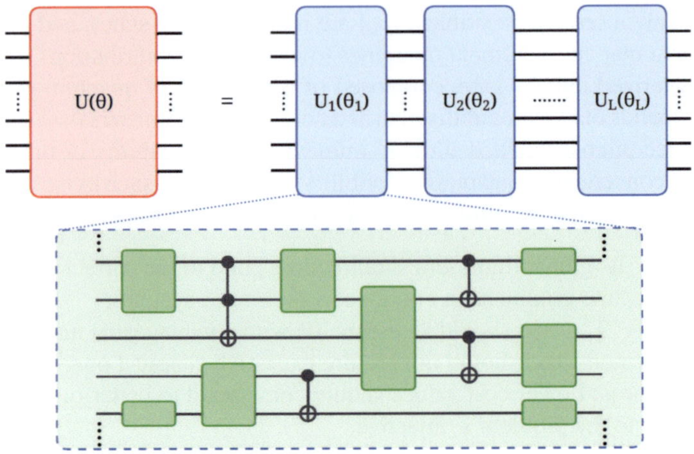

Fig. 3 Schematic of an ansatz, where the unitary $U(\boldsymbol{\theta})$, parameterized by the set $\boldsymbol{\theta}$, is represented as a product of L unitaries $U_l(\theta_l)$ applied sequentially to an input state. Each $U_l(\theta_l)$ can be further decomposed into a sequence of parameterized and fixed gates

with W_m being an unparameterized unitary, and H_m a Hermitian generator. This modular structure, as shown in Fig. 3, allows for flexible circuit design, enabling the encoding of problem-specific knowledge or generic strategies.

Since the choice of ansatz is critical, it is important to quantify its quality using rigorous metrics. Expressibility and the entanglement capability of an ansatz are two metrics that can be used to compare the quality of the ansatz.

Expressibility refers to the ability of a parameterized quantum circuit to explore the Hilbert space uniformly. An ansatz with high expressibility can represent a broad and diverse set of quantum states. The metric of expressibility can be formally quantified by the norm $\|A^{(t)}(U)\|$, where $A^{(t)}(U)$ compares the statistical distribution of quantum states produced by the ansatz to that of a Haar-random ensemble. It is defined as

$$A^{(t)}(U) := \int (|\psi\rangle\langle\psi|)^{\otimes t}\, d\psi - \int \left(U^{\otimes t}|\phi_0\rangle\langle\phi_0|(U^{\dagger})^{\otimes t}\right) dU \tag{4}$$

The first term averages over all pure states over Haar-random, while the second term averages over the states generated by the ansatz $U(\theta)$. Here, $|\phi_0\rangle$ is a reference state, usually set to $|0\rangle$. The closer $A^{(t)}(U)$ is to zero, the more expressive the ansatz is. While a highly expressive ansatz increases coverage of the state space, it may also make optimization more difficult and lead to phenomena such as barren plateaus. Conversely, a simpler ansatz may converge faster but fail to reach the desired solution. Therefore, expressibility must be carefully balanced against trainability when designing ansatz circuits.

Entangling capability quantifies the potential of an ansatz to generate entanglement among qubits when its parameters θ are randomly sampled. It serves as an important indicator of how effectively the circuit can represent highly entangled quantum states using shallow-depth architectures. This property is especially valuable in the context of NISQ devices. An ansatz with strong entangling capability can efficiently represent the solution space, making it particularly well suited for solving challenging problems within the VQA framework.

In the literature, several implementations of the ansatz can be found that fall into two main categories: problem-specific and generic. The problem-specific ansatzes are designed with prior knowledge of the specific problem. The structure of the ansatz often reflects the physical or computational properties of the problem. In contrast, generic (problem-agnostic) ansatzes can be applied without detailed knowledge of the underlying problem. A widely used example of a problem-specific ansatz is the Unitary Coupled Cluster (UCC) ansatz, which is particularly common in quantum chemistry. The UCC ansatz is designed to find the ground-state energy of a fermionic molecular Hamiltonian. It constructs the variational state using a motivated reference state $|\phi_0\rangle$, often the Hartree–Fock state, and applies unitary transformations.

Hardware-efficient ansatzes are considered generic and are designed to be easily implemented on NISQ devices. They minimize circuit depth and leverage the native gate sets and qubit connectivity of specific hardware platforms. Hardware-efficient ansatzes are also flexible enough to incorporate problem symmetries when needed. Despite their versatility, they may suffer from trainability issues, particularly when initialized with random parameters. The Quantum Alternating Operator Ansatz is another widely used generic ansatz, inspired by the Trotterized adiabatic transformation, and is specifically designed to solve combinatorial optimization problems. A key advantage of this ansatz is that, for certain problem classes, it restricts the quantum evolution to a smaller subspace of the full Hilbert space that is directly relevant to the solution. This constraint can lead to more efficient algorithms by reducing the effective search space while preserving problem-specific structure.

In addition to the commonly used ansatzes, there are several other ansatz classes, such as the variational Hamiltonian ansatz, variable structure ansatz, sub-logical ansatz, and hybrid ansatz. These designs offer unique flexibility in balancing key trade-offs among expressibility, circuit depth, and compatibility with NISQ-era hardware. Each class is actively being explored as a means to improve the efficiency, scalability, and trainability of variational quantum algorithms across diverse problem domains.

2.2.3 Classical Optimization

The classical optimization algorithm is used within the quantum–classical feedback loop to solve the variational optimization problem defined as

$$\boldsymbol{\theta}^* = \arg\min_{\boldsymbol{\theta}} C(\boldsymbol{\theta}). \tag{5}$$

These optimizers must operate over complex, non-convex cost landscapes that are often affected by noise, statistical uncertainty from finite sampling, and barren plateaus where gradients vanish. Classical optimizers in VQAs are broadly categorized into two groups: gradient-based and gradient-free methods.

Among the gradient-based optimizers, Stochastic Gradient Descent (SGD) and its adaptive variant Adam are commonly used. Adam dynamically adjusts learning rates during training, helping accelerate convergence. More quantum-aware approaches, such as quantum natural gradient descent, incorporate concepts from information geometry using a metric tensor to guide updates in directions more sensitive to changes in the quantum state. Another advanced gradient-based method is meta-learning, which leverages a neural network to model the update rule itself. This optimizer learns how to optimize by using the gradient and optimization history to generate parameter updates more effectively than fixed rule-based approaches.

One of the key advantages of many VQAs is the ability to compute gradients of the cost function analytically. A widely adopted approach is the parameter-shift rule, which enables the exact evaluation of gradients by shifting a single parameter while holding others fixed. For a cost function $C(\theta)$, the partial derivative with respect to the parameter θ_l is given by

$$\frac{\partial C}{\partial \theta_l} = \sum_k \frac{1}{2\sin\alpha} \left(\mathrm{Tr}\left[O_k U^\dagger(\theta^+)\rho_k U(\theta^+)\right] - \mathrm{Tr}\left[O_k U^\dagger(\theta^-)\rho_k U(\theta^-)\right] \right), \quad (6)$$

where $\theta^\pm = \theta \pm \alpha \mathbf{e}_l$, and $\mathbf{e}_l$ is the unit vector with a 1 at the l-th index and 0 elsewhere. The angle α is typically chosen as $\pi/4$, which minimizes numerical instability arising from the $1/\sin\alpha$ factor. By applying the chain rule and evaluating this expression for each parameter, the full gradient can be constructed.

On the gradient-free side, Simultaneous Perturbation Stochastic Approximation (SPSA) is often grouped with gradient-free methods, where it does not compute gradients directly. Instead, SPSA estimates gradients using finite differences along randomly chosen directions. Additionally, newer methods fit cost landscapes using trigonometric expansions to enable local updates without relying on gradients. Overall, optimizer selection in VQAs is a crucial step, as it directly impacts the success and performance of the VQA.

Having developed a foundational understanding of the cost function, ansatz design, and optimizer, the overall VQA process can be decomposed into four key phases, as illustrated in Fig. 4.

- **Initial state preparation** ($|\phi_0\rangle$)—Preparing a known fiducial quantum state, often $|0\rangle^{\otimes n}$, as the input for the ansatz circuit.
- **Trial state preparation**—Applying a sequence of parameterized quantum gates ($U(\theta)$) to generate a variational state.
- **Measurement and cost function evaluation** ($C(\theta)$)—Estimating the expectation value of the cost function using quantum measurements on the variational state.
- **Classical optimization loop**—Using a classical optimizer to update the parameters θ to minimize the cost function, forming a hybrid feedback loop.

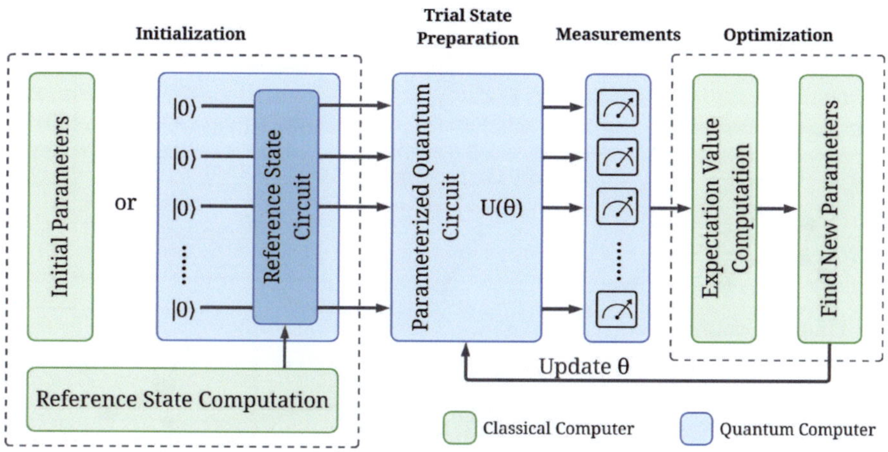

Fig. 4 Overview of the four core phases in a variational quantum algorithm: initial state preparation, trial state preparation, measurement and cost function evaluation, and classical optimization

2.3 Applications of Variational Quantum Algorithms

VQAs offer a flexible framework for addressing a diverse range of computational problems on near-term quantum hardware. Thanks to their hybrid quantum–classical structure, they have been utilized for diverse applications, spanning both inherently quantum mechanical problems and tasks of a more classical nature. Figure 5 shows applicability of VQAs across problem domains. In this section, we explore seven application areas of variational quantum algorithms.

2.3.1 Finding Ground States

One of the most prominent use cases for VQAs is finding low-lying eigenstates and eigenvalues of quantum Hamiltonians. Early approaches, such as adiabatic state preparation and quantum phase estimation, required circuit depths far beyond what is currently feasible. The Variational Quantum Eigensolver (VQE) emerged to fill this gap by leveraging the variational principle to approximate ground-state energies. Given a Hamiltonian H, the VQE seeks to minimize the expectation value $C(\theta) = \langle\psi(\theta)|H|\psi(\theta)\rangle$ over a parameterized state $|\psi(\theta)\rangle$, ensuring that $C(\theta)$ provides an upper bound on the true ground-state energy. Here $|\psi(\theta)\rangle = U(\theta)|\psi_0\rangle$, where $U(\theta)$ is the ansatz and $|\psi_0\rangle$ is the initial state.

Beyond ground states, extensions such as orthogonality-constrained VQE, subspace expansion methods, and subspace VQE enable the estimation of excited states. These approaches rely on modifying the cost function to penalize overlaps with previously found eigenstates or to explore a variational subspace spanned by states generated from the approximate ground state.

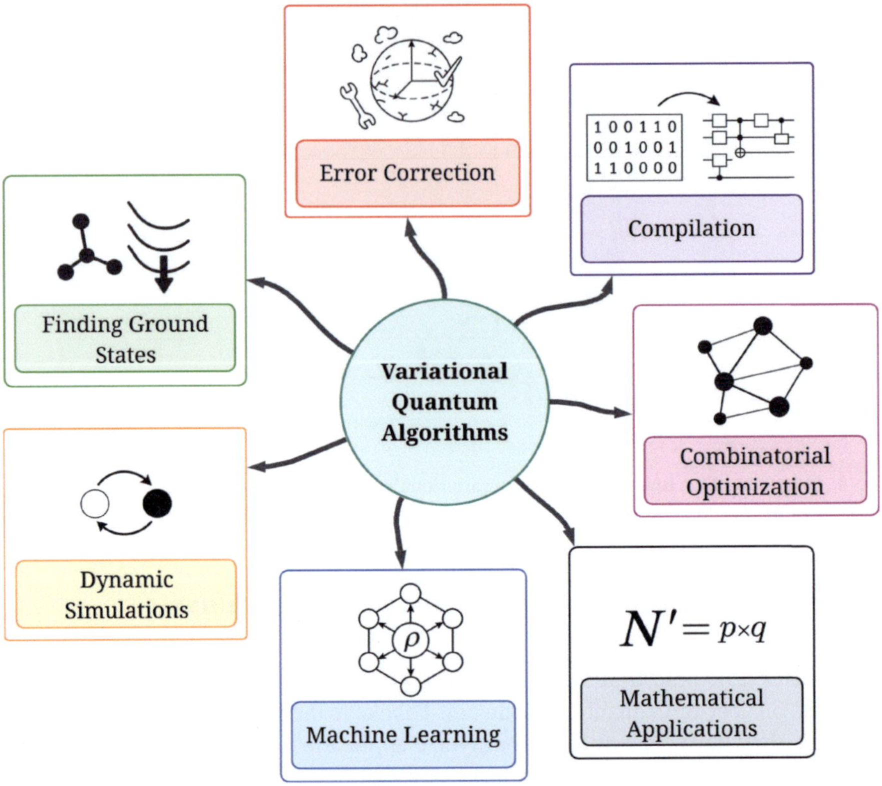

Fig. 5 Illustration of various applications of variational quantum algorithms across different problem domains

2.3.2 Dynamical Quantum Simulation

VQAs are also being utilized to simulate time evolution. Unlike traditional dynamic Hamiltonian simulation techniques that require deep circuits scaling with simulation time, variational methods can achieve approximate dynamics with shallow circuits. For example, iterative approaches update variational parameters (θ) according to differential equations derived from variational principles, while variational fast forwarding and subspace approaches approximate time evolution operators ($exp(-iHt)$) directly. Moreover, VQAs have been adapted to simulate open quantum systems by mapping the evolution of mixed states into parameter dynamics using principles, such as McLachlan's variational formulation. This enables the study of dissipative processes within a variational framework, often reducing the quantum resources required compared to direct approaches.

2.3.3 Optimization Problems

Beyond inherently quantum tasks, VQAs are promising for solving classical optimization problems. The Quantum Approximate Optimization Algorithm (QAOA) is the most famous example, designed to tackle combinatorial optimization problems like Max-Cut and Constraint-Satisfaction (SAT). QAOA alternates between applying problem-specific and mixing Hamiltonians, with variational parameters that are classically optimized to minimize an energy expectation corresponding to the cost function of the original classical problem. This framework directly leverages the quantum state space to encode and navigate complex optimization landscapes.

2.3.4 Mathematical Applications

Several VQAs have been developed for mathematical problems such as solving linear systems of equations or performing matrix-vector multiplications. For instance, the Quantum Linear Systems Problem (QLSP) involves preparing a quantum state $|x\rangle$ such that $A|x\rangle \propto |B\rangle$ for a given matrix A and $|B\rangle = b/||b||$, where b is a column vector. While the Harrow–Hassidim–Lloyd algorithm offers exponential speedup under certain conditions, its circuit depth is impractical for near-term devices. VQA approaches construct a Hamiltonian whose ground state encodes the solution, then use a variational optimization to approximate this state. These methods also extend to non-linear equations and problems, such as integer factoring, by translating them into optimization tasks over quantum states.

2.3.5 Quantum Compilation

VQAs are used to compile quantum programs into hardware-efficient circuits. Given a target unitary V, a VQA seeks a parameterized circuit $U(\theta)$ that approximates V, minimizing some fidelity-based cost function. Such variational compilation approaches are not only useful for executing specific algorithms but also play a role in error mitigation by shortening circuit depths [10, 11].

2.3.6 Quantum Error Correction

VQAs have been utilized for discovering or adapting quantum error-correcting codes tailored to the noise characteristics of specific hardware, thereby optimizing protection schemes for quantum memories.

2.3.7 Machine Learning

VQAs provide a natural way to implement quantum classifiers, autoencoders, generative models, and even quantum versions of neural networks. For example, parameterized quantum circuits can be trained to classify data by encoding inputs into quantum states and using expectation values to predict labels. The expressibility and entanglement structure of quantum circuits serve as powerful resources for capturing complex data patterns. In generative tasks, VQAs can implement quantum analogs of Generative Adversarial Networks (GANs), training quantum circuits to produce data distributions that match a target dataset. These applications exploit the large Hilbert space for representing intricate probability distributions, potentially surpassing classical methods in expressiveness.

2.4 Role of State Preparation in VQAs

As depicted in Fig. 4, state preparation is a fundamental component of VQAs. We can divide state preparation in VQAs into two main types: initial (reference) state preparation and trial state preparation.

The first is the **initial (reference) state preparation**, where the quantum circuit prepares a known fiducial state or an application-specific educated reference state. The choice of this initial state depends heavily on the target problem. For example, in VQE applications using a UCC ansatz, the initial state is typically chosen as the Hartree–Fock state. If the VQE is implemented as an adiabatically assisted scheme, finding the ground state of a complex problem by evolving from a simpler one, then the initial state is the ground state of that simpler problem. In QAOA, the initial state is prepared as the ground state of an appropriately chosen mixer Hamiltonian. In variational quantum error correction schemes, the initial state is often simply ($|0\rangle$).

The second type is **trial state preparation**, where a parameterized quantum circuit (the ansatz) prepares a trial state based on the current set of parameters (θ). These parameters are updated by a classical optimizer in response to the cost function evaluations from previous iterations. Note that state preparation in VQAs, including both the initialization and the trial state generated by the ansatz, directly influences the performance and accuracy of the algorithm.

3 State Preparation for Quantum Computing

One of the fundamental requirements in quantum computing is the ability to prepare an arbitrary quantum state. As shown in Fig. 6, this essential task is typically accomplished in two stages. First, the quantum computer is initialized to a known fiducial state, most commonly the all-zero state $|0\rangle^{\otimes n}$ for an n-qubit system. Then, a sequence of carefully designed discrete quantum gates are applied to this initial

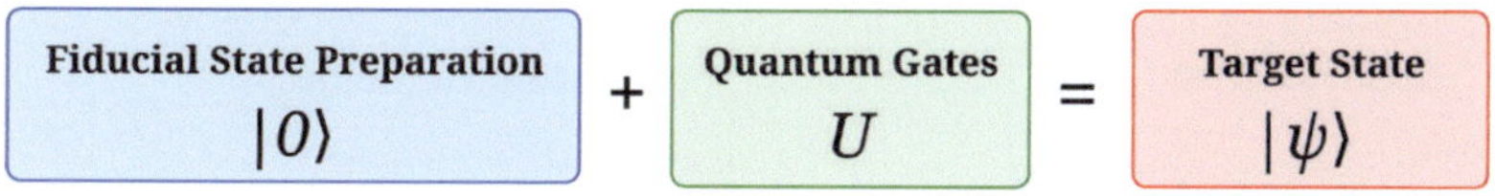

Fig. 6 Preparation of arbitrary quantum states consists of two major components, fiducial state preparation followed by application of quantum gates

state to generate the desired target state, formally expressed as $|\psi\rangle = U|0\rangle^{\otimes n}$, where U denotes the unitary operation realized by the quantum circuit. These steps enable the preparation of complex quantum states required for a wide range of quantum algorithms.

This same two-stage process of arbitrary quantum state preparation appears in VQAs, although with slight variations. Similar to standard approaches, VQAs begin by preparing either a known fiducial state or an application-specific educated reference state. However, instead of applying a fixed unitary operation U, VQAs employ a parameterized unitary operation $U(\theta)$, where the parameters θ are iteratively adjusted during the optimization process. Thus, state preparation stands at the core of VQAs, determining both the starting point and the landscape explored by the classical optimizer.

This section surveys various techniques for quantum state preparation. We begin by examining conventional methods such as active and passive reset protocols, which are fundamental for reliably initializing qubits to desired reference states. Next, we explore advanced strategies like steering techniques that enable the efficient construction of complex trial states within parameterized circuits.

3.1 *State Preparation Using Passive Reset*

Passive reset is one of the most fundamental and widely employed methods for initializing qubits in quantum computing systems. In this approach, qubits are simply allowed to thermalize to their ground state by waiting for a time on the order of their energy relaxation time, commonly characterized by the T_1 coherence time. This technique exploits the natural dissipation of energy from the qubit into its environment, driving it back to the lowest energy state without requiring explicit control pulses or engineered dissipative dynamics. For instance, in superconducting systems, passive reset involves waiting for the transmon or other superconducting qubit to spontaneously relax into its ground state. Figure 7a shows an overview of passive reset.

While passive reset is conceptually simple and requires no additional hardware or control complexity, it comes with important trade-offs. The main challenge among these is the direct dependence of the reset time on the qubit's T_1 coherence time. As qubit technologies advance and coherence times improve into the hundreds of microseconds or even seconds, the time required for passive thermalization cor-

(a) Passive reset (b) Active reset

Fig. 7 Two conventional state preparation methods

respondingly increases. For example, superconducting qubits embedded in three-dimensional waveguide cavities have demonstrated T_1 times approaching 70 μs [12], while trapped-ion qubits have memory coherence times exceeding 50 seconds [13]. In these technologies, simply waiting to ensure ground state initialization can easily dominate the total execution time of quantum programs, undermining the benefits of long coherence.

Moreover, passive reset is inherently limited to preparing fiducial ground states. This makes it unsuitable for algorithms that require initialization into more general or application-specific reference states. For example, preparing the Hartree–Fock state in VQE, or a specific mixer Hamiltonian ground state in QAOA, often demands active gate operations to generate the required superposition or entangled state from $|0\rangle^{\otimes n}$. Passive reset alone cannot achieve such non-trivial initializations, highlighting a fundamental restriction of this method.

In summary, while passive reset continues to play an important role in quantum computing due to its simplicity, its dependence on T_1 times poses significant scalability challenges. As coherence times improve, the reset overhead becomes increasingly costly relative to circuit execution times. Additionally, the inability of passive reset to prepare arbitrary initial states necessitates complementary techniques, such as active resets or sophisticated state engineering, to meet the demands of modern variational quantum algorithms and beyond. Thus, understanding both the advantages and the limitations of passive reset is essential when designing state preparation strategies for high-performance quantum applications.

3.2 State Preparation Using Active Reset

Active reset refers to protocols that deliberately drive a qubit from an arbitrary mixed or excited state back to a desired fiducial state, typically the ground state, using explicit control operations. Unlike passive reset, which relies on natural thermal relaxation governed by T_1 times, active reset uses mechanisms such as projective measurements or coupling to ancillary systems to rapidly extract entropy from the qubit. In the projective measurement approach, the target quantum state may not align with an eigenstate of the measurement operator, resulting in probabilistic measurement outcomes. Consequently, once the qubit state collapses upon measurement, single-qubit rotations are applied conditionally to correct the state based on the observed readout results. Figure 7b shows an overview of active reset.

A major advantage of active reset is speed. Because passive reset times scale with T_1, which is increasing in modern devices due to longer coherence times, waiting passively would quickly become the dominant overhead relative to computation. Active reset breaks this dependence. By leveraging strong qubit-ancilla interactions or carefully optimized control pulses, it is possible to reduce the reset time to tens of nanoseconds, independent of T_1. This capability not only accelerates the repetition of quantum circuits but also enables more frequent calibrations and characterizations, as demonstrated by restless measurements, where dynamic repetition rates can achieve speed-ups in benchmarking and calibration tasks [14–16].

However, active reset also introduces trade-offs. These methods typically demand precise control over coupling strengths, timing, and sometimes rely on assumptions about ancilla purity or the absence of residual entanglement. The work on fundamental bounds explicitly shows that there are limits on how fast and how accurately qubit reset can be performed, governed by the choice of interaction Hamiltonians and the dimension of the ancillary system [17]. While coupling to higher-dimensional ancillas can increase achievable purity, it does not reduce the minimum reset time below the one set by the optimal interaction strength. Moreover, errors introduced during reset such as residual excitations, imperfect swaps, or heating can propagate into subsequent quantum operations, potentially degrading overall algorithm fidelity.

In summary, active reset is an indispensable tool for modern quantum computing, providing the means to initialize qubits quickly and repeatedly, independent of their natural relaxation times. It underpins high-throughput experiments, frequent calibrations, and variational algorithms that demand repeated state preparation. Nonetheless, careful engineering is required to balance speed, fidelity, and complexity, as theoretical analyses make clear that there are unavoidable bounds on performance dictated by quantum control principles and system design. These considerations highlight the importance of choosing the right active reset strategy, whether via direct measurement, coupling to engineered reservoirs, or optimized pulse protocols to match the specific demands of quantum applications.

3.3 State Preparation Based on Dissipative Dynamics

State preparation based on dissipative dynamics intentionally utilizes interaction with the environment to guide the system into a desired quantum state, rather than being just a source of noise. Unlike traditional approaches that rely on purely coherent unitary evolutions, these methods utilize controlled dissipation, typically described by Lindblad master equations, to drive a quantum system toward a unique steady state that encodes the target state. This strategy enables the robust preparation of complex quantum states, including entangled and topologically ordered states, starting from arbitrary initial conditions, as the dissipation effectively "cools" the system into the target manifold regardless of its initial state [18, 19].

One of the fundamental advantages of dissipative state preparation is its inherent insensitivity to certain types of errors and initial state imperfections. Because the dissipative evolution is designed to have a unique dark state, which leads to the target state by the process. This has been demonstrated in the context of preparing stabilizer states, cluster states, and projected entangled pair states (PEPS). For example, in the work by Kraus et al. [18], dissipative processes were constructed for which the desired multipartite entangled state emerges as the unique stationary solution, independent of system size, with relaxation times determined primarily by local coupling strengths rather than by the total number of qubits.

These approaches have been extended to more intricate many-body scenarios. For instance, Rydberg atom platforms leverage strong dipole-dipole interactions in combination with optical pumping to implement stroboscopic sequences of coherent and dissipative operations. As shown by Weimer et al. [20], auxiliary Rydberg atoms can mediate multi-qubit interactions while optical pumping introduces controlled dissipation, effectively simulating master equations that combine Hamiltonian evolution with carefully crafted Lindblad terms. This allows for the dissipative preparation of ground states of models like Kitaev's toric code or lattice gauge theories, where the Liouvillian dynamics systematically annihilates excitations and cools the system into topologically ordered phases.

However, these strategies are not without limitations. While dissipation provides robustness against certain coherent errors, it also imposes fundamental constraints tied to the interplay between the system-reservoir coupling and the spectrum of the induced dynamics. For instance, the design of local quantum jump operators that uniquely stabilize a highly entangled target state can be intricate or even impossible for certain state classes without introducing non-local interactions. Furthermore, while dissipative processes can guarantee convergence to a pure steady state, the relaxation times may still depend unfavorably on parameters like the locality or overlap properties of the jump operators [18, 19].

Another important consideration is experimental implementation. Engineering dissipative couplings often requires auxiliary systems and precisely tuned interactions that preserve local structure while ensuring that only the target state remains unaffected (i.e., a dark state). In practice, realizing such control whether through Rydberg gates combined with optical pumping, or through Markovian reservoir engineering using ancillas, can be technologically demanding. Moreover, imperfections in gate operations or residual decoherence outside the engineered dissipative channels can introduce heating or unwanted steady states, as highlighted by detailed studies of gate errors in Rydberg lattice implementations [20].

In conclusion, state preparation strategies based on dissipative dynamics offer a compelling route to engineer complex quantum states with intrinsic robustness against certain initialization errors and coherent fluctuations. They expand the toolbox beyond purely unitary evolution, allowing the systematic preparation of entangled and topologically ordered states directly as attractors of the dynamics. Nonetheless, they also bring new challenges, including the need for careful design of quasi-local jump operators and precise experimental control over engineered dissipation. As experimental platforms mature, these methods are likely to play an increasingly central role in quantum simulation and in realizing robust quantum resources for computation and communication.

3.4 State Preparation Based on Algorithmic Cooling

Algorithmic cooling represents an innovative approach to quantum state preparation, especially vital for initializing qubits in highly pure states needed for scalable quantum computing. At its core, algorithmic cooling combines principles from information theory and thermodynamics, manipulating entropy within a system of spins to achieve effective temperatures much lower than the surrounding environment. Unlike traditional reversible entropy manipulations limited by Shannon's bound, algorithmic cooling pushes beyond this limit by offloading entropy into an external heat bath, thereby purifying qubits well past what closed-system compression allows [21, 22].

The earliest implementations of algorithmic cooling were motivated by challenges in nuclear magnetic resonance (NMR) quantum computing. In these systems, large ensembles of molecules act as parallel quantum computers, but their low thermal polarization severely limits signal strength. Traditional pseudo-pure state methods introduce an exponential scaling problem, where the pure state fraction decreases exponentially with the number of qubits. Algorithmic cooling emerged as a method to overcome this by using a set of rapidly thermalizing auxiliary spins that absorb the concentrated entropy from the slower-relaxing computation qubits. This process dramatically improves the polarization of selected qubits without requiring impractically large sample sizes [22, 23].

A typical algorithmic cooling protocol involves iterative rounds of entropy compression and reset. Initially, a data compression subroutine is applied among computation qubits to concentrate entropy into a subset of spins, which are then swapped out with reset spins that quickly thermalize back to the bath polarization. By repeatedly applying these steps each time starting from an increasingly polarized system, the target qubits can achieve significantly higher polarization. Variants include Heat Bath Algorithmic Cooling (HBAC), where these resets are actively managed through controlled interaction with a cold bath [24, 25].

One of the major advantages of algorithmic cooling is that it breaks the entropy conservation constraint inherent to closed-system data compression, achieving effective spin temperatures far below the bath temperature. This is especially critical for enabling fault-tolerant quantum error correction, where ancilla qubits require polarization levels exceeding certain thresholds. Techniques like the Partner Pairing Algorithm (PPA) have demonstrated optimal entropy extraction, theoretically allowing polarization enhancements up to factors of $2^{(n-2)}$ for n qubits, thereby providing exponential improvements over reversible approaches [24, 25].

However, algorithmic cooling is not without limitations. Its performance heavily depends on the ratio between the relaxation times of computation and reset qubits. In practice, reset spins do not instantly re-thermalize, and finite relaxation rates introduce heating of the computation qubits over long protocols. As explored through both theoretical bounds and numerical simulations, for realistic relaxation ratios ($R = T_1(\text{comp})/T_1(\text{reset})$), the exponential cooling advantage degrades sharply. Simulations for the 2PAC (Practicable Algorithmic Cooling) algorithm show that even for very large $R \approx 10^7$, cooling is not possible for more than about 20 spins. For $R \approx 10^4$, cooling is achievable in solid-state NMR, using spin diffusion degradation occurs at even smaller system sizes [23].

Another practical challenge is implementing the complex multi-qubit gates required for the compression steps, especially in larger systems. While simpler PAC or mPAC variants trade off optimal cooling depth for implementability using standard gates, they achieve less polarization than algorithms like the Partner Pairing Algorithm (PPA). Furthermore, in many realistic settings, the number of required reset operations grows quadratically with the targeted polarization level, which can impose significant time costs and potential decoherence risks in experimental systems [23].

Recent extensions of algorithmic cooling exploit correlations between qubits and their environment. For example, generalized state-reset operations leveraging correlated relaxation processes akin to the Nuclear Overhauser Effect can enable polarization transfer that surpasses traditional independent reset limitations. These methods, however, demand precise engineering of the system-environment interaction and introduce additional complexities in control and error management [25].

In summary, algorithmic cooling stands out as a powerful strategy for preparing highly pure quantum states, crucial for initializing quantum registers and enabling advanced quantum algorithms. While it offers clear theoretical advantages, including breaking Shannon's bound and providing pathways to scalable quantum computing, practical implementations must carefully address issues like finite relaxation, gate complexity, and error propagation. As experimental platforms evolve, particularly with advances in hybrid electron-nuclear systems and more robust control techniques, algorithmic cooling is poised to remain as a cornerstone in the quantum state preparation toolkit.

3.5 State Preparation Using Measurement-Induced Steering

Quantum measurements not only provide information but also fundamentally alter the state of the system being measured. This back-action of quantum measurements lies at the heart of measurement-induced quantum steering (MIQS), a process that leverages entanglement and measurement to drive a quantum system toward a desired target state. MIQS provides an alternative to traditional state preparation by utilizing open system dynamics implemented digitally on quantum computers.

In MIQS, a system of qubits is coupled to ancillary qubits through an entangling unitary operator U. After this entangling interaction, the ancilla is measured and reset to its initial state. The measurement induces a non-unitary transformation on

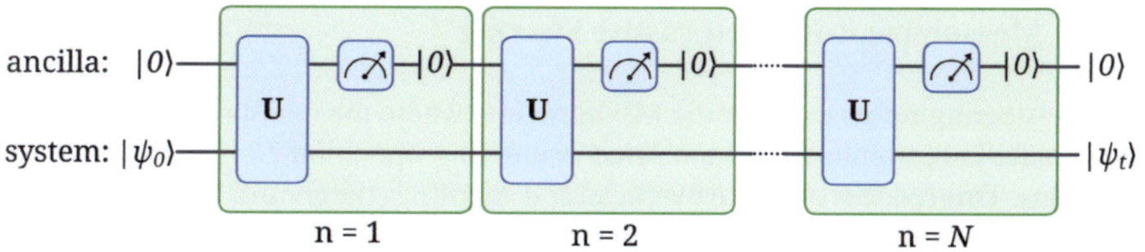

Fig. 8 Circuit diagram for measurement-induced steering protocol: the system qubit is repeatedly coupled to an ancilla through a unitary U, followed by measurement and reset of the ancilla. Over N iterations, the system converges to the target state $|\psi_t\rangle$ starting from the unknown initial state $|\psi_0\rangle$

the system, effectively "steering" it toward a desired target state. This process is repeated over multiple iterations to gradually refine the system's state.

Mathematically, if the system is in a state ρ_S^n after n iterations, and the ancilla is initialized in state $\rho_A = |\psi_A\rangle\langle\psi_A|$, then the update rule after applying U and tracing out the ancilla is given by

$$\rho_S^{n+1} = \mathrm{Tr}_A \left[U(\rho_A \otimes \rho_S^n) U^\dagger \right]. \tag{7}$$

Figure 8 depicts the quantum circuit corresponding to the steering process. The objective is to steer the system state toward the desired target state $|\psi_t\rangle$. The dynamics of U should be designed such that the following steering inequality is satisfied:

$$\langle\psi_t|\rho_S^{n+1}|\psi_t\rangle \geq \langle\psi_t|\rho_S^n|\psi_t\rangle. \tag{8}$$

This inequality implies that the fidelity with the target state $|\psi_t\rangle$ does not decrease over iterations, and under suitable conditions, the system converges asymptotically to the target state.

The entangling operator U must be designed to satisfy this condition. Typically, it is constructed as a time evolution of a Hamiltonian H over a short interval:

$$U = \exp(-iH\delta t), \tag{9}$$

where H has the structure:

$$H = \sum_n \left(O_A^{(n)} |\psi_A\rangle\langle\psi_A| \right) \otimes \Omega_S^{(n)} + \text{h.c.}, \tag{10}$$

where n labels the ancilla qubits. The Hamiltonian consists of a direct product of operators $O_A^{(n)}$, which rotate the ancillas from their initial state to an orthogonal subspace, and operators $\Omega_S^{(n)}$, which rotate the system to an orthogonal subspace, as proposed in [3]. The remainder of this section describes two variations of measurement-induced steering: passive steering and active steering.

3.5.1 Measurement-Induced Passive Steering

Passive steering refers to a specific MIQS regime where the measurement outcomes of the ancilla are ignored and the same entangling operation U is used across all iterations. This reduces classical overhead and simplifies the control flow on quantum hardware. As shown in Fig. 9, in passive steering, the ancilla is measured, but the result is discarded. The system state is updated using the average (mixed) post-measurement state.

A useful case study is the preparation of the $|+\rangle = \frac{1}{\sqrt{2}}(|0\rangle + |1\rangle)$ state. The Hamiltonian for steering to this state can be simplified as

$$H = \frac{J}{2}\left(\sigma_x^A \sigma_z^S - \sigma_y^A \sigma_y^S\right), \tag{11}$$

where J is a tunable coupling constant. The corresponding unitary is:

$$U = \exp\left(-i\frac{J}{2}(\sigma_x^A \sigma_z^S - \sigma_y^A \sigma_y^S)\right). \tag{12}$$

This operator is implemented as a small fixed-depth quantum circuit (see Fig. 10) and repeated across multiple iterations, yielding exponential convergence to the target state.

As shown in Fig. 11, this protocol was experimentally realized on IBM Quantum processors. Qiskit was used to implement mid-circuit measurement and active reset. For a given target state, multiple iterations of the steering circuit are applied, and state tomography is used to estimate the system's density matrix. Fidelity improves monotonically with the number of iterations, confirming the theoretical guarantees [3].

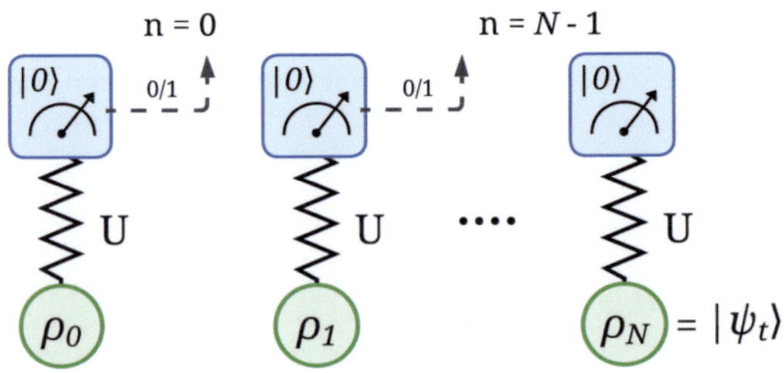

Fig. 9 An overview of measurement-induced passive steering. After each ancilla measurement, the readout is discarded

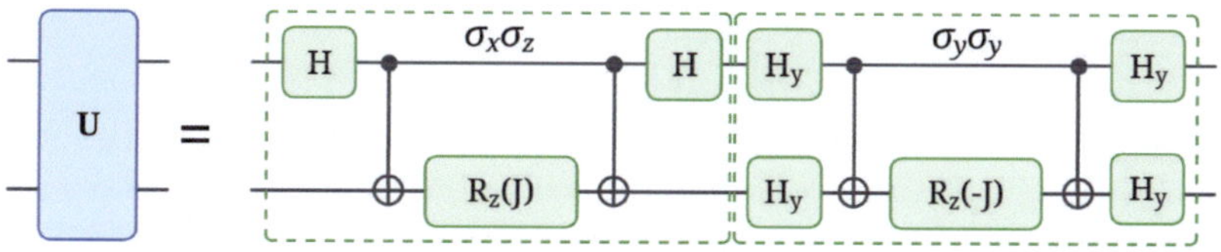

Fig. 10 Quantum circuit implementing the steering unitary U for preparing the $|+\rangle$ state via passive steering. The figure represents a Pauli-based quantum circuit applied to the system and ancilla qubits. The circuit consists of local rotations and CNOT gates applied to the system and ancilla qubits

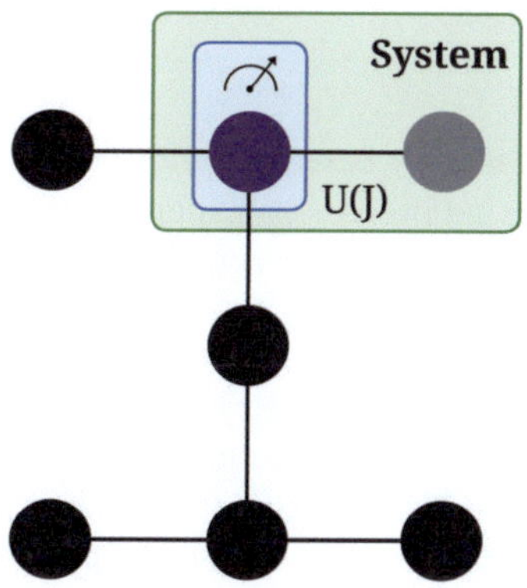

(a) Qubit connectivity layout of `ibm_perth`.

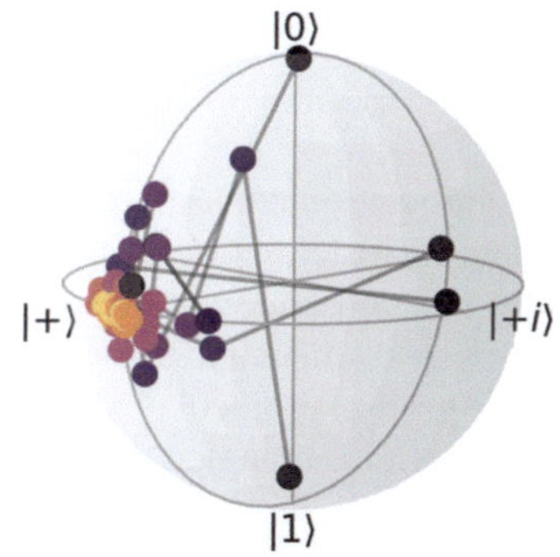

(b) Visualization of state convergence on the Bloch sphere of an arbitrary initial state to the $|+\rangle$ state under passive steering over multiple iterations [23].

Fig. 11 The passive steering protocol is demonstrated on IBM's superconducting quantum hardware, exemplified by `ibm_perth`, whose connectivity graph is shown in **a**. For the system qubit, the transmon is selected, offering the highest measurement contrast between computational states, while the ancilla is chosen as the nearest connected qubit according to the device topology. **b** Displays Bloch sphere data illustrating passive steering on `ibm_perth`, where arbitrary initial states (black dots) are driven to an equal superposition state (yellow dots)

Measurement-induced steering and its passive variant provide an elegant and hardware-friendly approach to quantum state preparation. By leveraging entanglement and measurement back-action, this method circumvents the challenges of deep unitary circuits and long passive resets, while offering guaranteed convergence under well-characterized dynamics. Its compatibility with NISQ hardware makes it particularly attractive for near-term quantum algorithm design.

3.5.2 Measurement-Induced Active Steering

Active steering extends the measurement-induced steering framework by incorporating real-time feedback into the evolution of the quantum system. Unlike passive steering, where the system evolves under a fixed coupling unitary regardless of

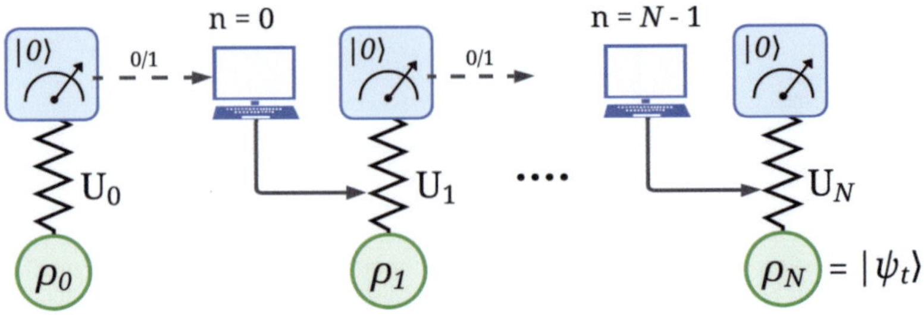

Fig. 12 Overview of feedback-based active steering. After each detector measurement, the readout is used to adapt the next steering unitary, forming a feedback loop

measurement outcomes, active steering adapts this unitary based on prior detector readouts. This closed-loop approach allows the steering process to respond dynamically to stochastic measurement outcomes, thereby accelerating convergence and improving resilience to noise.

The general protocol of active steering is illustrated in Fig. 12. Initially, the quantum system is entangled with a detector via a unitary operation U_i. The detector is then measured, resulting in an outcome $r_i \in \{0, 1\}$. The backaction induced by this measurement updates the system's state from ρ_S^i to ρ_S^{i+1}. Crucially, this readout is passed to a classical processor, which uses it to choose the next coupling operator U_{i+1}, effectively steering the system along a feedback-informed trajectory.

Mathematically, the update of the system state conditioned on measurement outcome r is given by

$$\rho_S^{i+1} = \frac{\mathrm{Tr}_D\left[\Pi_r U_i(\rho_D \otimes \rho_S^i)U_i^\dagger \Pi_r\right]}{p_r}, \tag{13}$$

where ρ_S is the density matrix of the system, ρ_D is the density matrix of the coupled detector, and Π_r is the measurement projector and p_r is the probability of observing outcome r. The next unitary U_{i+1} is then selected adaptively as

$$U_{i+1} = U_\theta(r_i, r_{i-1}, \ldots, r_1), \tag{14}$$

with parameters θ optimized to minimize a cost function based on the final state fidelity:

$$J(r) = 1 - \langle\psi_t|\rho_S^n(r)|\psi_t\rangle. \tag{15}$$

A practical implementation of this adaptive strategy is the gradient-based feedback method proposed in [4]. This method optimizes the sequence of unitary parameters via automatic differentiation by summing over all possible readout trajectories. Figure 13 demonstrates how a two-step feedback policy successfully prepares a target superposition state starting from a random initial state.

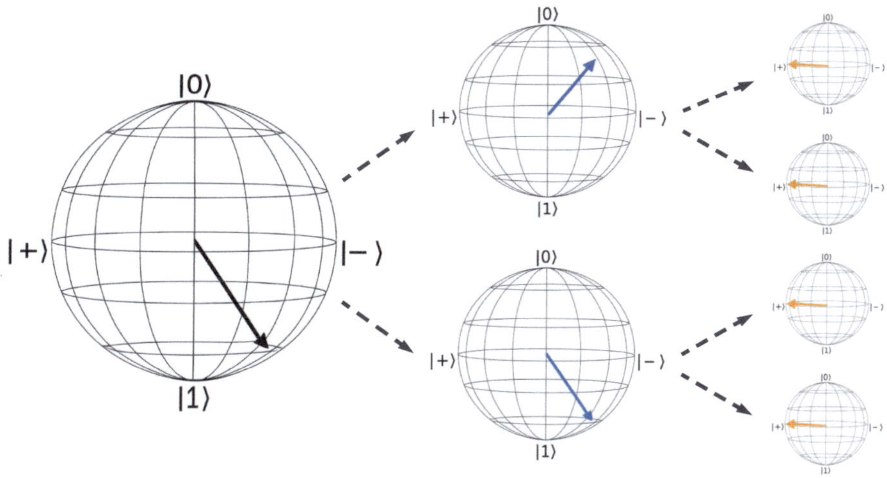

Fig. 13 Bloch sphere visualization of single-qubit active steering. Feedback after the first measurement allows for deterministic convergence to the target state after the second step

Despite its advantages, active steering introduces overhead in both control complexity and hardware requirements. Adaptive feedback demands fast classical processing, mid-circuit measurements, and low-latency updates to quantum operations, which may be challenging on current quantum architectures. Nevertheless, its ability to adaptively suppress decoherence and improve convergence makes it a powerful tool for high-fidelity state preparation.

In summary, active steering enhances the steering protocol by leveraging real-time feedback from measurements. Through strategies such as gradient-based optimization, it achieves state preparation with higher fidelity and robustness to noise, positioning it as a critical technique for future quantum technologies.

4 State Preparation for Variational Quantum Algorithms Using Measurement-Induced Passive Steering

VQAs exploit a hybrid quantum–classical workflow to solve complex problems, including finding ground state energies (using VQE) and combinatorial optimization tasks. A key stage in these algorithms is the preparation of a trial quantum state whose properties (such as energy) are evaluated and used to guide classical optimization.

The effectiveness of VQE depends heavily on the ability to prepare trial quantum states that sufficiently overlap with the ground state of a target Hamiltonian. Traditionally, these trial states are generated by applying a parameterized ansatz to a known fiducial state, such as $|0\rangle^{\otimes n}$, or to a reference state derived from classical approximations like Hartree–Fock. However, this process is susceptible to noise, requires precise initializations, and becomes impractical for large systems. To overcome these limitations, steering-based approaches have been introduced that leverage measurement-induced passive steering to prepare quantum states in a robust and scalable way.

Measurement-induced steering offers a powerful alternative to traditional approaches by enabling the preparation of arbitrary quantum states starting from unknown initial conditions. There are two key avenues for incorporating steering into the VQE pipeline:

- **Reference state preparation:** Passive steering can be used to prepare a reference state from an arbitrary initial state. This avoids the overhead of precisely initializing the quantum register to a specific fiducial state.
- **Replacing the ansatz:** More radically, passive steering itself can serve as the trial state preparation mechanism. Instead of tuning parameters inside a variational circuit, the classical optimization loop directly specifies the target state. Passive steering circuits are then constructed to steer the system toward this target state, effectively collapsing the traditional ansatz into the steering process.

4.1 Reference State Preparation via Passive Steering

In many VQE applications, particularly in quantum chemistry, the trial state is prepared by applying a parameterized ansatz circuit on top of a reference state, typically derived from mean-field methods such as Hartree–Fock. For example:

$$|\psi(\theta)\rangle = U(\theta)|\phi_0\rangle,$$

where $|\phi_0\rangle$ is the Hartree–Fock state mapped to the qubit register.

However, preparing such a reference state assumes the quantum register can be initialized to a known fiducial state, like $|0\rangle^{\otimes n}$, followed by a series of well-calibrated gates to construct $|\phi_0\rangle$. This becomes impractical as system sizes grow, due to increased gate errors and the difficulty of synchronizing multi-qubit operations.

Passive steering addresses this challenge by preparing the reference state directly from an arbitrary initial state, as shown in Fig. 14. The process involves:

1. Specifying the target reference state $|\phi_0\rangle$.
2. Constructing a steering operator U that couples system qubits to ancillas, designed such that iterative measurement should drive the system toward $|\phi_0\rangle$.
3. Repeatedly measuring and resetting the ancillas, thereby exploiting the back-action of quantum measurements to "steer" the system toward the reference state.

This guarantees that even starting from an unknown or noisy initial state, the system approximates $|\phi_0\rangle$ after sufficient iterations, eliminating the need for explicit

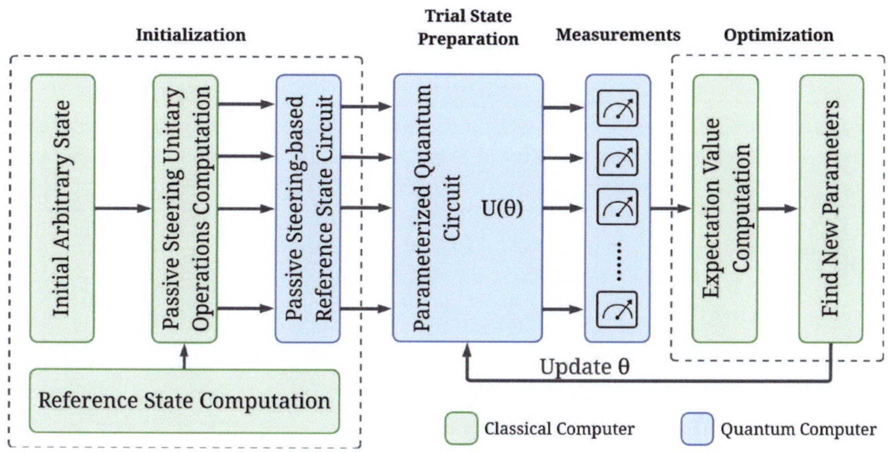

Fig. 14 Passive steering can replace traditional reference state preparation

initialization to $|0\rangle^{\otimes n}$. After the reference state is generated using passive steering, the conventional ansatz within the VQE pipeline is incorporated, yielding a hybrid strategy that enhances both the algorithm's performance and adaptability.

4.2 *Replacing the Ansatz with Passive Steering*

Beyond reference state preparation, steering can directly generate the trial state, removing the need for a parameterized ansatz, as depicted in Fig. 15. In standard VQE, the classical optimizer searches over parameters θ in a fixed circuit structure $U(\theta)$. This is prone to barren plateaus, where the gradients vanish in high-dimensional Hilbert spaces. In contrast, steering-centric VQE operates by

1. Letting the classical optimizer directly adjust the target state $|\psi_t\rangle$ itself, using Bloch sphere angles or more general wavefunction coefficients.
2. For each choice of $|\psi_t\rangle$, constructing a passive steering circuit that iteratively steers the system toward $|\psi_t\rangle$.
3. Measuring the expectation value $\langle\psi_t|H|\psi_t\rangle$ and feeding it back to the classical optimizer.

This process eliminates the need for deep parameterized circuits entirely. Instead of tuning internal gate parameters, it dynamically adjusts the target state and relies on the steering protocol to robustly realize this state through repeated measurement and reset. This approach is naturally resilient to noise, with each iteration partially "refreshing" the state through measurements and resets. This approach also enables scalable state preparation by decoupling from the exponential gate calibration required for complex ansatz circuits. Recent experimental studies have demon-

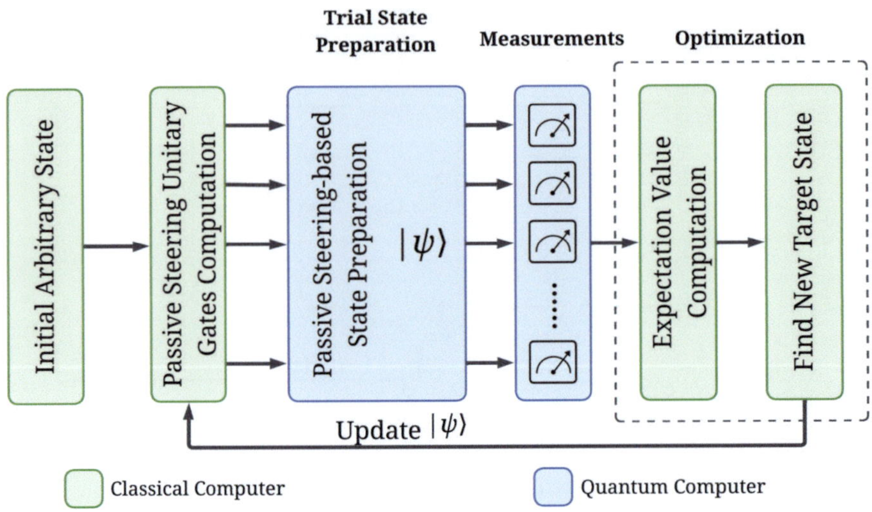

Fig. 15 Passive steering can replace traditional ansatz, steering the quantum system toward the desired trial state before energy evaluation and classical optimization

strated that passive steering used in this way achieves competitive or superior results compared to state-of-the-art approaches like UCCSD, particularly under noisy conditions [26].

4.3 Experimental Evaluation

In this section, we present the experimental results using the Qiskit [27] simulator and compared the passive steering based state preparation method against UCCSD (Unitary Coupled Cluster Singles and Doubles). We have evaluated the utility of passive steering using expressibility as well as effectiveness in finding the ground state energy of the H_2 molecule across a range of interatomic distances. Expressibility is an essential property of any state preparation technique, which describes the ability to generate a broad distribution of quantum states. Passive steering was evaluated using fidelity distributions and KL-divergence relative to Haar random states.

As shown in Fig. 16, steering circuits demonstrated nearly uniform state coverage with a low KL-divergence (0.0211), significantly better than UCCSD's (0.4147). This suggests that passive steering has the expressive power necessary for variational algorithms across a wide range of problem classes [26].

To evaluate the effectiveness of replacing reference state preparation and trial state preparation, experiments were performed on the hydrogen molecule (H_2) across varying bond lengths. In the noiseless setting, using passive steering to prepare the reference state and then applying UCCSD produced energy estimates nearly identical

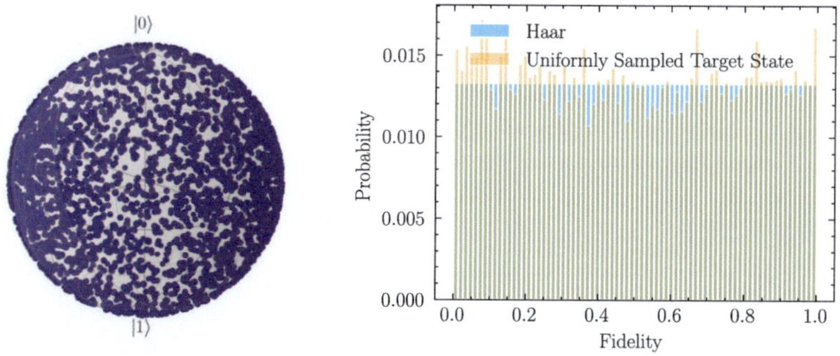

(a) Passive steering-based uniform state preparation

(b) Histogram of the estimated fidelities, overlaid with the distribution given by the Haar measure.

Fig. 16 Measured expressibility metric for a single-qubit parameterized circuit

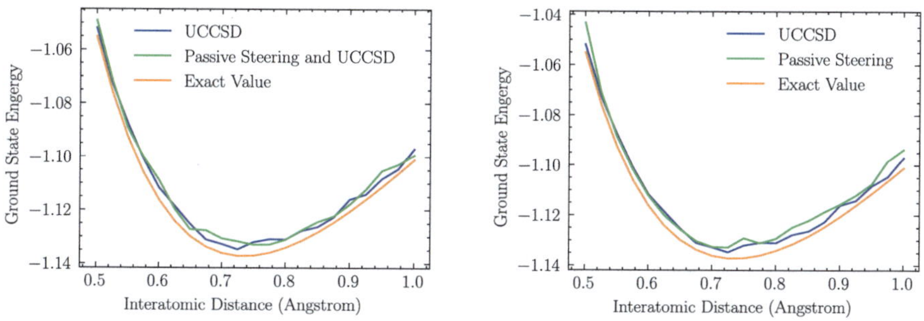

(a) VQE results with passive steering and UCCSD vs. pure UCCSD without noise

(b) VQE results with passive steering vs. UCCSD without noise

Fig. 17 Performance comparison of H_2 ground state energy produced by passive steering with UCCSD. [26]

to those of traditional UCCSD initialized with a clean Hartree–Fock state. Figure 17a confirms the validity of this combination. Moreover, using steering alone to generate the trial state (replacing the ansatz) also resulted in ground state energies comparable to UCCSD with known initial state as shown in Fig. 17b.

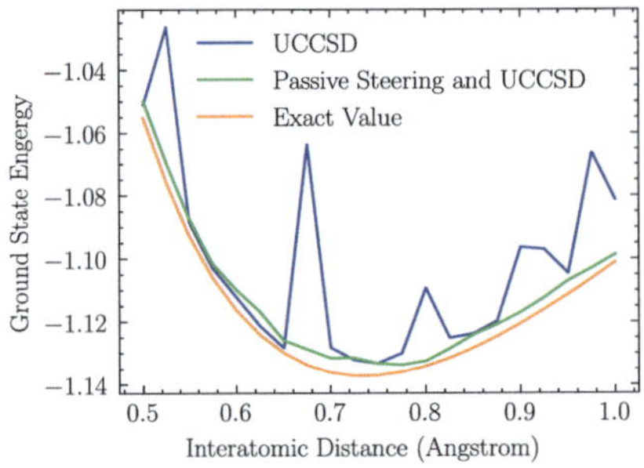

(a) VQE results with passive steering and UCCSD vs. UCCSD with noise

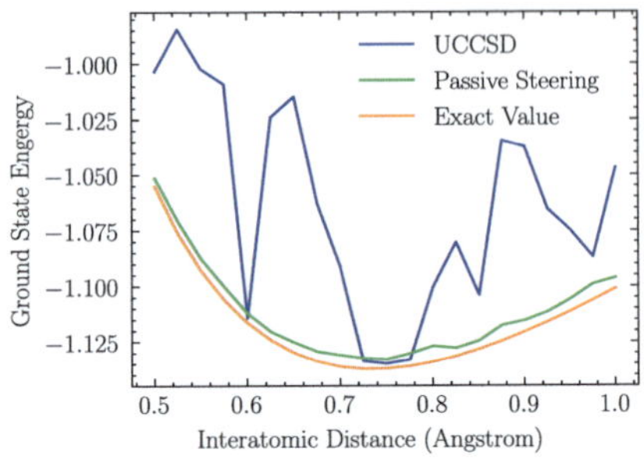

(b) VQE results with passive steering and UCCSD vs. UCCSD with random initial state

Fig. 18 Performance degradation of UCCSD under noise and random initial state compared to the passive steering-enhanced version [26]

The steering-based approaches particularly shine in noisy environments. As shown in Fig. 18a, traditional UCCSD degrades under noise due to circuit depth and lack of reset. However, passive steering by construction performs repeated resets and short sequences, naturally resisting decoherence. In all trials, steering-enabled methods achieved significantly lower energy errors compared to their ansatz-based counterparts when subjected to realistic noise models.

As depicted in Fig. 18b, passive steering exhibits a remarkable advantage in scenarios where no educated reference state is available for initializing the ansatz. In traditional VQE with UCCSD, the choice of initial state plays a critical role. Without a well-prepared Hartree–Fock reference, the optimizer may begin far from the true ground state, resulting in slow convergence or entrapment in local minima. In contrast, passive steering initializes the trial state directly through an iterative, measurement-driven process, without requiring a pre-constructed reference. This independence from fiducial states enables the steering protocol to effectively explore the relevant subspace of the Hilbert space, even from random or mixed initializations. The results show that, under such conditions, passive steering consistently achieves lower energy errors and better convergence compared to UCCSD initialized from uninformed starting points. This robustness highlights the versatility of steering as a state preparation tool, especially in settings where prior knowledge of the system is limited or inaccessible.

In summary, state preparation via steering in VQE enables two practical enhancements: (1) steering as a reference state generator when used with traditional ansatzes like UCCSD, and (2) steering as a complete replacement for the ansatz, enabling direct target state construction. Both paths reduce initialization overhead, resist noise, and eliminate circuit complexity. These advancements form a flexible and scalable solution suitable for near-term quantum applications.

5 Summary

This chapter described an end-to-end workflow of Variational Quantum Algorithms (VQAs) from fiducial initialization through ansatz application, measurement, cost evaluation, and classical updates. We provided an overview of conventional preparation strategies (active and passive reset), dissipative engineering, and algorithmic cooling, highlighting their resource costs, hardware dependencies, and robustness trade-offs. Building on this foundation, we discussed quantum steering as a measurement-induced, environment-assisted approach that can operate in two modes, replacing an educated reference state or substituting the trial (parametrized) circuit, thereby reducing depth and shot complexity without sacrificing expressivity. Empirically, steering-initialized VQE gives more accurate results and, when no strong prior such as Hartree–Fock is available, can surpass UCCSD baselines, providing a pragmatic recipe for state preparation in the next generation of VQAs.

References

1. Daniel Volya and Prabhat Mishra. Quantum spectral clustering of mixed graphs. In *2021 58th ACM/IEEE design automation conference (DAC)*, pages 463–468. IEEE, 2021.
2. Daniel Volya and Prabhat Mishra. Quantum steering of surface error correcting codes. In *2023 IEEE International Conference on Quantum Computing and Engineering (QCE)*, volume 1, pages 1394–1400. IEEE, 2023.
3. Volya, Daniel, and Prabhat Mishra. 2024. State preparation on quantum computers via quantum steering. *IEEE Transactions on Quantum Engineering (TQE)* 5: 1–14.
4. Daniel Volya, Zhixin Pan, and Prabhat Mishra. Feedback-based steering for quantum state preparation. In *2023 IEEE International Conference on Quantum Computing and Engineering (QCE)*, volume 1, pages 1308–1318. IEEE, 2023.
5. Zachery Utt, Daniel Volya, and Prabhat Mishra. Quantum measurement discrimination using cumulative distribution functions. In *2023 Design, Automation & Test in Europe Conference & Exhibition (DATE)*, pages 1–6. IEEE, 2023.
6. Utt, Zachery, Daniel Volya, and Prabhat Mishra. 2024. Quantum measurement classification using statistical learning. *ACM Transactions on Quantum Computing (TQC)* 5 (2): 1–16.
7. Daniel Volya and Prabhat Mishra. Impact of noise on quantum algorithms in noisy intermediate-scale quantum systems. In *IEEE International Conference on Computer Design (ICCD)*, pages 1–4. IEEE, 2020.
8. Daniel Volya and Prabhat Mishra. Modeling of noisy quantum circuits using random matrix theory. In *2022 IEEE 40th International Conference on Computer Design (ICCD)*, pages 132–138. IEEE, 2022.
9. Hasini Witharana, Daniel Volya, and Prabhat Mishra. Qcassert: Quantum device testing with concurrent assertions. In *Asia and South Pacific Design Automation Conference (ASP-DAC)*, pages 491–496. IEEE, 2024.
10. Daniel Volya and Prabhat Mishra. Quantum data compression for efficient generation of control pulses. In *Asia and South Pacific Design Automation Conference (ASPDAC)*, pages 216–221, 2023.
11. Daniel Volya and Prabhat Mishra. Qudcom: Towards quantum compilation for qudit systems. arXiv preprint arXiv:2311.07015, 2023.
12. Chad Rigetti, Jay M Gambetta, Stefano Poletto, Britton LT Plourde, Jerry M Chow, Antonio D Córcoles, John A Smolin, Seth T Merkel, Jim R Rozen, George A Keefe, et al. Superconducting

qubit in a waveguide cavity with a coherence time approaching 0.1 ms. *Physical Review B–Condensed Matter and Materials Physics*, 86(10):100506, 2012.
13. Harty, T. P., D. T. C. Allcock, C. Ballance, L. Guidoni, H. A. Janacek, N. M. Linke, D. N. Stacey, and D. M. Lucas. 2014. High-fidelity preparation, gates, memory, and readout of a trapped-ion quantum bit. *Physical review letters* 113 (22): 220501.
14. Caroline Tornow, Naoki Kanazawa, William E Shanks, and Daniel J Egger. Minimum quantum run-time characterization and calibration via restless measurements with dynamic repetition rates. *Physical Review Applied*, 17(6):064061, 2022.
15. Daniel Volya and Prabhat Mishra. Quantum benchmarking via random dynamical quantum maps. In *2024 IEEE International Conference on Quantum Computing and Engineering (QCE)*, volume 1, pages 1277–1283. IEEE, 2024.
16. Daniel Volya, Andrey Nikitin, and Prabhat Mishra. Fast quantum characterization via riemannian optimization. In *2024 IEEE International Conference on Quantum Computing and Engineering (QCE)*, volume 1, pages 1270–1276. IEEE, 2024.
17. Daniel Basilewitsch, Jonas Fischer, Daniel M Reich, Dominique Sugny, and Christiane P Koch. Fundamental bounds on qubit reset. *Physical Review Research*, 3(1):013110, 2021.
18. Barbara Kraus, Hans P Büchler, Sebastian Diehl, Adrian Kantian, Andrea Micheli, and Peter Zoller. Preparation of entangled states by quantum markov processes. *Physical Review A–Atomic, Molecular, and Optical Physics*, 78(4):042307, 2008.
19. Frank Verstraete, Michael M Wolf, and J Ignacio Cirac. Quantum computation and quantum-state engineering driven by dissipation. *Nature physics*, 5(9):633–636, 2009.
20. Hendrik Weimer, Markus Müller, Igor Lesanovsky, Peter Zoller, and Hans Peter Büchler. A rydberg quantum simulator. *Nature Physics*, 6(5):382–388, 2010.
21. Boykin, P. Oscar, Tal Mor, Vwani Roychowdhury, Farrokh Vatan, and Rutger Vrijen. Algorithmic cooling and scalable nmr quantum computers. *Proceedings of the National Academy of Sciences*, 99(6):3388–3393, 2002.
22. Jose M Fernandez, Seth Lloyd, Tal Mor, and Vwani Roychowdhury. Algorithmic cooling of spins: A practicable method for increasing polarization. *International Journal of Quantum Information*, 2(04):461–477, 2004.
23. Brassard, Gilles, Yuval Elias, Tal Mor, and Yossi Weinstein. 2014. Prospects and limitations of algorithmic cooling. *The European Physical Journal Plus* 129 (11): 258.
24. Daniel K Park, Nayeli A Rodriguez-Briones, Guanru Feng, Robabeh Rahimi, Jonathan Baugh, and Raymond Laflamme. Heat bath algorithmic cooling with spins: review and prospects. *Electron Spin Resonance (ESR) Based Quantum Computing*, pages 227–255, 2016.
25. Nayeli A Rodriguez-Briones, Jun Li, Xinhua Peng, Tal Mor, Yossi Weinstein, and Raymond Laflamme. Heat-bath algorithmic cooling with correlated qubit-environment interactions. *New Journal of Physics*, 19(11):113047, 2017.
26. Sahan Sanjaya, Daniel Volya, and Prabhat Mishra. Variational quantum algorithms via measurement-induced passive steering. In *2024 IEEE International Conference on Quantum Computing and Engineering (QCE)*, volume 1, pages 481–487. IEEE, 2024.
27. IBM Qiskit. https://qiskit.github.io/qiskit-aer/stubs/qiskit_aer.AerSimulator.html. Accessed: 2025-08-20.

Quantum Artificial Intelligence

Emma Andrews and Prabhat Mishra

Abstract Machine learning and artificial intelligence have risen in popularity, capable of performing a wide variety of tasks with great accuracy. In classical domains, machine learning models are restricted by the computational resources required, often taking days to train. Quantum computing provides a possible avenue to accelerate these machine learning models in diverse application domains. In this chapter, we analyze the state of quantum machine learning today and compare the machine learning models between classical and quantum domains. Specifically, we discuss the four possible intersections of classical and quantum domains based on data and models, including classical–classical (classical model processing classical data), classical–quantum (classical model processing quantum data), quantum–classical (quantum model processing classical data), and quantum–quantum (quantum model processing quantum data).

1 Introduction

Machine learning and artificial intelligence (AI) are widely used in diverse application domains, including classification [52, 53] and cybersecurity [38–40]. Machine learning was originally proposed for classical computing with classical data. However, with the rise of quantum computing, implementing these models in quantum space is promising to advance research in both computing areas, either accelerating classical models by running on quantum devices or proposing new architectures in quantum computing that can perform better on specific tasks. There is a great synergy between classical and quantum domains in terms of the origin of the data and underlying machine learning models. In other words, we can consider the following four ways of integrating models and data [3, 16]:

E. Andrews (✉) · P. Mishra
University of Florida, Gainesville, FL, USA
e-mail: e.andrews@ufl.edu

P. Mishra
e-mail: prabhat@ufl.edu

P. Mishra (ed.), *Design Automation for Quantum Computing*,
https://doi.org/10.1007/978-3-032-09303-5_4

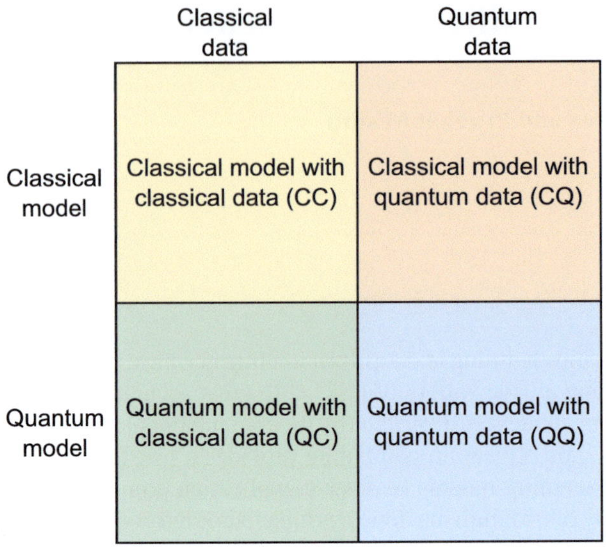

Fig. 1 Four ways of combining models and data from classical and quantum domains

- *Classical–Classical (CC)*, classical model processing classical data;
- *Quantum–Quantum (QQ)*, quantum model processing quantum data;
- *Classical–Quantum (CQ)*, classical model processing quantum data; and
- *Quantum–Classical (QC)*, quantum model processing classical data.

Figure 1 provides an overview of these four scenarios based on data and models. Each quadrant needs to consider domain-specific requirements. For example, *CC* and *QQ* need to consider the constraints of classical and quantum computing, respectively. However, the hybrid ones, *CQ* and *QC*, must carry out additional preprocessing steps to effectively map the data from one domain into a form that is readable by the other domain. This often involves extra computational resources that the pure combinations, *CC* and *QQ*, do not typically have to consider.

The remainder of this chapter is organized as follows. Section 2 presents an overview of machine learning models in the classical domain. Section 3 covers the machine learning models in the quantum domain. Section 4 describes different ways of combining classical/quantum machine learning models with classical/quantum data. Finally, Sect. 5 concludes the chapter.

2 Classical Machine Learning Models

This section provides an overview of different classical models, including linear regression, support vector machine, neural networks, convolutional neural networks, recurrent neural networks, autoencoders, and transformers.

2.1 Linear Regression

Linear regression is a supervised machine learning technique that models an input to an output by learning a linear function. This linear function takes the form of

$$f(\mathbf{x}) = \mathbf{W}^{\top}\mathbf{x} + b \tag{1}$$

where $\mathbf{x}$ is the input vector; $\mathbf{W}$ is the learnable weights, often in matrix form; and b is the bias or intercept.

2.2 Support Vector Machine

Support vector machine (SVM) is a supervised machine learning model that learns a decision boundary between the input data points, classifying them into two different groups. This decision boundary is also known as a hyperplane, following the linear function equation in Eq. 1 such that

$$\mathbf{w}^{\top}\mathbf{x} + b = 0. \tag{2}$$

Determining the best fitting hyperplane for a given input dataset amounts to learning the best fitting $\mathbf{w}$. Figure 2 shows an example linear hyperplane fit for the given data samples.

A drawback of SVMs is that the decision boundary must be linear, meaning that the data separation must also be linear for the best possible classification accuracy. The kernel trick can be used to map the linear, two-dimensional data into nonlinear,

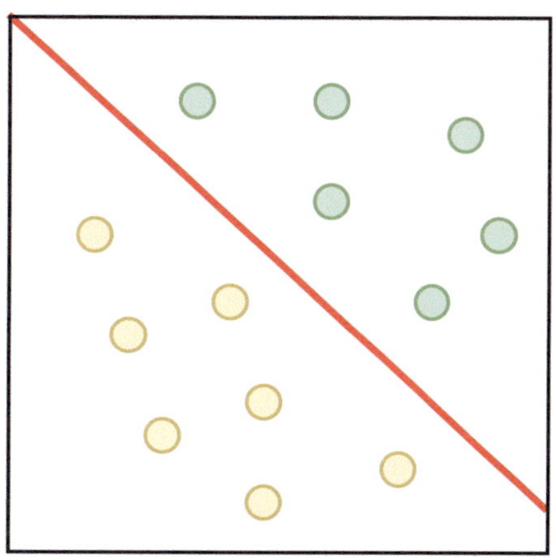

Fig. 2 Linear hyperplane in red as the best fit line to divide data points of two classes, one class colored yellow and the other green

higher-dimensional data spaces. This replaces part of the computation of the linear hyperplane with a kernel function $k(x_i, x_j)$, with the function determining the resulting dimensionality and linearity of the hyperplane decision boundary.

Kernel functions can be used to map nonlinear data into high-dimensional spaces where a linear mapping can be applied. A kernel function often takes the form of

$$k(x_i, x_j) = \langle f(x_i), f(x_j) \rangle, \tag{3}$$

where the kernel function k maps x_i and x_j using the function f.

2.3 *Neural Networks*

Neural networks (NNs), or artificial neural networks (ANNs), are a common machine learning model architecture that uses neurons and layers to process an input into a specified output type. NNs consist of an input layer, an output layer, and hidden layers, along with each layer containing several neurons and connections between the neurons of other layers. The input layer is responsible for processing the original input data to the model, then sending its output to one or more hidden layers, where additional processing takes place. Results from the hidden layers are used as input to the output layer, which produces the overall output of the model. This architecture is depicted in Fig. 3.

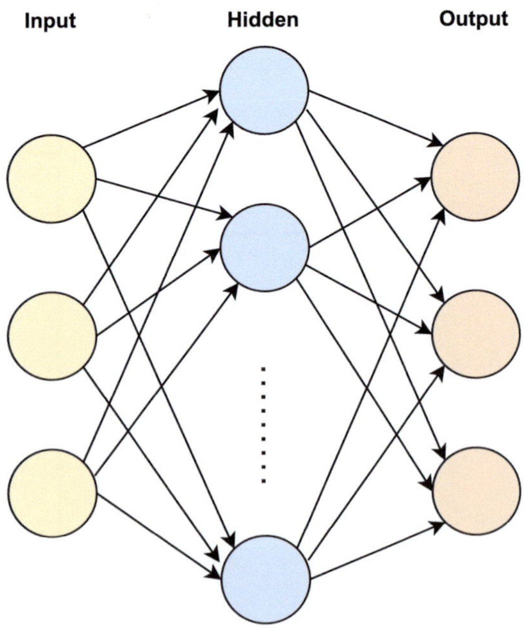

Fig. 3 Generalized architecture of a neural network, consisting of an input layer, a hidden layer, and an output layer. In this example, the input and output layers each have three neurons, while the hidden layer has more than three neurons

There are different variations of the neural network, varying in purpose, application, and resulting structure. The generic variants are described below, and more specific ones are described in the subsequent sections.

- **Deep neural network (DNN)**: Neural network architecture where the layers and neurons are structured such that there is more than one hidden layer.
- **Feedforward neural network (FNN)**: Neural network where the data is sent forward through the layers, without revisiting a previous or current layer.
- **Multi-layer perceptron (MLP)**: A feedforward neural network consisting of several layers containing nonlinear activation functions.

Activation functions are functions that run after a hidden layer to add non-linearity to the overall architecture. Without activation functions, neural networks are structured linearly with the input, output, and hidden layers. Depending on the data and application, it may be necessary to have non-linearity in the model architecture to best represent and establish dependencies in the data. There are several common activation functions:

- **ReLU**: The rectified linear unit (ReLU) maintains positive output values while zeroing out negative ones. It is expressed as

$$g(x) = \max(0, x). \tag{4}$$

- **Sigmoid**: The sigmoid activation function creates an output between 0 and 1, indicating a probability. It is expressed as

$$\sigma(x) = \frac{1}{1 + e^{-x}}. \tag{5}$$

- **Tanh**: The tanh activation function creates an output between -1 and 1, effectively centering it around 0. It is expressed as

$$\tanh(x) = \frac{e^x - e^{-x}}{e^x + e^{-x}}. \tag{6}$$

- **Softmax**: The Softmax function is similar to sigmoid, mapping the output to a probability between 0 and 1. However, this is done across multiple classes, instead of binary, like with sigmoid. It is defined by

$$\text{softmax}(x) = \frac{e^{x_i}}{\sum_{j=1}^{J} e^{x_j}}. \tag{7}$$

2.4 Convolutional Neural Networks

While feedforward neural networks can work well for one-dimensional data, they often have difficulties processing two-dimensional or grid-like data, such as images. To address this, convolutional neural networks (CNNs) are a type of feedforward neural network that uses convolutional layers in the model to process two-dimensional data effectively [32]. The hidden layers of a CNN consist of the convolutional layer and the pooling layer, as shown in Fig. 4.

Convolutional layers: Convolutional layers make use of kernels, or windows, on the data to observe a small portion of the input data at a time. As this kernel is moved across the input data, also known as the sliding window, the dot product between the kernel and the current view of the data is taken. This sliding window method learns features within the data. Convolutional layers are defined for each dimension of data. The 2D version, for example, is defined as

$$y_{ij} = \sum_{m=0}^{k_h-1} \sum_{n=0}^{k_w-1} x_{i+m,j+n} \cdot w_{mn}, \tag{8}$$

where k_h and k_w are the height and the width of the kernel, respectively, and w is the kernel.

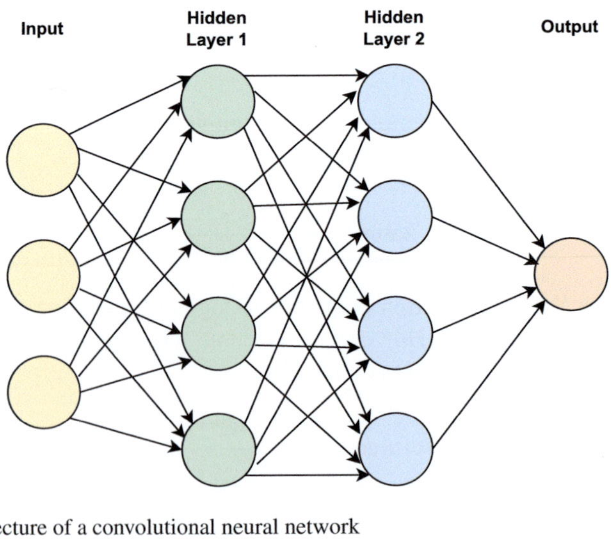

Fig. 4 Architecture of a convolutional neural network

Pooling layers: Typically, after a convolutional layer has completed, the output is sent as input to a pooling layer to reduce the data space even further. This reduction is achieved by collecting the input data into smaller tensors. The dimensionality of the input tensor is not changed.

2.5 Recurrent Neural Networks

While feedforward neural network architectures produce good results for a variety of applications, it may be beneficial for the machine learning model to learn better dependencies between input data. Instead of feeding the output of layers forward into the next layer in the model, it is possible to connect it in such a way to feed it back as input to the current layer. Establishing a memory in this manner is one of the key architectural features of recurrent neural networks and Boltzmann machines.

In recurrent neural networks (RNNs), the neurons are able to use their output as part of their input [17, 46]. This effectively maintains a memory for each neuron, using information from the previous execution in the next execution, as shown in Fig. 5. Sequences of data are the primary use case for RNNs. The inputs, each of size N, to the recurrent hidden layers are a sequence of vectors across time t defined by $\mathbf{x}_t = (x_1, x_2, \ldots, x_N)$. The outputs, each of size M, are similarly defined

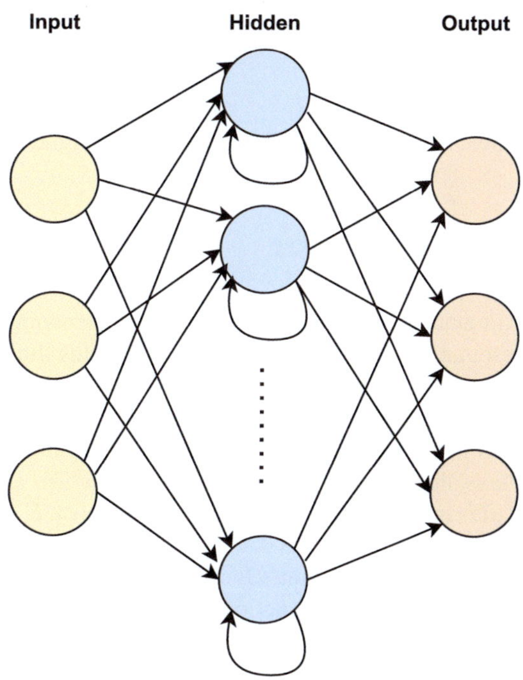

Fig. 5 RNN architecture. The hidden layer uses the output of neurons as input to the next timestep for the same neuron. This is the memory of the RNN

as a sequence of vectors across time t defined by $\mathbf{y}_t = (y_1, y_2, \ldots, y_M)$. To keep a memory, each layer has a hidden vector of size P corresponding to each time step t, defined by $\mathbf{h}_t = (h_1, h_2, \ldots, h_P)$. The hidden vector is the output of the current layer at the current time step t, and is used as information in processing the same hidden layer for the next time step $t + 1$ and its associated input $\mathbf{x}_{t+1}$. Therefore, the full recurrent hidden layer can be expressed as

$$\mathbf{y}_t = \mathbf{W}\mathbf{x}_t + \mathbf{W}\mathbf{h}_{t-1} + \mathbf{b}_h, \tag{9}$$

where $\mathbf{W}$ is the weight matrix and $\mathbf{b}_h$ is the bias vector.

2.6 *Boltzmann Machines*

Instead of using a deterministic process like in RNNs, Boltzmann machines (BMs) utilize a stochastic process to learn the distribution of a set of data points [21]. The probability P of a state vector $\mathbf{v}$ is defined as

$$P(\mathbf{v}) = \frac{e^{-E(\mathbf{v})}}{\sum_{\mathbf{u}} e^{-E(\mathbf{u})}}, \tag{10}$$

where the $E(\mathbf{v})$ is the energy of the state vector $\mathbf{v}$. The energy is given by

$$E(\mathbf{v}) = -\sum_i s_i^{\mathbf{v}} b_i - \sum_{i<j} s_i^{\mathbf{v}} s_j^{\mathbf{v}} w_{ij}, \tag{11}$$

where neuron i has the binary state $s_i^{\mathbf{v}}$ assigned to it by $\mathbf{v}$.

2.7 *Autoencoders*

Autoencoders [8] are a type of neural network that learns a low-dimensional representation of the original data. It does this through an encoder and a decoder. The encoder is responsible for reducing the high-dimensional original data into the low-dimensional data space, also known as the latent space. The decoder thus takes the data in the latent space and reconstructs it into the original data space. Figure 6 displays an overview of the encoder and decoder and their interactions.

Formally, this is defined as the encoder $\mathcal{E}$ learning the function $\mathcal{E}_\phi : \mathcal{X} \rightarrow \mathcal{Z}$, where $\mathcal{X}$ is the original data space and $\mathcal{Z}$ is the latent space. Conversely, the decoder $\mathcal{D}$ learns the function $\mathcal{D}_\theta : \mathcal{Z} \rightarrow \mathcal{X}$. $\mathcal{E}_\phi$ and $\mathcal{D}_\theta$ are parameterized by ϕ and θ, respectively. These functions aim to satisfy

$$\underset{\theta,\phi}{\operatorname{argmin}}\ L(\theta, \phi), \tag{12}$$

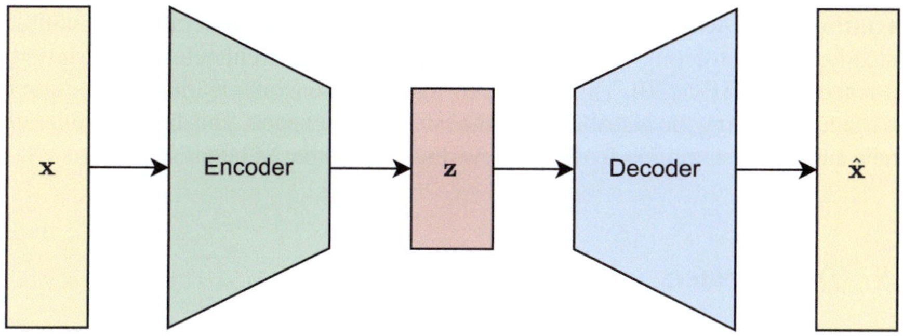

Fig. 6 Architecture of autoencoders. The input $\mathbf{x}$ is given to the encoder, producing $\mathbf{z}$ in the latent space. The decoder then processes $\mathbf{z}$ for reconstruction, resulting in $\hat{\mathbf{x}}$

where $L(\theta, \phi)$ is the loss function and is defined by

$$L(\theta, \phi) := \mathbb{E}_{x \sim \mu_{ref}}[d(x, \mathcal{D}_\theta(\mathcal{E}_\phi(x)))]. \tag{13}$$

Simplified, this results in the encoder producing a latent space vector such that

$$\mathbf{z} = \mathcal{E}_\phi(\mathbf{x}), \tag{14}$$

where $\mathbf{x}$ is the input vector and $\mathbf{z}$ is the latent space vector. Conversely, the decoder produces a reconstructed version of the original input data such that

$$\hat{\mathbf{x}} = \mathcal{D}_\theta(\mathbf{z}), \tag{15}$$

where $\hat{\mathbf{x}}$ is the reconstruction of the original input vector $\mathbf{x}$.

Dimensionality reduction, and thus compression, is the primary application of autoencoders, taking high-dimensional data and reducing it into the latent space of a lower dimension. With autoencoders learning the latent space representation of the data, it allows for the latent space representation of a data sample to be reconstructed back to the original high-dimensional space. Additionally, there are specific applications that the autoencoder structure has been modified to work well. This includes the *denoising autoencoder* and the *variational autoencoder*.

Denoising Autoencoders: Denoising autoencoders [55] are a specific type of autoencoder with the goal of denoising the data. The input data is noisy and the encoder is responsible for learning the data while reducing it to the latent space and learning the features of the data. The decoder reconstructs the original input from the latent space, cleaning up the noisy data in the original data with learned features.

Variational Autoencoders: Autoencoders can be used to generate new samples instead of reconstructing the input data sample with the architecture of variational autoencoders (VAEs) [30]. This is done by having the encoder learn a distribution of the input data, using the distribution as the latent vector space. The decoder generates a new sample by sampling from this new distribution-based latent space.

2.8 Transformers

Transformers are a neural network architecture that uses self-attention to learn the dependencies between data, unlike the memory structure of RNNs [54]. Attention mechanisms are a series of functions that enable the memory-like aspect in these models, helping to find and establish long-range dependencies between data [7]. Transformers are often used in natural language processing for handling text data. Large language models (LLMs) are a type of transformer model that can generate large amounts of text in response to a prompt, which is often also text-based.

There are several major components involved in the transformer architecture. These are the *input embeddings*, *self-attention mechanisms*, and *multi-head attention*.

- *Input embeddings* are the vector representation of the tokenized input. For example, with text input, words can be tokens, which are then processed into their embedding as a vector for use in the model.
- *Self-attention mechanism* is used to learn dependencies across sequence data. One such way to determine self-attention is with scaled dot product attention, defined as

$$\text{attention}(Q, K, V) = \text{softmax}\left(\frac{QK^T}{\sqrt{d_k}}\right)V, \tag{16}$$

  where Q is the queries, K is the keys, V is the values, and d_k is the dimension of the keys so that $\frac{1}{\sqrt{d_k}}$ acts as a scaling factor. Q, K, and V are derived from the input embedding.
- *Multi-head attention* calculates multiple self-attentions concurrently. Each one is specifically geared toward a specific portion of the sequence, allowing for robust learning of the features of the input embeddings.

The transformer architecture consists of two major sections of hidden layers, divided by the type of input to each.

1. *Input embedding section* takes in the input embeddings, processing them through multi-head attention and feedforward layers. The output of this section is used as input to layers within the second section.

2. *Output embedding section* takes in the output from the previous iteration as input. It also consists of several layers of multi-head attention and feedforward layers. One of the multi-head attention layers takes in the output from the input embedding section as additional input to the layer.

The output from the model is the output probabilities of tokens.

Vision transformers (ViTs) are an adjustment to the traditional transformer architecture that operates on text to process image data instead [15]. To create the notion of a token in images like words are in text, the image is broken into several patches, or rectangular regions of the image. Specifically, a 2D image $\mathbf{x} \in \mathbb{R}^{H \times W \times C}$ is reshaped into N patches $\mathbf{x}_p \in \mathbb{R}^{N \times (P^2 \cdot C)}$ where H, W, and C are the height, width, and number of channels of the original image, respectively; P is the height and width of each individual patch; and $N = \frac{HW}{P^2}$.

3 Quantum Machine Learning Models

This section presents the machine learning models utilized for quantum computing, including quantum support vector machines, quantum neural networks, quantum convolutional neural networks, quantum recurrent neural networks, quantum autoencoders, and quantum transformers. Note that most of these models are derived from their classical counterparts described in Sect. 2.

3.1 Quantum Support Vector Machines

Kernel functions are a class of mathematical functions that are used in supervised machine learning. The most popular example of kernel functions is the support vector machine (SVM) used for supervised classification tasks. Supervised binary classification of data is an important task and is the goal of quantum SVMs. Quantum SVMs aim to find the hyperplane to divide the data into two classes, like classical SVMs do (see Sect. 2.2) [42]. Like classical SVMs, quantum SVMs are restricted to linear hyperplanes. Kernel functions can also be used to increase the dimensionality of the hyperplane and observe non-linearity in the decision boundary.

3.2 Quantum Neural Networks

As neural networks serve as the backbone for many machine learning architectures, adapting them to quantum computing is promising in accelerating applications. Quantum neural networks (QNNs) are a category of models that contain a

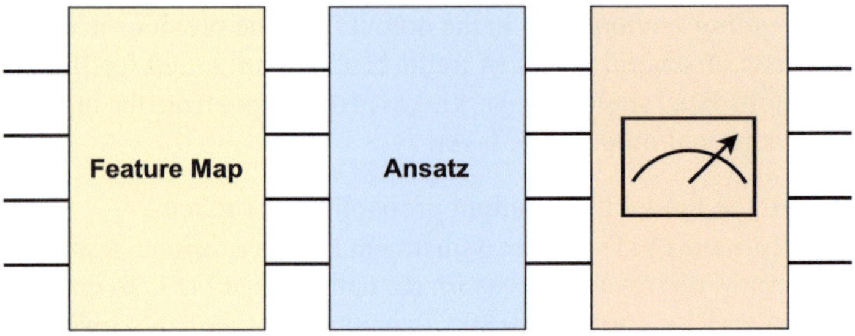

Fig. 7 Generalized QNN architecture with a feature map to prepare the data before sending to the ansatz for processing

neural network-like architecture. However, one major challenge is that the concepts in classical that define what makes a neural network, with its neurons, cannot be easily translated to quantum systems.

Quantum neural networks as a variational quantum circuit have risen in popularity [1, 25]. These variational quantum circuits are used to perform machine learning tasks similar to those in the classical domain, with the most popular training technique of quantum circuit learning [35]. This includes first loading the data, processing the data using an ansatz, performing a measurement to obtain the output, and, finally, updating the parameters of the ansatz with a classical optimizer. An overview of these processes is shown in Fig. 7 that consists of three major steps: data preparation, learning from data, and output processing.

Data Preparation: Data preprocessing prepares the data as input $|\psi\rangle$ to the ansatz of the QNN for full processing. Depending on the domain of the data, this preprocessing step can be different. Quantum data often does not need any additional preprocessing to prepare it. On the other hand, classical data must be converted into quantum state values. This is often done with a feature map. A feature map is a circuit consisting of rotation and unitary gates, placing the data into superposition and entanglement.

Learning from Data: The ansatz $U(\theta)$, parameterized by θ, is the portion of the overall quantum circuit representing the model that is responsible for the core processing of the input data. This is where both training and testing inputs occur after the classical input has been preprocessed using a feature map. The parameterized gates of the variational quantum circuit that compose the ansatz operate on the preprocessed data, such that the output is the result of $U(\theta)|\psi\rangle$.

Output Processing: Once the input data is processed by the ansatz, it is measured to obtain the output. This output can then have further processing on it as part of the training cycle to optimize the parameterized gates, updating the weights. With measuring, it is possible to run classical optimization techniques on the output to update the parameters in the ansatz, such as gradient descent and the Adam optimizer [29].

3.3 Quantum Convolutional Neural Networks

Quantum convolutional neural networks (QCNNs) [14] use convolutional layers for the hidden layers like in CNNs. Their implementation in quantum algorithms consists of a hidden layer where the convolutional layer is followed by a pooling layer, as shown in Fig. 8. This layer pair is repeated until the desired output size is reached. As quantum circuits cannot dynamically change the number of qubits used during execution, extra qubits are left unutilized as the convolutional and pooling layers reduce the dimensional space of the data. The structure of the layers and the extra qubits resulting from reducing the size of the data are shown in Fig. 8. The convolutional and pooling layers are often performed in the same circuit, with the output of

$$|\psi_i(\theta_i)\rangle\langle\psi_i(\theta_i)| = \mathrm{Tr}_{B_i}(U_i(\theta_i)|\psi_{i-1}\rangle\langle\psi_{i-1}|U_i(\theta_i)^\dagger), \tag{17}$$

where i is the layer, U_i is the parameterized circuit containing the convolution and pooling operations, B_i is the subsystem such that $B_i \in \mathbb{C}^{2^{n/2^i}}$, n is the number of qubits, and Tr_{B_i} performs a partial trace over B_i.

3.4 Quantum Boltzmann Machines

Quantum Boltzmann machines (QBMs) [4] are a class of quantum algorithms that learn a probability distribution of a given set of inputs. From QBMs, it is also possible to learn hidden features of the data. Formally, QBMs are defined as learning the probability distribution

$$P(\mathbf{v}) = \frac{\sum_{\mathbf{h}} e^{-E(\mathbf{z})}}{Z}, \tag{18}$$

where $E(\mathbf{z})$ is the energy function for combined state $\mathbf{z}$ and Z is the quantum partition function defined by

$$Z = \sum_{\mathbf{z}} e^{-E(\mathbf{z})}. \tag{19}$$

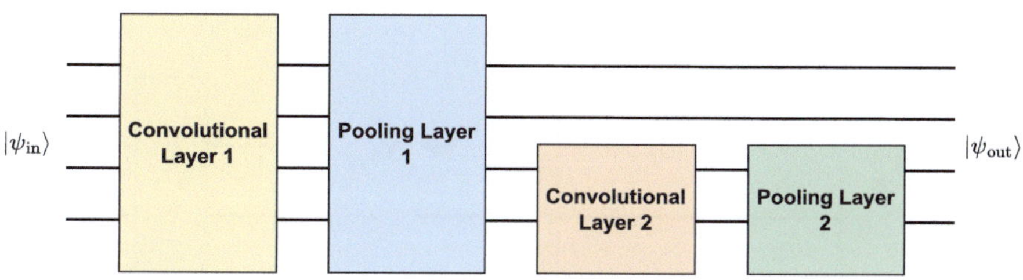

Fig. 8 QCNN architecture with blocks of convolutional and pooling layers

The energy function is given by

$$E(\mathbf{z}) = -\sum_{a} b_a z_a - \sum_{a,b} w_{ab} z_a z_b, \tag{20}$$

where b_a and w_{ab} are the trainable parameters.

3.5 *Quantum Recurrent Neural Networks*

Quantum recurrent neural networks (QRNNs) attempt to learn the relationship between sequence-based data [9, 49]. As with other neural network models, the goal of running RNNs on quantum devices is to learn stronger relationships in data due to placing the data into superposition and entanglement during the processing of the model. QRNNs exhibit the same memory structure as RNNs do, where the output of a current hidden layer is used as input to the current layer on subsequent iterations, as shown in Fig. 9. The hidden register $|\psi_{t-1}\rangle$ at timestep $t-1$ is sent into a layer of a parameterized unitary gate $U(x_t, \theta)$, resulting in output

$$|\psi_t\rangle = U(x_t, \theta)|\psi_{t-1}\rangle. \tag{21}$$

3.6 *Quantum Autoencoders*

Like classical autoencoders, quantum autoencoders encode data into a reduced latent space, then decode it back to the original data space [44]. However, careful consideration must be taken in the quantum algorithm for both the encoder and decoder to ensure correct functionality. Romero et al. [44] first proposed a quantum algorithm for autoencoders in 2017. Quantum autoencoders (QAEs) aim to learn which unitaries allow for the reduced-dimensional space with the latent space to be reconstructed correctly.

The ansatz is broken down into two variational quantum circuits, one for the encoder and one for the decoder. The encoder is $U(\theta)$ and thus the decoder is $U^\dagger(\theta)$,

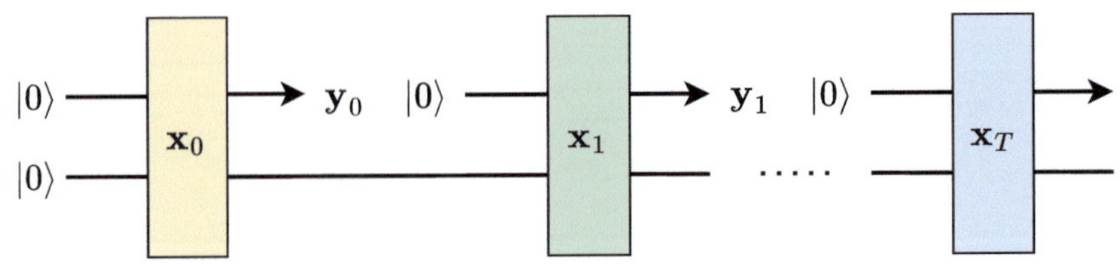

Fig. 9 Architecture of QRNN over timesteps 0 to T

both being parameterized by θ. For effective computation, including training the autoencoder, more qubits are required than the number of qubits required to handle the input state, unlike other QNN ansatz-based architectures. In addition to the input qubits, there are additional qubits for the reference state and auxiliary. The decoder also requires latent space qubits and trash qubits as inputs.

To effectively represent the computations of both the encoder and decoder, along with the specific qubits they use, these qubits are classified into subsystems. The latent space qubits are subsystem A, the trash space qubits are subsystem B, and the reference space qubits are subsystem B'. These qubits are chosen as the ones used in the circuit, such as $U_{AB}(\theta)$ operating using both the latent space and trash space qubits for system AB. Therefore, the output state from the full ansatz, containing both the encoder and decoder, is defined by

$$\rho_{\text{out}} = U_{AB'}^{\dagger}(\theta)\,\text{Tr}_B\left[U_{AB}(\theta)\left[|\psi_{in}\rangle\langle\psi_{in}|_{AB} \otimes |a\rangle\langle a|_{B'}\right] U_{AB'}(\theta)\right] U_{AB'}(\theta), \quad (22)$$

where $|\psi_{in}\rangle$ is the input state to the encoder, or the AB system, and $|a\rangle$ is the reference state contained in B'. Figure 10 provides an overview of the entire workflow.

Similar to classical autoencoders, there are quantum autoencoders for performing denoising functions and variational autoencoders.

Denoising with QAEs: While there is no specific denoising autoencoder in quantum domain, modifications can be made to the general QAE architecture to perform denoising tasks [10]. This includes adjusting the cost function within the optimization routine during training. Bondarenko and Feldmann [10] showed that using a cost function with the fidelity of the data in the QAE can cause noisy quantum states to be denoised. Specifically, the noisy quantum states were highly entangled Greenberger–Horne–Zeilinger (GHZ), W, Dicke, and cluster states. In addition to the denoising, spin-flip errors were also corrected.

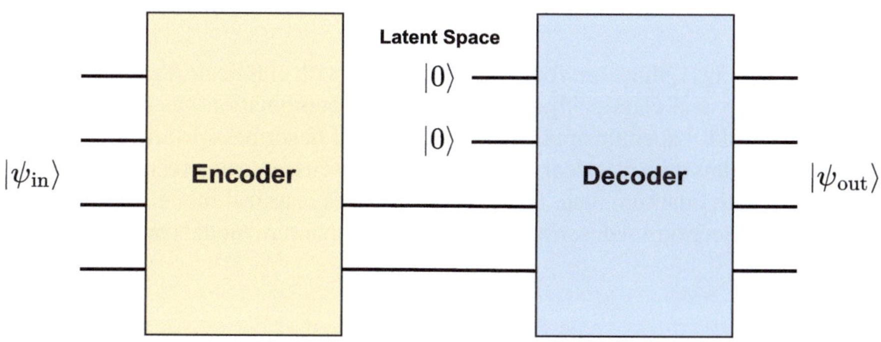

Fig. 10 Overview of the architecture of a quantum autoencoder

Quantum Variational Autoencoders: Quantum variational autoencoders (QVAEs) are the generative version of QAEs [27]. The key structure in VAEs is sampling from a learned distribution rather than using the latent space representation of an input sample during the reconstruction process. For QVAEs, this sampling step is implemented as preparing a quantum state of appropriate size of the latent space. This prepared state is then used as the input to the decoder, resulting in a generated output.

3.7 *Quantum Transformers*

Quantum transformers aim to implement the self-attention mechanism from classical domain for effective encoded representation learning of data [13]. While classical transformers begin by encoding the data, it is implemented as a quantum orthogonal layer in quantum domain. The quantum orthogonal layer processes input quantum state $|\mathbf{x}_i\rangle$ through unitaries $\mathbf{V}$, resulting in output $|\mathbf{V}\mathbf{x}_i\rangle$. Self-attention performs a similar computation, calculating $|\mathbf{x}_i \mathbf{W} \mathbf{x}_j|^2$ as the attention coefficient A_{ij} with $\mathbf{W}$ as a trainable weight matrix. The output from self-attention is thus defined by

$$\mathbf{y}_i = \sum_j A_{ij} \mathbf{V} \mathbf{x}_j, \tag{23}$$

and is computed classically with measurements from the attention coefficient and the quantum orthogonal layer. These attention calculations work with patches, providing an implementation of the vision transformers as well [13].

4 Combination of Quantum Computing with Machine Learning

As outlined in Fig. 1, there are four ways of dealing with classical/quantum machine learning models and classical/quantum data. Each combination has unique advantages and useful for specific applications. Section 4.1 describes specific applications of classical models with classical data. Section 4.2 discusses usage scenarios of quantum models with quantum data. Section 4.3 describes classical models operating on quantum data. Section 4.4 describes applications of quantum models processing classical data.

4.1 *Classical Models with Classical Data*

This section describes different application scenarios of classical models with classical data. Specifically, we provide an overview of how different classical machine learning models (e.g., neural networks or transformers) deal with different classical data types (e.g., text or image).

4.1.1 Convolutional Neural Networks

CNNs are widely used for dealing with images or videos, such as image classification and segmentation [34, 37]. U-Net is a popular CNN architecture for image segmentation, learning a feature map of the input image data, then producing different objects and their boundaries within the input image [45]. They have also been extended to work in diffusion models [43].

4.1.2 Recurrent Neural Networks

Seq2Seq is an RNN-based model that transforms an input sequence into another sequence [48]. This model is commonly used with text data, transforming one sentence into another sentence, and is the precursor to transformers and large language models. The memory allows for the generation task to produce coherent output by storing previous information during training and execution.

4.1.3 Autoencoders

Autoencoders can be used for a variety of tasks centered around feature learning. Modifications to the autoencoder architecture can lead to efficient anomaly detection within a given dataset [71]. This dataset can potentially contain noise and outliers, or anomalies in the data, which need to be identified for further usage of the overall dataset. The autoencoder can also accurately detect these anomalies without training on data where anomalies are present, making it robust.

Variational autoencoders are utilized in the generation of new data samples. Image generation is a popular task with VAEs, such as IntroVAE, which can generate high-resolution photographic images [23]. The VAE architecture includes an additional inference model to assess the quality of the generated image in an adversarial fashion. In addition to generation of images, the labels and captions for the images can also be generated [41]. The latent space of the image generation portion is given to the classifier handling the label and caption generation and prediction for accuracy.

4.1.4 Transformers

Vision transformers (ViTs) can be used for data types that are not images. The idea of transforming the images into patches to act as the token has inspired new research in using a similar strategy for their data space. For example, they can be adapted to work on satellite image time series [50].

4.2 Quantum Models with Quantum Data

This section describes different application scenarios of quantum models using quantum data. Specifically, we provide an overview of how different quantum machine learning models (e.g., neural networks or transformers) deal with different quantum data types. Quantum data are produced through various stages in quantum computing, including quantum algorithms [58], quantum state preparation [64, 66], quantum error correction [61], quantum measurement [52, 53], quantum noise analysis [57, 59], quantum characterization [65], quantum compilation [60, 62], quantum verification [67], and quantum benchmarking [63].

4.2.1 Quantum Neural Networks

QNNs and their ansatz often include basic gates. However, some specific types of quantum systems do not work well with these gates, or cannot be easily reversed. Continuous variables are one such system that requires all unitary gates for best performance. Killoran et al. [28] created a VQC for the QNN ansatz to use only unitary gates and elements to allow for training on continuous variable quantum state data. Continuous variables are represented with fields instead of qubit states, and thus require modification to the QNN architecture for specific processing. QNNs can also be used for specific tasks in quantum information processing, such as determining the quantum mutual information between two quantum states [47]. The QNN architecture can estimate this calculation by adjusting the loss function to account for quantum Donsker–Varadhan representations.

4.2.2 Quantum Convolutional Neural Networks

Quantum state characterization is an important task for analyzing the quantum state data produced by quantum circuits. Primarily, this is more important for quantum circuits executed on real quantum devices due to the presence of noise and errors that can be present, unlike simulators. A QCNN can be trained to help characterize quantum state data, specifically the symmetry-protected topological phases of a spin model [20]. The data is prepared using a variational state preparation circuit, parameterized with rotation gates and CZ gates.

4.2.3 Quantum Recurrent Neural Networks

The memory aspect of QRNNs and its subsequent learning of relationships between sequences of data make it a prime candidate for quantum process tomography, or learning the quantum process from its input and output data [11]. Memory acts as the quantum channels in quantum process tomography, able to detail certain aspects of the quantum process.

4.2.4 Quantum Autoencoders

One of the primary applications of autoencoders and QAEs is compression. In QAEs, the latent space is the learned representation of the input data, using fewer qubits than the original data. In fact, QAEs can provide a lossless data compression, reconstructing it from the latent space through the decoder with minimal information loss [22]. However, with quantum data, it is often noisy, which can cause autoencoders to not learn an effective latent space representation of the data due to the noise impacting the real features. Instead, the autoencoder can be updated to take into account the noise present in the data and from the underlying quantum device running the machine learning model to produce a better compression compared to other related techniques [12].

Error correction is a major task in ensuring that quantum circuits are run with minimal noise and error. Typically, error correction is done through the usage of error-correcting codes, but this can be laborious as it can require manual intervention by experts. An autoencoder can instead be trained to detect and correct errors present during the execution of a quantum circuit. Specifically, the encoder is able to learn the features of the underlying data that are present and reconstruct it as full data items with the decoder. Bondarenko et al. [10] showed that using a cost function with the fidelity of the data in the QAE caused noisy quantum states to be denoised. Specifically, the noisy quantum states were highly entangled Greenberger–Horne–Zeilinger (GHZ), W, Dicke, and cluster states. In addition to the denoising, spin-flip errors were also corrected.

4.2.5 Quantum Transformers

In quantum computing, quantum transformers can be used for a variety of tasks, from generating diagrams, predicting values from measurements, and more. Transformer quantum state is a specific implementation of the quantum transformer to carry out these tasks, all in one model [70]. This is done by learning the features of many similar types of quantum states, rather than just a single quantum state.

4.3 Classical Models with Quantum Data

There are various applications where it is advantageous to use a classical model despite having quantum data. This section describes different application scenarios of classical models with quantum data. Specifically, we provide an overview of how different classical machine learning models (e.g., neural networks or transformers) deal with different quantum data types.

4.3.1 Neural Networks

Quantum data often contains entanglement between qubits and noise from the quantum device that generated the data. In the case of entanglement, it can be difficult to attribute a full value to a specific qubit if its value depends on another qubit. Conversely, noise can exemplify errors in the data, and should be mitigated or removed to represent the quantum data generated from the quantum device accurately.

Entanglement detection is the process of analyzing quantum data and determining if it has entanglement present. If the entanglement is present, this can also include determining the qubits that are entangled. It is an important task in analyzing states from a quantum circuit, as measurement operators often lose this valuable information when reading the value of a qubit into a classical register. There are several neural networks for providing an accurate entanglement detection. Asif et al. [6] use a neural network to detect mixed entangled states. Specifically, the neural network examines the coherence and entropy of quantum state data to find the entanglement present. While quantum state data is often complete, sometimes it is incomplete, especially with measurement operators. Koutný et al. [31] use a deep neural network to determine the degrees of entanglement by learning the features of the incomplete data, filling in the missing data with learned features. Often, these models can only work in two-qubit systems. Ureña et al. [51] proposed an MLP model that can work in two- and three-qubit systems, detecting possible entanglement between the qubits of the system. In addition to detection, it can categorize which qubits are contributing to the present entanglement in the three-qubit system.

Gao et al. [19] present a classifier for quantum states to be used in quantum process tomography and other similar applications. This aims to learn through the presence of noise, which can often make reconstruction or learning of the quantum process difficult. Quantum devices should behave the same; however, environmental noise can often cause two quantum devices to exhibit different behavior. To examine if two quantum devices behave similarly, a neural network can analyze data from executing quantum circuits on both devices. The continuous variable states from this resulting data are of the greatest importance in determining the similarity. It is often necessary to characterize the quantum state data resulting from the execution of a quantum circuit, especially on real quantum devices. Zhu et al. [72] present a neural network that is trained on simulated state data to characterize data from real quantum devices.

The data from the real quantum devices differs from the simulated data due to the presence of noise and other environmental factors, and thus appears to the model as completely unknown data.

4.3.2 Convolutional Neural Network

While CNNs and QCNNs are heavily used to process image data, they can also be used for other data types. In this case, a CNN can be used to classify and reconstruct optical quantum states [2]. This works as the quantum states are 2D in nature with the density matrix representation, leading to an effective use of the convolution layers with kernels like images have. Follow-up work has also used a conditional GAN to reconstruct the density matrices to the same effect.

4.3.3 Recurrent Neural Networks

RNNs learn the relationship between data in sequence, which is beneficial for applications requiring characterizing or feature learning of quantum data. As such, RNNs have been shown to predict the result of a qubit and its evolution under specific environmental factors [18]. The relationship that the RNN learns between the input data of input states to a quantum circuit is equated to the tomography of the circuit, without knowing the circuit.

4.3.4 Autoencoders

A VAE can be utilized to learn the phase transitions in a parameterized quantum circuit [68]. With the optimal parameters for the parameterized circuit obtained, the VAE encodes these parameters to learn the relevant features. Multi-head self-attention over the feature maps resulting from the encoder thus relates the different features together. Generation from the decoder using these feature maps results in learning about the phase transitions of the original parameterized circuit.

4.3.5 Transformers

Transformers, while popular in their use with text and images, can be utilized to process quantum data. One such application is for quantum gate set tomography, characterizing a quantum process, including the full gate set [69]. To perform gate set tomography efficiently, the transformer encodes the input data consisting of gate sequence data, learning the relationships between the sequences and their elements. From these learned features, a feedforward neural network is used to predict error parameters as the final result of the given input data.

4.4 Quantum Models with Classical Data

Many classical ML models have been adapted to work on quantum computers, with varying success. These models are often adapted as a quantum algorithm for the best success. However, these models struggle to work with classical data, as for every bit of data, there needs to be one qubit. Models dealing with large data, like images, thus require thousands of qubits for the smallest image datasets like MNIST and CIFAR-10 to function. These quantum models thus often take advantage of tensor networks to compress the parameters of the model, allowing for more data to be processed with less memory.

To use classical data with quantum models, it must be mapped into quantum states. This is often done using a feature map, with the results used as inputs to the ML model in question. One of the more popular feature map techniques is the ZZ feature map, or the second-order Pauli rotation circuit, as shown in Fig. 11 for two qubits. This section describes different application scenarios of quantum models with classical data. Specifically, we provide an overview of how different quantum machine learning models (e.g., neural networks or transformers) deal with different classical data types (e.g., text or image).

4.4.1 Quantum Convolutional Neural Networks

QCNNs are largely used like their classical counterparts to process image data, primarily for image classification [14]. The original QCNN architecture can be modified to achieve better results or to work under different constraints. The QCNN, primarily the ansatz, can be modified to work with two-qubit circuit blocks to reduce the number of necessary parameters per layer to 15 [24]. These two-qubit circuit blocks represent U_i in the layer of QCNNs defined in Sect. 3.3. Adjusting to two-qubit circuits allows for image classification tasks to perform better and achieve better accuracy results. Li et al. [33] present a quantum deep convolutional neural network (QDCNN) for image recognition, modifying the traditional QCNN architecture. In QDCNN, there is an additional quantum classification layer inserted after the convolutional and pooling layers. This layer allows for the model to effectively use the feature mappings learned from the previous layers and interpret them into class labels.

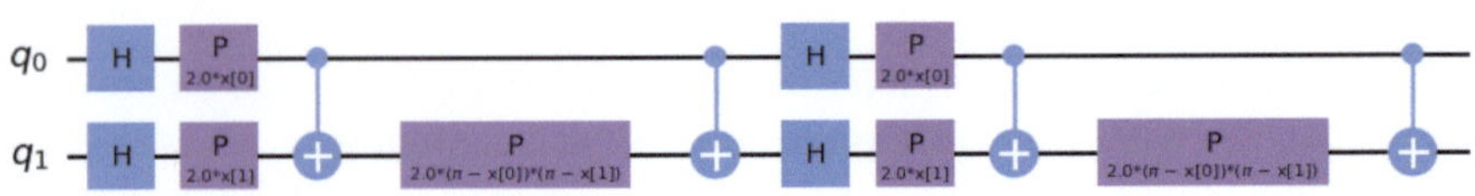

Fig. 11 Two-qubit ZZ feature map

4.4.2 Quantum Recurrent Neural Networks

The need to implement hidden vectors using several measurements in the middle of the QRNN ansatz can increase resource requirements and potentially cause bugs due to collapsing to the basis state. Viqueira et al. [56] instead formulate a QRNN using density matrices, replacing statevector simulation. This allows for operations, such as operator sum, to use fewer resources. With the reduced requirements and efficient structure without the measurements, the QRNN can be used to generate additional sequence elements for univariate and multivariate time series, specifically signal data.

4.4.3 Quantum Autoencoders

While QCNNs are traditionally used to carry out image classification, QAEs can also be used. Asaoka and Kudo [5] proposed a variation of the QAE architecture to classify images so that the features of the images are extracted through a quantum process. This is done with the addition of a circuit over the reference space B' to encode information about the class label. Thus, the result from this new circuit takes the place of the reference state $|a\rangle$ as used in Eq. 22. Figure 12 contains an overview of this resulting architecture, where $V(x)$ is the feature map mapping the classical image data into quantum states, $V_L(y)$ is the circuit for the class label, and $U(\theta)$ is the ansatz.

4.4.4 Quantum Transformers

Quantum transformers are able to process and learn over classical data when feature mapped into quantum states. Images can be processed with a quantum transformer for visual clustering with QClusformer [36]. QClusformer uses one VQC for processing

Fig. 12 QAE architecture for image classification [5]

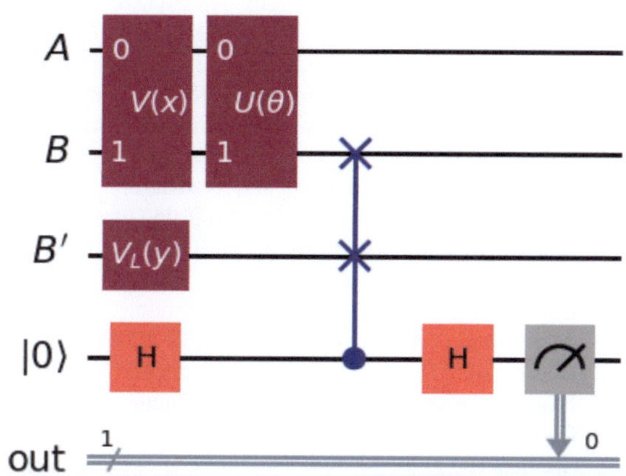

the query, key, and value by taking a linear combination of the Pauli matrices. The resulting architecture performs accurate clustering of images. Text is another popular data modality with transformers, and Quixer is an alternative version of transformers for word-level language modeling tasks [26]. Quixer utilizes unitary gates in linear combination along with quantum singular value transform to learn the relations between the input token data, which has been encoded into quantum states. Attention is thus implemented through the composition of the unitaries.

5 Summary

Machine learning is widely used in a variety of applications. With the onset of quantum computing, these classical models can be utilized to work as quantum circuits and quantum algorithms to varying success. This chapter surveyed various machine learning models in classical and quantum domains and explored their applicability in four categories based on models and data types, including classical model processing classical data, classical model processing quantum data, quantum model processing classical data, and quantum model processing quantum data.

References

1. Abbas, Amira, David Sutter, Christa Zoufal, Aurelien Lucchi, Alessio Figalli, and Stefan Woerner. 2021. The power of quantum neural networks. *Nature Computational Science* 1 (6): 403–409.
2. Ahmed, Shahnawaz, Carlos Sánchez Muñoz, Franco Nori, and Anton Frisk Kockum. 2021. Classification and reconstruction of optical quantum states with deep neural networks. *Physical Review Research* 3 (3): 033278.
3. Aïmeur, Esma, Gilles Brassard, and Sébastien Gambs. 2006. Machine Learning in a Quantum World. In *Advances in Artificial Intelligence*, ed. Luc Lamontagne and Mario Marchand, 431–442. Berlin, Heidelberg: Springer.
4. Amin, Mohammad H., Evgeny Andriyash, Jason Rolfe, Bohdan Kulchytskyy, and Roger Melko. 2018. Quantum Boltzmann Machine. *Physical Review X* 8 (2): 021050.
5. Asaoka, Hinako, and Kazue Kudo. 2025. Quantum autoencoders for image classification.
6. Asif, Naema, Uman Khalid, Awais Khan, Trung Q. Duong, and Hyundong Shin. 2023. Entanglement detection with artificial neural networks. *Scientific Reports* 13 (1): 1562.
7. Bahdanau, Dzmitry, Kyunghyun Cho, and Yoshua Bengio. 2016. Neural Machine Translation by Jointly Learning to Align and Translate.
8. Bank, Dor, Noam Koenigstein, and Raja Giryes. 2021. Autoencoders. arXiv, April 2021.
9. Bausch, Johannes. 2020. Recurrent Quantum Neural Networks. In *Advances in Neural Information Processing Systems*, vol. 33, 1368–1379. Curran Associates, Inc.
10. Bondarenko, Dmytro, and Polina Feldmann. 2020. Quantum autoencoders to denoise quantum data. *Physical Review Letters* 124 (13): 130502.
11. Bondarenko, Dmytro, Robert Salzmann, and Viktoria-S. Schmiesing. 2023. Learning Quantum Processes with Memory – Quantum Recurrent Neural Networks.
12. Cao, Chenfeng, and Xin Wang. 2021. Noise-Assisted Quantum Autoencoder. *Physical Review Applied* 15 (5): 054012.

13. Cherrat, El Amine, Iordanis Kerenidis, Natansh Mathur, Jonas Landman, Martin Strahm, and Yun Yvonna Li. 2024. *Quantum Vision Transformers. Quantum* 8: 1265.
14. Cong, Iris, Soonwon Choi, and Mikhail D. Lukin. 2019. Quantum convolutional neural networks. *Nature Physics* 15 (12): 1273–1278.
15. Dosovitskiy, Alexey, Lucas Beyer, Alexander Kolesnikov, Dirk Weissenborn, Xiaohua Zhai, Thomas Unterthiner, Mostafa Dehghani, Matthias Minderer, Georg Heigold, Sylvain Gelly, Jakob Uszkoreit, and Neil Houlsby. 2021. An Image is Worth 16x16 Words: Transformers for Image Recognition at Scale.
16. Dunjko, Vedran, Jacob M. Taylor, and Hans J. Briegel. 2016. Quantum-enhanced machine learning. *Physical Review Letters* 117 (13).
17. Elman, Jeffrey L. 1990. Finding structure in time. *Cognitive Science* 14 (2): 179–211.
18. Flurin, E., L. S. Martin, S. Hacohen-Gourgy, and I. Siddiqi. 2020. Using a recurrent neural network to reconstruct quantum dynamics of a superconducting qubit from physical observations. *Physical Review X* 10 (1): 011006.
19. Gao, Jun, Lu-Feng Qiao, Zhi-Qiang Jiao, Yue-Chi Ma, Hu Cheng-Qiu, Ruo-Jing Ren, Ai-Lin Yang, Hao Tang, Man-Hong Yung, and Xian-Min Jin. 2018. Experimental machine learning of quantum states. *Physical Review Letters* 120 (24): 240501.
20. Herrmann, Johannes, Sergi Masot Llima, Ants Remm, Petr Zapletal, Nathan A. McMahon, Colin Scarato, François Swiadek, Christian Kraglund Andersen, Christoph Hellings, Sebastian Krinner, Nathan Lacroix, Stefania Lazar, Michael Kerschbaum, Dante Colao Zanuz, Graham J. Norris, Michael J. Hartmann, Andreas Wallraff, and Christopher Eichler. 2022. Realizing quantum convolutional neural networks on a superconducting quantum processor to recognize quantum phases. *Nature Communications* 13 (1): 4144.
21. Hinton, Geoffrey E. 2002. Training products of experts by minimizing contrastive divergence. *Neural Computation* 14 (8): 1771–1800.
22. Huang, Chang-Jiang, Hailan Ma, Qi Yin, Jun-Feng Tang, Daoyi Dong, Chunlin Chen, Guo-Yong Xiang, Chuan-Feng Li, and Guang-Can Guo. 2020. Realization of a quantum autoencoder for lossless compression of quantum data. *Physical Review A* 102 (3): 032412.
23. Huang, Huaibo, Zhihang li, Ran He, Zhenan Sun, and Tieniu Tan. 2018. IntroVAE: Introspective Variational Autoencoders for Photographic Image Synthesis. In *Advances in Neural Information Processing Systems*, vol. 31. Curran Associates, Inc.
24. Hur, Tak, Leeseok Kim, and Daniel K. Park. 2022. Quantum convolutional neural network for classical data classification. *Quantum Machine Intelligence* 4 (1): 3.
25. Jeswal, S. K., and S. Chakraverty. 2019. Recent developments and applications in quantum neural network: A review. *Archives of Computational Methods in Engineering* 26 (4): 793–807.
26. Khatri, Nikhil, Gabriel Matos, and Luuk Coopmans, and Stephen Clark. 2024. Quixer: A Quantum Transformer Model.
27. Khoshaman, Amir, Walter Vinci, Brandon Denis, Evgeny Andriyash, Hossein Sadeghi, and Mohammad H Amin. 2018. Quantum variational autoencoder. *Quantum Science and Technology* 4 (1): 014001.
28. Killoran, Nathan, Thomas R. Bromley, Juan Miguel Arrazola, Maria Schuld, Nicolás Quesada, and Seth Lloyd. 2019. Continuous-variable quantum neural networks. *Physical Review Research* 1 (3): 033063.
29. Kingma, Diederik P., and Jimmy Ba. 2014. Adam: A Method for Stochastic Optimization.
30. Kingma, Diederik P., and Max Welling. 2013. Auto-Encoding Variational Bayes.
31. Koutný, Dominik, Laia Ginés, Magdalena Moczała-Dusanowska, Sven Höfling, Christian Schneider, Ana Predojević, and Miroslav Ježek. 2023. Deep learning of quantum entanglement from incomplete measurements. *Science Advances* 9 (29): eadd7131.
32. LeCun, Yann, Yoshua Bengio, and Geoffrey Hinton. 2015. Deep learning. *Nature* 521 (7553): 436–444.
33. Li, YaoChong, Ri-Gui Zhou, Xu RuQing, Jia Luo, and Hu WenWen. 2020. A quantum deep convolutional neural network for image recognition. *Quantum Science and Technology* 5 (4): 044003.

34. Li, Zewen, Fan Liu, Wenjie Yang, Shouheng Peng, and Jun Zhou. 2022. A survey of convolutional neural networks: analysis, applications, and prospects. *IEEE Transactions on Neural Networks and Learning Systems* 33 (12): 6999–7019.
35. Mitarai, K., M. Negoro, M. Kitagawa, and K. Fujii. 2018. Quantum circuit learning. *Physical Review A* 98 (3): 032309.
36. Nguyen, Xuan-Bac, Hoang-Quan Nguyen, Samuel Yen-Chi Chen, Samee U. Khan, Hugh Churchill, and Khoa Luu. 2024. QClusformer: A quantum transformer-based framework for unsupervised visual clustering. In *2024 IEEE International Conference on Quantum Computing and Engineering (QCE)*, vol. 02, 347–352.
37. O'Shea, Keiron, and Ryan Nash. 2015. An Introduction to Convolutional Neural Networks.
38. Pan, Zhixin, and Prabhat Mishra. 2022. A survey on hardware vulnerability analysis using machine learning. *IEEE Access* 10: 49508–49527.
39. Pan, Zhixin, and Prabhat Mishra. 2023. *Explainable AI for Cybersecurity*. Springer.
40. Pan, Zhixin, Jennifer Sheldon, and Prabhat Mishra. 2022. Hardware-assisted malware detection and localization using explainable machine learning. *IEEE Transactions on Computers* 71 (12): 3308–3321.
41. Pu, Yunchen, Zhe Gan, Ricardo Henao, Xin Yuan, Chunyuan Li, Andrew Stevens, and Lawrence Carin. 2016. Variational autoencoder for deep learning of images, labels and captions. In *Advances in Neural Information Processing Systems*, vol. 29. Curran Associates, Inc.
42. Rebentrost, Patrick, Masoud Mohseni, and Seth Lloyd. 2014. Quantum support vector machine for big data classification. *Physical Review Letters* 113 (13): 130503.
43. Rombach, Robin, Andreas Blattmann, Dominik Lorenz, Patrick Esser, and Björn Ommer. 2022. High-resolution image synthesis with latent diffusion models. In *Proceedings of the IEEE/CVF Conference on Computer Vision and Pattern Recognition*, 10684–10695.
44. Romero, Jonathan, Jonathan P Olson, and Alan Aspuru-Guzik. 2017. Quantum autoencoders for efficient compression of quantum data. *Quantum Science and Technology* 2 (4): 045001.
45. Ronneberger, Olaf, Philipp Fischer, and Thomas Brox. 2015. U-Net: Convolutional Networks for Biomedical Image Segmentation. In *Medical Image Computing and Computer-Assisted Intervention – MICCAI 2015*, Lecture Notes in Computer Science, 234–241. Cham: Springer.
46. Salehinejad, Hojjat, Sharan Sankar, Joseph Barfett, Errol Colak, and Shahrokh Valaee. 2018. Recent Advances in Recurrent Neural Networks.
47. Shin, Myeongjin, Junseo Lee, and Kabgyun Jeong. 2024. Estimating quantum mutual information through a quantum neural network. *Quantum Information Processing* 23 (2): 57.
48. Sutskever, Ilya, Oriol Vinyals, and Quoc V. Le. 2014. Sequence to Sequence Learning with Neural Networks.
49. Takaki, Yuto, Kosuke Mitarai, Makoto Negoro, Keisuke Fujii, and Masahiro Kitagawa. 2021. Learning temporal data with a variational quantum recurrent neural network. *Physical Review A* 103 (5): 052414.
50. Tarasiou, Michail, Erik Chavez, and Stefanos Zafeiriou. 2023. ViTs for SITS: Vision Transformers for Satellite Image Time Series. In *Proceedings of the IEEE/CVF Conference on Computer Vision and Pattern Recognition*, 10418–10428.
51. Ureña, Julio, Antonio Sojo, Juani Bermejo-Vega, and Daniel Manzano. 2024. Entanglement detection with classical deep neural networks. *Scientific Reports* 14 (1): 18109.
52. Utt, Zachery, Daniel Volya, and Prabhat Mishra. Quantum measurement discrimination using cumulative distribution functions. 2023. In *2023 Design, Automation & Test in Europe Conference & Exhibition (DATE)*, 1–6. IEEE.
53. Utt, Zachery, Daniel Volya, and Prabhat Mishra. 2024. Quantum measurement classification using statistical learning. *ACM Transactions on Quantum Computing (TQC)* 5 (2): 1–16.
54. Vaswani, Ashish, Noam Shazeer, Niki Parmar, Jakob Uszkoreit, Llion Jones, Aidan N Gomez, Łukasz Kaiser, and Illia Polosukhin. 2017. Attention is all you need. In *Advances in Neural Information Processing Systems*, vol. 30. Curran Associates, Inc.
55. Vincent, Pascal, Hugo Larochelle, Yoshua Bengio, and Pierre-Antoine Manzagol. 2008. Extracting and composing robust features with denoising autoencoders. In *Proceedings of the 25th International Conference on Machine Learning*, ICML '08, 1096–1103, New York, NY, USA, July 2008. Association for Computing Machinery.

56. Viqueira, J. D., D. Faílde, M. M. Juane, A. Gómez, and D. Mera. 2025. Density matrix emulation of quantum recurrent neural networks for multivariate time series prediction. *Machine Learning: Science and Technology* 6 (1): 015023.
57. Volya, Daniel, and Prabhat Mishra. 2020. Impact of noise on quantum algorithms in noisy intermediate-scale quantum systems. In *IEEE International Conference on Computer Design (ICCD)*, 1–4. IEEE.
58. Volya, Daniel, and Prabhat Mishra. 2021. Quantum spectral clustering of mixed graphs. In *2021 58th ACM/IEEE design automation conference (DAC)*, 463–468. IEEE.
59. Volya, Daniel, and Prabhat Mishra. 2022. Modeling of noisy quantum circuits using random matrix theory. In *2022 IEEE 40th International Conference on Computer Design (ICCD)*, 132–138. IEEE.
60. Volya, Daniel, and Prabhat Mishra. 2023. Quantum data compression for efficient generation of control pulses. In *Asia and South Pacific Design Automation Conference (ASPDAC)*, 216–221.
61. Volya, Daniel, and Prabhat Mishra. 2023. Quantum steering of surface error correcting codes. In *2023 IEEE International Conference on Quantum Computing and Engineering (QCE)*, vol. 1, 1394–1400. IEEE.
62. Volya, Daniel, and Prabhat Mishra. 2023. Qudcom: Towards quantum compilation for qudit systems. arXiv preprint arXiv:2311.07015.
63. Volya, Daniel, and Prabhat Mishra. 2024. Quantum benchmarking via random dynamical quantum maps. In *2024 IEEE International Conference on Quantum Computing and Engineering (QCE)*, vol. 1, 1277–1283. IEEE.
64. Volya, Daniel, and Prabhat Mishra. 2024. State preparation on quantum computers via quantum steering. *IEEE Transactions on Quantum Engineering (TQE)* 5: 1–14.
65. Volya, Daniel, Andrey Nikitin, and Prabhat Mishra. 2024. Fast quantum characterization via riemannian optimization. In *2024 IEEE International Conference on Quantum Computing and Engineering (QCE)*, vol. 1, 1270–1276. IEEE.
66. Volya, Daniel, Zhixin Pan, and Prabhat Mishra. 2023. Feedback-based steering for quantum state preparation. In *2023 IEEE International Conference on Quantum Computing and Engineering (QCE)*, vol. 1, 1308–1318. IEEE.
67. Witharana, Hasini, Daniel Volya, and Prabhat Mishra. 2024. Qcassert: Quantum device testing with concurrent assertions. In *Asia and South Pacific Design Automation Conference (ASPDAC)*, 491–496. IEEE.
68. Xin, Li, and Zhang-Qi Yin. 2025. Learning quantum phase transition in parametrized quantum circuits with an attention mechanism.
69. Yu, King Yiu, Aritra Sarkar, Maximilian Rimbach-Russ, Ryoichi Ishihara, and Sebastian Feld. 2025. Transformer models for quantum gate set tomography. *Quantum Machine Intelligence* 7 (1): 10.
70. Zhang, Yuan-Hang, and Massimiliano Di Ventra. 2023. Transformer quantum state: A multipurpose model for quantum many-body problems. *Physical Review B* 107 (7): 075147.
71. Zhou, Chong, and Randy C. Paffenroth. 2017. Anomaly detection with robust deep autoencoders. In *Proceedings of the 23rd ACM SIGKDD International Conference on Knowledge Discovery and Data Mining*, KDD '17, 665–674, New York, NY, USA, August 2017. Association for Computing Machinery.
72. Zhu, Yan, Ya-Dong Wu, Ge Bai, Dong-Sheng Wang, Yuexuan Wang, and Giulio Chiribella. 2022. Flexible learning of quantum states with generative query neural networks. *Nature Communications* 13 (1): 6222.

Quantum Compilation and Error Correction

Hardware-Aware Compilation for Different Quantum Computing Platforms

Ludwig Schmid, Daniel Schoenberger, Yannick Stade, Lukas Burgholzer, and Robert Wille

Abstract Quantum computing is rapidly advancing toward practical utility. But current hardware platforms face significant physical constraints. These include limited qubit connectivity, noise, and architectural restrictions, necessitating sophisticated compilation strategies. This chapter presents an overview of hardware-aware quantum circuit compilation tailored to each of the three major hardware platforms: superconducting qubits, neutral atoms, and trapped ions. We discuss key compilation stages—mapping, routing, and scheduling—emphasizing how platform-specific features influence algorithm design. Methods include optimization based on satisfiability solvers and heuristic approaches like A* search. All discussed techniques are available in the open-source *Munich Quantum Toolkit* (MQT), facilitating further community development. The discussion underlines the importance of architecture-specific compilation and suggests how these approaches can extend to further platforms such as photonic and spin-based systems.

L. Schmid (✉) · D. Schoenberger · Y. Stade · L. Burgholzer · R. Wille
Chair for Design Automation, Technical University of Munich, Munich, Germany
e-mail: ludwig.s.schmid@tum.de

D. Schoenberger
e-mail: daniel.schoenberger@tum.de

Y. Stade
e-mail: yannick.stade@tum.de

L. Burgholzer
e-mail: lukas.burgholzer@tum.de

R. Wille
e-mail: robert.wille@tum.de

L. Burgholzer · R. Wille
Munich Quantum Software Company GmbH, Garching near Munich, Germany

R. Wille
Software Competence Center Hagenberg GmbH, Hagenberg, Austria

P. Mishra (ed.), *Design Automation for Quantum Computing*,
https://doi.org/10.1007/978-3-032-09303-5_5

1 Introduction

Quantum computing is transitioning from theoretical promise to practical realization, with hardware platforms becoming increasingly capable and scalable. However, these systems still operate under significant constraints, including limited connectivity, noise, and architectural restrictions that hinder the direct execution of arbitrary quantum circuits [80].

To bridge the gap between high-level quantum programs and the physical realities of hardware, advanced software tools are indispensable [18, 98]. Among these, *quantum circuit compilation* plays a central role: It translates abstract quantum algorithms into concrete sequences of hardware-compatible operations, while optimizing for platform-specific constraints and performance objectives. This process must be tightly coupled with the physical characteristics of the target architecture, as different platforms offer distinct native operations and constraints.

Effective compilation strategies must therefore be hardware-aware. What is optimal for one platform may be inefficient or even infeasible for another. More precisely:

(1) **Superconducting systems** present challenges due to static and sparse qubit connectivity. Compilation must account for a fixed coupling map, often requiring careful insertion of `SWAP` gates to enable necessary interactions [19, 58, 106].
(2) **Neutral atom platforms** offer more flexibility through atom rearrangement, enabling dynamic qubit layouts [88]. However, compilation must respect physical constraints like non-crossing movements and designated zones [9, 94].
(3) **Trapped-ion devices**, particularly those using the Quantum Charge Coupled Device (QCCD) architecture [45], require careful scheduling of ion transport and gate execution, balancing latency and architectural limitations [50, 90].

These variations necessitate tailored compilation pipelines that exploit each platform's strengths while mitigating its limitations.

In this chapter, we present an overview of quantum circuit compilation across these three leading hardware platforms. We focus on three key stages of the compilation process that are particularly influenced by hardware constraints:

- **Mapping**—Assigns logical qubits to physical hardware qubits, ensuring that the circuit can be executed within the device's connectivity structure.
- **Routing**—Introduces additional operations—such as `SWAP` or `MOVE` operations—to enable necessary two-qubit interactions between non-adjacent qubits.
- **Scheduling**—Organizes operations over time to respect hardware-specific execution constraints and to optimize overall performance.

Together with **Synthesis** of the gate sequence and a final **Evaluation**, these steps form the core of hardware-aware quantum compilation.

All discussed techniques are implemented in open-source as part of the *Munich Quantum Toolkit* [109] (MQT), giving easy access to the community to use and improve the available tools.

The remainder of this chapter is structured as follows. Section 2 offers an overview of quantum circuit compilation, covering key subroutines: synthesis, mapping, routing, scheduling, and evaluation. Sections 3–5 focus on hardware-aware compilation for specific quantum platforms. Section 3 addresses superconducting chips, detailing routing challenges and symbolic SWAP optimization. Section 4 discusses neutral atom platforms, highlighting compilation strategies for hybrid and zoned architectures under physical and optical constraints. Section 5 explores trapped ion systems with a focus on time-optimal shuttling in shuttling architectures. Section 6 briefly surveys other quantum platforms and their compilation considerations. Finally, Sect. 7 concludes with a summary and future outlook.

2 Quantum Circuit Compilation

Broadly speaking, the term "compilation" refers to the transformation of a high-level, abstract quantum algorithm description into a lower-level representation comprising operations suitable for further processing or direct hardware execution [18, 27, 33, 64]. This process is typically realized through a layered software architecture, where each layer is dedicated to handling specific compilation tasks. Due to the interdisciplinary nature of quantum computing, which merges concepts from both computer science and physics, terminology—including the term "compilation" itself—may vary depending on the context. To mitigate any ambiguity of the term *compilation*, we provide a brief overview and clarification of the compilation subroutines considered in this chapter.

2.1 Compilation Subroutines

The compilation subroutines can be grouped into three principal categories, classified according to their level of abstraction and their dependence on a specific hardware architecture. These three categories are adapted from similar considerations for classical compilation for different hardware architectures [18].

- *Platform-independent compilation*
  This category comprises high-level optimization techniques that are applicable regardless of the target quantum platform. Examples include loop unrolling and function inlining [43], which simplify constructs such as for-loops and function invocations in a manner analogous to classical compiler optimizations. Additional techniques in this category involve gate-level optimizations, such as exploiting commutation rules to reduce gate count or depth [40, 65].

- *Platform-dependent compilation*
  These subroutines are tailored to the specific features and constraints of a given quantum hardware platform. The resulting output is a sequence of platform-specific instructions, which remain abstracted from the actual physical hardware implementation.
- *Device-dependent compilation*
  This lowest layer of compilation is customized for a concrete hardware configuration, or even a specific device. It converts platform-specific operations into hardware-level instructions executable by the quantum processing unit. It is sometimes referred to as *firmware*, emphasizing its close integration with the hardware.

This work focuses on the platform-dependent compilation subroutines, with a particular emphasis on the stages of *Mapping* and *Routing*, which will be explained in more detail below. We analyze these steps across different quantum hardware architectures—namely, superconducting qubits, neutral atoms, and trapped ions. These platforms are characterized by fundamentally distinct physical constraints and interaction topologies, which require tailored and platform-specific compilation strategies.

While this general structure above applies also to many other compilation schemes, such as compilation for measurement-based quantum computing (MBQC) and compilation for machines operating on a fault-tolerant level using quantum error correction (QEC), our focus here lies on compilation for gate-based computation on near-term intermediate-scale quantum (NISQ) devices. Nevertheless, many of the principles and hardware-specific considerations discussed in this chapter can inform and be adapted to MBQC or QEC-based compilation scenarios, where similar challenges manifest at different levels of abstraction.

From here on, we assume that both hardware-dependent and platform-independent optimization steps have been completed in advance. This can be achieved using, e.g., ZX-based tools such as `PyZX` [46]. Our aim now is to produce hardware-oriented instruction sequences while avoiding lower-level aspects such as signal control or pulse-level implementation, which depend to an even higher degree on the specific hardware setup. This setting naturally leads to five core tasks within the compilation chain, illustrated in Fig. 1.

(1) *Synthesis*, also commonly termed *decomposition*, concerns the breakdown of abstract quantum gates into operations that are natively supported by the target platform.

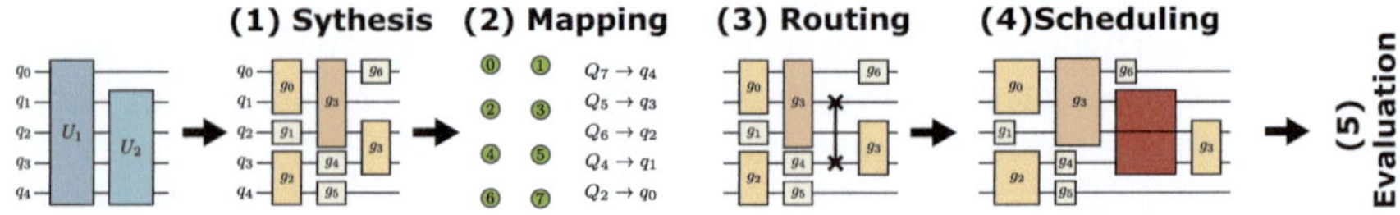

Fig. 1 The five steps of hardware-dependent compilation, adapted from [88]

(2) *Mapping* addresses the spatial allocation of circuit qubits onto physical hardware qubits.
(3) *Routing* ensures the necessary qubit connectivity is achieved by inserting connectivity-altering operations such as `SWAP` or `MOVE` operations. This enables two-qubit interactions across initially unconnected physical qubits.
(4) *Scheduling* involves the temporal organization of operations, ensuring that the platform's architectural constraints are fulfilled while achieving an optimized execution of the operations.
(5) *Evaluation* assesses the quality of the compilation output based on appropriate figures of merit. These metrics may vary depending on the abstraction level and the target hardware platform, and may include measures such as circuit depth, gate count, estimated fidelity, execution time, or resource overhead.

The subsequent sections provide more formal definitions and some examples of these five steps, as applicable to the context of this chapter.

2.1.1 Synthesis

In the *quantum circuit model*, any non-dissipative quantum computation $U \in \mathbb{C}^{2^n \times 2^n}$ can be represented as a finite sequence of unitary operations $g \in \mathbb{C}^{2^m \times 2^m}$, with $m < n$. These smaller unitary operations are known as *quantum gates*. Each gate acts on a subset of qubits, with n (m) denoting the number of qubits involved in the overall circuit (individual gate), respectively. These gates are restricted to reversible unitary transformations and therefore are closely related to classical reversible gates and the corresponding logic, such that many of the classical synthesis methods can be directly mapped to the quantum case [78].

But while in the classical world one only has to consider operations on the binary computational states, quantum algorithms exhibit arbitrary unitary rotations around the Bloch sphere, requiring some kind of discretization. The Solovay-Kitaev theorem [47] ensures that for a general unitary operation U, an approximation up to an arbitrarily small error can always be obtained efficiently using only a discrete set of finite size. Such a collection of gates is referred to as an *universal gate set* Σ^{univ}. On real hardware, however, only a limited set of operations can be implemented directly, which is specified by the *native gate set* Σ^{native}. This native set is typically chosen to perform operations that closely represent the physical behavior of the system. To have a *universal quantum computer*, one requires Σ^{nativ} to also be a universal gate set.

Typically, this is realized by Σ^{native} comprising single-qubit rotations around at least two different axes and at least one entangling two-qubit gate, such as CX, CZ, or the Mølmer–Sørensen gate [15]. To then enable execution on a particular quantum device, any quantum operation not included in the native gate set must be decomposed into a sequence the native gates, a process commonly referred to as *synthesis* or *decomposition* [1, 28, 66, 67].

Definition 1 (*Synthesis*) Given a unitary quantum operation $U \in \mathbb{C}^{2^n \times 2^n}$ and a native gate set Σ^{native}, the *synthesis* task consists in finding a gate sequence

$$\tilde{U} = g_{N-1} \circ \cdots \circ g_0$$

with all $g_0, \ldots, g_{N-1} \in \Sigma^{\text{native}}$ such that $\tilde{U}$ approximates U up to a specified precision.

Example 1 Examples of native gate sets for selected quantum platforms, depending mostly on the way the gates are realized on hardware:

- IBM (until 2021): $[U_3(\theta, \phi, \lambda), CX]$,
- IBM (until 2024): $[Id, R_Z(\theta), \sqrt{X}, X, CX]$,
- IBM (current): $[Id, R_Z(\theta), \sqrt{X}, X, CZ]$,
- IonQ/AQT: $[R_X(\theta), R_Y(\theta), R_Z(\theta), R_{XX}(\theta)]$.

As an example, none of these gate sets support operations acting on more than two qubits directly. Hence, all larger multi-qubit operations, including three-qubit gates such as the Toffoli gate, must be decomposed into single- and two-qubit gates.

Example 2 Consider the circuit depicted in Fig. 2a, which implements a simplified version of Grover's search algorithm. This circuit contains a three-qubit Toffoli gate, which cannot be implemented natively on any of the above platforms. Consequently, synthesis must be applied to decompose the Toffoli gate into native operations. Figure 2b (b) illustrates one such decomposition using the Clifford+T gate set.

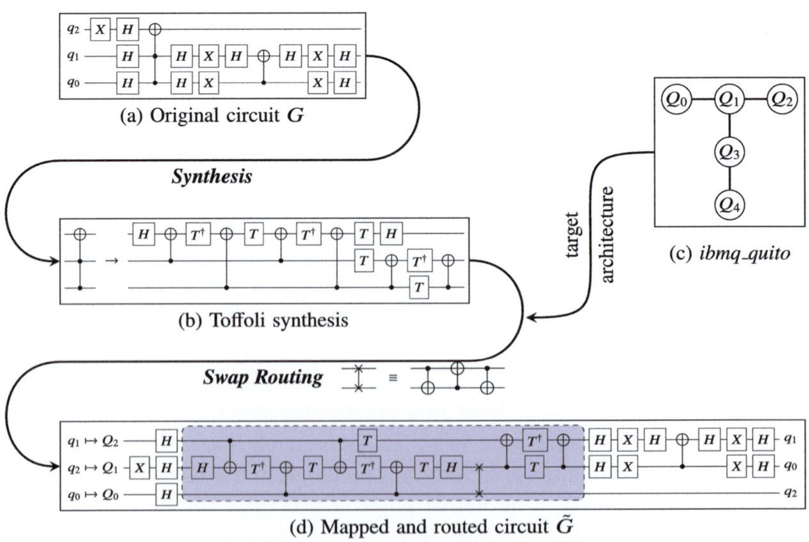

Fig. 2 Exemplary compilation-flow illustration for a Grover-type circuit G

2.1.2 Mapping

After synthesis, the next step in the compilation flow is *mapping*, which establishes a correspondence between the logical circuit qubits $\mathbf{Q} = \{q_0, \ldots, q_{n-1}\}$ and the physical hardware qubits $\mathbf{P} = \{Q_0, \ldots, Q_{|\mathbf{P}|}\}$ with $|\mathbf{P}| \geq n$. This relationship is formalized as a bijective mapping function $f : \mathbf{Q} \rightarrow \mathbf{P}$.

Some *dynamic quantum processors*, such as neutral atoms or trapped ions, allow for rearrangement of the physical qubits. Here, the physical qubits $\mathbf{P}$ are not fixed in space but must be placed on a coordinate system $\mathbf{C} = \{C_\alpha\}_{\alpha=0}^{|\mathbf{C}|}$, where $|\mathbf{C}| > |\mathbf{P}| \geq n$ represent sites where physical qubits can be placed.

A good assignment must account for the constraints imposed by the hardware's physical connectivity. The connectivity is typically encoded in the form of a *coupling graph* $G = (\mathbf{P}, \mathbf{E})$, where nodes correspond to physical qubits, and edges $\mathbf{E}$ represent pairs that can directly interact through native two-qubit gates. The goal is to place circuit qubits that frequently interact onto adjacent physical qubits in the coupling graph.

Example 3 The coupling graph of a 4-qubit superconducting chip is shown in Fig. 2c.

Definition 2 (*Mapping*) Given a quantum circuit $U = g_{N-1} \circ \cdots \circ g_0$ and a coupling graph $G = (\mathbf{P}, \mathbf{E})$, the task of *mapping* is to compute two bijective functions:

- **Qubit mapping** $f_\text{q} : \mathbf{Q} \rightarrow \mathbf{P}$ assigning circuit qubits to physical atoms.
- **Atom mapping** $f_\text{a} : \mathbf{P} \rightarrow \mathbf{C}$ placing physical qubits at coordinates.

such that as many two-qubit gates $g(q_i, q_j)$ as possible satisfy $(f_\text{a} \circ f_\text{q}(q_i), f_\text{c} \circ f_\text{q}(q_j)) \in \mathbf{E}$.

Example 4 Figure 3a illustrates how $n = 3$ circuit qubits are mapped to $|\mathbf{Q}| = 4$ physical qubits which are layed out on a 3×3 grid of $|\mathbf{C}| = 9$ coordinates.

However, due to the limited connectivity of most quantum hardware and the evolving interaction patterns of quantum circuits, a single static assignment rarely suffices to enable all required interactions, making dynamic adjustments of f_q or f_a necessary.

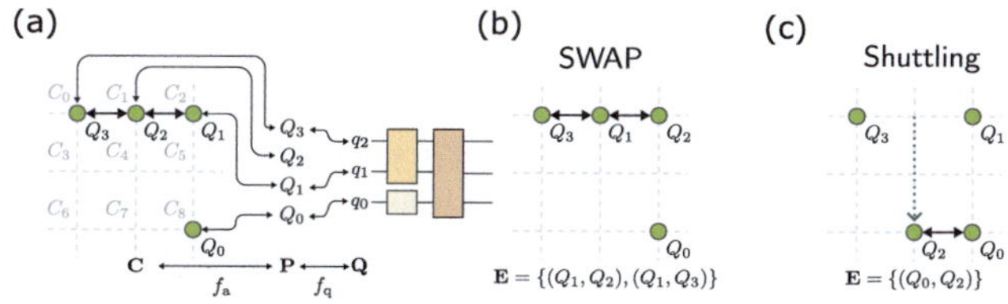

Fig. 3 Illustration [adapted from [89]] of the qubit and coordinate mapping (**a**) and the two routing strategies, swapping (**b**) and shuttling (**c**)

2.1.3 Routing

Because a static mapping typically does not fulfill all interactions, especially in circuits with complex entanglement structures, a dynamic adaptation step called *routing* is necessary.

Routing introduces auxiliary operations—SWAP and MOVE—to modify the mapping functions f_{q} and f_{a} from above throughout circuit execution. Inserting these operations throughout the circuit execution enables qubit interactions that are otherwise disallowed by the initial hardware topology.

A $\mathtt{SWAP}(Q_i, Q_j)$ operation exchanges the logical assignments of two physical qubits, effectively updating f_{q} at runtime. This is implemented on the circuit level via a sequence of three CX gates, introducing additional circuit depth and noise.

Example 5 Figure 3a shows an initial mapping of the circuit. (b) shows the mapping after applying a SWAP between qubit Q_1 and Q_2 changing the coupling graph edges **E**.

Dynamic systems offer an alternative: $\mathtt{MOVE}(Q_i)$, which physically repositions a qubit from one coordinate to another, altering the connectivity graph G without the use of entangling gates but by directly acting on f_{a}. While this avoids increased gate error, it can incur errors from the rearrangement operation, such as additional latency.

Example 6 In (c) of Fig. 3, the action of $\mathtt{MOVE}\,Q_2$ to position C_7 is illustrated, connecting now Q_2 with Q_0.

Definition 3 (*Routing*) Given a quantum circuit $U = g_{N-1} \circ \cdots \circ g_0$, initial mappings f_{q}^0 and f_{c}^0, with the corresponding coupling graph $G = (\mathbf{P}, \mathbf{E})$, *routing* is the task of inserting SWAP and/or MOVE operations such that each gate $g(q_i, q_j)$ satisfies $(f_{\mathrm{c}} \circ f_{\mathrm{q}}(q_i), f_{\mathrm{c}} \circ f_{\mathrm{q}}(q_j)) \in \mathbf{E}$ at the time of execution.

Example 7 After mapping and routing the final circuit $\tilde{G}$ of Fig. 2d fulfills the connectivity constraints of the architecture in Fig. 2c.

2.1.4 Scheduling

The scheduling of quantum gates refers to the temporal arrangement of gate executions after the routing step. A common constraint is that each physical qubit can typically participate in only one gate per timestep.

Further limitations arise, e.g., from the classical control infrastructure. In many quantum platforms, not all qubits can be independently addressed at the same time due to limitations in control electronics, such as shared signal lines or bandwidth constraints. Physical effects can also constrain parallelism. For instance, in neutral atom systems, the Rydberg blockade effect prevents nearby atoms from being simultaneously excited, effectively limiting parallel gate execution based on spatial locality [35, 85, 88].

Definition 4 (*Scheduling*) Given a routed quantum circuit $U = g_{N-1} \circ \cdots \circ g_0$, *scheduling* is the task of assigning a discrete or continuous timestep t_i to each gate g_i such that:

- no qubit is involved in more than one gate at any timestep,
- hardware control constraints are satisfied,
- physical interaction constraints (e.g., Rydberg blockade) are obeyed.

For this task, gates can be considered as blocks of different sizes and widths, as illustrated in Fig. 1(4). Typically minimizing circuit depth and time, this results in a Tetris-like arrangement of gates [59].

Example 8 The last step of Fig. 1 illustrates the temporal arrangement of different gates by representing them as differently sized blocks.

2.1.5 Evaluation

Evaluation assesses the quality of the compilation output through *metrics*, or *figures of merit* tailored to the specific constraints and capabilities of the target hardware. Typically, metrics such as gate count are used as indicators for the routing overhead, particularly for hardware with static connectivity, such as superconducting architectures. However, in the context of dynamic architectures, including systems capable of physically reconfiguring qubit connectivity via `SWAP` or `MOVE` operations, gate count alone is insufficient to capture performance trade-offs.

To better reflect the physical realities of such systems, we adopt a more comprehensive figure of merit: the *approximate success probability* P [88]. This metric incorporates in its definition both temporal and operational aspects of the circuit execution:

$$P = \exp\left(-\frac{t_{\text{idle}}}{T_{\text{eff}}}\right)\prod_{O}\mathcal{F}_O, \quad T_{\text{eff}} = \frac{T_1 T_2}{T_1 + T_2}, \tag{1}$$

where $\mathcal{F}_O \in [0, 1]$ denotes the fidelity of operation O, and T_1, T_2 are the characteristic relaxation and dephasing times of the system. The *effective coherence time* T_{eff} captures the decoherence time scale. The *idle time* t_{idle} accounts for the cumulative time during which qubits are not operated on and, therefore, suffer from decoherence. Intuitively, this metric incorporates two principal considerations:

- **Decoherence errors**: For idle qubits, amplitude damping and dephasing lead to idle errors. These are captured through T_1, T_2, and summarized in T_{eff}.
- **Gate infidelities**: Executing gates, particularly multi-qubit gates, exhibit errors characterized by an average gate fidelity $\mathcal{F}_O$.

Due to its simplicity, the approximate success probability $P(U)$ provides a compact, computationally tractable, and physically informed figure of merit that allows comparing circuits at various abstraction levels and with varying hardware capabilities. Its primary advantage lies in integrating scheduling and routing consequences directly into the metric, thereby enabling more informed and platform-specific optimization strategies.

In the following, these steps are discussed for different hardware platforms, including superconducting chips, neutral atoms, and trapped ions. The focus lies on the mapping and routing tasks, which are illustrated for the different hardware constraints, followed by a short summary of possible algorithmic solutions to these problems.

3 Superconducting Chips

Superconducting chips represent a leading hardware platform for constructing quantum processors [13, 37, 49]. Unlike trapped ions or neutral atoms that operate at microscopic scales, superconducting qubits are macroscopic circuits fabricated using superconducting materials. These circuits exploit the quantum mechanical properties of superconducting currents and Josephson junctions to define and control the qubits.

3.1 Background

The macroscopic design allows for detailed engineering of qubit characteristics such as energy levels, coupling strengths, and coherence times via circuit parameters [37, 82, 92, 105]. One widely used superconducting qubit design is the transmon, which minimizes charge noise and allows for relatively long coherence times. Transmons are typically implemented using a nonlinear LC oscillator formed by a Josephson junction shunted with a large capacitor, effectively reducing sensitivity to charge fluctuations [92]. The qubits are operated at millikelvin temperatures inside dilution refrigerators to maintain superconductivity and suppress thermal noise.

Qubit control and readout are achieved via microwave pulses. Single-qubit gates are performed by applying resonant microwave fields that drive Rabi oscillations between the qubit's ground and excited states. Two-qubit gates typically rely on fixed or tunable couplings mediated by microwave resonators or direct capacitive/inductive interactions. The most common gate mechanisms include cross-resonance, iSWAP, and CZ interactions, with platforms often choosing one based on gate fidelity, crosstalk, and implementation complexity [37].

In a typical architecture, qubits are connected through microwave resonators or direct couplings to enable two-qubit interactions. The specific arrangement of the coupling graph is dictated by architectural choices aiming to balance connectivity, frequency crowding, and cross-talk mitigation. Examples of common layouts are 2D topologies like square grids, IBM's heavy-hex, or Rigetti's octagonal lattice.

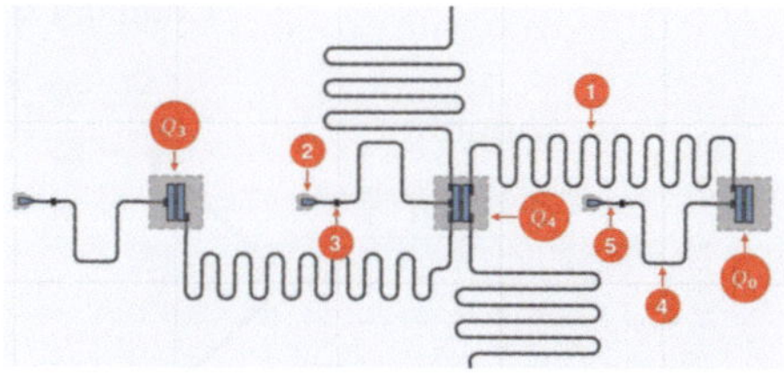

Fig. 4 Chip design with different chip components, adapted from [51]

Example 9 Figure 4 shows a chip section created using the open-source tool Qiskit Metal [68]. It shows three trasmon qubits Q_3, Q_4, Q_0, connected by resonators ①. Each qubit also has a readout/control ② with a resonator ③ merged using a connection between capacitors and readout/control⑤.

The available gate set on superconducting platforms is restricted to a native gate set native gate set Σ^{native} as discussed in Sect. 2.1.1. For instance, IBM devices support single-qubit gates such as $R_Z(\theta)$, $\sqrt{X}$, X, and two-qubit gates like CNOT, CZ.

Importantly, the connectivity of a fabricated chip is fixed, and as a result, two-qubit gates can only be directly applied between physically connected qubits. If a gate in the quantum circuit requires interaction between unconnected qubits, additional routing operations are necessary. This gives rise to the original `SWAP`-insertion problem of quantum circuit compilation.

3.2 *Problem Formulation and Motivation*

Superconducting quantum hardware imposes severe restrictions on qubit connectivity and allows two-qubit interactions (e.g., CNOT or CZ gates) only between specific pairs of physical qubits defined by a coupling graph. Consequently, executing arbitrary quantum circuits often requires the insertion of *SWAP gates*, which permute the positions of circuit qubits on physical hardware to enable necessary interactions.

However, each `SWAP` operation introduces a substantial overhead. In most gate libraries, a `SWAP` must be decomposed into three native two-qubit gates (e.g., CNOTs), increasing circuit depth and amplifying the impact of decoherence and gate errors. This problem of inserting the minimal number of `SWAP` gates during compilation, while respecting hardware connectivity constraints, is known to be *NP-complete* [11].

Example 10 Consider the example from Fig. 5. The initial circuit to the left is mapped and routed for two different coupling graphs corresponding to two differently designed chips. The output contains the minimal number of `SWAP` gates, which can differ significantly depending on the input circuit and the coupling graph.

Existing methods typically assume a static hardware architecture, focusing on optimizing the logical-to-physical qubit mapping and gate scheduling [53, 57, 76,

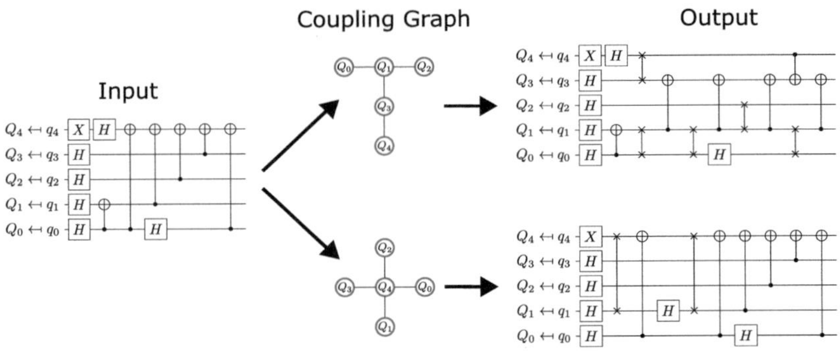

Fig. 5 Output of minimal SWAP gate insertion for two different coupling graph architectures

99, 100, 114]. In the following, we illustrate how this can be done using satisfiability solvers and show how to explore a broader design space that not only seeks optimal mappings for a fixed architecture but also considers potential modifications to the hardware layout.

3.3 *Solution*

To address the `SWAP` insertion problem optimally, it was proposed to use a formal and symbolic approach that fully captures the mapping constraints of a quantum circuit to a given superconducting architecture [107]. Our methodology consists of two main components: (1) an *exhaustive symbolic formulation* of the mapping problem, and (2) the *use of powerful reasoning engines*, such as SAT or SMT solvers, to efficiently navigate the resulting search space.

Given an n-qubit quantum circuit $G = g_0, g_1, \ldots, g_N$ composed of two-qubit gates, and a device described by a coupling graph $G = (\mathbf{P}, \mathbf{E})$, the goal is to find a mapping from logical to physical qubits that enables the execution of each gate as described in Definition 2 and Definition 3. We focus on minimizing the number of additional gates introduced, particularly those arising from `SWAP` insertions.

3.3.1 Symbolic Formulation

The mapping of logical to physical qubits is captured through *Boolean variables*. Specifically, for each gate g_k in the circuit, a variable x_{ij}^k indicates whether circuit qubit q_j is mapped to physical qubit Q_i before gate g_k. This approach accounts for the possibility of remapping (via `SWAP`s) at any point in the circuit.

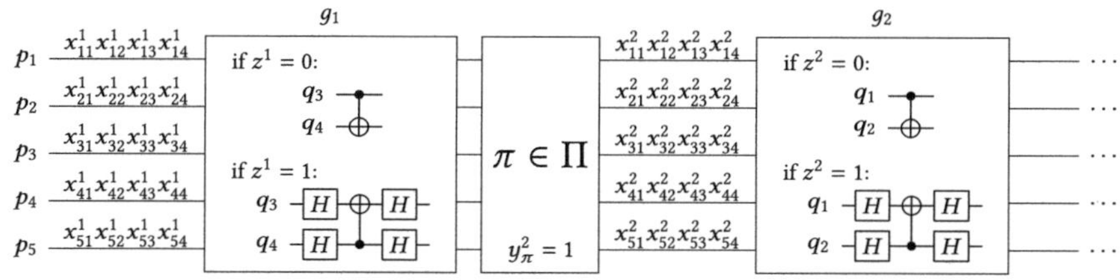

Fig. 6 Illustration of the symbolic encoding of the mapping and routing problem into binary variables, taken from [107]

To ensure only valid mappings are produced, we add constraints to enforce:

(1) **One-to-one mapping**: Each circuit qubit must be mapped to exactly one physical qubit, and no two circuit qubits may occupy the same physical qubit at any given time. This corresponds to the scheduling constraint from Definition 4.
(2) **Connectivity constraints**: Each two-qubit gate $g_k(q_c, q_t)$ must operate on qubits mapped to connected physical qubits in **E**.

In addition, we define *permutation variables* y_π^k to model the application of `SWAP` operations that change the mapping between consecutive gates. These variables represent permutations $\pi \in \Pi$ of the physical qubits and are used to track when and how the mapping changes throughout the circuit.

The overall cost of a mapping, in terms of `SWAP` overhead, is expressed as

$$\mathcal{F} = \sum_{k=1}^{N} \sum_{\pi \in \Pi} \text{swaps}(\pi) \cdot y_\pi^k$$

where swaps(π) denotes the number of `SWAP`s needed to realize permutation π. The symbolic formulation can be passed to a solver with an optimization objective of minimizing $\mathcal{F}$.

Example 11 Figure 6 illustrates such an encoding for a sequence of three CX gates. The z variables are only required if the architecture is directed, as discussed below.

3.3.2 Performance Considerations

While this approach is complete and optimal, the associated computational complexity remains exponential in the worst case. To alleviate this, several performance optimizations are proposed:

- **Search space reduction** by pruning permutations that do not impact upcoming gates.
- **Constraint tightening** to eliminate invalid or redundant mappings early.

- **Omitting intermediate variables** like y_π^k in the optimization process and deriving them post hoc from x_{ij}^k assignments.

For details on these optimizations, we refer to the literature [16, 99, 107]. In the following, we want to briefly discuss some additional considerations connected to this SWAP insertion problem. All these improvements substantially lower the runtime of the symbolic mapping procedure, making it feasible to apply the method to circuits and architectures of moderate size, while still achieving near-optimal solutions [16, 107].

3.3.3 Considering Sub-architectures

When the number of circuit qubits n is less than the number of hardware qubits m, the mapping may be restricted to a subset of n physical qubits. This would require solving $\binom{m}{n}$ subproblems, each corresponding to a different n-qubit sub-architecture. To avoid this additional overhead, higher efficiency is achieved by (among other optimizations) only considering connected and non-isomorphic subarchitectures [77].

3.3.4 Layering Strategies

Instead of allowing permutations before every gate, mappings may only change before selected subsets $G' \subseteq G \setminus \{g_0\}$. This significantly reduces the complexity to $2^{nm(N)}$. Several strategies for selecting G' include

- *Disjoint qubits*: grouping gates that act on disjoint qubit sets into layers [42, 114].
- *Odd gates*: only allowing permutations before gates with odd indices.
- *Qubit triangle*: clustering gates acting on up to three qubits to exploit triangle structures in the device.

Depending on the selected G' this can significantly increase the computation at the cost of sacrificing optimality guarantees.

3.3.5 Improved Mappings for Undirected Architectures.

For devices with unidirectional coupling constraints, additional freedom is obtained by allowing direction-reversing transformations (e.g., replacing a reversed CNOT by surrounding it with Hadamard gates). The symbolic model introduces Boolean variables z^k to track direction reversals and extends the cost function to account for both SWAP and reversal operations.

3.4 Discussion

The compilation of quantum circuits for devices based on superconducting chips with a static coupling map is predominantly challenged by routing constraints, namely the insertion of SWAP gates due to limited qubit connectivity. These SWAP operations introduce significant depth and error overhead, making their minimization essential.

This section presented a symbolic and optimal approach to tackle the SWAP insertion problem. By encoding the mapping problem as a Boolean function and utilizing powerful reasoning engines, one can obtain provably optimal solutions [107]. To mitigate the high computational cost associated with this approach, a range of performance improvements can be employed, including sub-architecture selection, layering strategies, and direction-aware mappings [16].

It is important to note that this symbolic method is one of many strategies in the compilation landscape. Numerous heuristic approaches—such as those based on greedy algorithms, A*-search, reinforcement learning, or stochastic optimization [61, 76, 114] provide faster, albeit approximate, alternatives. These are especially relevant as the number of qubits in current quantum architectures has surpassed the size trackable by optimal approaches.

This section focused on superconducting hardware as a representative and widely used platform. The next sections will extend this investigation to other leading hardware platforms, namely neutral atom and trapped ion devices.

4 Neutral Atom Quantum Computing

Neutral atom quantum computing (NAQC) [35, 54, 87, 88, 110] utilizes individual atoms as qubits, stored in programmable arrays of optical traps. These atoms can be manipulated and entangled using lasers, allowing for scalable quantum computation. Due to their long coherence times, high-fidelity gate operations, and the ability to dynamically rearrange atomic positions, neutral atoms are a promising platform for quantum information processing.

4.1 Background

In NAQC, qubits are realized using individual atoms held in optical dipole traps, such as optical lattices or optical tweezer arrays. These traps are generated by tightly focused laser beams, creating localized potential wells in which atoms can be confined and cooled to their motional ground states.

Atom trapping is typically realized using either static optical lattices or dynamically configurable tweezer traps generated via Spatial Light Modulators (SLMs).

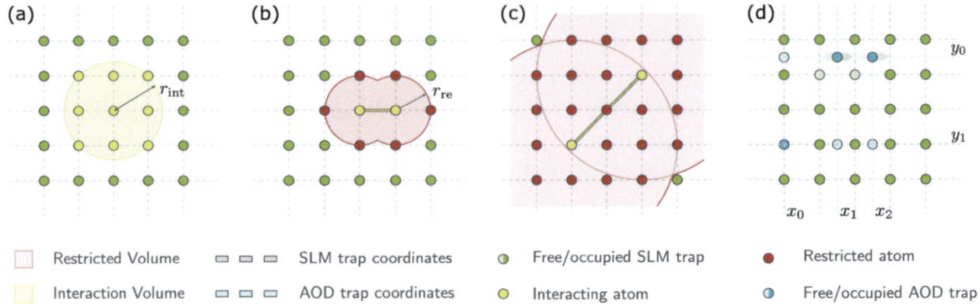

Fig. 7 Illustration of Rydberg interaction and atom rearrangements

These allow for initial stochastic loading followed by defect-free rearrangements [5, 17, 31, 32, 74].

Example 12 Figure 7 shows such optical tweezer arrays arranged in a square grid using Spatial Light Modulators (SLM). While other geometries are possible, for computational tasks, regular patterns are typically advantageous.

The qubit encoding relies on long-lived internal atomic states. In alkali atoms like Rb or Cs, the qubit states $|0\rangle$ and $|1\rangle$ are commonly represented by hyperfine ground states [30, 55]. Alkaline-earth(-like) atoms such as Sr and Yb allow for alternative encodings using metastable states, nuclear spins [4, 44, 63], or fine-structure levels [81, 102].

Single-qubit gates are implemented via laser-driven transitions between qubit states. These transitions can be induced globally across the entire array or targeted locally to individual atoms using focused beams [9, 87]. This allows for high-fidelity state preparation and arbitrary single-qubit rotations.

Two-qubit gates leverage the Rydberg blockade mechanism [30, 86], which involves transient excitation of atoms to high-lying Rydberg states that exhibit strong dipole-dipole or van der Waals interactions [26, 70]. When one atom is excited to a Rydberg state, it shifts the energy levels of nearby atoms, preventing their simultaneous excitation. This conditional dynamic is exploited to implement entangling gates such as the controlled-Z (CZ) gate [25, 29, 39, 41, 56].

A necessary condition for this mechanism to work is that the involved qubits lie within an *interaction radius* r_{int}. Additionally, to mitigate crosstalk during parallel operations, distinct gates must be separated by at least a *restriction radius* $r_{\text{re}} \geq r_{\text{int}}$.

Example 13 The yellow region in Fig. 7a indicates the possible interaction candidates for two-qubit gates for a certain interaction radius r_{int}. Figure 7b shows a possible choice and the corresponding restriction radius r_{restr}. The red region corresponds to the restricted volume for other multi-qubit gates to be executed simultaneously. This restricted volume increases for long-range interactions as illustrated in Fig. 7c.

Atom shuttling is realized by loading atoms from static SLM traps to traps created using two Acousto-Optic Deflectors (AODs) in horizontal and vertical directions.

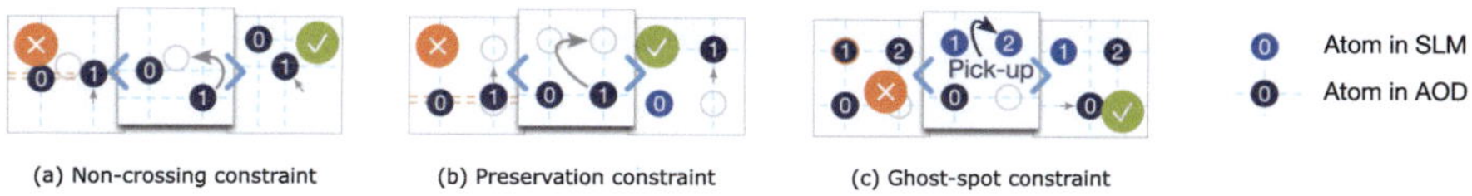

Fig. 8 Constraints on the atom movements using Acousto-Optic Deflectors (AODs)

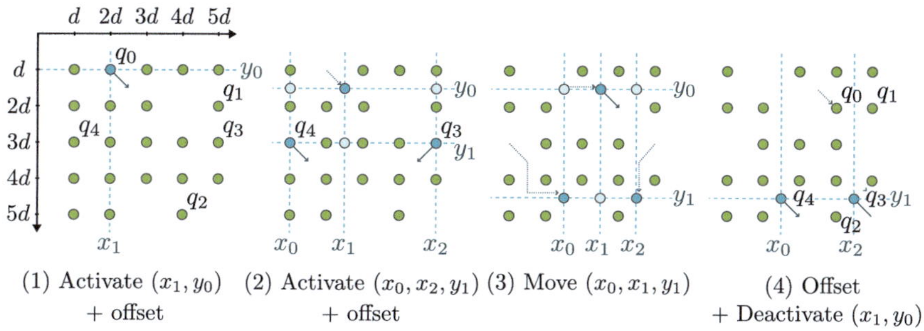

Fig. 9 Loading and unloading sequences due to AOD constraints, taken from [89]

These AOD traps can be described using row and column indices with trap sites at the intersections [88, 101]. By modulating the trapping lasers, one can move rows/columns, resulting in mobile trap sites which can be *rearranged* or *shuttled*. Several qubits can be moved simultaneously by moving multiple row/column intersections. However, three constraints, illustrated in Fig. 8 must be respected [88, 94, 101]:

(a) *Non-crossing constraint*: Row and column movements must preserve ordering, i.e., they cannot cross.
(b) *Preservation constraint*: Atoms trapped in the same AOD row/column must remain in the same row/column.
(c) *Ghost-spot constraint*: Inactive AOD sites still form traps, creating *ghost spots* that must not intersect with existing qubit paths.

These are mitigated through careful loading sequences and offset maneuvers that isolate ghost spots from occupied paths.

Example 14 Figure 9 illustrates a scenario where q_0 needs to be shuttled to q_1 and q_3, q_4 next to q_2. Due to the above three constraints, this needs to be done in multiple loading and movement steps. As a first step (1), q_0 is loaded by activating AOD coordinates at $x_1 = 2d$ and $y_0 = d$. To prevent problematic ghost spots with the following activations, a small offset move is applied. Then (2), q_3 and q_4 can be loaded simultaneously in the same row by activating $x_0 = d$, $x_2 = 5d$ and row $y_1 = 3d$. Note that due to the previous offset, all resulting ghost spots (light blue) are in the empty inter-qubit regions. The qubits can then be moved (3) to their destinations without crossing any other AOD coordinate and placed sequentially using offset movements (4).

4.2 Problem Motivation and Description

Compilation for NAQC systems introduces novel challenges due to the physical and architectural flexibility of neutral atom platforms. Unlike static architectures such as the previous superconducting chips with fixed connectivity, NAQC requires a dynamic determination of qubit placement and routing paths, particularly under the above AOD-movement constraints. This section explores this compilation problem for two different neutral atom-based architectures: hybrid and zoned.

4.2.1 Hybrid Architectures

In *hybrid architectures*, atoms are initially arranged in static optical traps formed by a Spatial Light Modulator (SLM) and are loaded into dynamic Acousto-Optic Deflector (AOD) traps only when necessary [87, 89]. Furthermore, all atoms are assumed to be individually addressable. Two-qubit gate execution depends on the inter-atomic distance: two atoms only interact and execute the two-qubit gate if they are within the interaction radius r_{int} as discussed in Sect. 4.1.

We want to focus on the mapping and routing task, which is a trade-off between two routing possibilities:

- *Gate-based routing*, where connectivity is achieved by inserting `SWAP` operations to move circuit qubits closer together (similar to superconducting hardware).
- *Shuttling-based routing*, where atoms are physically moved to new coordinates using AOD-based traps.

In particular, considering Definition 2, gate-based mapping acts on the circuit qubit to hardware qubit mapping f_{q}, while shuttling-based routing acts on the mapping f_{a} from hardware qubits to positional coordinates.

This creates a two-fold mapping challenge with an expanded search space and the risk of mapping conflicts. Some recent works have focused exclusively on one method [2, 59, 75, 101], but a hybrid approach is necessary to fully exploit NAQC capabilities.

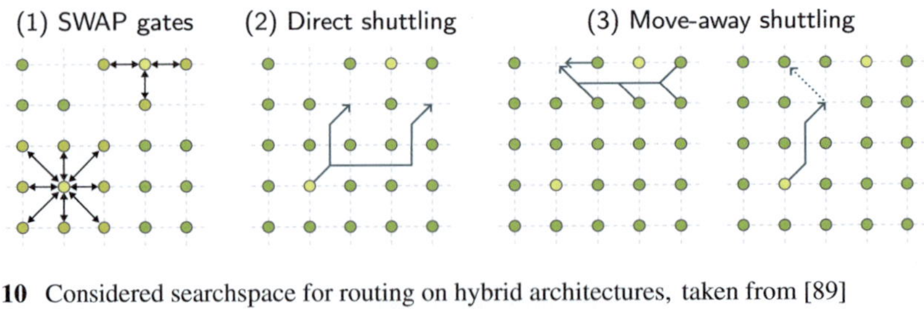

Fig. 10 Considered searchspace for routing on hybrid architectures, taken from [89]

Example 15 Figure 10 illustrates the considered search space to connect two qubits. Shown are the possible operations for the three cases: (1) gate-based `SWAP`s, (2) directly shuttling to one of the free coordinates, and (3) requiring a move-away operation beforehand.

The individual addressability of hybrid architectures imposes a challenge on the control and focus of the Rydberg laser beam. To circumvent this, architectures with global control over a subregion of the atom registers were proposed as discussed in the following.

4.2.2 Zoned Architectures

Figure 11 depicts a schematic of a *zoned neutral atom architecture* based on the experimental setup of Bluvstein et al. [9]. Zoned architectures divide the hardware layout into distinct regions, typically a *storage zone* for idle qubits and an *entanglement zone* where active two-qubit operations are performed. These can be complemented by additional zones, e.g., for readout or a reservoir of atoms to compensate for losses [10, 17]. Atoms are initially loaded into dense, static traps in the storage zone using SLMs. AODs are then used to shuttle atoms to and from the entanglement zone where global Rydberg beams perform CZ operations on selected atom pairs [25, 29]. In particular, the distance between atoms is chosen such that r_{int} and r_{restr} only include the paired atom to perform a two-qubit gate. While this avoids parallelism constraints due to blocking, it also necessitates atom rearrangements before every two-qubit layer, increasing the shuttling overhead.

Example 16 In Fig. 11 an illustration of a two-zone architecture is given. The entanglement zone consists of trap pairs, with qubits placed in adjacent traps (here Q_0 and Q_1) perform a CZ interaction during the global Rydberg beam. Qubits Q_2 and Q_3 are placed in the storage zone, unaffected by the Rydberg laser.

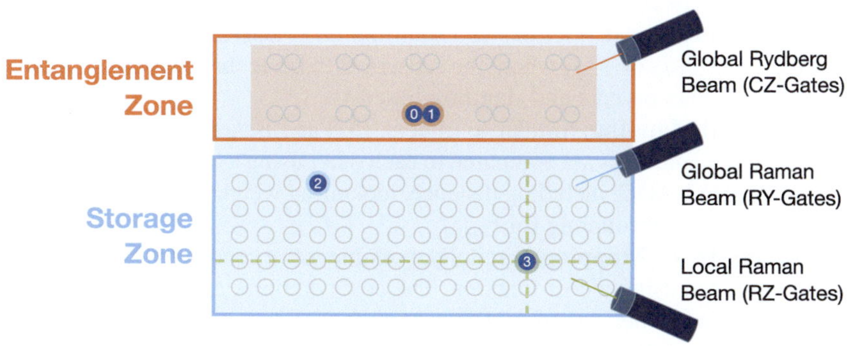

Fig. 11 Schematic of a zoned neutral atom architecture illustrating global and local operations on atoms (blue)

Because only atoms in the entanglement zone are affected by Rydberg excitations, all other atoms remain coherent and shielded from unintended interactions. Multiple two-qubit operations can thus be executed in parallel in a spatially isolated manner.

The compilation problem here focuses on mapping circuit operations to physical movements constrained by AOD limitations, due to the shuttling-only structure of the architecture. During a *rearrangement step*, qubits can be transferred using AODs under the constraints reviewed in Sect. 4.1.

4.3 Proposed Solutions

Here, we present two approaches to address the compilation challenges in neutral atom quantum computing, focusing separately on hybrid and zoned architectures. These solutions consider the physical constraints, most importantly, gate restriction due to r_{restr} and shuttling parallelism due to the AOD constraints. The aim is to optimize the approximate success probability of Eq. 1.

4.3.1 Hybrid Architectures

The routing problem for hybrid architectures is dominated by the choice between `SWAP`-gate insertion and AOD-shuttling. To balance this gate-based and shuttling-based routing, a hybrid compilation pipeline composed of five key stages was proposed, illustrated in Fig. 12 and discussed in more detail in [89]:

(1) **Layer Creation**: Iterate over the input circuit and identify executable gates based on gate commutativity. This defines a frontier layer f and a lookahead layer l, where the lookahead contains gates which become executable once f is empty.
(2) **Mapping Capability Decision**: Here, the costs for `SWAP`s and shuttling operations are estimated based on the arrangement and Euclidean distance of the atoms on the grid. The cost function is based on the approximate success probability to balance between gate fidelities and shuttling time. Based on that, the gates are assigned to f_g, l_g or f_s, l_s where the subscripts indicate the mapping and routing capability (gate-based s and shuttling-based s).
(3) **Gate-based Mapping**: Similar to Li et al. [57] and Baker et al. [2] all possible `SWAP` gate candidates are compared to each other using a cost function. The cost is adapted to NAQC, including distance, lookahead weight w_l, and a time-based decay factor λ^t accounting for r_{restr}.
(4) **Shuttling-based Mapping**: Similarly, possible move candidates are determined and evaluated using a cost function. The cost combines distance reduction and parallelism penalties due to AOD constraints.
(5) **Translation to Hardware Operations**: This layer decomposes `SWAP`s into native CZ and single-qubit gates and schedules AOD movements adhering, again, to the AOD constraints.

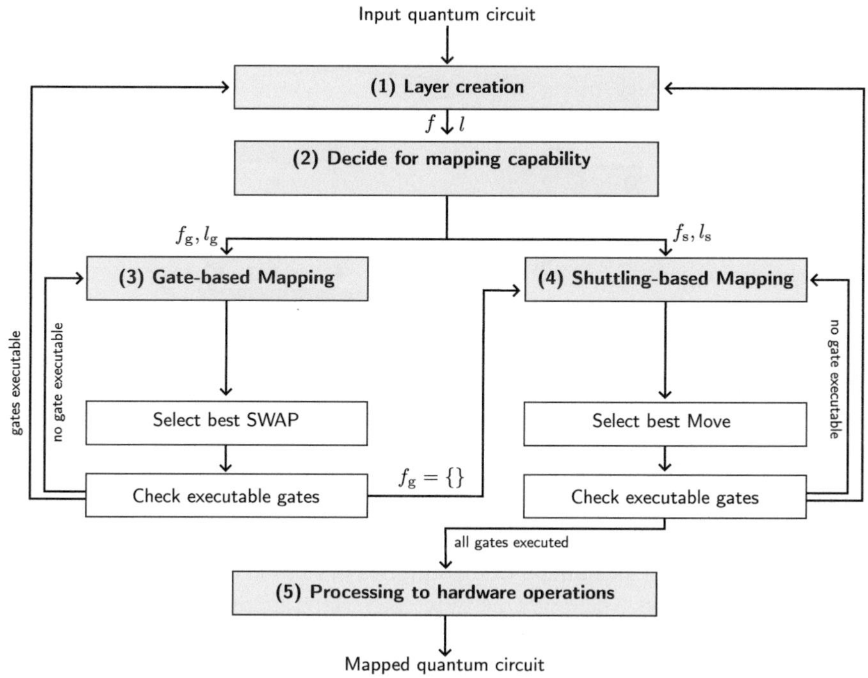

Fig. 12 Resulting hybrid mapping process, adapted from [89]

As the cost functions take gate fidelities and shuttling speeds as input, the compilation will adapt to possibly device-specific characteristics. This way, the framework selects operations that are either more cost-effective or more parallelizable, depending on the scenario, providing a general and flexible compilation for hybrid architectures. Further details and implementation are provided in [89].

4.3.2 Zoned Architectures

To manage the rearrangement complexity in zoned architectures, we can employ *routing-aware placement* [95]. Here, gates are assigned to different layers, and within each layer, the qubits are mapped anew such that the routing overhead is minimized. This optimization is done using an A^*-based search [97].

The input circuit is split into two types of subsequent layers: (1) Layers containing only single-qubit gates (2) layers containing independent two-qubit gates, i.e., two-qubit gates that can be executed in parallel. For every two-qubit gate layer, a gate and an intermediate placement are computed. The *gate placement* facilitates the execution of the corresponding two-qubit gates in the entanglement zone. An *intermediate placement* returns the qubits into the storage zone, where also the single-qubit gates are executed. The overall compilation flow is illustrated in Fig. 13 below.

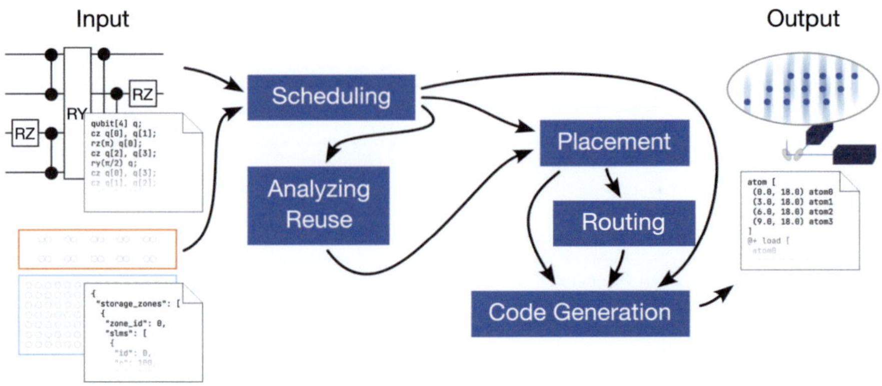

Fig. 13 The compilation flow of the routing-aware compiler introduced in [95]. The compiler takes a quantum circuit as input and produces a sequence of target-specific instructions

The placements are retrieved following a layer-by-layer strategy and are optimized in a way that minimizes the number of rearrangement steps in the subsequent routing stage of the compilation. The fewer rearrangement steps are required, the shorter the circuit duration and the lower the idle errors. As this routing strategy never introduces additional gate overhead, the optimization of the approximate success probability of Eq. 1 reduces to an optimization of t_{idle}. Since the number of rearrangement steps and the atoms' traveling distance are the main contributions to t_{idle}, the cost function employed by the compiler is defined as the sum over the maximum distance traveled by any atom in each rearrangement step.

$$\text{cost} = \sum_{s \in \{\text{rearr. steps}\}} \max_{d \in \{\text{travel dist. in } s\}} d$$

For finding good placements, the compiler employs an A* search algorithm that explores the placement space of the atoms. Central to A* is a suitable heuristic estimating the actual cost of placements that, in this case, combines three components [97]:

(1) An **admissible component**, estimating the lower bound of the cost.
(2) An **accelerating component**, favoring placements preserving gaps for later atoms, accelerating the search toward a good placement of all atoms.
(3) A **look-ahead component**, encouraging placements that reduce cost in future layers.

This method allows routing-aware placement to efficiently navigate an exponential placement space, minimizing the total number of rearrangement steps and, hence, the quantum circuit duration.

4.4 Discussion

Quantum circuit compilation for neutral atoms (NAQC) is significantly more complex than for static platforms like superconducting qubits due to the increased search space. Unlike fixed connectivity in superconducting chips, NAQC introduces a dynamic qubit layout via atom rearrangement, which expands the compilation problem beyond logical `SWAP`-insertion. NAQC compilation demands architecture-aware, constraint-respecting strategies, including physical constraints such as non-crossing, ghost spots, and row/column preservation of AOD rearrangements.

The best compilation strategy depends heavily on the architecture. Hybrid systems must choose between inserting `SWAP`s or shuttling atoms, balancing fidelity and latency. Zoned architectures prioritize minimizing rearrangement steps between storage and entanglement zones, where heuristic placement and scheduling (e.g., A*) can be very useful.

Finally, there have been alternative approaches to NAQC compilation for both the hybrid architecture [2, 38, 59, 75] and the zoned architecture [23, 60, 94, 101, 103]. Also, novel compilation schemes using neutral atom-specific capabilities have been proposed, such as "flying ancillas" [104], where entanglement is mediated by additional ancillary qubits, and the idea of continuous placement of atoms instead of regular grids [62]. Continuous improvements of existing approaches and development of new methods are necessary to keep track with the rapid development of the hardware.

5 Trapped Ions

Trapped ions are among the most promising physical platforms for quantum computing. Sharing several desirable features with neutral atoms, such as their identical nature and the long coherence times, trapped-ion systems currently achieve the highest two-qubit gate fidelities of any quantum computing architecture [14, 22, 34]. Simultaneously, they suffer from scalability issues with one of the possible paths to large-scale computation, namely *Quantum Charge Coupled Devices* (QCCDs), discussed below.

5.1 Background

Trapped-ion qubits are realized using two-level quantum systems within the ions, typically based on the ions' electronic states [14, 96]. The most common encodings include hyperfine qubits [3, 8], optical qubits [79], and Zeeman-encoding based on an energy splitting induced by an external magnetic field [84]. Initialization can be achieved via optical pumping: a laser selectively excites transitions from all but one qubit state, causing the ion population to accumulate in the non-coupled state with fidelities exceeding 99.9% [12]. Measurement is performed by coupling laser light to only one of the qubit states; if the ion is in that state, it fluoresces, while the absence of fluorescence indicates the ion is in the other state [73]. This method allows for measurement fidelities exceeding 99.99% [14].

Universal *single-qubit rotations* are implemented using stimulated Raman transitions (for hyperfine qubits) or electric quadrupole transitions (for optical qubits). These operations are modeled using the Jaynes-Cummings Hamiltonian under the rotating wave approximation and are expressed as rotations on the Bloch sphere, allowing arbitrary control of the qubit state [14].

Two-qubit entangling gates rely on the Mølmer–Sørensen (MS) interaction [69], which exploits shared vibrational modes within a trap. This scheme supports *all-to-all* connectivity in an ion-chain and achieves the highest known entangling-fidelities [6, 71].

Ions are confined using RF (radio-frequency) and DC electric fields within so-called Paul traps, which can be fabricated either as three-dimensional structures or as planar surface-electrode traps [14].

Example 17 Figure 14 illustrates Paul traps in both 3D and 2D (surface) configurations. The ions (blue) are confined by carefully shaped electric fields produced by the RF (pink) and DC (purple) control elements. Gates can be realized by focused lasers shining onto the chain either axially or orthogonally to address single ions.

While single linear traps are sufficient for small-scale devices, increasing the number of ions to more than tens of ions within one chain is not feasible. Among other consequences, overcrowding an ion chain leads to slower gate speeds $R_{\text{gate}} \sim 1/\sqrt{N}$ and increased motional mode crowding. These limitations motivate scalable trap architectures, where multiple individual traps are combined while ion manipulation remains independently within a single trap.

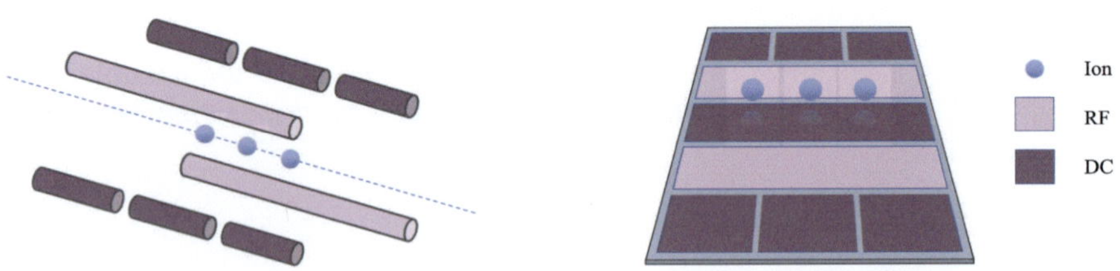

(a) 3D Paul trap (b) Surface trap

Fig. 14 An illustration of two possible realizations of a Paul trap. A combination of RF and DC electric fields produced by control electronics (pink and purple) confines the ions (blue) at their current position. The figure is adapted from [91]

5.1.1 Quantum Charge Coupled Device (QCCD)

To enable scalable computation, the *Quantum Charge Coupled Device* (QCCD) architecture was proposed by Kielpinski et al. [45]. Similarly to the zoned neutral atom architecture, this design segments the trap into specialized functional zones: *processing zones* (PZs) for high-fidelity gate operations, *memory zones* (MZs) for storing idle qubits isolated from noise and control lasers, *measurement zones* for rapid and accurate state readout, and *loading zones* for introducing new ions. Advanced fabrication techniques, including CMOS-compatible processes, have enabled the realization of compact and scalable QCCD devices with integrated control elements.

Since ions can be moved (shuttled) throughout a trap, the ions move between the different zones as needed for the desired quantum computation [36, 72].

Example 18 Figure 15a shows a linear QCCD architecture, where each site (six purple squares) can confine and shuttle ion chains using segmented control electrodes.

To increase flexibility, linear zones can be interconnected via junctions (e.g., T-, Y-, or X-junctions) to form 2D architectures [7]. This allows parallel ion transport and circumvents the bottleneck of swap operations within a single linear trap.

Example 19 Figure 15b displays a possible 2D QCCD layout. Memory and processing zones are arranged in a grid interconnected by X-junctions, enabling scalable ion movement between zones.

While ion transport is a key advantage of trapped-ion systems, it incurs both physical and architectural costs. Linear shuttling is relatively fast, but junction crossings are slower and contribute significantly to overall latency [7, 111]. During motion, ions acquire motional energy and must be cooled, typically using Doppler and sideband cooling techniques [12], to prevent decoherence. This further increases the necessary time to move ions, making time the main optimization focus.

In the next section, we explore how quantum circuits are compiled to such QCCD architectures while optimizing for fast circuit execution.

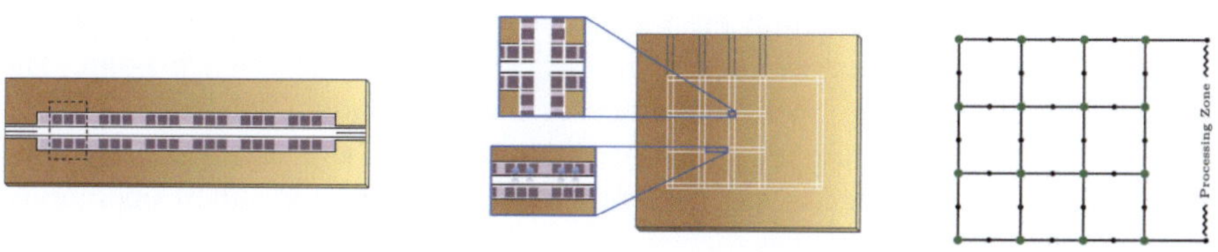

(a) Linear QCCD architecture (b) 2D QCCD architecture (c) Graph abstraction

Fig. 15 An illustration of two possible QCCD architectures and a corresponding graph abstraction, adapted from [90]

5.2 Problem Motivation and Description

Executing quantum algorithms on trapped-ion quantum computers requires precise and efficient control of ion movement within a QCCD architecture. In this context, the considered problem is to determine a time-optimal shuttling schedule for executing a given quantum circuit on a QCCD-based architecture.

As previously discussed, the architecture consists of multiple different zones interconnected by a system of linear traps and junctions [45]. Each ion chain holds multiple qubits and acts as a single transportable entity. In the following, we focus on an architecture consisting of a grid-type memory zone, which is connected to a linear processing zone. Furthermore, we assume that ions can not swap places directly with each other. This means that ions may block each other, which necessitates careful coordination of the ions' movement. To work with the discussed architecture, we abstract its design as a graph.

Definition 5 Consider a graph $G = (V, E)$ (note that this is different from the coupling graph from Sect. 2.1.2) representing the memory zone and the interface to the processing zone. Nodes V consist of major nodes (junctions) and minor nodes (intermediate linear segments). Edges $E = \{e_0, \ldots, e_k\}$ denote all positions available to ion chains, including special inbound and outbound edges for entering and exiting the processing zone. A set $C = \{i_0, \ldots, i_l\}$ contains all ion chains relevant to the circuit.

Example 20 In the QCCD layout shown in Fig. 15b, memory zones are graph-abstracted with major nodes forming a 4×4 grid, connected by edges. Minor nodes split regions between junctions, and a processing zone is connected via designated inbound and outbound edges. The corresponding graph is shown in Fig. 15c.

To model the system dynamically, time is discretized into individual steps. At each time step, the state of all ion chains is described by their positions on the graph. Transitions between time steps capture movement, where no more than one ion chain may pass through a given junction at a time. This constraint ensures physical realism while maintaining abstraction from hardware-specific timing details.

Simultaneously, the quantum circuit dictates temporal constraints by specifying which ion chains must be present in the processing zone at each point. The required sequence of gate operations imposes a corresponding sequence S of ion chains derived from the circuit structure. According to that sequence, ions must move to the outbound edge to enter the processing zone, perform their operations, and return via the inbound edge.

Given these constraints, the problem addressed can be formally described as

- **Given**: A graph layout G representing the memory zone and processing zone interface, and a sequence S derived from a quantum circuit.
- **Goal**: Determine a schedule with the minimum number of time steps $\hat{T}$ to execute all required shuttling operations for the circuit.

Because cooling and movement control infrastructure remain difficult to scale, minimizing the duration and number of shuttling operations is crucial for future large-scale devices. This motivates the search for optimized, hardware-agnostic shuttling strategies that respect architectural and operational constraints while minimizing circuit execution time.

5.3 Proposed Solution

To address the shuttling problem introduced above, two complementary methods have been proposed: an *exact solution* based on Boolean satisfiability (SAT) [91] and a *scalable heuristic* tailored to the topology of QCCD architectures [90]. The exact SAT-based method ensures optimality, while the heuristic enables efficient shuttling on larger-scale devices.

5.3.1 Exact Solution

The problem of finding a time-optimal shuttling schedule for a given quantum circuit can be formulated as a classical combinatorial optimization problem. This is a similar reasoning as for the optimal approach as the compilation for superconducting chips from Sect. 3.3. Although complex, this type of problem can be tackled using SAT solvers, which have become powerful tools in fields such as logic synthesis, verification, and test generation.

To apply SAT to the shuttling problem, we reduce the optimization problem to a sequence of decision problems. Specifically, we ask: "Can the circuit be executed with a shuttling schedule of T time steps?" Starting from $T = 1$, we increment T until a satisfiable solution is found. The first such T is guaranteed to be the minimal solution $\hat{T}$.

The problem is encoded as a SAT instance Φ, with variables representing the position and movement of ion chains at each time step, and constraints ensuring correctness, such as exclusivity of ion positions, valid transitions across junctions, and adherence to the circuit-derived sequence S [91].

Definition 6 Let Φ be a Boolean formula over a set of variables. The *Boolean satisfiability* problem is to determine if there exists an assignment of truth values such that Φ evaluates to true.

Definition 7 The Boolean variables employed to describe the position and movement of ion chains can be expressed as $x^t_{e,i} \in \{0, 1\}$ with $0 \leq t \leq T$, $i \in C$, and $e \in E$. Each such variable represents whether there is an ion chain i present on site e at time step t. The value 0 ("false") means absent, whereas 1 ("true") means present.

With this representation, further constraints can be specified to capture the full dynamics of the system. Using these constraints, an equation can be formulated, which can then be given to a SAT-solver to find a satisfiable solution in a minimal number of time steps.

While SAT is NP-complete, modern solvers are capable of solving large-scale instances efficiently. Initial work has demonstrated the potential of SAT in quantum computing contexts [108]. This work continues this trend by applying SAT solving to the domain of trapped-ion circuit compilation.

While this approach yields shuttling sequences that are optimal in the number of timesteps, the problem becomes infeasible for larger graphs. Also, in dense memory zones, many ion chains may block each other, making optimal shuttling impractical. To enable scalability, another approach is necessary.

5.3.2 Heuristic and Scalable Solution

To scale to larger QCCD architectures, we also propose a heuristic based on cycle-based shuttling. The core idea is to resolve conflicts of chains blocking each other by forming cycles along chains.

A *cycle* is defined as a closed path in the graph representation of the memory zone. If a desired path is blocked, a cycle encompassing the target and blocking chains can be constructed. Moving all chains one edge forward on this cycle avoids collisions and shuttles the target ion in the right direction.

Example 21 In Fig. 16a, chains c_0 and c_2 are blocked by c_5 and c_7. By forming a cycle around the blocked paths and rotating all chains, c_0 and c_2 can proceed without conflict (Fig. 16b).

For grid-based QCCD layouts composed of X-junctions, cycles can be formed using rectangular regions. Vertical moves require single-rectangle cycles, while horizontal moves span two rectangles.

Example 22 Figure 16c illustrates two cases: (top) c_0 is blocked vertically by c_1, resolved with a single-rectangle cycle; (bottom) c_0 is blocked horizontally, resolved with a two-rectangle cycle.

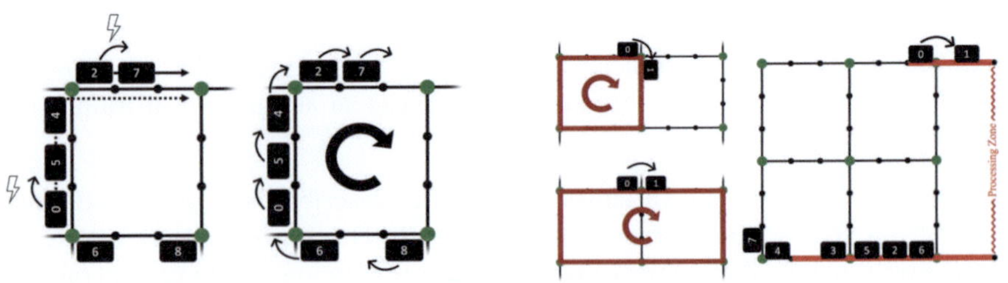

(a) Blocked paths (b) Cycle (c) In memory (d) Through processing

Fig. 16 Illustration of individual cycle construction and cycles across different zones. Composed and adapted from [90]

Cycle-based movement is efficient because all involved junctions can be used in parallel, requiring only a single time step per full rotation. Combined, all memory zone movements can be expressed as sequences of such cycle operations.

Processing zones are accessed via a one-way pass as illustrated in Fig. 16d. Direct cycles are avoided to prevent unintended chain movements. Instead, a chain is routed through the processing zone to a free edge in the memory zone.

Example 23 In Fig. 16d, c_0 needs access to the processing zone, but the inbound edge is occupied by c_1. A search through the processing zone identifies a path to an unoccupied memory edge for exit.

Combining this memory zone and processing zone shuttling strategies, an arbitrary quantum circuit can be compiled to hardware QCCD operations, and efficient shuttling schedules can be created.

5.4 Discussion

Quantum circuit compilation for trapped-ion platforms, particularly with QCCD architectures, differs fundamentally from that of superconducting or neutral atom systems. Unlike static connectivity models, trapped ions require coordinated physical movement of qubits along the QCCD graph. In QCCD devices, the time overhead of shuttling ion chains between memory and processing zones makes temporal scheduling a central challenge.

These approaches demonstrate how SAT solvers can be leveraged to compute exact ion transport schedules, and the abstraction of QCCD devices as graph layouts, along with strategies such as cycle-based routing, enable us to tackle the scheduling constraints involved in compiling for large-scale devices. Together, these exact and heuristic approaches provide a foundation for scalable and constraint-aware compilation tailored to the demands of trapped-ion devices.

6 Alternative Platforms

While superconducting qubits, trapped ions, and neutral atoms currently concentrate a large part of the experimental efforts in quantum computing, a diverse set of alternative hardware platforms is actively being explored. These systems often feature fundamentally different physical principles, control mechanisms, and architectural constraints, each requiring specialized compilation approaches tailored to their native operations, connectivity, and noise characteristics. Many methods discussed above, such as SAT-based encoding and efficient heuristics based on A* search, could be promising candidates to develop optimal and efficient compilation software for these alternative platforms.

Photonic and Measurement-Based Quantum Computing Photonic quantum computing [48, 93] leverages single photons—typically represented in specific optical modes—as qubits. Quantum gates can be implemented using linear optical elements such as beam splitters, phase shifters, and interferometers, which perform unitary transformations on the photon modes. Measurement-based models [83], such as those relying on cluster states, shift the computational effort to the generation of entangled resource states followed by adaptive measurements, posing distinct compilation challenges related to resource optimization and feed-forward logic [112, 113].

Quantum Dots and Spin Shuttling Architectures In architectures based on spin qubits, qubits are encoded into the spin states of electrons in quantum dots [21]. Operations are implemented through controlled exchange interactions with proposals that introduce qubit shuttling along 1D or 2D arrays [52] to dynamically alter the connectivity graph, imposing unique temporal and spatial constraints on circuit compilation [24].

Silicon-Based Quantum Computing Silicon is a promising platform owing to its compatibility with existing semiconductor fabrication technologies. Qubits can be implemented using donor-based approaches, such as phosphorus atoms. Silicon qubits benefit from long coherence times—especially in isotopically purified silicon-28—and high-fidelity gate operations. However, achieving high connectivity and scaling remains difficult. The compilation process must accommodate layout constraints and the typically sparse or dynamic connectivity, as well as limited parallelism in control [20].

Together, each platform introduces its own architectural primitives and noise models, underscoring the importance of hardware-aware quantum compilation strategies beyond those developed for the mainstream platforms.

7 Conclusion

As quantum hardware platforms grow in scale and maturity, they continue to face fundamental resource constraints such as limited qubit connectivity, restricted mobility, and fidelity limitations.

Different hardware platforms impose distinct challenges that require tailored compilation strategies. Superconducting qubits necessitate careful `SWAP` gate insertion due to static coupling maps, with symbolic and heuristic methods balancing optimality and scalability. Neutral atom systems demand dynamic qubit layout management, addressing rearrangement constraints like non-crossing and zone-based operations. Trapped-ion platforms with QCCD architectures introduce temporal scheduling challenges tied to ion shuttling, solvable via both SAT-based and heuristic techniques.

This work demonstrates that hardware-specific strategies are key to efficient quantum compilation. All the presented methods above are available in open-source as part of the *Munich Quantum Toolkit* [109] (MQT), giving easy access to potential users and developers.

While superconducting qubits, trapped ions, and neutral atoms currently dominate experimental efforts, a range of alternative quantum hardware platforms is rapidly emerging. Systems such as photonic and measurement-based quantum computing, quantum dots with spin shuttling, and silicon-based qubits introduce fundamentally different operational principles and constraints. Extending the techniques discussed in this chapter—including SAT-based encodings and efficient heuristic methods like A^* search—offers a promising direction for developing compilation tools tailored to these next-generation platforms.

Acknowledgements The authors acknowledge funding from the European Research Council (ERC) under the European Union's Horizon 2020 research and innovation program grant agreement No. 101001318 and No. 101114305 ("MILLENION-SGA1" EU Project), and the Munich Quantum Valley (MQV K5 + K7), which is supported by the Bavarian state government with funds from the Hightech Agenda Bayern Plus. Furthermore, this work was supported by the BMFTR under grant number 13N17298 (SYNQ) and the Deutsche Forschungsgemeinschaft (DFG, German Research Foundation) under grant numbers 563402549 and 563436708.

References

1. Amy, M., et al. 2013. A meet-in-the-middle algorithm for fast synthesis of depth optimal quantum circuits. *IEEE Transactions on CAD of Integrated Circuits and Systems* 32 (6): 818–830.
2. Baker, J. M., et al. 2021. Exploiting long-distance interactions and tolerating atom loss in neutral atom quantum architectures. In *Int'l Symposium on Computer Architecture*, 818–831.
3. Ballance, C. J., et al. 2016. High-fidelity quantum logic gates using trapped-ion hyperfine qubits. *Physical Review Letters* 117 (6): 060504.
4. Barnes, K., et al. 2022. Assembly and coherent control of a register of nuclear spin qubits. *Nature Communications* 13 (1): 2779.
5. Barredo, D., et al. 2016. An atom-by-atom assembler of defect-free arbitrary two dimensional atomic arrays. *Science* 354 (6315): 1021–1023.
6. Benhelm, J., et al. 2008. Towards fault-tolerant quantum computing with trapped ions. *Nature Physics* 4 (6): 463–466.
7. Blakestadetal, R. B. 2009. High-fidelity transport of trapped-ion qubits through an x-junction trap array. *Physical Review Letters* 102 (15): 153002.
8. Blinovetal, B. B. 2004. Quantum computing with trapped ion hyperfine qubits. *Quantum Information Processing* 3 (1): 45–59.
9. Bluvstein, D., et al. 2022. A quantum processor based on coherent transport of entangled atom arrays. *Nature* 604 (7906): 451–456.
10. Bluvstein, D., et al. 2023. Logical quantum processor based on reconfigurable atom arrays. *Nature* 1–3.
11. Botea, A., et al. 2018. On the complexity of quantum circuit compilation. In *Int'l Symposium on Combinatorial Search.*
12. Brandl, M. F. 2017. A quantum von neumann architecture for large-scale quantum computing. arXiv: 1702.02583.
13. Bravyi, S., et al. 2022. The future of quantum computing with superconducting qubits. *Journal of Applied Physics* 132 (16): 160902.
14. Bruzewicz, C. D., et al. 2019. Trapped-ion quantum computing: progress and challenges. *Applied Physics Reviews* 6 (2): 021314.

15. Brylinski, J.-L., et al. Universal quantum gates, 2001. arXiv: quant-ph/0108062.
16. Burgholzer, L., et al. 2022. Limiting the search space in optimal quantum circuit mapping. In *Asia and South Pacific Design Automation Conference*, 466–471. arXiv: 2112.00045.
17. Chiu, N.-C., et al. 2025. Continuous operation of a coherent 3,000-qubit system. arXiv: 2506.20660.
18. Chong, F. T., et al. 2017. Programming languages and compiler design for realistic quantum hardware. *Nature* 549 (7671): 180–187.
19. Cowtan, A., et al. 2019. On the Qubit Routing Problem. In *14th Conference on the Theory of Quantum Computation, Communication and Cryptography (TQC 2019), volume 135 of Leibniz International Proceedings in Informatics (LIPIcs)*, ed. W. van Dam et al., 5:1–5:32, Dagstuhl, Germany. Schloss Dagstuhl–Leibniz-Zentrum fuer Informatik, eprint: 1902.08091.
20. Crawford, O., et al. 2023. Compilation and scaling strategies for a silicon quantum processor with sparse two-dimensional connectivity. *npj Quantum Information* 9 (1): 1–11.
21. De Smet, M., et al. 2025. High-fidelity single-spin shuttling in silicon. *Nature Nanotechnology* 1–7.
22. Debnath, S., et al. 2016. Demonstration of a small programmable quantum computer with atomic qubits. *Nature* 536 (7614): 63–66.
23. Decker, E. 2024. Arctic: A Field Programmable Quantum Array Scheduling Technique. arXiv: 2405.06183.
24. Escofet, P., et al. 2025. Compilation Techniques for Spin Qubits in a Shuttling Bus Architecture. arXiv: 2502.06263.
25. Evered, S. J., et al. 2023. High-fidelity parallel entangling gates on a neutral-atom quantum computer. *Nature* 622 (7982): 268–272.
26. Evered, S. J., et al. 2023. High-fidelity parallel entangling gates on a neutral-atom quantum computer. *Nature* 622 (7982): 268–272.
27. Ge, Y., et al. 2024. Quantum Circuit Synthesis and Compilation Optimization: Overview and Prospects. arXiv: 2407.00736.
28. Giles, B., et al. 2013. Exact synthesis of multiqubit Clifford+T circuits. *Physical Review A* 87 (3): 032332.
29. Giudici, G., et al. 2024. Fast entangling gates for Rydberg atoms via resonant dipoledipole interaction. arXiv: 2411.05073
30. Graham, T. M., et al. 2019. Rydberg-mediated entanglement in a two-dimensional neutral atom qubit array. *Physical Review Letters* 123 (23): 230501.
31. Grimmetal, R. 2000. Optical dipole traps for neutral atoms. In Advances in Atomic, Molecular, and Optical Physics, vol. 42, ed. B. Bederson et al., 95–170. Academic Press.
32. Gyger, F., et al. 2024. Continuous operation of large-scale atom arrays in optical lattices. *Physical Review Research* 6: 033104, 3.
33. Häner, T., et al. 2018. A software methodology for compiling quantum programs. *Quantum Science and Technology* 3 (2): 020501.
34. Harty, T. P., et al. 2016. High-fidelity trapped-ion quantum logic using near-field microwaves. *Physical Review Letters* 117 (14).
35. Henriet, L., et al. 2020. Quantum computing with neutral atoms. *Quantum* 4: 327.
36. Hensinger, W. K. 2021. Quantum computer based on shuttling trapped ions. *Nature* 592 (7853): 190–191.
37. Huang, H.-L., et al. 2020. Superconducting quantum computing: a review. *Science China Information Sciences* 63 (8): 180501.
38. Huang, Y., et al. 2025. DasAtom: A Divide-and-Shuttle Atom Approach to Quantum Circuit Transformation. arXiv: 2409.03185.
39. Isenhower, L., et al. 2010. Demonstration of a neutral atom controlled-NOT quantum gate. *Physical Review Letters* 104 (1): 010503.
40. Itoko, T., et al. 2020. Optimization of quantum circuit mapping using gate transformation and commutation. *Integration* 70: 43–50, 1.
41. Jandura, S., et al. 2022. Time-optimal two- and three-qubit gates for Rydberg atoms. *Quantum* 6: 712.

42. Javadi-Abhari, A., et al. 2024. Quantum computing with Qiskit. arXiv: 2405.08810.
43. Javadi Abhari, A., et al. 2014. ScaffCC: a framework for compilation and analysis of quantum computing programs. In *Conference on Computing Frontiers*, 1–10.
44. Jenkins, A., et al. 2022. Ytterbium nuclear-spin qubits in an optical tweezer array. *Physical Review X* 12 (2): 021027.
45. Kielpinski, D., et al. 2002. Architecture for a large-scale ion-trap quantum computer. *Nature* 417 (6890): 709–711.
46. Kissinger, A., et al. 2020. PyZX: Large scale automated diagrammatic reasoning. *Electronic Proceedings in Theoretical Computer Science* 318: 229–241. arXiv: 1904.04735.
47. Kitaev, A. Y. 1997. Quantum computations: Algorithms and error correction. *Russian Mathematical Surveys* 52 (6): 1191.
48. Kok, P., et al. 2007. Linear optical quantum computing with photonic qubits. *Reviews of Modern Physics* 79 (1): 135–174.
49. Krantz, P., et al. 2019. A quantum engineer's guide to superconducting qubits. *Applied Physics Reviews* 6: 021318.
50. Kreppel, F., et al. 2023. Quantum circuit compiler for a shuttling-based trapped ion quantum computer. *Quantum* 7: 1176.
51. Kunasaikaran, J., et al. 2024. A Framework for the Design and Realization of Alter native Superconducting Quantum Architectures. arXiv: 2305.07052.
52. Künneetal, M. 2024. The SpinBus architecture for scaling spin qubits with electron shuttling. *Nature Communications* 15 (1): 4977.
53. Lao, L., et al. 2022. Timing and resource-aware mapping of quantum circuits to superconducting processors. *IEEE Transactions on CAD of Integrated Circuits and Systems* 41 (2): 359–371.
54. Levine, H. 2021. Quantum Simulation and Quantum Information Processing with Programmable Rydberg Atom Arrays. https://dash.harvard.edu/handle/1/37368497.
55. Levine, H., et al. 2019. Parallel implementation of high-fidelity multiqubit gates with neutral atoms. *Physical Review Letters* 123 (17): 170503.
56. Levine, H., et al. 2019. Parallel implementation of high-fidelity multiqubit gates with neutral atoms. *Phys. Rev. Lett.* 123 (17): 170503.
57. Li, G., et al. 2019. Tackling the qubit mapping problem for NISQ-era quantum devices. In *Int'l Conference on Architectural Support for Programming Languages and Operating Systems*.
58. Li, G., et al. Tackling the qubit mapping problem for NISQ-era quantum devices. In *Int'l Conference on Architectural Support for Programming Languages and Operating Systems*, 1001–1014.
59. Li, Y., et al. 2023. Timing-aware qubit mapping and gate scheduling adapted to neutral atom quantum computing. *IEEE Transactions on CAD of Integrated Circuits and Systems*, 1–1.
60. Lin, W.-H., et al. 2025. Reuse-aware compilation for zoned quantum architectures based on neutral atoms. In *Int'l Symposium on High-Performance Computer Architecture*. arXiv: 2411.11784.
61. Lin, W.-H., et al. 2023. Scalable optimal layout synthesis for NISQ quantum processors. In *Design Automation Conference* 1–6.
62. Ludmiretal, J. 2024. Parallax: A compiler for neutral atom quantum computers under hardware constraints. In *Int'l Conference for High Performance Computing*, SC '24, 1–17, Atlanta, GA, USA. IEEE Press.
63. Ma, S., et al. 2022. Universal gate operations on nuclear spin qubits in an optical tweezer array of 171Yb atoms. *Physical Review X* 12 (2): 021028.
64. Maronese, M., et al. 2022. Quantum compiling. In *Quantum Computing Environments*, ed. S. S. Iyengar, et al., 39–74. Cham: Springer.
65. Maslov, D., et al. 2008. Quantum circuit simplification and level compaction. *IEEE Transactions on CAD of Integrated Circuits and Systems* 27 (3): 436–444.
66. Maslov, D. 2016. On the advantages of using relative phase Toffolis with an application to multiple control Toffoli optimization. *Physical Review A* 93 (2): 022311, 10.

67. Miller, D. M., et al. 2011. Elementary quantum gate realizations for multiple-control Toffoli gates. In *Int'l Symposium on Multi-valued Logic*.
68. Minev, Z. K., et al. 2021. Qiskit Metal: An Open-Source Framework for Quantum Device Design & Analysis.
69. Mølmer, K., et al. 1999. Multiparticle entanglement of hot trapped ions. *Physical Review Letters* 82 (9): 1835–1838.
70. Morgado, M., et al. 2021. Quantum simulation and computing with Rydberg interacting qubits. *AVS Quantum Science* 3 (2): 023501.
71. Moses, S. A., et al. 2023. A race track trapped-ion quantum processor. arXiv: 2305.03828.
72. Murali, P., et al. 2020. Architecting noisy intermediate-scale trapped ion quantum computers. In *Int'l Symposium on Computer Architecture*, ISCA '20, 529–542, Virtual Event. IEEE Press.
73. Myerson, A. H., et al. 2008. High-fidelity readout of trapped-ion qubits. *Physical Review Letters* 100 (20): 200502.
74. Norcia, M. A., et al. 2024. Iterative assembly of ^{171}Yb atom arrays with cavity enhanced optical lattices. *PRX Quantum* 5: 030316, 3.
75. Nottingham, N., et al. 2023. Decomposing and Routing Quantum Circuits Under Constraints for Neutral Atom Architectures. arXiv: 2307.14996.
76. Park, S., et al. 2022. A fast and scalable qubit-mapping method for noisy intermediate-scale quantum computers. In *Design Automation Conference*, 13–18.
77. Peham, T., et al. 2023. On optimal subarchitectures for quantum circuit mapping. *ACM Transactions on Quantum Computing*. arXiv: 2210.09321.
78. Peres, A. 1985. Reversible logic and quantum computers. *Physical Review A* 32 (6): 3266–3276.
79. Pogorelov, I., et al. 2021. Compact ion-trap quantum computing demonstrator. *PRX Quantum* 2 (2): 020343.
80. Preskill, J. 2018. Quantum computing in the NISQ era and beyond. *Quantum* 2: 79.
81. Pucher, S., et al. 2024. Fine-Structure Qubit Encoded in Metastable Strontium Trapped in an Optical Lattice. arXiv: 2401.11054.
82. Rasmussen, S., et al. 2021. Superconducting circuit companion-an introduction with worked examples. *PRX Quantum* 2 (4): 040204.
83. Raussendorf, R., et al. 2003. Measurement-based quantum computation on cluster states. *Physical Review A* 68 (2): 022312.
84. Ruster, T., et al. 2016. A long-lived Zeeman trapped-ion qubit. *Applied Physics B* 122 (10): 254.
85. Saffman, M. 2016. Quantum computing with atomic qubits and Rydberg interactions: Progress and challenges. *Journal of Physics B: Atomic, Molecular and Optical Physics* 49 (20): 202001.
86. Saffman, M., et al. 2010. Quantum information with Rydberg atoms. *Reviews of Modern Physics* 82 (3): 2313–2363.
87. Saffman, M. 2019. Quantum computing with neutral atoms. *National Science Review* 6 (1): 24–25.
88. Schmid, L., et al. 2024. Computational capabilities and compiler development for neutral atom quantum processors—connecting tool developers and hardware experts. *Quantum Science and Technology* 9 (3): 033001.
89. Schmid, L., et al. 2024. Hybrid circuit mapping: Leveraging the full spectrum of computational capabilities of neutral atom quantum computers. In *Design Automation Conference*. arXiv: 2311.14164.
90. Schoenberger, D., et al. 2024. Shuttling for scalable trapped-ion quantum computers. *IEEE Transactions on CAD of Integrated Circuits and Systems* 1–1.
91. Schoenberger, D., et al. 2024. Using Boolean satisfiability for exact shuttling in trapped-ion quantum computers. In *Asia and South Pacific Design Automation Conference*, 127–133.
92. Siddiqi, I. 2021. Engineering high-coherence superconducting qubits. *Nature Reviews Materials* 6 (10): 875–891.
93. Slussarenko, S., et al. 2019. Photonic quantum information processing: A concise review. *Applied Physics Reviews* 6 (4): 041303.

94. Stade, Y., et al. 2024. An abstract model and efficient routing for logical entangling gates on zoned neutral atom architectures. In *Int'l Conference on Quantum Computing and Engineering*. arXiv: 2405.08068.
95. Stade, Y., et al. 2025. Routing-aware placement for zoned neutral atom-based quantum computing. arXiv: 2505.22715.
96. Steane, A., et al. 2000. Quantum computing with trapped ions, atoms and light. *Fortschritte der Physik* 48 (9–11): 839–858.
97. Stefan Edelkamp, S. J. Cost-Algebraic Heuristic Search.
98. Svore, K., et al. 2006. A layered software architecture for quantum computing design tools. *Computer* 39 (1): 74–83.
99. Tanet, B., al. 2020. Optimal layout synthesis for quantum computing. In *Int'l Conference on CAD*, 1–9.
100. Tan, B., et al. 2021. Optimality study of existing quantum computing layout synthesis tools. *IEEE Transactions on Computers* 70 (9): 1363–1373.
101. Tan, D. B., et al. 2024. Compiling quantum circuits for dynamically field-programmable neutral atoms array processors. *Quantum* 8: 1281.
102. Unnikrishnan, G., et al. 2024. Coherent Control of the Fine-Structure Qubit in a Single Alkaline-Earth Atom. arXiv: 2401.10679.
103. Wang, H., et al. 2024. Atomique: A quantum compiler for reconfigurable neutral atom arrays. In *Int'l Symposium on Computer Architecture*, 293–309. arXiv: 2311.15123 [quant-ph].
104. Wang, H., et al. 2024. Q-Pilot: Field programmable quantum array compilation with flying ancillas. In *Design Automation Conference*. arXiv: 2311.16190.
105. Wendin, G. 2017. Quantum information processing with superconducting circuits: A review. *Reports on Progress in Physics* 80 (10): 106001.
106. Wille, R., et al. 2019. Mapping quantum circuits to IBM QX architectures using the minimal number of SWAP and H operations. In *Design Automation Conference*, 1–6.
107. Wille, R., et al. 2019. Mapping quantum circuits to IBM QX architectures using the minimal number of SWAP and H operations. In *Design Automation Conference*, 2.
108. Wille, R., et al. 2023. MQT QMAP: Efficient quantum circuit mapping. *CoRR*, abs/2301.11935. arXiv: 2301.11935.
109. Wille, R., et al. 2024. The MQT handbook: A summary of design automation tools and software for quantum computing. In *IEEE International Conference on Quantum Software (QSW)*. arXiv: 2405.17543. A live version of this document is available at https://mqt.readthedocs.io.
110. Wintersperger, K., et al. 2023. Neutral atom quantum computing hardware: Performance and end-user perspective. *EPJ Quantum Technology* 10 (1): 32.
111. Wright, K., et al. 2013. Reliable transport through a microfabricated X-junction surface-electrode ion trap. *New Journal of Physics* 15 (3): 033004.
112. Zhang, H., et al. 2023. OneQ: A compilation framework for photonic one-way quantum computation. In *Int'l Symposium on Computer Architecture*, 1–14.
113. Zilk, F., et al. 2022. A compiler for universal photonic quantum computers. In *2022 IEEE/ACM Third International Workshop on Quantum Computing Software (QCS)*, 57–67. arXiv: 2210.09251.
114. Zulehner, A., et al. 2019. An efficient methodology for mapping quantum circuits to the IBM QX architectures. *IEEE Transactions on CAD of Integrated Circuits and Systems*.

Design Automation in Quantum Error Correction

Archisman Ghosh, Avimita Chatterjee, and Swaroop Ghosh

Abstract Quantum error correction (QEC) underpins practical fault-tolerant quantum computing (FTQC) by addressing the fragility of quantum states and mitigating decoherence-induced errors. As quantum devices scale, integrating robust QEC protocols is imperative to suppress logical error rates below threshold and ensure reliable operation, though current frameworks suffer from substantial qubit overheads and hardware inefficiencies. Design automation in the QEC flow is thus critical, enabling automated synthesis, transpilation, layout, and verification of error-corrected circuits to reduce qubit footprints and push fault-tolerance margins. This chapter presents a comprehensive treatment of design automation in QEC, structured into four main sections. The first section delves into the theoretical aspects of QEC, covering logical versus physical qubit representations, stabilizer code construction, and error syndrome extraction mechanisms. In the second section, we outline the QEC design flow, detailing the areas highlighting the need for design automation. The third section surveys recent advancements in design automation techniques, including algorithmic T-gate optimization, modified surface code architecture to incorporate lesser qubit overhead, and machine-learning-based decoder automation. The final section examines near-term FTQC architectures, integrating automated QEC pipelines into scalable hardware platforms and discussing end-to-end verification methodologies. Each section is complemented by case studies of recent research works, illustrating practical implementations and performance trade-offs. Collectively, this chapter aims to equip readers with a holistic understanding of design automation in QEC system design in the fault-tolerant landscape of quantum computing.

A. Ghosh (✉) · A. Chatterjee · S. Ghosh
Pennsylvania State University, University Park, PA, USA
e-mail: apg6127@psu.edu

A. Chatterjee
e-mail: amc8313@psu.edu

S. Ghosh
e-mail: szg212@psu.edu

P. Mishra (ed.), *Design Automation for Quantum Computing*,
https://doi.org/10.1007/978-3-032-09303-5_6

1 Introduction

Quantum error correction (QEC) is a foundational component in the realization of fault-tolerant quantum computing (FTQC) [1]. Unlike classical bits, quantum bits (qubits) are inherently fragile and susceptible to a wide range of noise processes, including bit-flip, phase-flip, decoherence, and crosstalk errors [2]. These challenges are exacerbated in the Noisy Intermediate-Scale Quantum (NISQ) era [3], where quantum hardware is still in its nascent stages of scalability and stability. QEC techniques aim to protect logical quantum information by encoding it into entangled states of multiple physical qubits, thereby enabling detection and correction of errors without direct measurement of the quantum state. While the theoretical foundations of QEC, such as stabilizer codes [4] and surface codes [5], are well developed, practical implementation remains resource-intensive and operationally complex. This has motivated a growing interest in automating QEC workflows, from circuit-level optimization and code selection to decoder design and syndrome-based correction. As quantum systems scale, automation becomes not merely beneficial but essential to ensure the efficiency, scalability, and adaptability of QEC strategies in modern quantum architectures.

1.1 What Is Quantum Error Correction?

QEC essentially protects quantum information against decoherence, gate infidelities, and measurement errors by encoding logical qubits into entangled states of multiple physical qubits [1]. This is essential because quantum states are inherently fragile and susceptible to both bit-flip (X) and phase-flip (Z) errors, which cannot be mitigated by classical redundancy due to the no-cloning theorem [6]. QEC employs syndrome measurement and conditional correction without disturbing the encoded state, enabling reliable quantum operations [1]. As we scale beyond the NISQ era, QEC becomes indispensable for achieving large-scale FTQC.

Why does classical error correction not work?

Classical error correction codes (ECC), such as Hamming [7], Bose–Chaudhuri–Hocquenghem (BCH) [8], and Reed-Solomon codes [9], function by introducing redundancy through the *duplication* of classical bits and performing majority voting or algebraic checks to detect and correct transmission errors. These codes rely on two critical assumptions: (1) data can be freely copied, and (2) data can be measured repeatedly without altering the underlying state. Both assumptions fail in the quantum domain due to fundamental principles of quantum mechanics.

- **No-Cloning Theorem:** A central principle of quantum theory is the no-cloning theorem, which states that it is impossible to create an independent and identical copy of an arbitrary unknown quantum state. Mathematically, there exists no unitary operation U such that $U(|\psi\rangle \otimes |0\rangle) = |\psi\rangle \otimes |\psi\rangle, \forall |\psi\rangle \in \mathcal{H}$; $\mathcal{H}$ being the Hilbert space of qubit states [6, 10]. This precludes the use of direct repetition codes (e.g., $0 \rightarrow 000$) in the quantum setting, thereby invalidating the core mechanism of classical ECC.

- **Measurement-induced collapse:** Classical ECCs perform detection and correction by directly reading bit values at the receiver and applying logical inference. However, in quantum mechanics, measurement is a non-unitary, non-reversible operation that collapses the superposition of a qubit state into one of its computational basis states [11]. For example, if a qubit is in a state $|\psi\rangle = \alpha|0\rangle + \beta|1\rangle$, any standard measurement in the computational basis results in the state irreversibly collapsing to either $|0\rangle$ or $|1\rangle$, with respective probabilities $|\alpha|^2$ and $|\beta|^2$. This destructive property of measurement prevents direct syndrome extraction, as it would annihilate the very quantum information we intend to protect.
- **Complexity of the Quantum Errors:** In classical systems, the error model is typically limited to bit-flip errors ($0 \leftrightarrow 1$), which can be handled by checking parity or Hamming distance [1]. In contrast, quantum systems are subject to a richer error space spanned by the Pauli group $\mathcal{P} = \{I, X, Y, Z\}$. A general error on a qubit can be expressed as a linear combination of Pauli operators: $E = \alpha I + \beta X + \gamma Y + \delta Z$, where X (bit-flip), Z (phase-flip), and $Y = iXZ$ (combined) all act non-trivially on the state. Therefore, a quantum error-correcting code (QECC) must be able to detect and correct not only X and Z-type errors but also their simultaneous occurrence, even when occurring coherently. Furthermore, errors in quantum systems are not necessarily discrete events but may be continuous or coherent due to interactions with the environment (decoherence), imprecise gate implementations (gate errors), or imperfect readouts (measurement errors). Classical ECC assumes a discrete and probabilistic error model, which is not sufficient to model quantum noise accurately.

1.2 Stabilizer Formalism

A stabilizer [1, 4, 12] is an operator that preserves a quantum state under its action. Formally, for a quantum state $|\psi\rangle$, a stabilizer g satisfies $g|\psi\rangle = |\psi\rangle$. Stabilizers are elements of the *Pauli group* $\mathcal{P}_n$, which is defined as the group of n-fold tensor products of the single-qubit Pauli matrices I, X, Y, Z, including multiplicative factors $\{\pm 1, \pm i\}$. That is, each element of $\mathcal{P}_n$ is of the form $i^\alpha P_1 \otimes P_2 \otimes \cdots \otimes P_n$, where each $P_j \in \{I, X, Y, Z\}$ and $\alpha \in \{0, 1, 2, 3\}$. The group has cardinality $2 \cdot 4^n$, and is closed under multiplication. A *stabilizer code* is defined by an abelian subgroup $\mathcal{S} \subset \mathcal{P}_n$ that does not contain $-I$, and is generated by a set of $m = n - k$ independent, commuting operators $G = \{g_1, g_2, \ldots, g_m\}$. The codespace C is the subspace of the Hilbert space $\mathbb{C}^{2^n}$ that is stabilized by all elements of $\mathcal{S}$, i.e.,

$$C = \left\{ |\psi\rangle \in \mathbb{C}^{2^n} \;\middle|\; g_i|\psi\rangle = |\psi\rangle; \forall g_i \in G \right\}$$

Because each generator constrains the state to a $+1$-eigenspace, and the generators are independent and commuting, the dimension of C is 2^k, corresponding to k logical qubits encoded in n physical qubits. To qualify as valid stabilizer generators, the set $\{g_1, \ldots, g_m\}$ must satisfy three structural rules:

1. **Each generator is a Pauli string**: That is, each g_i must be an element of the Pauli group $\mathcal{P}_n$.
2. **Commutativity**: All generators must mutually commute, i.e., $g_i \otimes g_j = g_j \otimes g_i$ for all i, j.
3. **Independence**: No generator may be written as a product of others, i.e., there must be no nontrivial subset $\{g_{i_1}, \ldots, g_{i_r}\}$ such that their product equals another generator in the set.

Logical operations on encoded qubits are performed using *logical Pauli operators*, denoted $\bar{X}_j, \bar{Z}_j$, which commute with all elements of $\mathcal{S}$ but are not themselves in $\mathcal{S}$. These operators act non-trivially on the codespace and are used to implement quantum gates in a fault-tolerant manner. The overall quantum error correcting code is denoted as $[[n, k, d]]$, where d is the minimum weight of any Pauli operator that commutes with all elements of $\mathcal{S}$ but is not in $\mathcal{S}$, i.e., the minimum weight of a logical error. The stabilizer formalism thus forms both a mathematical and operational framework for defining quantum error-correcting codes.

1.3 Constructing a Stabilizer Generator

The construction of stabilizer generators is guided by the principle that an error must anticommute with at least one stabilizer to be detected. That is, for a stabilizer operator S and an error operator $E \in \mathcal{P}_n$, the stabilizer must satisfy $\{S, E\} = SE + ES = 0$. In such a case, the presence of the error flips the eigenvalue of the stabilizer measurement from $+1$ to -1, thereby providing a detectable syndrome. Consequently, to detect bit-flip (Pauli-X) errors, the stabilizer must involve Pauli-Z operators; to detect phase-flip (Pauli-Z) errors, the stabilizer must involve Pauli-X operators. This design criterion underpins the structural logic of quantum parity-check circuits.

In practical implementations, stabilizers are measured using ancilla-based circuits. Consider a system of three data qubits $|\psi_0\rangle, |\psi_1\rangle, |\psi_2\rangle$ (Fig. 1). A typical Z-type stabilizer generator acts as the parity check $Z_0 \otimes Z_1 \otimes Z_2$, which anticommutes with X-type errors on any of the three qubits. The syndrome is extracted using an ancilla qubit initialized to $|0\rangle$, which interacts with each data qubit via a CNOT gate (with data as control, ancilla as target). After the sequence of CNOTs, the ancilla is measured in the computational basis. If no error is present, the measurement yields $+1$; if a bit-flip error has occurred on one of the data qubits, the ancilla's outcome flips to -1, thus enabling error detection without directly measuring the data qubits. Detection of Z-type errors requires an analogous structure using X-type stabilizers. Since hardware platforms often do not support direct controlled-X interactions, these stabilizers are constructed by conjugating the circuit with Hadamard gates. Specifically,

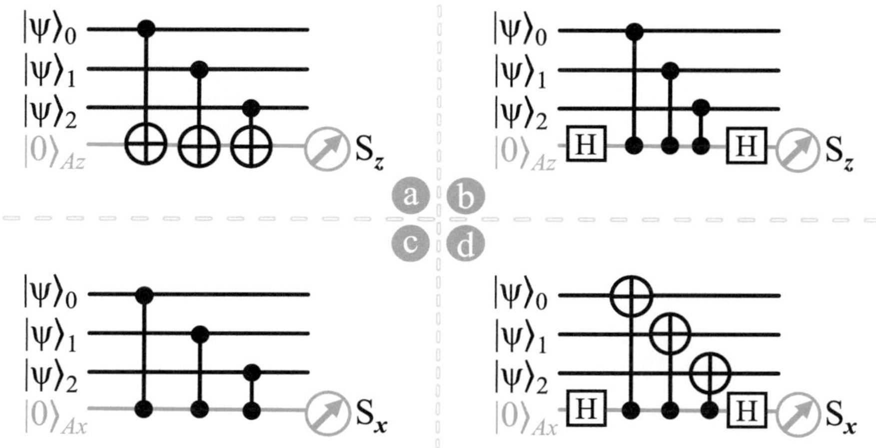

Fig. 1 Different approaches to generate a Z-*stabilizer* and X-*stabilizer* operating on 3 qubits, namely $|\psi\rangle_0$, $|\psi\rangle_1$ and $|\psi\rangle_2$ [1]. The ancilla qubit, on which the stabilizer output is being projected, is represented as $|0\rangle_{Az}$ and $|0\rangle_{Ax}$ for Z-*stabilizers* and X-*stabilizers*, respectively. The circuits are presented as follows: **a** This circuit exhibits a Z-*stabilizer*. The ancilla qubit, $|0\rangle_{Az}$ measures $Z\,|\psi\rangle_0 \otimes Z\,|\psi\rangle_1 \otimes Z\,|\psi\rangle_2$, commonly denoted as $Z_0 \otimes Z_1 \otimes Z_2$. **b** This circuit demonstrates the same Z-*stabilizer* as in (**a**). The stabilizer is regenerated using H gates and CZ gates instead of $CNOT$ gates. The ancilla qubit, $|0\rangle_{Az}$ also measures $Z_0 \otimes Z_1 \otimes Z_2$. **c** This circuit represents an X-*stabilizer*. The ancilla qubit, $|0\rangle_{Ax}$ measures $X\,|\psi\rangle_0 \otimes X\,|\psi\rangle_1 \otimes X\,|\psi\rangle_2$, commonly denoted as $X_0 \otimes X_1 \otimes X_2$. **d** This circuit portrays the same X-*stabilizer* as mentioned in (**c**). The stabilizer is recreated using H gates and $CNOT$ gates instead of CZ gates. The ancilla qubit $|0\rangle_{Ax}$ also measures $X_0 \otimes X_1 \otimes X_2$

to measure $X_0 \otimes X_1 \otimes X_2$, Hadamard gates are first applied to each data qubit and the ancilla (converting the X-basis to the Z-basis), followed by the same CNOT sequence as for a Z-stabilizer, and finally another layer of Hadamard gates is applied to restore the original basis. This method transforms a parity check in the X-basis into a standard Z-basis measurement on the ancilla.

A general and universal circuit for measuring stabilizer generators of the form $P_0 \otimes P_1 \otimes \cdots \otimes P_{n-1}$, where $P \in \{X, Z\}$, uses a sequence of controlled-P gates mediated by a single ancilla qubit (Fig. 2). The circuit begins by preparing the ancilla in the state $|0\rangle$, followed by a Hadamard gate to place it in the superposition $(|0\rangle + |1\rangle)/\sqrt{2}$. Then, a sequence of controlled-P gates is applied such that each control is the ancilla and each target is a data qubit. Finally, a second Hadamard gate is applied to the ancilla before measurement. The resulting quantum state evolves as follows:

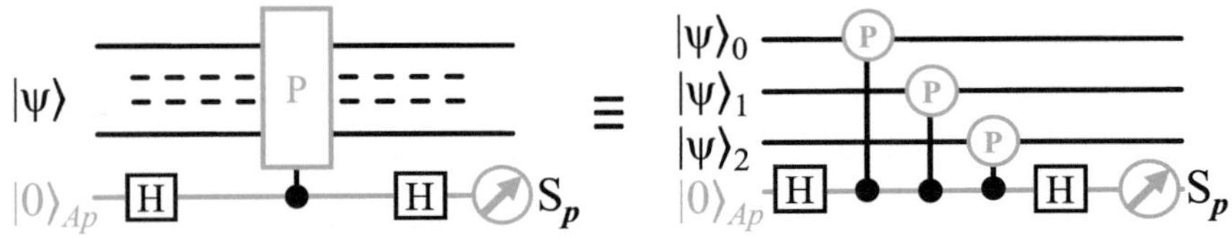

Fig. 2 Illustration of a universal circuit of a stabilizer generator, in which P represents a Pauli-gate operation. This circuit measures $P\,|\psi\rangle_0 \otimes P\,|\psi\rangle_1 \otimes P\,|\psi\rangle_2 \equiv P_0 \otimes P_1 \otimes P_2$. Usually for a stabilizer generator, $P \in \{X, Z\}$ [1]

$$\begin{aligned}
\text{After first Hadamard:}\quad & \frac{1}{\sqrt{2}}\left(|0\rangle + |1\rangle\right) \otimes |\psi\rangle, \\
\text{After controlled-}P:\quad & \frac{1}{\sqrt{2}}\left(|0\rangle \otimes |\psi\rangle + |1\rangle \otimes P|\psi\rangle\right), \\
\text{After second Hadamard:}\quad & \frac{1}{2}\left(|0\rangle \otimes (I + P)|\psi\rangle + |1\rangle \otimes (I - P)|\psi\rangle\right).
\end{aligned}$$

Measuring the ancilla in the computational basis projects the system onto one of two branches: the outcome $|0\rangle$ corresponds to the eigenvalue $+1$, with probability proportional to $\|(I + P)|\psi\rangle\|^2$; the outcome $|1\rangle$ corresponds to the eigenvalue -1, indicating that an error has occurred. Notably, if an error commutes with the stabilizer (i.e., $[P, E] = 0$), the measurement outcome remains unaffected; if it anticommutes, the measurement outcome flips, yielding an eigenvalue of -1. This technique ensures that even if the physical error occurs at an arbitrary angle (i.e., not precisely a Pauli operator), the stabilizer circuit projects the state onto a well-defined eigenstate of the stabilizer operator, suppressing coherence across the error subspace, rendering the error syndrome detectable and classically interpretable.

1.4 Circuit Architecture of a QECC

QECCs are not merely algebraic constructs—they manifest physically as quantum circuits comprising well-defined components that collectively facilitate the encoding, monitoring, and preservation of logical quantum information [13]. In this section, we describe the circuit-level structure of a QECC, focusing on the practical instantiation of the encoding and stabilizer modules, while deferring the decoding and correction logic to subsequent sections. A typical quantum error correction circuit can be decomposed into three interdependent stages:

- **Encoding:** The initial logical state, often denoted $|\psi\rangle_k \in \mathcal{H}_{2^k}$, is mapped to a higher-dimensional subspace $|\psi\rangle_L \in \mathcal{H}_{2^n}$, known as the codespace, via a unitary encoding circuit. This stage establishes logical redundancy by entangling the logical qubits with additional physical qubits.

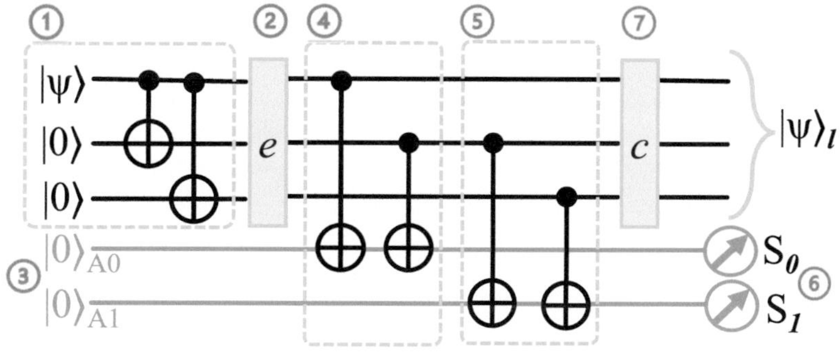

Fig. 3 Demonstrating the quantum circuit that implements a 3-*qubit* repetition code. The circuit comprises the following components: (1) The state preparation circuit. (2) A single bit-flip error that may occur on any of the 3 qubits. (3) Two ancilla qubits that are initialized to the state, $|0\rangle$ are employed for parity checking purposes. (4) The first stabilizer circuit, $Z \otimes Z \otimes I$, is responsible for measuring the parity between the first two qubits. (5) The second stabilizer circuit, $I \otimes Z \otimes Z$, measures the parity between the last two qubits. (6) The ancilla qubits are utilized to obtain the syndrome bits, S_0, S_1, which enable the detection and deduction of errors. (7) To rectify the errors, a correction operator, c, comprising a sequence of self-inverse Pauli-gates, is applied to the qubit that requires correction [1]

- **Stabilizer Interaction Layer:** A set of ancilla-assisted subcircuits are employed to perform parity checks—each associated with a stabilizer generator. These checks involve entangling gates (typically CNOTs or controlled-Zs) between the data qubits and auxiliary ancilla qubits.
- **Syndrome Projection:** Each ancilla qubit is measured in the computational basis to obtain a binary outcome corresponding to the eigenvalue of its associated stabilizer ($+1$ or -1). Crucially, this projection extracts syndrome information without collapsing the logical state, thereby enabling non-demolition error detection.

Case Study 1.4.1: The 3-Qubit Repetition Code [1]:
The fundamental ideas above are best illustrated through the 3-qubit repetition code, which protects against single-qubit bit-flip (Pauli-X) errors as seen in Fig. 3. The logical qubit $|\psi\rangle = \alpha|0\rangle + \beta|1\rangle$ is encoded into a three-qubit entangled state $|\psi\rangle_L = \alpha|000\rangle + \beta|111\rangle$ via a cascade of CNOT gates, where the original qubit acts as the control and two ancillary physical qubits serve as the targets. After encoding, the stabilizer layer implements two parity-check operations:

- $Z \otimes Z \otimes I$, which checks the parity between the first and second qubit.
- $I \otimes Z \otimes Z$, which checks the parity between the second and third qubit.

These stabilizers are implemented using two ancilla qubits, initialized in the state $|0\rangle$, which interact with the data qubits via controlled-NOT operations. The resulting ancilla states are measured to yield syndrome bits S_0

Table 1 Detection, deduction, and correction of errors with respect to the syndrome measurements [1]

Detection		Deduction		Correction
S_0	S_1	Error location	Erroneous state	Correction operator
+1	+1	No error	$\|000\rangle$	III
−1	+1	Qubit 1	$\|100\rangle$	XII
−1	−1	Qubit 2	$\|010\rangle$	IXI
+1	−1	Qubit 3	$\|001\rangle$	IIX

and S_1. The entire circuit layout is shown in Fig. 3, where each functional component—encoding, error insertion, stabilizer subcircuits, ancilla readout, and correction—is modularized. This design enables the detection of single bit-flip errors on any of the three physical qubits. A lookup table (Table 1) associates each syndrome pattern (S_0, S_1) with the location of the error, allowing an appropriate Pauli-X correction (to be addressed in the decoding stage). Importantly, because stabilizer measurements are performed via ancilla projections, the encoded state remains coherent and unmeasured.

While the 3-qubit repetition code serves as a pedagogical example, the architectural template generalizes naturally to more sophisticated QECCs. In a generic $[[n, k, d]]$ stabilizer code, a logical register of k qubits is encoded into a block of n physical qubits. A set of $m = n - k$ stabilizer generators $\{g_1, g_2, \ldots, g_m\}$ is employed to define the codespace and detect errors. The corresponding circuit-level realization, illustrated in Fig. 4, consists of the following components: A **Unitary encoder** that maps k logical qubits into n entangled physical qubits forming the initial codeword $|\psi\rangle_L$; **Ancilla qubits** $\{|0\rangle_{A_1}, \ldots, |0\rangle_{A_m}\}$, each dedicated to measuring one stabilizer generator g_i; **Stabilizer measurement circuits**, wherein each g_i is decomposed into a sequence of Pauli operations acting conditionally on the ancilla qubit. The interaction between data and ancilla qubits is governed by entangling gates, and measurement of the ancilla yields the syndrome bit S_i; and the **Final logical state**, now possibly afflicted by correctable errors, is routed to downstream modules for decoding and correction. This architectural abstraction underlies modern quantum codes, including surface codes, Bacon-Shor codes, and color codes, all of which instantiate stabilizer checks through structured ancilla-mediated circuits.

1.5 QEC Codes

Beyond the general stabilizer formalism, quantum error correction codes can be broadly categorized based on their structural principles, encoding strategies, and

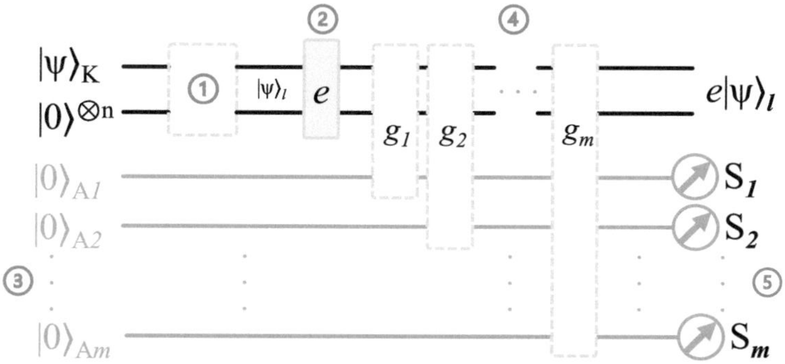

Fig. 4 Illustration of a generalized stabilizer circuit that encodes k physical qubits with n logical qubits and employs m stabilizers to detect potential errors in the logical codeword, $|\psi\rangle_l$, where $m = n - k$. The circuit comprises the following components: (1) The encoder circuit that encodes k physical qubits with n logical qubits; (2) A bit-flip or phase-flip or both errors that may occur on one or more qubits within the logical codeword, $|\psi\rangle_l$; (3) m ancilla qubits that are initialized to $|0\rangle$ to facilitate the projection of the measurements of the stabilizer generators; (4) m stabilizer generators, designated as $g_1, g_2, \ldots, g_m$, are employed; and (5) Each stabilizer g_i projects its measurements onto the ancilla qubit $|0\rangle_{Ai}$. The ancilla qubits are subsequently measured as syndrome measurements, S_i. Based on these syndrome values, error deduction and correction mechanisms are executed. It is essential to highlight that the provided diagram primarily illustrates a generalized stabilizer circuit designed for detecting errors. This representation does not encompass the execution of correction operators; consequently, the final basis state persists as the erroneous state, $e\,|\psi\rangle_l$ [1]

implementation constraints. In what follows, we outline four key families of QEC codes that have played pivotal roles in the development of fault-tolerant quantum architectures: concatenated codes, topological codes, subsystem codes, and bosonic codes.

- **Concatenated Codes:** These codes recursively compose quantum codes to achieve exponential suppression of logical error rates [14–16]. Given a base code $C_1 = [[n, k, d]]$ encoding k logical qubits into n physical qubits with code distance d, one can encode each of the n physical qubits of C_1 using another code C_2, and so forth. The resultant code has an effective distance that grows multiplicatively across levels: for L levels of concatenation, the overall distance becomes d^L, assuming identical base codes. Fault-tolerant gadgets for encoded gates, syndrome extraction, and error correction can be systematically constructed within this recursive structure. Classical control complexity and decoding remain tractable due to the tree-like modularity. A notable example is the concatenation of the [[7, 1, 3]] Steane code [17], which allows for transversal implementation of the Clifford group at each level. However, the total number of physical qubits scales as n^L, leading to prohibitive overhead for large L, especially in architectures constrained by spatial locality.
- **Topological Codes:** These encode logical information in the global degrees of freedom of a spatially extended lattice of qubits [18]. These codes are defined via

local stabilizer generators that act on a small, fixed number of neighboring physical qubits. Logical operators are non-local strings or loops that span the code lattice and commute with all stabilizers but are not part of the stabilizer group itself. Let $\mathcal{S} \subset \mathcal{P}_n$ be the stabilizer group generated by local check operators. A topological code defines logical Pauli operators $\bar{X}_i, \bar{Z}_i \in \mathcal{P}_n$ such that

$$\forall g \in \mathcal{S}, \quad [g, \bar{X}_i] = [g, \bar{Z}_i] = 0, \quad \text{and} \quad \bar{X}_i \bar{Z}_i = -\bar{Z}_i \bar{X}_i$$

The most studied example is the *surface code* [6], defined on a 2D square lattice with two types of stabilizers: *star operators* (X-type) and *plaquette operators* (Z-type), each acting on four qubits. Logical qubits are encoded via nontrivial homology cycles across the lattice (in the toric code [19]) or via boundaries and defects (in the planar code [20]). The code distance d scales with the linear dimension of the lattice, and local syndrome measurements suffice for error detection. Topological codes exhibit high threshold error rates (e.g., $\sim$1% for surface codes under depolarizing noise), and only require nearest-neighbor interactions, making them highly favorable for 2D hardware implementations [21]. However, realizing a universal gate set necessitates either lattice surgery, magic state distillation, or code switching, as not all logical operations can be implemented transversally.

- **Subsystem Codes:** In *subsystem codes* [22], the codespace is partitioned as $\mathcal{H}_{\text{code}} = \mathcal{H}_{\text{logical}} \otimes \mathcal{H}_{\text{gauge}}$, where $\mathcal{H}_{\text{gauge}}$ is a gauge subsystem that carries no logical information [23]. The stabilizer group $\mathcal{S}$ is a subgroup of the (not necessarily abelian) gauge group $\mathcal{G}$. Error correction is performed by measuring only a subset of $\mathcal{G}$ (typically low-weight operators), using these to infer syndromes for the stabilizer subspace. An important example is the *Bacon–Shor code* [24–26], which protects against both bit-flip and phase-flip errors using two-dimensional repetition codes along orthogonal axes. The gauge operators are two-qubit Pauli operators along rows and columns, while logical operators span full rows or columns. Subsystem surface codes similarly introduce gauge degrees of freedom into surface codes to simplify syndrome extraction, particularly enabling the use of weight-2 checks instead of weight-4 or weight-6 stabilizers. The main advantage of subsystem codes lies in reduced measurement complexity and increased fault-tolerance in certain hardware settings. However, gauge freedom complicates decoding, and tradeoffs exist between gauge simplicity and code distance.
- **Bosonic Codes:** In these codes [27], the quantum information is encoded into continuous-variable modes rather than qubit-based systems. They exploit the infinite-dimensional Fock space of harmonic oscillators (e.g., photonic or phononic modes) and are particularly suited to superconducting cavity architectures or optical systems. The *Gottesman–Kitaev–Preskill (GKP)* code [28, 29] encodes a qubit into periodic grid-like superpositions of position eigenstates, enabling protection against small displacement errors in phase space. Logical operators are implemented as translations in quadrature space. The code stabilizers are displacement operators that define a lattice in the phase plane, and error correction involves measuring modular position and momentum. Other bosonic codes include *cat codes* [30, 31], which use superpositions of coherent states (e.g., $|\alpha\rangle \pm |-\alpha\rangle$), and *bino-*

mial codes [32], which encode logical qubits using weighted superpositions of Fock states [33] chosen to cancel specific error channels (e.g., photon loss). These codes allow for hardware-efficient error correction, particularly in platforms where bosonic modes exhibit longer coherence times than transmon qubits. Decoding bosonic codes involves a combination of continuous-variable measurement techniques and phase estimation, often integrated with feedback control. While they show promising error-correcting performance against dominant noise channels (e.g., amplitude damping), their performance under general noise remains an area of active research.

Among the various classes of quantum error correction codes, *surface codes* have emerged as the leading candidate for scalable, fault-tolerant quantum computation [21]. Their construction on a 2D lattice of physical qubits with strictly local stabilizer measurements makes them well-suited for implementation in current quantum hardware architectures. Surface codes exhibit high error thresholds under realistic noise models and support flexible encoding schemes through boundaries and defects. Furthermore, their compatibility with efficient decoding algorithms and modular gate constructions, such as lattice surgery [34], renders them a practical foundation for large-scale FTQC. The following subsection explores the structure and operation of surface codes in detail.

1.6 The Surface Code

Surface codes (Fig. 5) constitute a class of topological stabilizer codes defined on a two-dimensional square lattice [1, 6]. Each physical qubit is associated with an edge of the lattice, and the code's stabilizer group is generated by two sets of local opera-

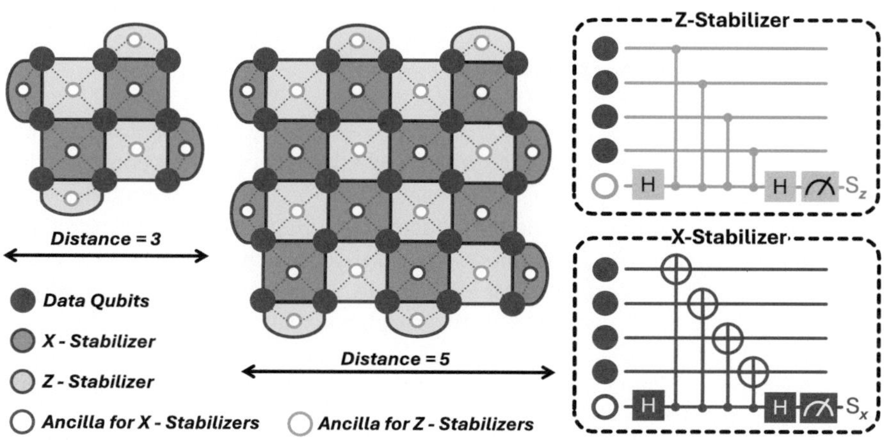

Fig. 5 The diagrammatic representation of the structure of a rotated surface code

tors: *star operators* and *plaquette operators*. For a given vertex s, the corresponding star operator is defined as

$$A_s = \prod_{i \in \text{star}(s)} X_i,$$

where the product ranges over the four edges incident to s, and X_i denotes the Pauli-X operator acting on qubit i. Similarly, for each face or plaquette p, the plaquette operator is defined as

$$B_p = \prod_{i \in \text{plaquette}(p)} Z_i,$$

with the product taken over the four edges surrounding the plaquette and Z_i the Pauli-Z operator. The set of all such A_s and B_p defines an abelian subgroup $\mathcal{S} \subset \mathcal{P}_n$, where $\mathcal{P}_n$ is the n-qubit Pauli group. These stabilizers mutually commute and can be measured locally, as each acts on at most four qubits. By design, the stabilizer structure imposes a gauge symmetry on the lattice: the valid codespace is the simultaneous $+1$-eigenspace of all stabilizer generators. Errors anticommute with a subset of the generators and thus produce detectable syndrome changes without collapsing the encoded logical information. To enable local measurement, each stabilizer is assigned a dedicated ancilla qubit located at the center of the corresponding vertex or plaquette. These ancilla qubits interact with the surrounding data qubits through a fixed entangling circuit during syndrome extraction. The overall layout ensures that all operations required for stabilizer measurement involve only nearest-neighbor interactions, making the architecture highly compatible with planar hardware designs. At the boundaries of the lattice, stabilizer generators become truncated due to the reduced number of neighboring edges. This leads to the definition of two boundary types: *smooth boundaries*, which terminate X-type stabilizers, and *rough boundaries*, which terminate Z-type stabilizers. These boundary conditions not only affect the stabilizer structure but also play a crucial role in the definition of logical operators, as they determine the permissible endpoints of nontrivial Pauli operator strings.

Logical Qubits and Code Distance

Logical qubits in a surface code are encoded by exploiting the global topological features of the lattice. In the planar surface code, logical operators correspond to nontrivial paths of Pauli operators that connect specific types of boundaries. Two types of boundaries are defined on the lattice: *rough boundaries*, which terminate Z-type plaquette stabilizers, and *smooth boundaries*, which terminate X-type star stabilizers. A logical $\overline{Z}$ operator is implemented by a chain of Pauli-Z operators acting on qubits along a path that connects two rough boundaries. Similarly, a logical $\overline{X}$ operator is a chain of Pauli-X operators that connect two smooth boundaries. These logical operators commute with all stabilizers but are not themselves part of the stabilizer group, and they satisfy the requisite anticommutation relation $\overline{X}\,\overline{Z} = -\overline{Z}\,\overline{X}$. The *code distance* d is defined as the minimum weight (i.e., the number of qubits acted upon) of any Pauli operator that implements a nontrivial logical operation. For

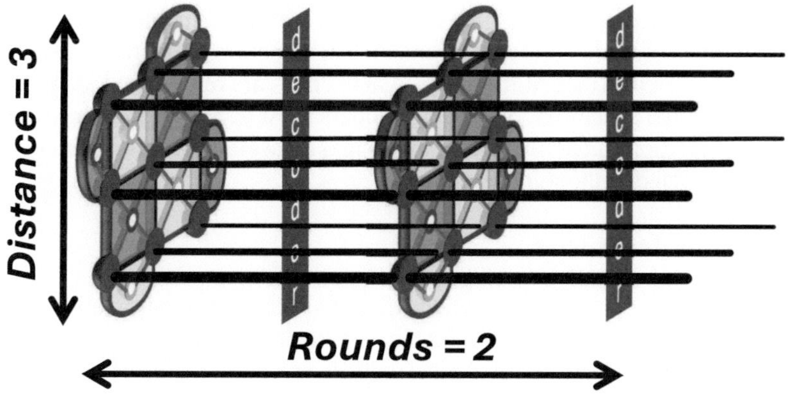

Fig. 6 A diagrammatic representation of running a surface code of distance 3 for 2 *rounds*

a square lattice of linear size d, the shortest logical operator spans d physical qubits, and the code can thus correct up to $\lfloor (d-1)/2 \rfloor$ arbitrary single-qubit errors. Surface codes perform stabilizer measurements in repeated *rounds* to extract syndromes over time. Multiple rounds—typically d—are needed to distinguish data errors from measurement noise. These rounds define the temporal dimension of decoding graphs, enabling identification of error chains via algorithms such as minimum-weight perfect matching on space-time syndrome data (Fig. 6). The number of logical qubits encodable depends on the geometry of the patch: for a planar code with an appropriate boundary configuration, a single logical qubit is typically encoded per patch.

Syndrome Extraction

Syndrome extraction in surface codes is performed by measuring the stabilizer generators A_s and B_p repeatedly over time using dedicated ancilla qubits. Each ancilla qubit is associated with a specific stabilizer and interacts with the data qubits on the corresponding star or plaquette via a fixed sequence of entangling gates. To extract the eigenvalue of a star operator A_s, the corresponding ancilla qubit is initialized in the $|+\rangle$ state, entangled with the neighboring data qubits through a series of $CNOT$ gates (with the ancilla as control), and then measured in the X-basis. A similar procedure applies for plaquette operators B_p, using ancilla initialization in the $|0\rangle$ state, $CNOT$ gates with the ancilla as target, and measurement in the Z-basis. A nontrivial measurement outcome (i.e., -1) indicates that an error has occurred on one or more of the neighboring data qubits that anticommute with the corresponding stabilizer. Since errors affect the measured eigenvalue rather than the data state itself, the extraction process is nondestructive. Repeating syndrome measurements over multiple cycles enables temporal redundancy, which is essential for identifying and correcting time-correlated or measurement-induced errors. To achieve full fault tolerance, syndrome extraction must be repeated over multiple time steps. For a surface code of distance d, it is necessary to perform d rounds of syndrome measurements in order to reliably detect and correct time-correlated errors, including faults arising during ancilla preparation and measurement. This repetition effectively extends the decoding problem into three dimensions, where errors are represented as strings in a $2 + 1$ dimensional space-time volume. Decoders must account for both spatial and temporal correlations, ensuring that physical and measurement errors can be simultaneously corrected without logical failure.

Decoding in Surface Codes

The decoding process [35, 36] operates on the space-time syndrome history generated by repeated rounds of stabilizer measurements, as described in the previous section, involving the identification of the most likely configuration of physical errors that could have produced the observed stabilizer syndrome. Since each stabilizer generator either commutes or anticommutes with a single-qubit Pauli error, a local error manifests as a violation (i.e., a -1 measurement outcome) of one or more stabilizer checks. These violations, also called *syndrome defects*, are detected at the endpoints of error chains. In this framework, errors are naturally represented as *strings* of Pauli operators acting on connected paths of qubits along the lattice [37]. A single Pauli error introduces a pair of syndrome defects at its adjacent stabilizers. More generally, a chain of single-qubit errors will produce syndrome defects at the endpoints of the string, while the internal stabilizers along the path remain satisfied due to cancellation of anticommuting effects. Logical errors correspond to nontrivial homologically inequivalent strings that span the code and connect distinct boundary types, potentially changing the encoded logical state if undetected. The task of the decoder is to infer the most probable set of error strings that match the observed syndrome data, ideally correcting all errors without inducing a logical fault. Under the assumption of independent and identically distributed noise, the decoding problem reduces to a minimum-weight matching problem on a graph where nodes represent syndrome defects and edge weights reflect error probabilities [36].

Logical Operations and Code Variants

Surface codes support fault-tolerant implementation of logical operations through topological manipulations of the encoded qubits. While transversal gates are severely limited by the Eastin–Knill theorem [38], the surface code architecture admits alternative schemes such as *lattice surgery* [34] and *code deformation* [39]. In lattice surgery, logical operations like $CNOT$ are realized by dynamically merging and splitting code patches via stabilizer reconfiguration along shared boundaries. Clifford group operations can thus be executed in a fault-tolerant manner using purely local interactions and syndrome measurements [40]. *Rotated surface codes* [41, 42] modify the lattice geometry to achieve the same code distance with fewer physical qubits. The $XZZX$ *code* [43–45] alternates the type of Pauli operators in stabilizers across the lattice, making it particularly effective under biased noise conditions. More advanced schemes based on defects [46], holes [47], or twist operators [48] allow the encoding of multiple logical qubits within a single patch and support braiding-based gate constructions. These variants retain the essential features of locality and high threshold while offering different trade-offs in encoding density and decoding complexity.

1.7 Lattice Surgery

Surface codes operate under the constraint of a two-dimensional nearest-neighbor (2DNN) interaction topology, wherein physical qubits are arranged in a planar grid and interact exclusively with their adjacent neighbors [48]. This constraint aligns naturally with the physical architecture of many leading quantum hardware platforms, particularly superconducting qubits, and enables fault-tolerant quantum computation via local stabilizer measurements. In the surface code framework, each logical qubit is encoded across a two-dimensional lattice of physical qubits. Executing a quantum algorithm thus entails assigning a distinct surface to every logical qubit, each supporting stabilizer measurement cycles for error detection and correction. As the logical width of a quantum algorithm increases, so too does the spatial demand for separate, concurrently active logical surfaces. While certain logical gates—such as X, Z, or H—can be implemented transversally within a single logical surface, two-qubit gates such as CNOT generally necessitate interactions between distinct logical patches. In a strict 2DNN architecture, the transversal application of two-qubit gates across distant surfaces violates locality constraints and imposes significant overheads in connectivity and error suppression. To address this, *lattice surgery* provides an alternative paradigm that facilitates logical operations via local boundary manipulations rather than global transversal gates. The key idea is to dynamically modify the topology of logical patches through operations known as *merging* and *splitting*. These operations, when executed along appropriate stabilizer boundaries (e.g., X-type or Z-type), enable the implementation of entangling gates and joint measurements entirely within the 2DNN model [34, 49].

Merge and Split Operations

The fundamental operations in lattice surgery are defined as follows (Fig. 7):

- **Merge:** Two adjacent logical surfaces are merged along a common boundary to perform a joint stabilizer measurement. For instance, a merge along the X-boundary enables the measurement of the joint $X \otimes X$ operator. This is achieved by modifying the stabilizer configuration across the boundary—specifically, by introducing new ancillary qubits between the patches and redefining the stabilizers to include these ancillae and boundary qubits. The outcome of the merge corresponds to a parity measurement between the two logical qubits, enabling fault-tolerant implementation of certain Clifford operations and state preparations.
- **Split:** The inverse of merging, splitting separates a previously merged surface into two independent logical patches. This is effected by measuring the ancillary

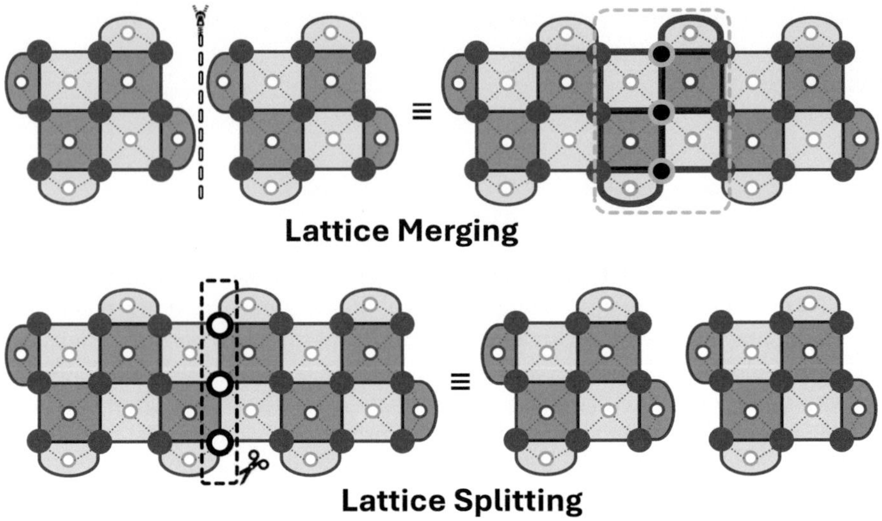

Fig. 7 Diagrammatic representation of the merging and splitting of the lattices in lattice surgery

qubits between the two logical regions in a basis (typically Z) that disentangles their stabilizer structure. The measurement outcomes are used to update the Pauli frame accordingly. After splitting, each surface resumes independent stabilizer measurement cycles, thus restoring individual logical qubits.

Merge and split operations can be performed along either X- or Z-type boundaries, depending on the desired logical operation. The process is entirely measurement-based, and all operations are confined to local stabilizer redefinitions and ancilla-mediated interactions. This locality is preserved at every step, thereby maintaining full compatibility with the 2DNN constraint and ensuring fault tolerance throughout.

Implications for Modular Architectures

Lattice surgery underpins many modern proposals for modular, patch-based quantum architectures [40, 48]. In particular, the ability to entangle, disentangle, and measure logical qubits through controlled reshaping of code patches facilitates the design of dedicated *data blocks* for computation and *distillation blocks* for magic state preparation. Communication between these blocks is mediated by lattice surgery operations, enabling scalable and efficient resource management without necessitating qubit transport or long-range interactions.

2 The QEC Design Flow

This subsection outlines the design flow of a quantum error correction (QEC) protocol, tracing the transformation from logical circuit synthesis through Pauli product rotations, Clifford gate pruning, and Clifford+T decomposition, culminating in magic state distillation and physical realization. The flow captures key transformations enabling FTQC.

2.1 Logical Circuit Synthesis

Logical circuit synthesis constitutes the first transformation in the QEC design flow. It bridges the gap between a high-level algorithmic description of quantum computation and a low-level, gate-based representation suitable for encoding under fault-tolerant quantum error correction schemes. The core objective is to represent the entire computation using a universal set of operations that are compatible with fault-tolerant implementation, particularly under stabilizer codes such as the surface code.

Pauli Product Rotation Formalism

Rather than immediately targeting a conventional gate set such as Clifford+T, recent advances propose an algebraically structured representation of quantum circuits using *Pauli product rotations* [40]. Each quantum gate is expressed as a unitary of the form

$$P_\varphi = \exp(-i\varphi P),$$

where P is a tensor product of Pauli operators (I, X, Y, Z) acting on one or more qubits, and φ is a rotation angle. This representation naturally subsumes all standard single- and multi-qubit gates:

$$\begin{aligned} S &= Z_{\pi/4}, \quad T = Z_{\pi/8}, \\ H &= Z_{\pi/4} X_{\pi/4} Z_{\pi/4}, \\ \text{CNOT} &= (Z \otimes X)_{\pi/4} \cdot (I \otimes X)_{-\pi/4} \cdot (Z \otimes I)_{-\pi/4}. \end{aligned}$$

More generally, a controlled-P gate, $C(P_1, P_2)$ can be written as

$$C(P_1, P_2) = (P_1 \otimes P_2)_{\pi/4} \cdot (I \otimes P_2)_{-\pi/4} \cdot (P_1 \otimes I)_{-\pi/4}.$$

This algebraic formalism unifies gate synthesis, gate commutation, and cancellation within a single operator-theoretic framework that is particularly well-suited for fault-tolerant execution, aligning with surface-code lattice surgery, which naturally implements such rotations via multi-qubit Pauli product measurements [40, 50, 51].

Clifford Elimination and Gate Commutation

Clifford gates, by virtue of mapping Pauli operators to Pauli operators under conjugation, can be commuted to the end of the circuit [40, 52, 53]. This is a key transformation step, as Clifford operations can typically be implemented transversally or classically tracked, whereas non-Clifford gates (e.g., T) must be realized through magic state injection and distillation [54]. For any Clifford operator C and Pauli product rotation P_φ, one has

$$C P_\varphi C^\dagger = P'_\varphi,$$

where P' is again a Pauli product. Hence, any quantum circuit U can be transformed into a canonical form:

$$U = C_f \cdot \left(\prod_j P^{(j)}_{\pi/8} \right) \cdot \left(\prod_k P^{(k)}_{\pi/4} \right),$$

where C_f is the final Clifford layer. This transformation has two practical consequences—(1) All *Clifford gates* are deferred to the end of the quantum circuit and eventually absorbed into the measurement bases by the following rules:

$$P_{\pi/4} \cdot P'_\varphi = \begin{cases} P'_\varphi P_{\pi/4}, & \text{if } [P, P'] = 0, \\ (i P P')_\varphi P_{\pi/4}, & \text{if } \{P, P'\} = 0. \end{cases}$$

and (2) the transformed circuit consists exclusively of *Pauli product rotations*, organized according to their rotation angles. All non-Clifford operations are executed via magic-state-assisted Pauli product measurements, while the residual Clifford action is accounted for through a classical update of the final measurement bases.

Clifford+T Formalism

After commutation, the resulting synthesized circuit adopts a structure tailored for QEC-aware transpilation:

- A sequence of $\pi/8$ rotations, i.e., T gates, each expressible as $P_{\pi/8}$,
- A sequence of $\pi/4$ Clifford rotations, i.e., $P_{\pi/4}$,
- A terminal layer of Pauli product measurements, where the final Clifford gates have been absorbed.

This *canonical Clifford+T form* [55–57] ensures that all operations beyond the Clifford group are explicitly isolated and ready for translation into fault-tolerant primitives. Each T gate is implemented via ancillary magic states, while all Clifford operations can be transversally encoded or classically managed.

Therefore, the original quantum circuit, when formulated via the Pauli product rotations and Clifford commutation, yields a representation of the form:

$$\text{Input circuit} \quad \Rightarrow \quad \{\pi/8\text{-rotations}\} + \{\pi/4\text{-rotations}\} + \{\text{Pauli measurements}\}$$

2.2 *Magic State Distillation*

The Clifford+T decomposition forms a universal gate set for quantum computation, wherein Clifford operations can typically be implemented fault-tolerantly and transversally under stabilizer codes. However, the non-Clifford T gate presents a fundamental challenge: due to the Eastin-Knill theorem [38], no quantum error-correcting code that admits universal quantum computation allows all gates, including T, to be implemented transversally. As a result, the standard method for implementing the T gate in fault-tolerant architectures is via *magic state injection*. A *magic state* is a specially prepared non-stabilizer ancilla qubit in the state

$$|m\rangle = \frac{1}{\sqrt{2}}(|0\rangle + e^{i\pi/4}|1\rangle),$$

which, when consumed in an ancilla-assisted measurement circuit, allows for the indirect application of the T gate on a data qubit, followed by a classically controlled Clifford correction. However, such magic states, when prepared at the physical level, are highly susceptible to noise and decoherence, rendering them unsuitable for direct use. Their error rates typically exceed the fault-tolerance threshold of the code and must be reduced via a process known as *magic state distillation* [58–63].

Distillation Protocols

Magic state distillation employs quantum codes that admit transversal T gates to recursively suppress noise in faulty ancillae. Classical protocols, such as the Bravyi-Kitaev 15-to-1 scheme based on the [15, 1, 3] punctured Reed-Muller code [40, 54, 64], take 15 noisy magic states as input and output a single, higher-fidelity magic state. The procedure involves the following steps:

1. Encoding the input states into the logical subspace of a distillation code.
2. Applying a transversal T gate across all qubits.
3. Decoding and measuring stabilizers to detect the presence of uncorrectable errors.
4. Postselecting on syndrome outcomes to accept or reject the output state.

This protocol reduces the error probability from p to approximately $35p^3$ under an independent depolarizing noise model [54, 61]. More generally, distillation protocols are characterized by triples (n, k, d), indicating n input states, k outputs, and distance d, often realized through triorthogonal matrices and CSS constructions. Following the magic state distillation process, the distilled states are placed and routed on the data patches on the 2D lattice of qubits, optimizing for the space-time cost [40].

2.3 Decoding

The decoding stage constitutes the classical inference process responsible for identifying and correcting quantum errors based on stabilizer measurement outcomes [35–37]. Given that direct measurement of data qubits is prohibited in fault-tolerant protocols to preserve quantum coherence, QEC architectures employ ancillary measurements of stabilizer generators to indirectly reveal the presence and type of errors. The decoding problem, then, is to map these syndrome outcomes to appropriate correction operations, with the goal of recovering the original logical state without introducing further logical errors. Mathematically, decoding can be viewed as a function

$$D : \{0, 1\}^m \to \mathcal{P}_n,$$

where $\{0, 1\}^m$ is the binary syndrome space generated by m stabilizer measurements, and $\mathcal{P}_n$ is the n-qubit Pauli group, from which a recovery operator is chosen. A decoder is thus a classical algorithm that infers an error coset representative from a syndrome string, ideally identifying the most probable equivalence class of errors consistent with the measured syndrome. The decoding process is intimately tied to the error model and the structure of the quantum code. Under a stochastic Pauli noise model, each error $E \in \mathcal{P}_n$ commutes or anticommutes with each stabilizer generator S_i, and the syndrome bit s_i is defined as

$$s_i = \begin{cases} 0 & \text{if } [E, S_i] = 0, \\ 1 & \text{if } \{E, S_i\} = 0. \end{cases}$$

Since multiple errors can yield the same syndrome—forming a coset of the stabilizer group—the decoding task is fundamentally probabilistic. A decoder aims to return a correction operator C such that $CE \in \mathcal{S} \cdot \mathcal{L}_0$, where $\mathcal{S}$ is the stabilizer group and $\mathcal{L}_0$ is the identity logical operator. Any deviation from this results in a logical error.

Code-Specific Decoding Strategies

The decoding algorithm employed depends heavily on the structure of the underlying quantum code.

- **Repetition Code:** In the n-qubit bit-flip repetition code, the syndrome reveals the parity of adjacent qubits [65, 66]. Decoding is performed by majority voting: the value shared by the majority of physical qubits is taken as the logical value.

- **Surface Code:** The most widely used code in practical architectures, surface codes define syndrome extraction in terms of vertex (X-type) and plaquette (Z-type) stabilizers on a two-dimensional lattice. Decoding is performed by matching the syndrome defects (syndrome bits equal to 1) using the *minimum-weight perfect matching* (MWPM) algorithm [36, 67]. This algorithm pairs detection events via a graph matching procedure that minimizes total edge weight, which corresponds to error likelihoods.
- **Color Code:** These codes, defined on trivalent lattices with three-colorable faces, present additional decoding challenges due to higher-weight stabilizers and overlapping supports. Decoders for color codes often rely on generalizations of matching techniques or employ message passing and belief propagation [68].
- **Subsystem Codes:** These codes introduce gauge degrees of freedom, further complicating decoding. In such cases, decoding typically proceeds in two stages: gauge fixing and then syndrome interpretation on the resulting stabilizer code [69].

Performance Criteria and Noise Models

The performance of a quantum decoder is evaluated along several critical dimensions, including accuracy, latency, scalability, and fault tolerance [36, 37]. Accuracy refers to the probability that the decoder returns a correction operator lying in the same coset as the true error, thereby preserving the logical state. Latency measures the computational delay incurred during decoding, which is particularly important for real-time feedback in fault-tolerant quantum computing. Scalability assesses the algorithm's complexity and feasibility as the code distance increases, especially relevant for large-scale codes such as surface or LDPC codes. Fault tolerance encompasses the decoder's robustness against ancillary imperfections, such as measurement errors or correlated faults, ensuring that the decoding process does not itself become a source of logical failure. These performance metrics are inherently dependent on the assumed noise model, which may include depolarizing noise (uniform Pauli errors), biased noise (e.g., dephasing-dominated models), erasure noise (common in photonic architectures), and spatially or temporally correlated noise. Additionally, practical implementations must often incorporate measurement errors, necessitating syndrome extraction over multiple rounds and the deployment of temporal decoding strategies that track syndrome histories across time to distinguish data and measurement faults.

The decoding procedure must be adapted to the specific characteristics of the underlying noise model, as the structure and correlations in physical errors directly influence syndrome patterns and decoder performance. Common noise models include depolarizing noise, where all single-qubit Pauli errors occur with equal probability; biased noise, such as dephasing-dominated channels where one Pauli error (typically Z) is more prevalent; erasure noise, wherein qubit loss is detectable and localized—frequently encountered in photonic systems; and correlated noise, arising from crosstalk, leakage, or environmental coupling, which induces multi-qubit error events. Additionally, measurement errors compromise the reliability of stabilizer readouts, necessitating repeated syndrome extraction across multiple error-correction cycles. To address this, modern decoders often employ *temporal decoding*, wherein sequences of syndromes are jointly analyzed to differentiate between data and measurement faults. In topological codes such as the surface code, this results in a space-time decoding framework, typically modeled as a three-dimensional syndrome graph on which algorithms like minimum-weight perfect matching (MWPM) are applied to infer the most likely error trajectories through both space and time.

Classical Decoding Algorithms

Several classical decoding schemes are employed in practice:

- **Lookup Table Decoders:** These decoders enumerate all possible syndromes and their corresponding optimal corrections in a precomputed table. For each observed syndrome, the decoder simply returns the associated correction operator. This method guarantees exact decoding under the assumed noise model and is computationally trivial at runtime [70]. However, the space complexity grows exponentially with the number of stabilizers, making this approach infeasible beyond small codes such as the 3-qubit repetition code or the 5-qubit perfect code. Lookup table decoders are primarily useful for benchmarking purposes and for pedagogical illustration.
- **Minimum-Weight Perfect Matching (MWPM):** Widely employed in decoding surface codes, MWPM [35, 36, 67] operates on a graph constructed from syndrome data, where each node represents a detection event (typically a nontrivial stabilizer measurement outcome) and weighted edges correspond to the likelihood or cost of error paths connecting them. Under the assumption of independent errors, the decoder seeks a perfect matching that minimizes the total weight, i.e., the sum of edge costs. This is classically solved using Edmonds' blossom algorithm in $O(n^3)$ time for n nodes, though optimizations and parallel implementations exist. The error chains corresponding to the matched paths are applied as corrections. MWPM is effective under low noise rates and local stochastic noise models but does not naturally generalize to codes with non-planar structures or correlated noise.
- **Union-Find Decoders:** These decoders offer a computationally efficient alternative to MWPM, operating in nearly linear time. The core idea is to treat syndrome defects as seeds of growing clusters and perform disjoint-set operations to identify and merge connected components. When clusters overlap or reach a predefined radius, corrections are inferred by identifying minimal connecting paths between defects. The union-find data structure supports efficient merging and path compression, significantly reducing decoding latency [71–73]. While union-find decoders may not always match the accuracy of MWPM, especially near the code threshold, they scale better with lattice size and are promising candidates for hardware-level integration.
- **Belief Propagation (BP):** Particularly suited to quantum low-density parity-check (LDPC) codes, belief propagation is an iterative algorithm that operates on a Tanner graph representing stabilizer checks and physical qubits [74, 75]. Messages

are passed between variable nodes (qubits) and check nodes (stabilizers), updating beliefs about the marginal probability of errors on each qubit conditioned on neighboring syndromes. The decoder converges to a configuration with maximum posterior probability under the assumption of a factorized prior. While BP is efficient for sparse graphs, it can fail to converge or yield suboptimal results in the presence of short cycles or strong correlations. Nevertheless, it provides a practical decoding strategy for large, sparse codes where exact inference is intractable.

Each decoding method makes specific assumptions about the noise model, code structure, and computational resources available. While optimal decoding (e.g., maximum likelihood decoding) is in general NP-hard [36, 37], these heuristic or approximate decoders offer practical trade-offs for real-world fault-tolerant quantum computing architectures.

3 Design Automation and Optimization in QEC

The design and implementation of large-scale quantum error-corrected computations demand rigorous optimization across multiple abstraction levels. In classical computing, Electronic Design Automation (EDA) [76–79] encompasses a suite of algorithmic tools and methodologies for the automated synthesis, placement, routing, and verification of digital circuits, whether targeting ASICs or FPGAs. Classical EDA frameworks are driven by a well-defined set of design objectives—primarily area, timing (latency), power consumption, and functional correctness—subject to physical constraints imposed by the technology node and fabrication process. Over several decades, EDA has matured into a highly structured and automated domain, with established toolchains and optimization strategies. In quantum computing, particularly in the context of fault-tolerant quantum error correction (QEC), the design landscape is still nascent [80]. However, the need for optimization is no less critical. Quantum resources are scarce, noisy, and expensive to operate; thus, optimizing their usage is essential for the viability of any large-scale quantum computation. The QEC design and compilation flow, although structurally different from its classical counterpart, inherits several optimization objectives. These can be broadly categorized as follows:

- **Resource Minimization:** The total number of physical qubits required—including data, ancilla, and magic state qubits—must be minimized to conform to the limited quantum hardware available. This also includes reducing the demand for magic state distillation, which dominates the space-time overhead in fault-tolerant architectures.
- **Latency and Depth Reduction:** Minimizing logical circuit depth directly impacts the overall time for computation and reduces cumulative decoherence and operational error. In surface code-based realizations, this translates to minimizing the number of sequential lattice surgery steps or measurement rounds.

- **Fidelity Preservation under Formal Verification:** Employing quantum model checking or SMT-based equivalence to ensure the optimized circuit maintains or improves logical-state fidelity and introduces no new uncorrectable error paths.

These objectives are constrained by the fault-tolerance requirements of the quantum code: gates must respect transversal or code-compatible implementations, ancilla states must be prepared and verified without measurement collapse, and operations must preserve the logical subspace. Unlike classical logic synthesis, quantum design workflows must also manage non-Clifford gate costs, non-determinism in measurement-based logic (e.g., teleportation or lattice surgery), and hardware-induced connectivity restrictions. As a result, the optimization of QEC workloads poses novel algorithmic challenges that require specialized automation strategies.

3.1 Reducing Qubit Overhead

The robustness afforded by QEC comes at the cost of significant physical resource overhead, often orders of magnitude larger than the logical information being protected [1]. This overhead manifests in multiple forms, including the spatial footprint of encoded data qubits, ancillary qubits required for stabilizer measurements, additional patches for routing and code deformation, and the redundancy needed to tolerate hardware defects or adversarial noise [40]. In surface code-based QEC architectures, which are widely regarded as a leading candidate for practical implementation due to their high logical error thresholds and local interactions, the qubit overhead scales approximately quadratically with the code distance. For a code of distance d, the number of data qubits is $O(d^2)$, and the number of ancilla qubits is typically of the same order [81]. Furthermore, to ensure reliable computation over large programs, the architecture must accommodate both spatial and temporal variations—such as fabrication-induced defects, cosmic ray-induced transient faults, and workload-induced congestion—which further increase the qubit requirement via dynamic deformation or redundancy provisioning [82].

As the field transitions from theoretical code design to practical realization, the challenge of qubit overhead has emerged as a critical bottleneck. This has motivated a growing body of research focused on automating overhead-aware QEC design. Recent research has explored the areas widely, from modifying the structure of the surface code by forming superstabilizers around fault regions [83] to enabling localized code deformation and dynamic patch resizing to recover lost code distance while minimizing physical footprint [84] as well as attempting to reduce the total qubit usage by reusing ancilla qubits across $X-$ and $Z-$type stabilizer measurements through temporally interleaved sub-rounds [81].

Case Study 3.1.1: Adaptive Surface Code [83]
The adaptive surface code framework introduces a dynamic method to preserve surface code functionality in the presence of temporary or permanent defective qubits by replacing stabilizers that intersect the defect region with superstabilizers. This enables the code to continue operating without relocating logical information or restarting computation.

Algorithmic Overview: The framework employs a defect-detection strategy using DBSCAN clustering over syndrome mismatches to identify faulty regions during execution. Upon locating a defect, the surrounding stabilizers are merged into a superstabilizer encircling the faulty patch. Two variants are used: a basic approach where the superstabilizer is directly inferred, and a shell-based approach that uses repeated gauge operator measurements for higher fidelity. The decoding graph is updated accordingly to incorporate the new stabilizer geometry.

While this method does not recover the original code distance, it avoids full logical block discard and circumvents the need for preallocated redundant logical qubits. For example, when applied to a 3×3 defect in a surface code of distance $d = 15$, the adaptive patch deformation preserved logical functionality with a threshold only marginally lower than the defect-free baseline (2.7% vs. 2.9%), using fewer qubits than static overprovisioning would require. Thus, it reduces qubit overhead by avoiding global redundancy in favor of localized code deformation.

Case Study 3.1.2: Surf-Deformer [84]
Surf-Deformer introduces a composable instruction set that enables fine-grained patch resizing and defect excision through gauge fixing. Unlike the adaptive surface code framework, it not only removes defective qubits but also restores lost code distance and supports dynamic routing.

Algorithmic Overview: The core technique models patch manipulation as a set of gauge-preserving transformations: stabilizer-to-gauge (S2G), gauge-to-stabilizer (G2S), stabilizer-to-stabilizer (S2S), and gauge-to-gauge (G2G) operations. At runtime, the controller applies $PatchQ\_RM$ to remove defective zones and $PatchQ\_ADD$ to regrow the patch boundary. A layout generator estimates inter-patch spacing (Δd) based on defect rate and expected growth, allowing compilation-time resource planning. Read Section IV from [84] for a detailed understanding.

Compared to Q3DE [85], which doubles the patch size to recover distance, Surf-Deformer achieves equivalent error suppression with $2\times$ fewer qubits. On QFT-100-20 benchmarks under burst faults, Surf-Deformer reduced retry risk by up to $6\times$ compared to the adaptive surface code framework while consuming 50% fewer physical qubits than Q3DE. Its ability to dynamically restore code integrity while preserving communication links makes it highly efficient for fault-aware layouts.

Case Study 3.1.3: Quantum Prometheus
This framework proposes a compile-time strategy for reducing ancilla overhead by reusing the same ancilla qubits to measure both X- and Z-type stabilizers, scheduled in interleaved sub-rounds instead of simultaneously.

Algorithmic Overview: The method divides each QEC round into two sequential sub-rounds. In the first, X-stabilizers are measured using a fixed set of ancilla qubits; after reset, the same ancillas are reused to measure Z-stabilizers in the second sub-round. This reuse is compiler-inserted and does not require geometric patch modification. Logical decoding is applied only after a full round, preserving syndrome integrity.

This simple reuse strategy surprisingly reduces ancilla requirements by 50% and achieves a total qubit reduction of 25%. Crucially, for large code distances ($d \geq 13$), a modified surface code of distance $d + 2$ using this framework consumes fewer qubits than a standard code of distance d, while achieving better logical error suppression. For instance, a $d = 31$ modified surface code achieved a 1.5×10^{-27} logical error rate with 1441 qubits, compared to a $d = 27$ original code requiring 1457 qubits for a higher error rate of 4.27×10^{-24}.

3.2 *T-Gate Optimization*

In two-dimensional topological codes such as the surface code, all logical Clifford operations ($H, S, CNOT$) admit low-overhead, transversal implementations or lattice-surgery realizations [40]; by contrast, the non-Clifford T-gate (a $\pi/8$ phase rotation) is *not* transversal [80] and must be effected by injecting a high-fidelity "magic" state into the code via a dedicated lattice-surgery protocol. Each such injection consumes one distilled $|m\rangle = \frac{|0\rangle + e^{i\pi/4}|1\rangle}{\sqrt{2}}$, magic state, and because raw injections at physical error rate p_{phys} remain faulty at $O(p_{\text{phys}})$, recursive magic-state distillation is required to reach logical error rates below 10^{-10} for classically intractable algorithms [54]. Two natural resource metrics then arise:

- T-**Count**: the total number of T (and $\text{T}^{\dagger}$) gates in the circuit, each corresponding to one $|m\rangle$ consumption and thus setting the *volume* (qubit footprint) of distillation factories [80, 86–89].

– T**-Depth**: the minimum number of sequential layers of commuting T-gates; because each layer incurs a full round of magic-state injection and surface-code syndrome extraction (on the order of one code cycle), T-depth directly controls the *latency* of the logical operation [80, 86–89].

Moreover, each distilled $|m\rangle$ must itself be error-corrected by a distance—d patch, occupying $d \times d$ data qubits plus ancillas, so that achieving high fidelity often requires hundreds of magic states in parallel—amounting to tens or hundreds of thousands of physical qubits devoted solely to distillation. Thus, any reduction in *T-count* directly shrinks the factory footprint, while any collapse of *T-depth* proportionally reduces the number of distillation rounds (and hence overhead qubits) needed. This key optimization can be performed both in the early-stage circuit synthesis by exploiting polynomial methods and rewriting gates graphically, as well as in later, post-Clifford pruned stages of circuit synthesis by merging T-gate layers that commute mathematically.

Early-Stage Circuit Synthesis

Since each logical T-gate requires its own distilled magic state, and hence a full round of surface code syndrome extraction, any reduction in the T-count or T-depth before backend packing directly shrinks both the qubit footprint of magic-state factories and the overall execution latency. Early-stage (front-end) synthesis, therefore, seeks to eliminate non-Clifford rotations as early as possible, at the algebraic, graphical, or tensor-network level, so that downstream scheduling and resource allocation never "see" redundant T-gates. We present a taxonomy of the current state-of-the-art circuit synthesizers specifically designed to reduce the T-gates.

– *Phase-Polynomial Solvers:* Extract the CNOT+T subcircuit into a Boolean phase polynomial $\phi : \mathbb{Z}_2^n \to \mathbb{Z}_8$ plus a gate-synthesis matrix, then formulate T-depth minimization as a matroid-partition problem. Representative tools include Tpar [86], TODD [88], and Faster-TODD[87].
– *ZX-Calculus Rewriters:* Translate the circuit into a graph-like ZX-diagram [90] of Z-spiders and Hadamard edges, apply local complementation and gadget-fusion rules to collapse phase gadgets, and perform phase-teleportation to eliminate redundant $\pi/4$ rotations. Representative tools include PyZX used for T-gate optimization [89].
– *Tensor-Network Unitary Synthesis.* Represent single-qubit gate sequences as a compact tensor network (MPS), compute overlaps with the target unitary via efficient tensor contractions, and sample high-quality sequences under a T-budget. A representative tool is TRASYN [91].

All three frameworks isolate a residual "phase gadget" structure, whether as third-order tensor entries $\mathcal{T}_{ijk} = \frac{\partial^3 \phi}{\partial x_i\, \partial x_j\, \partial x_k}$, as degree-one spiders in a ZX-diagram, or as MPS tensors indexing gate blocks. Optimization then reduces to a combinatorial problem—low-rank tensor decomposition or minimum matroid cover for T-depth, graph rewrites for T-count, or sampling from a tensor network under T-budget. Therefore, on integrating the circuit synthesis ideas in the QEC design workflow, it looks somewhat like this:

1. *Decomposition:* parse the user's target unitary into phase-polynomial, ZX, and tensor-network forms.
2. *Optimization Passes:* run the matroid-partition solver, apply ZX rewrite rules, and/or perform tensor-network sampling.
3. *Extraction:* reconstruct minimized Clifford+T (or Clifford+U_3) circuits from each optimized representation.
4. *Selection and Hybridization:* compare resulting T-count, T-depth, and ancilla requirements; optionally chain or interleave passes (e.g., PyZX followed by TODD and/or methods in [89]).

This automated early-stage pipeline ensures that the bulk of expensive magic-state resources are never allocated to redundant rotations and provides a rich design space of algebraic, graphical, and tensor-network "knobs" to tune for any hardware and fidelity constraint.

Case Study 3.2.1: Tpar [86]
The T-depth optimization process for $CNOT$+T circuits proceeds through four key steps. First, in the *phase-polynomial extraction* phase, the $CNOT$+T subcircuit is identified and its corresponding Boolean phase polynomial is computed as

$$\phi(\mathbf{x}) = \sum_{i=1}^{L} c_i\, f_i(\mathbf{x}), \quad c_i \in \mathbb{Z}_8, \ f_i : \{0, 1\}^n \to \{0, 1\}.$$

Next, during *matroid construction*, a matroid $(S, \mathcal{I})$ is formed over the set $S = \{(c_i, f_i)\}$, with the collection of independent sets defined by

$$\mathcal{I} = \{T \subseteq S \mid \dim(V) - \text{rank}(T) \leq n - |T|\}, \quad V = \mathbb{F}_2^n.$$

The third stage involves *partitioning* this set using Edmonds' algorithm, which splits S into the minimum number k of independent subsets:

$$S = I_1 \cup I_2 \cup \cdots \cup I_k, \quad I_j \in \mathcal{I}.$$

Finally, the *circuit re-synthesis* step is performed for each subset I_j: a $CNOT$ network is synthesized via Gaussian elimination to compute all $f_i \in I_j$, followed by a single layer of T-gates corresponding to the coefficients c_i, and then the $CNOT$ network is uncomputed. This results in a quantum circuit with

T-depth exactly k, which is provably minimal under the given matroid-based model.

On m-bit Galois-field multipliers, the T-depth was reduced by approximately 79% and the T-count by approximately 42% without the use of ancillas. When $O(L)$ ancillas were allowed, the T-depth collapsed from $12(m-1)$ layers to just 2 layers. For $\Lambda_k(X)$ gates, the T-count was reduced from $7(4k-8)$ to $3(4k-8)+4$, and the T-depth improved from $3(4k-8)$ to $4k-8$ using $k-2$ ancillas.

Case Study 3.2.2: ZX-Calculus Rewriting with PyZX [87–89]
The optimization process based on ZX-calculus proceeds through a sequence of structured transformations. First, in the *ZX-diagram translation* step, the Clifford+T circuit is converted into a graph-like ZX-diagram composed of Z-spiders connected via Hadamard edges. Next, during *Clifford simplification*, local complementation and Pauli-spider removal are applied to eliminate all residual Clifford structure. In the *phase-gadget fusion* stage, Identity Fusion and Gadget Fusion rules are applied iteratively to merge $\pi/4$ phase gadgets that share identical neighborhoods. This is followed by *phase teleportation*, where symbolic phase variables are tracked during fusion; each fusion zeroes one variable and sums its phase into another, thereby recording cancellations. After this, the *residual matroid pass* [88] is performed by extracting the updated phase polynomial and invoking the TODD algorithm to partition the remaining terms into layers of minimal T-depth. Finally, a *faster enhancement* [87] step improves the efficiency of the residual solver through advanced rank-decomposition routines, enabling the handling of larger sets of phase terms with improved computational performance.

PyZX alone matches or improves the then state-of-the-art (Tpar and/or TODD) T-count on approximately 72% of reversible-arithmetic benchmarks, achieving up to a 50% reduction on circuits rich in Hadamard gates. Incorporating the TODD algorithm provides an additional average T-count reduction of approximately 15%, thereby enabling efficient synthesis of circuits with more than 10^4 phase terms.

Case Study 3.2.3: TRASYN [91]
TRASYN focuses on the synthesis of the Clifford +$U3$ gate set, aiming to reduce the overall T-count as well as the Clifford gate count in the resulting circuit. The gate synthesis method proceeds as follows. First, a *gate-block library* is precomputed, consisting of tensor-network representations

of short sequences of Clifford+U_3 operations. Next, the target single-qubit unitary—or a small multi-qubit block—is encoded as a Matrix Product State (MPS) by evaluating its overlap with elements of the gate-block library. Then, under a *T-budget* constraint, tensor-network contractions are used to compute fidelity overlaps for candidate sequences, and sampling is performed to identify those sequences that satisfy the prescribed T-count limit. Finally, the *sequence extraction* step selects the sequence with the highest fidelity and translates it into a native Clifford+U_3 circuit. This process yields implementations that require fewer T gates than standard R_z-only decomposition techniques.

For arbitrary single-qubit rotations, TRASYN achieves up to $2\times$ fewer T-gates compared to conventional three-gate R_z decompositions, while maintaining equivalent error bounds. On benchmark sets consisting of random unitaries, the average T-count is reduced by $30 - 50\%$, with end-to-end synthesis times of under one second per rotation.

Late-Stage Circuit Synthesis

At this stage of the design flow, all Clifford gates have been commuted to the end of the circuit or absorbed into measurement bases [40, 54], leaving a sequence of non-Clifford $\pi/8$ rotations organized into m layers:

$$C = L_1 \to L_2 \to \cdots \to L_m, \quad L_j = \exp\Big(i\frac{\pi}{8}\sum_{q=1}^{n} P_q^{(j)}\Big),$$

where each $P_q^{(j)} \in \{I, X, Y, Z\}$ is a Pauli operator on qubit q. Because each layer L_j requires a fresh batch of distilled magic states and a full round of surface-code syndrome extraction, the total T-depth m dictates both the number of distillation rounds and the overall latency of the logical computation. The objective of late-stage synthesis is to minimize m by merging layers whose Pauli-product rotations commute, without introducing additional T-gates. Since the search space of all possible mergings grows combinatorially with m, practical approaches employ heuristic or meta-heuristic search to identify near-optimal packings of commuting layers.

Case Study 3.2.4: Genetic-Algorithm for T-depth optimization [80]
In the GA-based approach for optimization in the late-circuit synthesis, the post-Clifford circuit is represented as a list of length-n Pauli vectors $\{L_j\}_{j=1}^{m}$. Optionally, high-density layers may be split into finer sub-columns along low-density qubit subsets to expose additional merge candidates. Compute for each layer L_j its T-count T_j and density $D_j = T_j/n$. For every pair (i, j) whose Pauli products commute, assign a score

$$\mathrm{score}(i, j) = 1 - |D_i - D_j| + \beta\left(T_{\max} - (T_i + T_j)\right),$$

with tunable weight $\beta \in [0, 1)$. Select a maximal non-overlapping matching of highest-scoring pairs to seed the Genetic-Algorithm population. **Genetic-Algorithm Framework:** The layer-collapsing algorithm proceeds through an evolutionary optimization framework. It begins with an *initialization* step, wherein an initial population of merge-sets is created; each merge-set is a disjoint collection of commutable gate pairs. Next, during *fitness evaluation*, the fitness of each individual is defined as the total number of merges it contains. The *selection and elitism* phase retains the top k individuals without modification and generates new offspring through crossover, by combining merge-sets, and occasional mutation, which resets portions of the chromosome. Finally, in the *iteration* phase, the best merge-set is applied to collapse the paired circuit layers into single layers. The circuit representation is then updated, and the process is repeated until no further merges are possible.

On synthetic circuits with 90–100 qubits and maximal T-density, the genetic algorithm reduced T-depth by up to 79.2%, e.g., from approximately ~1024 layers to ~214—and T-count by 41.9%, outperforming lookahead heuristics by a factor of approximately 2.6× in T-depth reduction. On standard reversible-arithmetic benchmarks [86, 89], such as adders and Galois-field multipliers, the average T-depth decreased by a factor of 2.58× when compared to zero-ancilla matroid-based methods, all achieved without the use of additional ancilla qubits.

3.3 *Optimizing the Surface Code Layout*

Distillation Protocols in Surface-Code Architectures

Magic-state distillation in the surface code can be mapped onto *distillation blocks* [40, 54], each of which implements a fault-tolerant multi-qubit Pauli product measurement via lattice surgery on patches of code distance d. A distillation block consists of

- Logical qubit patches of size $d \times d$,
- Ancillary patches for multi-patch Pauli product measurements (to realize $\pi/8$ rotations),
- Classical feed-forward paths for measurement outcome processing.

The geometric cost of each block is measured by its space–time volume

$$V = A \times T,$$

where A is the number of $d \times d$ tiles and T the number of code cycles. For example, a single-level 15-to-1 protocol occupies an $11d \times 5d$ tile region over $12d$ cycles, yielding

$$V_{15\to 1} = 660\, d^3.$$

Alternative protocols include

- **20-to-4:** occupies $\approx 14d^2$ tiles over $4d$ cycles, producing four outputs with $O(p^2)$ error [54],
- **Higher-rate codes** (e.g., 116-to-12, 225-to-1): trade larger $A \times T$ for multi-output or improved error scaling.

In a full-scale computation, the data-block's demand—set by the circuit's T-count and T-depth—must be matched by the throughput of parallel distillation blocks. The key trade-offs in magic state distillation and factory design are as follows. *Underprovisioning* leads to execution stalls due to insufficient availability of distilled states. In contrast, *overprovisioning* results in increased qubit overhead, consuming more hardware resources than necessary. Additionally, while *higher-level* distillation protocols can reduce the per-state cost in terms of T-gates or time, they typically require a larger factory.

Case Study 3.3.1: Protocol Selection by Workload [54]
We analyze three representative T-gate workloads (physical error rate $p = 10^{-4}$) and identify the protocol minimizing space–time cost per magic state.

Workload	T-depth	T-depth	Best Protocol	Cost per state
High-rate streaming	10^8	10^6	20-to-4	$27\, d^3$
Ultra-low error	10^{10}	10^4	15-to-1 × 15-to-1	$25.9\, d^3$
Resource-constrained	10^6	10^6	15-to-1	$6.3\, d^3$

High-rate streaming. When T-count $\gg$ T-depth, multi-output blocks (e.g., 20-to-4) maximize throughput per tile—ideal for pipelined factories.
Ultra-low error. Moderate depth with stringent fidelity demands favors cascading two 15-to-1 stages, achieving $O(p^9)$ error at $\approx 25.9\, d^3$ per output.
Resource-constrained. Low-depth, small-scale machines use a single-level 15-to-1 block to minimize footprint ($\approx 6.3\, d^3$) at the cost of higher latency.
These examples demonstrate that matching protocol choice to (T-count, T-depth) yields orders-of-magnitude reductions in space–time overhead.

Case Study 3.3.2: Analysis of algorithms to choose Magic State Distillation Protocols [92]
Each magic state distillation (MSD) protocol is characterized by three primary metrics: the number of surface-code tiles D_p required, the number of distillation steps S_p, and the number of output magic states k_p produced per round. For a given physical error rate p, the success probability of protocol p is $P_s = (1-p)^{N_p}$, where N_p is the total number of raw noisy states consumed. Consequently, the effective latency per distilled state is expressed as $\frac{S_p}{k_p P_s}$. Representative examples include the 15-to-1 protocol with $D_p = 11$, $S_p = 11$, $k_p = 1$, and the 20-to-4 protocol with $D_p = 14$, $S_p = 17$, $k_p = 4$. The Brute Force algorithm exhaustively enumerates all sequences of distillation rounds up to a fixed length L, simulating resource consumption and pipeline stalls to compute exact total tile-time products. Despite its optimality guarantee, the exponential search complexity ($O(|P|^L)$) renders it impractical for large L. Dynamic Programming (DP) mitigates this by defining a cost function $C(i, s)$ over protocol index i and cumulative step s, and applying the recurrence

$$C(i, s) = \min_{p \in P} \{C(i-1, s-S_p) + D_p\},$$

yielding an exact tile-minimal solution in time $O(L \cdot S_{\max} \cdot |P|)$. Greedy selection iteratively chooses the protocol $p^* = \arg\min_p(\frac{D_p}{k_p} + S_p)$, updating available magic-state counts and accounting for stalls; this yields near-optimal latency with linear complexity $O(L \cdot |P|)$. A Random baseline assigns a fixed protocol per optimization objective without adaptation.

When minimizing tile usage, DP reproduces the Brute Force optimum with 0% tile overhead, whereas Greedy incurs an average overhead of 41.3% and Random 14.7%. For latency minimization, Greedy deviates by only 7.6% from the Brute Force optimum, substantially outperforming DP (64.6% overhead) and Random (148.5%). In a balanced multi-objective scenario, DP and Greedy yield distinct Pareto-optimal configurations: DP points minimize tiles at the cost of extra steps, while Greedy points minimize steps with higher tile counts.

3.4 ML in Decoder Automation

Machine-learning (ML) modules can be introduced into the QEC decoding pipeline at carefully chosen junctures to accelerate decision making and to capture complex noise correlations that classical decoders struggle to model. At its core, decoding begins with the syndrome extraction process, which yields a binary vector

$$\mathbf{s} \in \{0, 1\}^m$$

of length m per QEC cycle. Traditionally, this vector is fed directly into a matching or propagation algorithm (e.g., MWPM, union-find, belief propagation) to propose a recovery operator [37, 93–96]. In an automated flow, the first opportunity for ML intervention lies in *syndrome preprocessing*, where one learns a mapping

$$\mathbf{s} \longmapsto \mathbf{z} \in \mathbb{R}^d, \quad d \ll m,$$

via an autoencoder or embedding network. By compressing syndrome information into a low-dimensional latent space $\mathbf{z}$, one both reduces the memory footprint and highlights the most salient error features—whether spatially clustered bit-flips or time-correlated measurement faults—before any costly graph search. Once a compact representation is available, the next "injection point" is the *syndrome-to-recovery mapping* stage. Here, one replaces, or augments, the classical lookup table (LUT) or brute-force search by training a neural classifier f to predict the appropriate Pauli correction

$$R_{\text{pred}} = f(\mathbf{z})$$

from among the set $\mathcal{R} = \{I, X, Z, Y\}^{\otimes k}$ of k logical-qubit operators. This reframing transforms the combinatorial decoding problem into a constant-time inference task, yielding an $O(1)$ decision latency per cycle. In practice, a shallow feed-forward or convolutional network can achieve logical-error rates on par with minimum-weight perfect matching for small code distances, but with vastly reduced—and hardware-friendly—runtime. Beyond single-round decoding, real QEC experiments demand resilience to *temporal correlations* arising from repeated measurements. By treating the syndrome record as a time series $\{\mathbf{s}_t\}_{t=1}^{T}$, one can employ sequence models, e.g., attention-based encoders, to distinguish between transient measurement flips and persistent data errors [37]. The result is a decoder that adaptively weights historical syndrome events when proposing corrections, yielding enhanced suppression of correlated error strings without sacrificing the modular structure of the classical decoder. Finally, practical deployment on FPGA or ASIC platforms imposes stringent area and latency constraints. *Resource-aware model compression* (via pruning or quantization) and *co-design* of neural architectures with digital logic pipelines ensure that syndrome acquisition, ML inference, and Pauli-frame updates form a seamless, low-latency hardware flow. Hybrid schemes—such as learned edge-weight prediction within a classical matching algorithm, preserve polynomial-time guarantees while infusing context-sensitive intelligence into each stage of the decoder.

Case Study 3.4.1: Multi-layer Perceptron (MLP) in QEC Decoding [95]
A pure-error peel-off routine deterministically removes stabilizer contributions from the raw syndrome, after which a shallow MLP infers the remaining logical Pauli correction.

MLP Design and Training: The neural classifier takes as input the $d^2 - 1$ independent defect bits remaining after pure-error cancellation for a distance-d surface code, and outputs two logits ℓ_X, ℓ_Z corresponding to an uncompensated logical X and Z flip. The network comprises two fully connected hidden layers of size 256 and 64, each followed by a square-nonlinear (SQNL) activation. Training data are sampled on-the-fly under a depolarizing noise model at physical error rate $p \approx p_{\text{pseudo}}$ (the MWPM pseudo-threshold for each d), ensuring coverage of hard, high-weight error patterns. Parameters θ are optimized via ADAM to minimize

$$\begin{aligned}\mathcal{L}(\theta) = -\mathbb{E}_{(\mathbf{s},y)}\Big[& y_X \log \sigma(\ell_X(\mathbf{s})) + (1 - y_X) \log\big(1 - \sigma(\ell_X(\mathbf{s}))\big) + \\ & y_Z \log \sigma(\ell_Z(\mathbf{s})) + (1 - y_Z) \log\big(1 - \sigma(\ell_Z(\mathbf{s}))\big)\Big] + \\ & \lambda \|\theta\|_2^2 + \gamma\, Q(\theta)\end{aligned}$$

where $y_X, y_Z \in \{0, 1\}$ are the true logical flips, σ is the sigmoid, and $Q(\theta)$ is a quantization-penalty driving weights toward small integer levels.

The logical-error rate $p_L(d, p)$ is defined as the probability that the combined pure-error + MLP correction yields $CE \notin \mathcal{S}$. In the low-p regime, one observes the characteristic scaling

$$p_L(d, p) \approx A\, p^{\frac{d+1}{2}}, \quad d = 3, 5, 7, 9,$$

with an exponent matching that of MWPM. The MLP high-level decoder achieves a pseudo-threshold $p_{\text{th}} \approx 1.2\%$ (comparable to blossom matching) and maintains the $O(1)$ inference cost per cycle. Incorporating fourfold rotational weight-sharing reduces the variance in p_L across random noise realizations and slightly increases the exponent coefficient A at larger d. Under uniform b-bit quantization of weights ($b = 3 - 7$), the threshold degrades by only $\sim$0.1%, demonstrating that

$$p_L(d, p; b\text{-bit}) - p_L(d, p; \infty\text{-bit}) = O\big(p^{\frac{d+1}{2}}\big)$$

remains small even for aggressive finite-precision constraints.

Hardware Implementation: A custom ASIC implementation of the MLP inference stage achieves sub-10 ns latency per cycle—well within a typical 440 ns error-correction window—and occupies less than 0.1 mm^2 at $d = 7$. On FPGA prototypes, pipelined execution of the two-layer network and the pure-error XOR tree sustains a $>$ 100 MHz decoding clock. Power consumption scales as $\Theta(d^2)$, indicating that beyond $d \approx 11$, resource trade-offs must be carefully balanced against fidelity gains.

Case Study 3.4.2: Graph Neural Networks (GNNs) in QEC Decoding [96]
Leveraging the detector graph topology via a GNN enables direct learning of spatial–temporal error correlations beyond local matching heuristics.

GNN Design and Data Representation: A space–time graph $G = (V, E)$ is constructed from syndrome events: each node i holds features (b_i, x_i, y_i, t_i) encoding stabilizer type and coordinates; edges connect nearest neighbors within radius r, weighted by $e_{ij} = 1/\sqrt{\Delta x^2 + \Delta y^2 + \Delta t^2}$ and pruned to max degree k. Seven graph-convolutional layers

$$\mathbf{h}_i^{(l+1)} = \mathrm{ReLU}\big(W_1 \mathbf{h}_i^{(l)} + W_2 \sum_{j \in \mathcal{N}(i)} e_{ij}\, \mathbf{h}_j^{(l)} + b\big)$$

A mean-pooling $\mathbf{h}_G = |V|^{-1} \sum_i \mathbf{h}_i^{(L)}$ feeds two MLP heads with sigmoid outputs for logical-flip probabilities $\hat{p}_X, \hat{p}_Z$. On training on Stim [97] and experimental data for repetition codes at $d = 3, 5, 25$ using Adam on binary cross-entropy. On surface-code simulation ($p = 10^{-3}$, $d \leq 9$), the GNN achieves logical-error rate scaling

$$p_L(d, p) \approx A\, p^{(d+1)/2}$$

with threshold $p_{\mathrm{th}} \approx 1\%$, outperforming MWPM in low-d regimes and matching its exponent. Inference complexity grows as $O(|V| + |E|) \approx O(d^2 t)$, substantially below the $O(d^3)$ of blossom matching.

Hardware Implementation: While GPU-accelerated implementations sustain real-time decoding for $d \leq 9$, ASIC or FPGA deployment requires fixed-size, quantized GNN kernels and sparse message-passing units to meet sub-100 ns latency per QEC cycle. Edge-pruning and degree capping are critical to bounding on-chip memory and compute.

Case Study 3.4.3: Transformers in QEC Decoding [37]
By leveraging the global receptive field of self-attention, the approach jointly processes space–time syndrome information, facilitating effective transfer learning across different quantum code distances. Syndromes are encoded on a $(D+1)^2 \times (D+1) \times T$ cubic grid with 6-dimensional feature vectors (X/Z check flags, syndrome bits, temporal markers). Tokens are projected to embeddings, 3-dimensional sinusoidal positional encodings are added, and then flattened to a 1-dimensional sequence. A Transformer encoder of L layers (MHSA + FFN) yields syndrome representations. A Transformer decoder ingests data-qubit positional embeddings and attends to encoder outputs, producing per-qubit error logits. Finally, an auxiliary global parity head pools

encoder features to predict overall syndrome parity. It is trained on simulated and experimental datasets at distance $d = 5$ using a mixed loss

$$\mathcal{L} = \mathcal{L}_{\text{local}} + \alpha\,\mathcal{L}_{\text{global}},$$

where $\mathcal{L}_{\text{local}}$ is binary cross-entropy per data qubit and $\mathcal{L}_{\text{global}}$ from pooled parity. Adam is used with warmup + cosine decay for 100 epochs. For new distances ($d = 7, 9, 11, 13$), fine-tuned pretrained weights for 10 epochs at reduced learning rate, saves $> 10\times$ training cost. The logical-error rate $p_L(d, p)$ is evaluated under a phenomenological noise model for distances $3 \leq d \leq 13$ and $p = 10^{-3} - 5 \times 10^{-2}$. Transformer-QEC achieves thresholds $p_{\text{th}} \approx 3.8\%$ and consistently outperforms UF, MWPM [36], and MLP [95] across all benchmarks. The logical scaling follows

$$p_L(d, p) \sim A\, p^{\lfloor (d+1)/2 \rfloor},$$

matching theoretical exponents while maintaining $O(N)$ attention complexity (with $N \approx D^2 T$). Transfer learning further reduces the on-chip training overhead by an order of magnitude.

3.5 Formal Verification in QEC

The development of large-scale FTQC systems depends critically on the correctness of the QEC infrastructure [98]. While simulation-based testing remains useful for empirical exploration, it fails to provide formal guarantees of correctness or to scale to systems involving hundreds or thousands of physical qubits. Formal verification offers a principled alternative: a mathematically rigorous methodology to prove that QEC codes and operations preserve the intended logical behavior despite the presence of noise and implementation imperfections. A robust verification methodology for QEC codes is generally structured around three foundational pillars: a domain-specific language for encoding QEC procedures, a formal semantics and logical assertion framework, and a deductive proof system for establishing correctness properties [99].

- **Domain-Specific Representation Language:** The first requirement is a structured language capable of expressing the components of a QEC protocol, viz., encoding, stabilizer measurements, syndrome-dependent branches, recovery procedures, and logical gate implementations. The language must support quantum data operations alongside quantum-classical control constructs such as conditionals and loops. Also, the language should allow stabilizers to be treated as primary objects, abstracting the hardware-level details of measurement circuits while enabling logical reasoning over code behavior. This abstraction makes it possible to model code families like repetition and surface codes independently of specific architectures.

- **Formal Semantics and Logical Assertions:** To reason about correctness, the language must be paired with a well-defined semantic model. Operational semantics describe how program states evolve during execution, typically over a combination of quantum states and stabilizer configurations. Denotational semantics provide a higher-level view of program behavior, enabling reasoning about loops and composition via fixed-point interpretations. Assertions are specified using logical predicates defined over stabilizers or Pauli expressions, serving as preconditions and postconditions for program operations. This approach avoids the intractable cost of general Hermitian-based predicate reasoning while preserving the ability to express meaningful correctness conditions for stabilizer codes.
- **Deductive Proof System for QEC Programs:** The final component is a sound, syntax-directed proof system for deriving correctness judgments over QEC programs. Structured as a quantum analog of Hoare logic [100, 101], the system defines how assertions are propagated through each language construct—initializations, unitary operations, stabilizer measurements, branching, and loops. The logic supports compositional reasoning, enabling the decomposition of large programs into verifiable subcomponents. Importantly, for operations within the Clifford group, the stabilizer-based formulation yields polynomial-time verification procedures, in contrast to the exponential complexity required by general-purpose quantum verification tools.

These aspects of formal verification with respect to QEC allow rigorous certification of code behavior under fault models, support reasoning about logical gate implementations and syndrome-based corrections, and enable scalable verification workflows for realistic quantum codes.

Case Study 3.5.1: QECV [99]
Language Design: The QECV framework introduces a domain-specific programming language, referred to as `QECV-Lang`, explicitly designed to model QEC routines using the stabilizer formalism as the core abstraction. In this language, stabilizers are treated as first-class program variables, and quantum operations are expressed in a high-level syntax that encompasses initialization, Clifford unitaries, Pauli measurements, and quantum-classical control constructs. Each program is composed of statements such as

$$\texttt{q} := |0\rangle, \quad \texttt{s} := +Z_1Z_2, \quad \texttt{if}\ M[s]\ \texttt{then}\ P_1\ \texttt{else}\ P_2,$$

where $s \in \mathcal{S}_n$ denotes a stabilizer expression over n qubits. This abstraction avoids low-level circuit specification and enables formal reasoning over logical operations, syndrome extraction, and feedback control. A companion assertion language, `QECV-Assn`, defines predicates over stabilizer relations, allowing properties of the encoded state to be captured compactly.

Encoding Procedures: Using `QECV-Lang`, canonical QEC protocols such as the 3-qubit repetition code and distance-3 surface code are encoded in a compositional manner. For example, the 3-qubit bit-flip code is described via the following high-level steps:

1. Initialize physical qubits: `q1 :=` $|0\rangle$, `q2 :=` $|0\rangle$, `q3 :=` $|0\rangle$,
2. Encode logical state with CNOTs: `q2 := CNOT(q1, q2), q3 := CNOT(q1, q3)`,
3. Define and measure stabilizers: $s_1 = Z_1 Z_2$, $s_2 = Z_2 Z_3$,
4. Apply syndrome-based correction: conditional application of X gates based on measured outcomes of s_1 and s_2.

Stabilizer assignments and correction branches are represented symbolically and reasoned over collectively, enabling verification under all syndrome outcomes in a unified symbolic execution trace. The abstraction naturally extends to more complex codes, such as surface codes, where stabilizers are represented by local Pauli products over 2D lattices and syndrome extraction is modeled through iterative conditional logic.

Proof Strategy: The QECV framework defines an operational semantics over configurations (ρ, σ), where ρ is a quantum state and σ maps stabilizer variables to Pauli strings. A corresponding denotational semantics defines the behavior of programs as superoperators acting on such configurations. The correctness of QEC programs is verified using a syntax-directed Hoare-style proof system over stabilizer assertions:

$$\{A\}\ P\ \{B\},$$

where A and B are assertions in `QECV-Assn`, such as logical stabilizers or error syndromes, and P is a `QECV-Lang` program. Correctness proofs were constructed for logical initialization, Clifford gates, and error correction routines for both the repetition code and the surface code. For instance, in the 3-qubit code, the system verifies that an injected single-qubit X error is successfully corrected and the logical information is preserved:

$$\{X_i \cdot \mathcal{L}\}\ P_{\text{corr}}\ \{\mathcal{L}\},$$

where $\mathcal{L}$ denotes the logical stabilizer and P_{corr} is the correction program. Importantly, QECV demonstrates polynomial-time verification of these protocols, enabled by the restriction to stabilizer operations and Clifford semantics. This contrasts with prior frameworks requiring exponential resources to represent general quantum states and predicates. Empirical benchmarks confirm that verifying distance-d surface code routines scales as $O(d^3)$, highlighting the tractability of the stabilizer-based formalism and its suitability for design automation in realistic fault-tolerant QEC pipelines.

Case Study 3.5.2: Verification of FTQC [102]
Language Design: Surface codes are a class of topological stabilizer codes defined on a planar lattice, where logical qubits are encoded in global degrees of freedom, and stabilizer checks are local and geometrically constrained. The error correction (EC) gadgets in surface codes consist of repeated syndrome measurements, ancilla-assisted Pauli measurements (e.g., using cat states or flag qubits), and classical feedback via decoders. The behavior of an EC gadget is described as a classical-quantum program (`cq-prog`) S, comprising quantum initializations, Clifford gates, mid-circuit measurements, classical branching, and repeat-until-success constructs. The fault-tolerance specification follows Definition 6(4): for any logical input state $|\psi\rangle$, any Pauli error of weight r, and any execution with s faults such that $r + s \leq t$, the output must remain within the space $\mathcal{S}_s(|\psi\rangle)$, where $t = \lfloor \frac{d-1}{2} \rfloor$ is the code's correctable threshold.

Encoding Procedure: The symbolic execution framework encodes all cq-prog transitions as symbolic stabilizer updates. Each configuration is represented by a tuple:

$$\langle S, \tilde{\sigma}, \tilde{\rho}, p, \varphi, \tilde{F} \rangle,$$

where $\tilde{\sigma}$ is the symbolic classical state, $\tilde{\rho}$ is a symbolic stabilizer state parameterized over Boolean symbols, φ is the accumulated path condition, and $\tilde{F}$ is a symbolic counter representing the number of injected faults.

Error injection is modeled symbolically by modifying stabilizer generators. For a symbolic state $\tilde{\rho} = \langle (-1)^{g_1} P_1, \ldots, (-1)^{g_m} P_m \rangle$, a fault injection on qubit q produces:

$$\tilde{\rho}' = \langle (-1)^{g_1 \oplus e \cdot c_1} P_1, \ldots, (-1)^{g_m \oplus e \cdot c_m} P_m \rangle,$$

where e is a new fault symbol and $c_i = 1$ if P_i anticommutes with the injected Pauli and 0 otherwise. This allows symbolic tracking of error propagation without enumeration.

At the end of each execution path, the symbolic output $\tilde{\rho}'$ is compared with the ideal fault-free output ρ_{ideal}. Fault-tolerance is certified if, for all satisfying valuations $\mathbf{s}$, the Pauli distance $D(\tilde{\rho}', \rho_{\text{ideal}})$ is bounded:

$$\forall \mathbf{s}, \quad \varphi(\mathbf{s}) \Rightarrow D(\tilde{\rho}', \rho_{\text{ideal}}) \leq \tilde{F}(\mathbf{s}).$$

Here, $D(\cdot, \cdot)$ computes the minimal weight of a Pauli operator transforming one stabilizer state into another. This constraint is encoded into SMT queries using symplectic vector representations of the Pauli group.

Proof Strategy: The framework provides a complete and automated method for certifying fault-tolerance of surface code EC gadgets. For memoryless loops, the method is both sound and complete; for conservative loops, soundness is preserved under local non-propagation of errors. Verification produces either a formal certificate or a counterexample trace highlighting a violation of fault-tolerance. This capability allows both design-time validation of error correction protocols and comparative analysis of EC strategies, such as differing ancilla schemes or decoder behaviors. The framework also supports symbolic abstraction of decoder logic, making it scalable to realistic surface-code gadgets with dozens of physical qubits.

4 FTQC Architectures

The next step to establishing QEC codes is the integration of the entire design workflow in an FTQC framework. There have been recent studies showing how this can be achieved [103, 104]. Based on the existing research, we can say that any near-term FTQC architecture built upon surface codes should integrate the following structural layers:

- **Logical-Qubit Fabric:** A 2D lattice of distance-d surface-code patches, each comprising a $d \times d$ array of data qubits interleaved with ancilla qubits for X- and Z-stabilizer extraction every QEC cycle. Logical Clifford operations are performed via lattice-surgery merges and splits of adjacent patches, implementing $CNOT$, H, and S gates with $O(d)$ time overhead and local syndrome updates, while the fabric's native primitives for Pauli-product measurements—realized through correlated ancilla circuits or twist-defect sequences—enable arbitrary non-Clifford rotations when combined with magic-state injection.
- **Magic-State Factory:** Dedicated distillation blocks (e.g., Bravyi–Kitaev 15-to-1 magic state distillation protocol [54]) consume noisy $|T\rangle$ injections and output states with error $O(p^3)$, where p is the physical gate error rate; these blocks occupy an $O(d) \times O(d)$ footprint of surface-code patches arranged in a pipelined sequence of distillation rounds to produce one high-fidelity $|T\rangle$ every $O(d)$ code cycles; injection channels—surface-code ancilla chains—then ferry distilled $|T\rangle$ states into the compute fabric with bounded routing latency.
- **Interconnects/Routing Layer:** A fault-tolerant ancilla network, realized either as a fixed "bus" of qubits [104] or as a mesh of surface-code ancilla patches [103], supports logical-state teleportation and multi-qubit Pauli-product measurements; teleportation primitives comprise Bell-pair preparation, joint Bell-basis measure-

ment, and feed-forward Pauli corrections, each executed in $O(d)$ cycles; and configurable routing regions, either statically provisioned for predictable bandwidth or dynamically reconfigurable to adapt ancilla pools to runtime congestion, ensure balanced connectivity across compute, factory, and memory segments.

- **Classical Compilation, Mapping and Scheduling Layer:** A compiler front end transpiles high-level Clifford+T circuits into sequences of native Pauli-based operations and lattice-surgery primitives; mapping algorithms embed the logical-qubit interaction graph onto the 2D patch fabric, minimizing SWAP or braiding overhead under connectivity constraints; and a scheduler allocates time slots to each operation, respecting magic-state supply, ancilla bandwidth, and asynchronous logical-clock domains—and resolves resource contention via critical-path or list-scheduling heuristics.
- **Qubit Memory Layout:** This is a layer specific to qLDPC codes introduced in [104–106]. A high-density logical-qubit storage layer employs qLDPC codes (e.g., an $[[n, k, d]]$ gross code) to encode k qubits with low overhead, supporting only Clifford-type operations; a fault-tolerant teleportation interface between memory and compute regions is realized by preparing and measuring Bell pairs across the surface-code and qLDPC patches in $O(d)$ cycles; and a trade-off analysis between data-movement cost and qubit overhead places cold data in the qLDPC tier to minimize total patch count while keeping hot data in the surface-code fabric for rapid gate access.

We discuss two of the most recent heterogeneous and homogeneous architectures proposed for near-term FTQC in the following case studies.

Case Study 4.1: HetEC [104]
Logical-Qubit Fabric: At the heart of HetEC lies a checkerboard of planar surface-code patches, each of dimension $d \times d$, encoding one logical qubit of distance d. Nearest-neighbor stabilizer measurements enforce error correction with a pseudo-threshold around 0.7%. Clifford gates are realized via lattice surgery—merging and splitting patches—while non-Clifford T gates are executed by injecting distilled magic states into these patches. Logical error rates under surface-code operations scale as $\sim 0.03\,(p/0.01)^{(d+1)/2}$, and each Pauli-rotation requires between d and $3d$ measurement rounds depending on its weight.

Magic-State Factory: A dedicated magic-state factory sits adjacent to the compute fabric, continuously distilling $|T\rangle$ ancillas. By decoupling magic-state distillation from the main surface-code region, HetEC maintains a steady supply of high-fidelity non-Clifford resources without bloating the compute area. Distilled $|T\rangle$ states are teleported into compute patches via lattice-surgery protocols, preserving universality while containing overhead within the factory block.

Ancilla Bus for Inter-Code Data Movement: Compute and memory regions communicate through a 103-qubit ancilla bus, which interfaces with a single "port" qubit on each qLDPC memory block. Within the gross code, logical automorphism gates permute any target qubit to this port in just two syndrome cycles. Thereafter, a joint XX (or ZZ) measurement over the bus teleports the qubit state fault-tolerantly between memory and compute. This bi-directional teleportation mechanism underpins all inter-code data transfers.

HetEC Transpiler and Scheduler: The HetEC transpiler accepts high-level Clifford+T circuits and performs:

1. Decomposition into weighted Pauli-product rotations.
2. Constraint-aware pruning and commuting of Clifford rotations to fit surface-code capacity.
3. Scheduling of logical operations, data-movement calls, and in-memory Clifford executions to minimize teleportation steps and respect asynchronous syndrome clocks.

This compilation flow balances compute-region limits against memory-region efficiency, reducing costly inter-code transfers.

Memory layout: Quantum memory is provided by a [[144, 12, 12]] qLDPC CSS gross code, encoding 12 logical qubits with a 1/24 overhead and a ~0.65% physical threshold. One logical qubit serves as the bus port; the remaining 11 support native Clifford gates via in-block automorphism circuits (parallel CNOT layers), each completing in ~14 syndrome cycles. By executing most Clifford operations in memory, HetEC offloads the compute fabric and reduces overall qubit pressure.

Integration: By co-designing heterogeneous codes, an ancilla bus, and a specialized compiler, HetEC achieves up to a 6.42× reduction in physical qubits at the cost of a 3.43× slower logical-clock depth for benchmark circuits.

Case Study 4.2: SPARO [103]

Logical-Qubit Fabric: SPARO assumes a uniform array of rotated surface-code patches, each of code distance d, supporting nearest-neighbor stabilizer checks and lattice-surgery operations for both Clifford and Pauli-product measurements (PPMs). Multi-qubit PPM errors are modeled as

$$P_{\mathrm{op}}^{(t)} = f\big(n_q^{(t)},\ \ell_{\mathrm{anc}}^{(t)},\ \mathrm{op\_type}\big),$$

where $n_q^{(t)}$ is the number of qubits involved in layer t, $\ell_{\mathrm{anc}}^{(t)}$ is the ancilla-path length, and op_type distinguishes X/Z versus Y measurements. Single- and two-qubit depolarizing channels are incorporated (with two-qubit error probability $p_{\mathrm{dep2}}/16$), and patch rotations to realign boundaries incur additional error and latency.

Magic-State Factory: The baseline SPARO layout includes a single magic-state factory consuming 11 tiles to run a 15-to-1 distillation protocol, plus up to 2 ancillary tiles for Y-measurement decomposition if twist defects are unavailable. Magic states are produced at discrete timesteps (every 11 syndrome cycles) and qubits requiring a T-gate may idle while waiting, contributing a layer-wise idle error

$$P_{\mathrm{idle}}^{(t)} = 1 - \prod_{i=1}^{n_{\mathrm{idle}}^{(t)}} (1 - p_{i,t}),$$

with $p_{i,t}$ the physical idling error rate for qubit i in layer t.

Routing Region and Ancilla Pathfinding: Rather than a fixed bus, SPARO embeds a routing subregion next to compute patches. For each PPM, ancilla qubits traverse this region; SPARO formulates ancilla-path minimization as a Steiner-tree problem on the routing graph, pruning candidate Steiner vertices by Manhattan distance before invoking a classical solver to yield the shortest path length ℓ_{anc}. Shorter ℓ_{anc} directly reduces both the operation error and latency.

End-to-End PBC Pipeline: SPARO's compiler proceeds in three key phases:

1. **PPM Decomposition and Pathfinding**: Translates Clifford+T circuits into weighted Pauli rotations and computes minimal ancilla paths via Steiner trees.
2. **Operation Scheduling**: Rotation scheduling minimizes basis-shift overhead by reordering Y-measurements within each layer. Measurement parallelization builds an interaction graph of PPMs and applies graph-coloring to maximize concurrent measurements.
3. **Qubit Mapping**: Performs a greedy placement of high-use qubits near factories, then refines via simulated annealing to minimize a cost function based on total Manhattan distance between interacting qubits.

Memory Layout: In SPARO, "memory" is implicit: qubits not participating in a given layer simply idle on the surface code. The comprehensive logical-error model then computes for each layer t:

$$p_L^{(t)} = 1 - \left(1 - P_{\mathrm{op}}^{(t)}\right)\left(1 - P_{\mathrm{idle}}^{(t)}\right), \quad p_{\mathrm{total}}^{L} = 1 - \prod_{t=1}^{T} \left(1 - p_L^{(t)}\right),$$

capturing failures from active PPMs, rotations, and idling.

Integration: SPARO first quantifies total time and error contributions per category over all layers—$(T_{\mathrm{total}}^{\mathrm{WaitT}}, E_{\mathrm{total}}^{\mathrm{WaitT}})$, $(T_{\mathrm{total}}^{\mathrm{Rotation}}, E_{\mathrm{total}}^{\mathrm{Rotation}})$, and $(T_{\mathrm{total}}^{\mathrm{Op}}, E_{\mathrm{total}}^{\mathrm{Op}})$.

It then estimates the marginal error-reduction yield of adding one unit of resource area to either factories or routing:

$$\frac{|\Delta E_{\text{total}}^{\text{WaitT}}|}{A_{\text{factory}}} \quad \text{vs.} \quad \frac{|\Delta E_{\text{total}}^{\text{Op}} + \Delta E_{\text{total}}^{\text{Rotation}}|}{A_{\text{routing}}}$$

and greedily allocates resources to whichever option maximizes $\Delta p_{\text{total}}^{L}/\Delta A$. After each allocation step, SPARO re-runs affected compilation phases to update bottleneck statistics. Applying this dynamic approach under a fixed total-tile budget, SPARO achieves up to 51.11% reduction in logical-error rate for a 433-qubit adder—outperforming both the Minimal baseline and an Intermediate static layout ($\sim$30% extra tiles) across multiple benchmarks.

5 Future of FTQC Research

Future work should focus on bringing improvement to the existing techniques of design automation, as well as exploring other angles on reducing the runtime for current optimization ideas in QEC implementation, and modifying the near-term FTQC architecture to accommodate multiple qubit technologies to improve the flow of design. Additionally, incorporating modifications in the modern QEC codes for accounting for defects introduced due to radiation and calibration errors in quantum computers and accelerating the decoding algorithms using Application-Specific Integrated Circuits (ASICs) and Field-Programmable Gate Arrays (FPGAs) are plausible directions for progressing toward an FTQC future.

Acknowledgements The work is supported in parts by NSF (CNS-2129675, CCF-2210963), gifts from Intel and IBM Quantum Credits.

Competing Interests The authors have no conflicts of interest to declare that are relevant to the content of this chapter.

References

1. Chatterjee, A., K. Phalak, and S. Ghosh. 2023. Quantum error correction for dummies. In *2023 IEEE International Conference on Quantum Computing and Engineering (QCE)*, vol. 01, 70–81.
2. Clerk, A. A., M. H. Devoret, S. M. Girvin, F. Marquardt, and R. J. Schoelkopf. 2010. Introduction to quantum noise, measurement, and amplification. *Reviews of Modern Physics* 82: 1155–1208. [Online]. https://link.aps.org/doi/10.1103/RevModPhys.82.1155.
3. Preskill, J. 2018. Quantum computing in the nisq era and beyond. *Quantum* 2: 79. [Online]. http://dx.doi.org/10.22331/q-2018-08-06-79.

4. Gottesman, D. 1997. Stabilizer codes and quantum error correction. [Online]. https://arxiv.org/abs/quant-ph/9705052.
5. Fowler, A. G., M. Mariantoni, J. M. Martinis, and A. N. Cleland. 2012. Surface codes: Towards practical large-scale quantum computation. *Physical Review A* 86 (3). [Online]. http://dx.doi.org/10.1103/PhysRevA.86.032324.
6. Wootters, W. K., and W. H. Zurek. 1982. A single quantum cannot be cloned. *Nature* 299 (5886): 802–803. [Online]. https://doi.org/10.1038/299802a0.
7. Hamming, R. W. 1950. Error detecting and error correcting codes. *Bell System Technical Journal* 29 (2): 147–160.
8. Bose, R., and D. Ray-Chaudhuri. 1960. On a class of error correcting binary group codes. *Information and Control* 3 (1): 68–79. [Online]. https://www.sciencedirect.com/science/article/pii/S0019995860902874.
9. Reed, I. S., and G. Solomon. 1960. Polynomial codes over certain finite fields. *Journal of the Society for Industrial and Applied Mathematics* 8 (2): 300–304. [Online]. https://doi.org/10.1137/0108018.
10. Jozsa, R. 2002. A stronger no-cloning theorem. [Online]. https://arxiv.org/abs/quant-ph/0204153.
11. von Neumann, J. 2018. *Mathematical Foundations of Quantum Mechanics: New Edition*, vol. 53. Princeton, NJ: Princeton University Press.
12. Dauphinais, G., D. W. Kribs, and M. Vasmer. 2024. Stabilizer formalism for operator algebra quantum error correction. *Quantum* 8: 1261. [Online]. http://dx.doi.org/10.22331/q-2024-02-21-1261.
13. Reed, M. D., L. DiCarlo, S. E. Nigg, L. Sun, L. Frunzio, S. M. Girvin, and R. J. Schoelkopf. 2012. Realization of three-qubit quantum error correction with superconducting circuits. *Nature* 482 (7385): 382–385. [Online]. http://dx.doi.org/10.1038/nature10786.
14. Knill, E., and R. Laflamme. 1996. Concatenated quantum codes. [Online]. https://arxiv.org/abs/quant-ph/9608012.
15. Gottesman, D. 1996. Pasting quantum codes. [Online]. https://arxiv.org/abs/quant-ph/9607027.
16. Grassl, M., P. Shor, G. Smith, J. Smolin, and B. Zeng. 2009. Generalized concatenated quantum codes. *Physical Review A* 79 (5). [Online]. http://dx.doi.org/10.1103/PhysRevA.79.050306.
17. Steane, A. M. 1996. Simple quantum error-correcting codes. *Physical Review A* 54 (6): 4741–4751. [Online]. http://dx.doi.org/10.1103/PhysRevA.54.4741.
18. Bombin, H. 2013. An introduction to topological quantum codes. [Online]. https://arxiv.org/abs/1311.0277.
19. Kitaev, A. 2003. Fault-tolerant quantum computation by anyons. *Annals of Physics* 303 (1): 2–30. [Online]. http://dx.doi.org/10.1016/S0003-4916(02)00018-0.
20. Freedman, M. H., and D. A. Meyer. 1998. Projective plane and planar quantum codes. [Online]. https://arxiv.org/abs/quant-ph/9810055.
21. Acharya, R., et al. 2024. Quantum error correction below the surface code threshold. *Nature* 638 (8052): 920–926. [Online]. http://dx.doi.org/10.1038/s41586-024-08449-y.
22. Aly, S. A., A. Klappenecker, and P. K. Sarvepalli. 2006. Subsystem codes. [Online]. https://arxiv.org/abs/quant-ph/0610153.
23. Poulin, D. 2005. Stabilizer formalism for operator quantum error correction. *Physical Review Letters* 95 (23). [Online]. http://dx.doi.org/10.1103/PhysRevLett.95.230504.
24. Bacon, D. 2006. Operator quantum error-correcting subsystems for self-correcting quantum memories. *Physical Review A* 73 (1). [Online]. http://dx.doi.org/10.1103/PhysRevA.73.012340.
25. Shor, P. W. 1995. Scheme for reducing decoherence in quantum computer memory. *Physical Review A* 52: R2493–R2496. [Online]. https://link.aps.org/doi/10.1103/PhysRevA.52.R2493.
26. Bacon, D. 2003. Decoherence, control, and symmetry in quantum computers. [Online]. https://arxiv.org/abs/quant-ph/0305025.

27. Grimsmo, A. L., J. Combes, and B. Q. Baragiola. 2020. Quantum computing with rotation-symmetric bosonic codes. *Physical Review X* 10 (1). [Online]. http://dx.doi.org/10.1103/PhysRevX.10.011058.
28. Conrad, J. 2024. The fabulous world of gkp codes. Ph.D. dissertation. [Online]. https://refubium.fu-berlin.de/handle/fub188/45792.
29. Conrad, J., J. Eisert, and F. Arzani. 2022. Gottesman-kitaev-preskill codes: A lattice perspective. *Quantum* 6: 648. [Online]. http://dx.doi.org/10.22331/q-2022-02-10-648.
30. Dodonov, V., I. Malkin, and V. Man'ko. 1974. Even and odd coherent states and excitations of a singular oscillator. *Physica* 72 (3): 597–615. [Online]. https://www.sciencedirect.com/science/article/pii/0031891474902158.
31. Castaños, O., R. López-Peña, and V. I. Man'ko. 1995. Crystallized schrödinger cat states. *Journal of Russian Laser Research* 16: 477–525.
32. Albert, V. V., K. Noh, K. Duivenvoorden, D. J. Young, R. T. Brierley, P. Reinhold, C. Vuillot, L. Li, C. Shen, S. M. Girvin, B. M. Terhal, and L. Jiang. 2018. Performance and structure of single-mode bosonic codes. *Physical Review A* 97 (3). [Online]. http://dx.doi.org/10.1103/PhysRevA.97.032346.
33. Rivera, N., J. Sloan, Y. Salamin, J. D. Joannopoulos, and M. Soljačić. 2023. Creating large fock states and massively squeezed states in optics using systems with nonlinear bound states in the continuum. *Proceedings of the National Academy of Sciences* 120 (9). [Online]. http://dx.doi.org/10.1073/pnas.2219208120.
34. Chatterjee, A., S. Das, and S. Ghosh. 2025. Lattice surgery for dummies. *Sensors* 25 (6): 1854. [Online]. http://dx.doi.org/10.3390/s25061854.
35. Wu, Y., N. Liyanage, and L. Zhong. 2024. Lego: Qec decoding system architecture for dynamic circuits. [Online]. https://arxiv.org/abs/2410.03073.
36. Wu, Y., and L. Zhong. 2023. Fusion blossom: Fast mwpm decoders for qec. [Online]. https://arxiv.org/abs/2305.08307
37. Wang, H., P. Liu, K. Shao, D. Li, J. Gu, D. Z. Pan, Y. Ding, and S. Han. 2023. Transformer-qec: Quantum error correction code decoding with transferable transformers. [Online]. https://arxiv.org/abs/2311.16082
38. Eastin, B., and E. Knill. 2009. Restrictions on transversal encoded quantum gate sets. *Physical Review Letters* 102 (11). [Online]. http://dx.doi.org/10.1103/PhysRevLett.102.110502.
39. Dua, A., A. Kubica, L. Jiang, S. T. Flammia, and M. J. Gullans. 2024. Clifford-deformed surface codes. *PRX Quantum* 5 (1). [Online]. http://dx.doi.org/10.1103/PRXQuantum.5.010347.
40. Litinski, D. 2019. A game of surface codes: Large-scale quantum computing with lattice surgery. *Quantum* 3: 128. [Online]. http://dx.doi.org/10.22331/q-2019-03-05-128.
41. Bombin, H., and M. A. Martin-Delgado. 2007. Optimal resources for topological two-dimensional stabilizer codes: Comparative study. *Physical Review A* 76 (1). [Online]. http://dx.doi.org/10.1103/PhysRevA.76.012305.
42. Anderson, J. T. 2011. Homological stabilizer codes. [Online]. https://arxiv.org/abs/1107.3502.
43. Wen, X.-G. 2003. Quantum orders in an exact soluble model. *Physical Review Letters* 90 (1). [Online]. http://dx.doi.org/10.1103/PhysRevLett.90.016803.
44. Bonilla Ataides, J. P., D. K. Tuckett, S. D. Bartlett, S. T. Flammia, and B. J. Brown. 2021. The xzzx surface code. *Nature Communications* 12 (1). [Online]. http://dx.doi.org/10.1038/s41467-021-22274-1.
45. Kovalev, A. A., I. Dumer, and L. P. Pryadko. 2011. Design of additive quantum codes via the code-word-stabilized framework. *Physical Review A* 84 (6). [Online]. http://dx.doi.org/10.1103/PhysRevA.84.062319.
46. Nagayama, S., A. G. Fowler, D. Horsman, S. J. Devitt, and R. V. Meter. 2017. Surface code error correction on a defective lattice. *New Journal of Physics* 19 (2): 023050. [Online]. http://dx.doi.org/10.1088/1367-2630/aa5918.
47. Brown, B. J., K. Laubscher, M. S. Kesselring, and J. R. Wootton. 2017. Poking holes and cutting corners to achieve Clifford gates with the surface code. *Physical Review X* 7 (2). [Online]. http://dx.doi.org/10.1103/PhysRevX.7.021029.

48. Litinski, D., and F. v. Oppen. 2018. Lattice surgery with a twist: Simplifying Clifford gates of surface codes. *Quantum* 2: 62. [Online]. http://dx.doi.org/10.22331/q-2018-05-04-62.
49. Horsman, D., A. G. Fowler, S. Devitt, and R. V. Meter. 2012. Surface code quantum computing by lattice surgery. *New Journal of Physics* 14 (12): 123011. [Online]. http://dx.doi.org/10.1088/1367-2630/14/12/123011.
50. Lai, C.-Y., and H.-C. Cheng. 2022. Learning quantum circuits of some t gates. *IEEE Transactions on Information Theory* 68 (6): 3951–3964. [Online]. http://dx.doi.org/10.1109/TIT.2022.3151760.
51. Coecke, B., and Q. Wang. 2018. Zx-rules for 2-qubit Clifford+t quantum circuits. [Online]. https://arxiv.org/abs/1804.05356.
52. Pllaha, T., K. Volanto, and O. Tirkkonen. 2021. Decomposition of Clifford gates. In *2021 IEEE Global Communications Conference (GLOBECOM)*. IEEE, 01–06. [Online]. http://dx.doi.org/10.1109/GLOBECOM46510.2021.9685501.
53. Patel, K. N., I. L. Markov, and J. P. Hayes. 2003. Efficient synthesis of linear reversible circuits. [Online]. https://arxiv.org/abs/quant-ph/0302002.
54. Litinski, D. 2019. Magic state distillation: Not as costly as you think. *Quantum* 3: 205. [Online]. http://dx.doi.org/10.22331/q-2019-12-02-205.
55. Dawson, C. M., and M. A. Nielsen. 2005. The Solovay-Kitaev algorithm. [Online]. https://arxiv.org/abs/quant-ph/0505030.
56. Ross, N. J., and P. Selinger. 2016. Optimal ancilla-free Clifford+t approximation of z-rotations. [Online]. https://arxiv.org/abs/1403.2975.
57. Bravyi, S., and A. Kitaev. 2005. Universal quantum computation with ideal Clifford gates and noisy ancillas. *Physical Review A* 71 (2). [Online]. http://dx.doi.org/10.1103/PhysRevA.71.022316.
58. Campbell, E. T., B. M. Terhal, and C. Vuillot. 2017. Roads towards fault-tolerant universal quantum computation. *Nature* 549 (7671): 172–179. [Online]. http://dx.doi.org/10.1038/nature23460.
59. Howard, M., J. Wallman, V. Veitch, and J. Emerson. 2014. Contextuality supplies the 'magic' for quantum computation. *Nature* 510 (7505): 351–355. [Online]. http://dx.doi.org/10.1038/nature13460.
60. Knill, E. 2004. Fault-tolerant postselected quantum computation: Schemes. [Online]. https://arxiv.org/abs/quant-ph/0402171.
61. Bravyi, S., and J. Haah. 2012. Magic-state distillation with low overhead. *Physical Review A* 86 (5). [Online]. http://dx.doi.org/10.1103/PhysRevA.86.052329.
62. Bravyi, S., and A. Kitaev. 2005. Universal quantum computation with ideal clifford gates and noisy ancillas. *Physical Review A* 71 (2). [Online]. http://dx.doi.org/10.1103/PhysRevA.71.022316.
63. Bartlett, S. D. 2014. Powered by magic. *Nature* 510 (7505): 345–347. [Online]. https://doi.org/10.1038/nature13504.
64. Guruswami, V., L. Jin, and C. Xing. 2017. Efficiently list-decodable punctured reed-muller codes. [Online]. https://arxiv.org/abs/1508.00603.
65. Park, M., N. Maskara, M. Kalinowski, and M. D. Lukin. 2025. Enhancing quantum memory lifetime with measurement-free local error correction and reinforcement learning. *Physical Review A* 111: 012419. [Online]. https://link.aps.org/doi/10.1103/PhysRevA.111.012419.
66. Cirel'son, B. S. 1978. Reliable storage of information in a system of unreliable components with local interactions. In *Locally Interacting Systems and Their Application in Biology*, ser. Lecture Notes in Mathematics, vol. 653, ed. R. L. Dobrushin, V. I. Kryukov, and A. L. Toom, 15–30. Berlin, Heidelberg: Springer.
67. Wu, Y., N. Liyanage, and L. Zhong. 2025. Micro blossom: Accelerated minimum-weight perfect matching decoding for quantum error correction. In *Proceedings of the 30th ACM International Conference on Architectural Support for Programming Languages and Operating Systems, Volume 2*, ser. ASPLOS '25. ACM, 639–654. [Online]. http://dx.doi.org/10.1145/3676641.3716005.

68. Lee, S.-H., A. Li, and S. D. Bartlett. 2025. Color code decoder with improved scaling for correcting circuit-level noise. *Quantum* 9: 1609. [Online]. http://dx.doi.org/10.22331/q-2025-01-27-1609.
69. Sarvepalli, P. K., and A. Klappenecker. 2008. Encoding subsystem codes. [Online]. https://arxiv.org/abs/0806.4954.
70. Das, P., A. Locharla, and C. Jones. 2021. Lilliput: A lightweight low-latency lookup-table based decoder for near-term quantum error correction. [Online]. https://arxiv.org/abs/2108.06569.
71. Chan, T., and S. C. Benjamin. 2023. Actis: A strictly local union–find decoder. *Quantum* 7: 1183. [Online]. http://dx.doi.org/10.22331/q-2023-11-14-1183.
72. Griffiths, S. J., and D. E. Browne. 2024. Union-find quantum decoding without union-find. *Physical Review Research* 6 (1). [Online]. http://dx.doi.org/10.1103/PhysRevResearch.6.013154.
73. Valentino, F., B. Branchini, D. Conficconi, D. Sciuto, and M. D. Santambrogio. 2025. Quekuf: an fpga union find decoder for quantum error correction on the toric code. *ACM Transactions on Reconfigurable Technology and Systems*, just Accepted. [Online]. https://doi.org/10.1145/3733239.
74. Yao, H., W. A. Laban, C. Häger, A. G. i Amat, and H. D. Pfister. 2024. Belief propagation decoding of quantum ldpc codes with guided decimation. [Online]. https://arxiv.org/abs/2312.10950.
75. Kuo, K.-Y., and C.-Y. Lai. 2024. Fault-tolerant belief propagation for practical quantum memory. [Online]. https://arxiv.org/abs/2409.18689.
76. Huang, G., J. Hu, Y. He, J. Liu, M. Ma, Z. Shen, J. Wu, Y. Xu, H. Zhang, K. Zhong, X. Ning, Y. Ma, H. Yang, B. Yu, H. Yang, and Y. Wang. 2021. Machine learning for electronic design automation: A survey. *ACM Transactions on Design Automation of Electronic Systems* 26 (5). [Online]. https://doi.org/10.1145/3451179.
77. MacMillen, D., R. Camposano, D. Hill, and T. Williams. 2000. An industrial view of electronic design automation. *IEEE Transactions on Computer-Aided Design of Integrated Circuits and Systems* 19 (12): 1428–1448.
78. Pan, J., G. Zhou, C.-C. Chang, I. Jacobson, J. Hu, and Y. Chen. 2025. A survey of research in large language models for electronic design automation. *ACM Transactions on Design Automation of Electronic Systems* 30 (3). [Online]. https://doi.org/10.1145/3715324.
79. Chen, L., et al.. 2024. Large circuit models: Opportunities and challenges. *Science China Information Sciences* 67 (10). [Online]. http://dx.doi.org/10.1007/s11432-024-4155-7.
80. Ghosh, A., A. Chatterjee, and S. Ghosh. 2025. Survival of the optimized: An evolutionary approach to t-depth reduction. [Online]. https://arxiv.org/abs/2504.09391.
81. Chatterjee, A., A. Ghosh, and S. Ghosh. 2025. Quantum prometheus: Defying overhead with recycled ancillas in quantum error correction. In *2025 26th International Symposium on Quality Electronic Design (ISQED)*, 1–7.
82. Shor, P. W. 1997. Fault-tolerant quantum computation. [Online]. https://arxiv.org/abs/quant-ph/9605011.
83. Siegel, A., A. Strikis, T. Flatters, and S. Benjamin. 2023. Adaptive surface code for quantum error correction in the presence of temporary or permanent defects. *Quantum* 7: 1065. [Online]. http://dx.doi.org/10.22331/q-2023-07-25-1065.
84. Yin, K., X. Fang, Y. Shi, T. Humble, A. Li, and Y. Ding. 2024. Surf-deformer: Mitigating dynamic defects on surface code via adaptive deformation. [Online]. https://arxiv.org/abs/2405.06941.
85. Suzuki, Y., T. Sugiyama, T. Arai, W. Liao, K. Inoue, and T. Tanimoto. 2022. Q3de: A fault-tolerant quantum computer architecture for multi-bit burst errors by cosmic rays. In *2022 55th IEEE/ACM International Symposium on Microarchitecture (MICRO)*. IEEE, 1110–1125. [Online]. http://dx.doi.org/10.1109/MICRO56248.2022.00079
86. Amy, M., D. Maslov, and M. Mosca. 2014. Polynomial-time t-depth optimization of Clifford+t circuits via matroid partitioning. *IEEE Transactions on Computer-Aided Design of Integrated Circuits and Systems* 33 (10): 1476–1489. [Online]. http://dx.doi.org/10.1109/TCAD.2014.2341953.

87. Vandaele, V. 2024. Lower t-count with faster algorithms. [Online]. https://arxiv.org/abs/2407.08695.
88. Heyfron, L., and E. T. Campbell. 2018. An efficient quantum compiler that reduces t count. [Online]. https://arxiv.org/abs/1712.01557.
89. Kissinger, A., and J. van de Wetering. 2020. Reducing the number of non-clifford gates in quantum circuits. *Physical Review A* 102 (2). [Online]. http://dx.doi.org/10.1103/PhysRevA.102.022406.
90. Coecke, B., and R. Duncan. 2008. Interacting quantum observables. In *Automata, Languages and Programming*, ed. L. Aceto, I. Damgård, L. A. Goldberg, M. M. Halldórsson, A. Ingólfsdóttir, and I. Walukiewicz, 298–310. Berlin, Heidelberg: Springer.
91. Hao, T., A. Xu, and S. Tannu. 2025. Reducing t gates with unitary synthesis. [Online]. https://arxiv.org/abs/2503.15843.
92. Chatterjee, A., A. Ghosh, and S. Ghosh. 2025. The q-spellbook: Crafting surface code layouts and magic state protocols for large-scale quantum computing. [Online]. https://arxiv.org/abs/2502.11253.
93. Bhoumik, D., R. Majumdar, D. Madan, and S. Sur-Kolay. 2024. Machine learning based decoding of heavy hexagonal qecc for asymmetric quantum noise. In *IEEE Computer Society Annual Symposium on VLSI (ISVLSI)* 2024: 246–251.
94. Wang, Z., and H. Tang. 2024. Artificial intelligence for quantum error correction: A comprehensive review. [Online]. https://arxiv.org/abs/2412.20380.
95. Overwater, R. W. J., M. Babaie, and F. Sebastiano. 2022. Neural-network decoders for quantum error correction using surface codes: A space exploration of the hardware cost-performance tradeoffs. *IEEE Transactions on Quantum Engineering* 3: 1–19. [Online]. http://dx.doi.org/10.1109/TQE.2022.3174017.
96. Gong, A., S. Cammerer, and J. M. Renes. 2023. Graph neural networks for enhanced decoding of quantum ldpc codes. [Online]. https://arxiv.org/abs/2310.17758.
97. Gidney, C. 2021. Stim: A fast stabilizer circuit simulator. *Quantum* 5: 497. [Online]. https://doi.org/10.22331/q-2021-07-06-497.
98. Huang, Q., L. Zhou, W. Fang, M. Zhao, and M. Ying. 2025. Efficient formal verification of quantum error correcting programs. *Proceedings of the ACM on Programming Languages* 9 (PLDI), 1068–1093. [Online]. http://dx.doi.org/10.1145/3729293.
99. Wu, A., G. Li, H. Zhang, G. G. Guerreschi, Y. Xie, and Y. Ding. 2021. Qecv: Quantum error correction verification. [Online]. https://arxiv.org/abs/2111.13728.
100. Hoare, C. A. R. 1969. An axiomatic basis for computer programming. *Communications of the ACM* 12 (10): 576–580. [Online]. https://doi.org/10.1145/363235.363259.
101. Apt, K. R., and E.-R. Olderog. 2019. Fifty years of hoare's logic. *Form. Asp. Comput.* 31 (6): 751–807. [Online]. https://doi.org/10.1007/s00165-019-00501-3.
102. Chen, K., Y. Liu, W. Fang, J. Paykin, X.-C. Wu, A. Schmitz, S. Zdancewic, and G. Li. 2025. Verifying fault-tolerance of quantum error correction codes. [Online]. https://arxiv.org/abs/2501.14380.
103. Kan, S., Z. Du, C. Liu, M. Wang, Y. Ding, A. Li, Y. Mao, and S. Stein. 2025. Sparo: Surface-code pauli-based architectural resource optimization for fault-tolerant quantum computing. [Online]. https://arxiv.org/abs/2504.21854.
104. Stein, S., S. Xu, A. W. Cross, T. J. Yoder, A. Javadi-Abhari, C. Liu, K. Liu, Z. Zhou, C. Guinn, Y. Ding, Y. Ding, and A. Li. 2025. Hetec: Architectures for heterogeneous quantum error correction codes. In *Proceedings of the 30th ACM International Conference on Architectural Support for Programming Languages and Operating Systems, Volume 2*, ser. ASPLOS '25. New York, NY, USA: Association for Computing Machinery, 515–528. [Online]. https://doi.org/10.1145/3676641.3716001.
105. Bravyi, S., A. W. Cross, J. M. Gambetta, D. Maslov, P. Rall, and T. J. Yoder. 2024. High-threshold and low-overhead fault-tolerant quantum memory. *Nature* 627 (8005), 778–782. [Online]. http://dx.doi.org/10.1038/s41586-024-07107-7.
106. He, Z., A. Cowtan, D. J. Williamson, and T. J. Yoder. 2025. Extractors: Qldpc architectures for efficient pauli-based computation. [Online]. https://arxiv.org/abs/2503.10390.

Quantum Verification and Debug

Verification Paradigms for Quantum Computing

Navnil Choudhury, Ameya Bhave, and Kanad Basu

Abstract As quantum computing progresses toward practical deployment, ensuring that quantum circuits execute both correctly and reliably has emerged as a critical concern. Quantum programs manipulate delicate superposition and entangled states that are inherently sensitive to noise, calibration drift, and compilation artifacts. This chapter examines the field of quantum verification, with emphasis on two core objectives: establishing fidelity, which quantifies how closely a physical system approximates its ideal behavior, and ensuring functional correctness, which confirms that a quantum circuit performs its intended computational task. Modern quantum devices face a range of error sources, including coherent gate miscalibrations, stochastic decoherence processes, readout infidelities, and transformations introduced during circuit compilation. These factors introduce significant verification challenges, such as exponential state-space complexity, a lack of standardized metrics, limited expressiveness in specification languages, and fragmentation across toolchains. A variety of verification paradigms have been proposed to address these issues. These include formal equivalence checking using graphical and algebraic methods, simulation-based testing for small or structured circuits, empirical characterization techniques such as randomized benchmarking and gate-set tomography, and state/process tomography for detailed characterization. Recent developments also point toward the use of assertion-based reasoning frameworks and hardware-in-the-loop testing strategies, including quantum fuzzing. Emerging directions in automated specification inference, fault localization, and compiler-assisted circuit repair reflect the growing need for scalable, integrated verification workflows. This chapter provides an overview of these paradigms, offering both conceptual foundations and practical insights into the evolving landscape of quantum verification.

N. Choudhury (✉) · K. Basu
Rensselaer Polytechnic Institute, Troy, USA
e-mail: choudn3@rpi.edu

K. Basu
e-mail: basuk@rpi.edu

A. Bhave
The University of Texas at Dallas, Richardson, USA
e-mail: ameya.bhave@utdallas.edu

P. Mishra (ed.), *Design Automation for Quantum Computing*,
https://doi.org/10.1007/978-3-032-09303-5_7

1 Quantum Verification: Background and Motivation

1.1 Introduction

Quantum computing leverages uniquely quantum-mechanical phenomena such as superposition and entanglement to perform computational tasks intractable for classical architectures [1]. Advances across superconducting, trapped-ion, and photonic architectures—including order-of-magnitude extensions of qubit coherence times, two-qubit gate fidelities, reconfigurable connectivity, and sub-microsecond classical control loops—have collectively propelled quantum hardware performance and scalability [2, 3]. High-fidelity gates coupled with low-latency measurement and control now support the parameterized ansatz circuits and iterative classical optimization routines required by algorithms such as the variational quantum eigensolver (VQE) and the quantum approximate optimization algorithm (QAOA), enabling their execution on devices with tens of qubits [4, 5]. Unlike classical bits, which occupy a well-defined state $\{0, 1\}$ and can be observed nondestructively, qubits inhabit a 2^n-dimensional Hilbert space and evolve under reversible, unitary operations. Quantum gates, including single-qubit rotations like

$$R_x(\theta) = e^{-i\theta X/2}, \quad R_z(\phi) = e^{-i\phi Z/2},$$

and multi-qubit entangling operations such as the controlled-NOT,

$$\text{CNOT} : |c, t\rangle \mapsto |c, t \oplus c\rangle,$$

must preserve normalization and are inherently reversible. Unlike classical logic, there is no "textitfan-out" primitive, and copying arbitrary quantum states is forbidden by the no-cloning theorem [6]. Moreover, measurement projects qubits onto a basis state, collapsing superpositions and potentially destroying entanglement [7]. Because quantum gates operate coherently across exponentially many computational paths and measurement inherently perturbs the state, traditional verification methods such as exhaustive input-output testing or full state tomography are infeasible [8, 9].

Quantum hardware today deviates significantly from the ideal, exhibiting imperfections that introduce both coherent and incoherent disturbances [10]. Coherent errors arise from systematic control miscalibrations [11]. For example, a nominal $R_x(\pi/2)$ pulse might execute a slightly over or under-rotation, and these small offsets accumulate predictably across successive gates. Incoherent noise originates from unwanted coupling to the environment [12]. Energy relaxation shortens excited-state lifetimes, and random fluctuations in ambient fields scramble phase information, causing a gradual loss of quantum coherence. Readout itself is also imperfect as measurement signals can misidentify qubit states unless corrected by calibration matrices. Together, these hardware non-idealities erode the precision of individual operations, distort the intended quantum dynamics, and ultimately diminish the reliability of end-to-end computations.

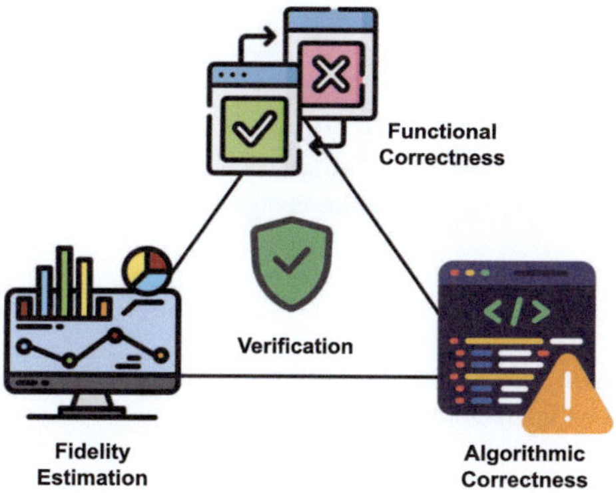

Fig. 1 The three pillars of quantum circuit verification

Additionally, quantum processors have intrinsic hardware limitations. These include sparse qubit connectivity dictated by physical coupling constraints [13], fixed native gate sets determined by the device's control electronics [15], and finite coherence windows limited by environmental decoherence [14]. Compilers must insert additional operations (e.g., SWAP networks) and reorder gates to align with the device's topology [16]. These adaptations not only increase circuit depth but also introduce idle periods during which qubits continue to lose coherence. As a result, compilation-and-routing overhead exacerbates the underlying error processes and highlights the necessity of verification strategies that are aware of hardware constraints and scheduling effects.

Quantum verification therefore broadly encompasses three complementary goals, as depicted in Fig. 1, and summarized as follows:

- **Algorithmic correctness**, which proves that a high-level quantum program or mathematical specification, such as a variational ansatz or phase-estimation routine, indeed implements the intended transformation or sampling task under ideal conditions.
- **Functional correctness**, which ensures that a compiled or optimized circuit faithfully realizes its specification—whether an exact unitary or target output distribution.
- **Fidelity estimation**, which quantifies the distance between implemented and ideal operations on real hardware.

Addressing these challenges has given rise to a spectrum of verification paradigms: formal equivalence checking using graphical calculi, simulation-based unit and integration tests, randomized benchmarking protocols, quantum state and process tomography, and emerging assertion languages grounded in quantum Hoare logic. This chapter provides a unified treatment of these methodologies, highlighting their assumptions, scalable implementations, and open limitations as quantum hardware progresses toward fault-tolerant thresholds.

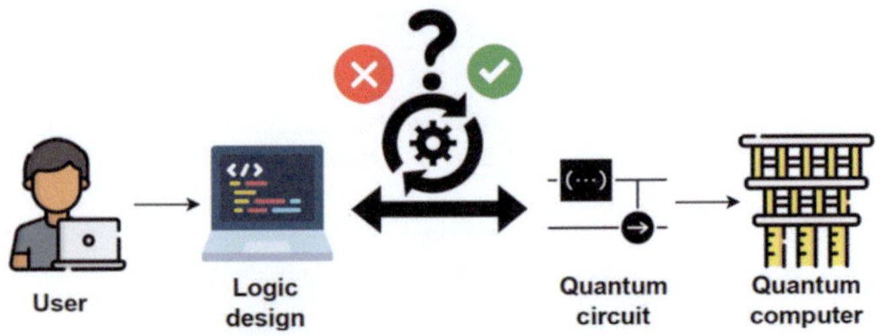

Fig. 2 Quantum verification is needed to check if the algorithm designed by a user is correctly and faithfully being executed on the quantum hardware

1.2 Why Verification Matters in Quantum Computing

Quantum computing operates at the frontier of physical and computational limits, harnessing superposition, entanglement, and interference to process information in ways unattainable by classical systems. However, the very features that make quantum computation powerful also render it extraordinarily fragile. Small perturbations, whether from environmental decoherence, imperfect control pulses, or software miscompilation, can have amplified consequences, especially as circuit depth and complexity increase. In this context, Fig. 2 depicts that *verification* is not merely a performance metric but a foundational requirement for correctness, reproducibility, and trust in quantum systems.

1.2.1 The Challenge of Trust in NISQ Devices

In the era of Noisy Intermediate-Scale Quantum (NISQ) devices, quantum processors are limited by short coherence times, gate infidelities, and imperfect measurements. Physical qubits today typically achieve gate fidelities in the range of 99% or lower, which, when compounded across hundreds of gates, results in non-negligible degradation of the final quantum state [17]. For example, in Google's 2019 demonstration of quantum supremacy, cross-entropy benchmarking revealed overall fidelities on the order of 10^{-3} for deep random circuits, a clear indication of the gap between ideal and practical computation using quantum hardware [18].

In practical applications such as variational quantum algorithms (e.g., VQE and QAOA), verification plays a critical role in separating algorithmic behavior from hardware-induced artifacts. A slight miscalibration in pulse amplitude or phase can bias energy landscapes, leading to convergence toward incorrect minima. In early VQE experiments on superconducting platforms, shifts in gate fidelity as small as 2–3% were shown to result in energy estimation errors that exceeded the threshold for chemical accuracy [19]. Without diagnostic tools such as direct fidelity estimation, randomized benchmarking, and circuit-level simulations, such discrepancies could easily go unnoticed.

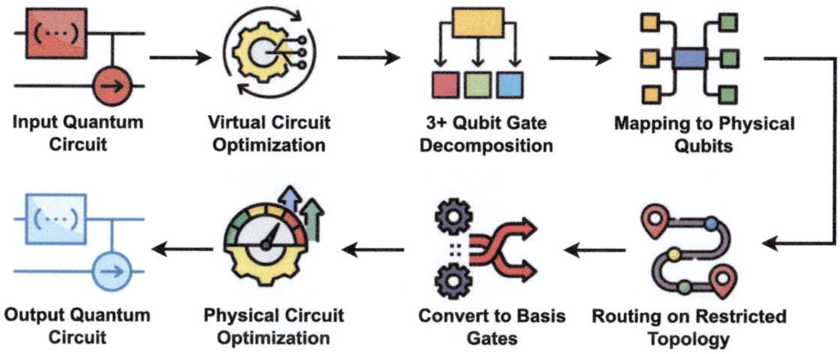

Fig. 3 Overview of the multiple transformations occurring during transpilation of a quantum circuit

1.2.2 Compiler Transformations and Logical Equivalence

Quantum compilers are responsible for mapping abstract circuit descriptions into hardware-specific instruction sequences that comply with both qubit connectivity and native-gate constraints. This compilation process involves gate scheduling, decomposition into basis operations, and qubit routing, each stage capable of introducing logical discrepancies if not performed with rigorous validation [20]. For example, an anomaly in a widely adopted transpiler caused the control and target qubits to be inadvertently reversed during SWAP network synthesis under certain topology constraints; the defect remained latent until uncovered by formal equivalence checking within a regression-testing framework. This illustrates a broader principle: verification must extend beyond hardware fidelity and into the software stack. Equivalence checking techniques based on reversible computing logic, or tensor contraction, can be used to formally prove that the transformed circuit remains semantically identical to its specification. In the absence of such tools, subtle software-level errors risk corrupting results silently (Fig. 3).

1.2.3 Reproducibility and Cross-Platform Comparisons

As quantum computing transitions from proof-of-concept experiments to scientific tools and commercial services, reproducibility becomes essential. However, identical circuits executed on different hardware platforms often yield inconsistent results due to differing noise profiles, connectivity graphs, and calibration schedules. A 2021 multi-platform study, for instance, ran the same QAOA circuit on Rigetti, IBM, and IonQ devices, observing energy differences of up to 10% for the same ansatz and parameters [21–23]. By embedding standardized verification protocols, such as randomized benchmarking [24], researchers are able to disentangle algorithmic behavior from hardware variability. Verification, in this sense, acts as an equalizer and ensures that cross-platform comparisons are meaningful and that results reported from one platform can be interpreted, reproduced, and trusted across others. In a scientific landscape increasingly concerned with reproducibility, the presence or absence of verification determines whether quantum advantage claims are robust or merely anecdotal.

1.2.4 The Role of Verification in Fault-Tolerant Architectures

As the field moves toward fault-tolerant quantum computing (FTQC), verification evolves from validating noisy, small-scale circuits to certifying the integrity of complex error-corrected systems. In this regime, correctness relies not only on physical qubits and gates but also on the proper execution of error-detecting codes, syndrome extraction cycles, and classical decoding algorithms. Modern FTQC schemes often incorporate lightweight "assertion checks", internal consistency tests that flag anomalies such as failed ancilla resets or unexpected syndrome patterns [37]. For instance, in surface-code architectures, any fault in ancilla preparation, stabilizer measurement, or decoder logic can introduce correlated logical errors [25]. Google's demonstration of a distance-5 surface code showed that omitting a simple post-measurement ancilla reset check increased logical error rates by over 15% [26]. Re-integrating that assertion check into the cycle reduced logical errors substantially, underscoring that even inner-loop checks are critical for fully verified fault-tolerant operation.

Moreover, in FTQC, verification must confirm that logical operations satisfy the required fault-tolerance thresholds. This involves simulating fault injection scenarios, estimating logical error rates under stochastic noise models, and validating that error syndromes map correctly to physical faults. These steps form the foundation of certified quantum computation, where users can rely on correctness guarantees backed by both physical error correction and formal verification.

1.2.5 System-Level Verification and Continuous Integration

Finally, at the full-stack level, verification encompasses every component, from quantum hardware and classical control electronics to firmware, drivers, and software toolchains. Organizations such as Riverlane and Quantinuum have adopted verification strategies drawn from electronic design automation (EDA), applying formal methods and hardware testbenches to quantum control stacks [27, 28]. Continuous integration pipelines now frequently include quantum "smoke tests," where benchmark circuits like GHZ states, Bell pairs, or teleportation protocols are run nightly to detect regressions in gate fidelity or routing logic [31]. These system-level verification strategies reflect a growing recognition: quantum computing is not just a novel physical platform, but a tightly integrated system of hardware, software, and theory. Only through layered, continuous verification can such systems scale from prototypes to mission-critical computing platforms.

Fig. 4 Algorithmic correctness ensures the specified Unitary (U_{spec}) is faithfully executed (U_{exec})

1.3 Algorithmic Correctness Versus Functional Correctness Versus Fidelity

Quantum verification must address three intertwined dimensions of correctness. **Algorithmic correctness** establishes that a high-level quantum program or specification, whether a variational ansatz, phase-estimation routine, or sampling protocol, indeed encodes the intended transformation or probability distribution in the ideal, noiseless setting (Fig. 4). **Functional correctness** then ensures that, after compilation and optimization (including gate decomposition, qubit mapping, and scheduling), the concrete circuit still realizes that same unitary or sampling behavior. Finally, **fidelity** quantifies how closely the physical implementation on real hardware matches its ideal counterpart, using metrics such as state fidelity $F(\rho, \sigma)$, process fidelity via Choi matrices, or average-case error rates from randomized benchmarking. Only by jointly proving algorithmic soundness, verifying circuit equivalence, and measuring implementation fidelity can we cultivate the trustworthiness required for scalable quantum computation.

1.3.1 Algorithmic Correctness

Algorithmic correctness asserts that a high-level quantum program or mathematical specification indeed encodes the intended transformation or sampling task in the ideal, noiseless model. It provides proof that, under perfect gates and no decoherence, the algorithm produces the desired output distribution or unitary action for all valid inputs. An example is illustrated in Sect. 2.4.2, where a single Toffoli gate is deconstructed into a set of simpler gates during transpilation. The correctness of such decompositions is typically validated as part of functional (compiler-level) correctness, assuming the underlying algorithmic specification is sound. It is imperative to verify whether this transformation is mathematically sound, in the absence of any errors.

In quantum computing, this means verifying that constructs such as variational ansätz, phase-estimation routines, or oracle-based subroutines truly implement their theoretical specifications. Any logical error—missing uncompute steps, incorrect parameterization, or improper entanglement structure—can invalidate downstream results regardless of hardware fidelity. Ensuring algorithmic correctness thus prevents wasted resources on flawed designs and serves as the foundation for all subsequent compilation and hardware testing. It also underpins reproducibility, since a provably correct algorithm should yield consistent behavior across different platforms and simulators.

Common methods for establishing algorithmic correctness include

- *Quantum Hoare Logic* and related program logics, which generalize pre-/post-condition reasoning to quantum states [29].
- *Interactive theorem proving* in proof assistants (e.g., Coq/SQIR, Isabelle/QBricks) to mechanize equivalence proofs between a circuit and its specification [30].
- *Symbolic simulation* and *assertion checks*, where intermediate state properties (e.g., fidelity to a Bell state) are verified analytically or via noise-free emulation.

Key metrics at this level include *operator norm distance* between the implemented and ideal unitaries, and the *success probability* (or overlap) of the final state with its target.

1.3.2 Functional Correctness

Functional correctness requires that, in the ideal (noiseless) setting, the observable input-output behavior of a compiled or optimized quantum circuit coincides exactly with its specification. Concretely, a circuit is functionally correct if, for every computational-basis input $|x\rangle$ and every measurement outcome y, the implemented circuit's probability

$$\Pr\big[y \mid M_{\text{impl}}\big(U_{\text{impl}}|x\rangle\big)\big]$$

matches the specification's probability

$$\Pr\big[y \mid M_{\text{spec}}\big(U_{\text{spec}}|x\rangle\big)\big].$$

Equivalently, when viewed as a classical function $f : \{0, 1\}^n \to \{0, 1\}^m$, the circuit must satisfy

$$\Pr\big[y = f(x)\big] = 1$$

(or $\geq 1 - \varepsilon$ in an approximate setting) for all x. This definition applies to basis-preserving or oracle-style circuits; for general quantum circuits with intrinsic randomness, functional correctness is instead defined in terms of equivalence between the induced output probability distributions.

Functional correctness, (Fig. 5) guards against semantic errors introduced by compiler optimizations, qubit routing, and gate decomposition. It guarantees that transformations–such as gate cancellations, SWAP insertions, and basis-gate lowering–preserve the intended classical input-output mapping on computational-basis states. Without functional correctness, even a perfectly calibrated device could execute the wrong computation, wasting quantum resources and classical post-processing effort. Verifying this property is therefore essential before hardware execution.

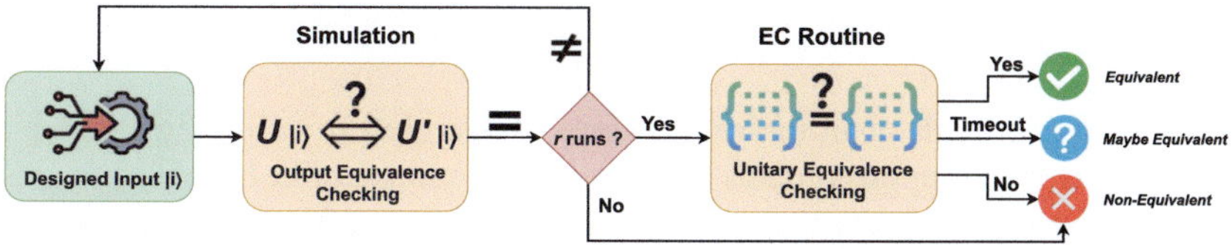

Fig. 5 Functional correctness guarantees that, for every valid input state, the unitary applied by the circuit is the same as the target unitary

Common verification methods and their scopes include:

- *Equivalence Checking:* Proving exact or observational equality of the specification and implementation. Graphical rewrite systems (e.g., ZX-calculus) or decision-diagram techniques may establish

$$U_{\text{impl}} \,|x\rangle = U_{\text{spec}} \,|x\rangle \quad \forall \,|x\rangle,$$

  which implies perfect functional correctness [32, 33].
- *Simulation-Based Testing:* Emulating both original and compiled circuits on classical hardware for a representative set of basis inputs and comparing either state vectors or output probability distributions [34, 35].
- *Assertion-Based Verification:* Embedding projective predicates or invariants at key circuit locations and checking that they hold throughout execution, thereby catching violations of expected behavior [37].

Key metrics for quantifying functional deviations include

- **Total Variation Distance (TVD):**

$$\max_x \tfrac{1}{2} \sum_y \Big| \Pr[y \mid U_{\text{impl}}|x\rangle] \;-\; \Pr[y \mid U_{\text{spec}}|x\rangle] \Big|,$$

  which directly measures worst-case distributional error over basis inputs.
- **Operator-Norm Distance:** $\|U_{\text{impl}} - U_{\text{spec}}\|$, a stronger measure that upper-bounds worst-case functional error but in practice often corresponds to algorithmic (full-unitary) correctness.

Completeness versus practicality: Exact equivalence checking (e.g., via ZX or decision diagrams) yields complete guarantees but can scale poorly. Simulation- and assertion-based methods are more scalable yet inherently sample- or assertion-limited, potentially missing corner-case mismatches.

1.3.3 Fidelity Metrics

Fidelity metrics provide quantitative measures of how closely an implemented quantum operation or state approximates its ideal specification. Unlike binary equivalence checks, which yield pass/fail outcomes, fidelity metrics offer graded, continuous assessments of accuracy. These scores are essential in the hardware execution setting, where small imperfections accumulate and must be managed, diagnosed, and bounded. In the compiler domain, where transformations are assumed to be noise-free, verification typically focuses on logical correctness rather than quantitative error. Common techniques include:

- *Exact Equivalence Checking*, which formally proves that the implemented circuit unitary U_{impl} matches the specification U_{spec}, either globally or on basis inputs.
- *Simulation-Based Testing*, where outputs of both circuits are compared across a representative input set using classical simulation.

These approaches are deterministic and sufficient in the absence of noise, but they do not capture how small deviations affect the circuit's behavior in realistic settings.

In the presence of noise, fidelity metrics become indispensable. They enable quantification of gate performance, circuit reliability, and overall system behavior under realistic execution. The following are the most widely used fidelity measures, each suited to different goals and resource constraints:

- **Average Gate Fidelity** Measures how closely a noisy gate approximates its ideal version, averaged over all pure input states:

$$F_{\text{avg}}(\mathcal{E}, \mathcal{U}) = \int d\psi \ \langle\psi|\,\mathcal{U}^{\dagger} \circ \mathcal{E}(|\psi\rangle\langle\psi|)\,|\psi\rangle.$$

  This metric is efficiently estimated via randomized benchmarking (RB) and is the standard for gate-level calibration. It correlates well with average-case behavior but does not capture worst-case deviations.
- **Total Variation Distance (TVD)** Quantifies the maximum deviation in output distributions on basis inputs:

$$\text{TVD} = \max_{x} \ \frac{1}{2} \sum_{y} \Big| \Pr_{\text{impl}}(y \mid x) - \Pr_{\text{ideal}}(y \mid x) \Big|.$$

  TVD is directly relevant for sampling-based tasks and functional correctness, indicating the maximum probability of observing an incorrect output string. It can be estimated via repeated measurements and aligns well with practical use cases in the NISQ regime.
- **Cross-Entropy Difference** Used in quantum advantage experiments, this metric compares the predicted and observed output distributions:

$$H_{\text{ideal}} - H_{\text{impl}} = \sum_y \Pr_{\text{ideal}}(y) \log\left(\frac{1}{\Pr_{\text{impl}}(y)}\right).$$

Cross-entropy difference emphasizes mismatches on low-probability (i.e., rare but important) outcomes, making it more sensitive than TVD in certain regimes.

- **State Fidelity** Evaluates the closeness of an experimentally prepared state σ to an ideal pure state $\rho = |\psi\rangle\langle\psi|$:

$$F(\rho, \sigma) = \left(\text{Tr}\sqrt{\sqrt{\rho}\sigma\sqrt{\rho}}\right)^2.$$

This is useful for validating small-scale state preparations (e.g., GHZ or Bell states) via tomography.

- **Process Fidelity** Compares quantum processes using their Choi representations $\chi_{\mathcal{E}}$ and $\chi_{\mathcal{U}}$:

$$F_p = \text{Tr}\big(\chi_{\mathcal{E}}\, \chi_{\mathcal{U}}\big).$$

Process fidelity is a foundation for direct fidelity estimation (DFE) protocols and is used when partial tomography of a quantum channel is feasible.

- **Diamond-Norm Distance** Provides a worst-case bound on how distinguishable two quantum channels are, even when acting on entangled inputs:

$$\|\mathcal{E} - \mathcal{U}\|_\diamond = \max_{\rho_{RA}} \|(\mathcal{E} \otimes \mathcal{I}_A)(\rho_{RA}) - (\mathcal{U} \otimes \mathcal{I}_A)(\rho_{RA})\|_1 .$$

While it offers the strongest theoretical guarantees, computing it requires full process tomography or semidefinite programming, making it impractical for large systems.

Selecting the appropriate fidelity metric depends on the verification objective:

- We would use *average gate fidelity* (via RB) for scalable gate-level benchmarking.
- We can use *TVD* or *cross-entropy difference* for functional correctness in sampling tasks, where classical output distributions are the target.
- *State fidelity* and *process fidelity* can be used when validating small systems or specific subroutines via tomography or DFE.

Together, these fidelity metrics provide a versatile toolkit for analyzing and validating quantum circuits across simulation, compilation, and hardware execution contexts.

1.3.4 Interplay and Trade-offs

Quantum verification must balance three orthogonal dimensions–*algorithmic correctness*, *functional correctness*, and *fidelity*–each addressing distinct failure modes and requiring different methods. Optimizing one does not guarantee the others, so effective workflows layer complementary checks.

First, **logical accuracy versus physical accuracy**: a circuit proven correct symbolically (algorithmic and functional correctness) can still fail on noisy hardware, while high gate fidelities alone do not ensure the right computation is being performed. Thus, equivalence-checking (ZX-calculus, decision diagrams) and compiler-level tests must be paired with randomized benchmarking or tomography.

Second, **error mitigation versus new failure modes**: techniques like dynamical decoupling or zero-noise extrapolation improve fidelity but increase depth, ancilla counts, and transpiler complexity—potentially introducing fresh logical errors. Verification must therefore revalidate functional correctness after each mitigation pass.

Third, **coverage versus scalability**: exhaustive proofs or full tomography offer deterministic guarantees but scale poorly. Scalable protocols (e.g., randomized benchmarking, cross-entropy tests) provide coarse-grained metrics for large systems yet cannot localize specific logical faults. A pragmatic strategy should aim to combine lightweight global benchmarks with targeted formal or simulation-based checks on critical subcircuits.

Finally, **deterministic versus statistical assurance**: Formal methods yield absolute correctness in ideal models but ignore noise, whereas statistical benchmarks capture real device behavior but only probabilistically. Integrating both approaches yields tight bounds on worst-case errors alongside confidence levels for expected performance.

By compressing these trade-offs into a unified workflow—incremental formal proofs, selective simulation tests, and layered fidelity benchmarks—practitioners can navigate the verification landscape efficiently, scaling rigorously from small NISQ experiments to future fault-tolerant architectures.

2 Sources of Error in Quantum Circuits

Quantum circuits must contend with a diverse range of error sources that are fundamentally different from faults in classical systems. Unlike digital logic gates, quantum operations are analog and continuous, making them susceptible to both systematic and stochastic imperfections. *Coherent control-pulse errors*, such as amplitude miscalibrations, frequency detunings, and residual crosstalk, introduce consistent but unintended unitary deviations that accumulate predictably across gates [38, 39]. *Incoherent noise channels*, including relaxation (T_1) and dephasing (T_2), arise from uncontrolled interactions with the environment and degrade quantum states in a probabilistic, irreversible manner [40]. *Measurement and readout errors* further distort the classical outcomes, while *compilation-induced transformations*, such as gate decompositions, qubit routing, and scheduling, can inadvertently alter the intended logic or exacerbate noise accumulation [41, 42]. Additionally, *idle-time decoherence* due to limited quantum resources causes qubits to degrade while waiting for operations [43]. These errors are particularly insidious in quantum systems because

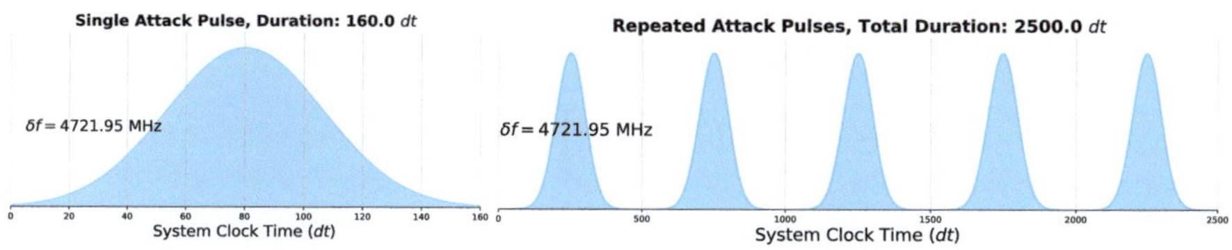

(a) Extended single gate pulse.

(b) Repeated gate pulses.

Fig. 6 An illustration of the pulse-level description of quantum gates. Please note the different cycle times on the x-axis as the figures are not drawn to scale

quantum information is fragile: it cannot be copied, observed non-invasively, or error-corrected trivially. Small inaccuracies can interfere constructively or destructively, leading to emergent failure modes that defy classical intuitions. As such, understanding the distinct physical and algorithmic origins of quantum errors is not only critical for mitigation, but also central to the design of verification protocols. Effective quantum verification—whether symbolic, simulation-based, or statistical—relies on accurately modeling how and where quantum operations deviate from their idealized behavior. This section provides a structured overview of these error mechanisms, laying the groundwork for the verification strategies discussed in subsequent chapters.

2.1 Coherent Control-Pulse Imperfections

Coherent control-pulse imperfections constitute a fundamentally distinct category of gate errors in quantum information processors. These errors originate from systematic deviations in the amplitude, phase, or timing of the applied control pulses (Fig. 6), such as miscalibrations or waveform distortions, and therefore implement unintended but unitary transformations on the quantum register. In contrast to incoherent noise, which irreversibly degrades information via stochastic, thermodynamic processes, coherent pulse errors accumulate deterministically over successive gate applications and may exhibit temporal drifts and spatial correlations across qubits. Consequently, their characterization and mitigation demand techniques that explicitly address systematic offsets—most notably pulse-shaping methods, routine hardware recalibration, and tailored benchmarking protocols that disentangle coherent and stochastic contributions.

Ideal single-axis rotations are specified by unitaries of the form:

$$U_{\text{ideal}} = \exp\Big[-\tfrac{i}{2}\,\theta\,P\Big] \quad , \quad P \in \{X, Y, Z\}.$$

In practice, a nominal

$$R_x\Big(\tfrac{\pi}{2}\Big) = \exp\Big[-\tfrac{i\pi}{4}\,X\Big]$$

pulse may instead realize

$$U = \exp\Big[-\tfrac{i}{2}\big(\tfrac{\pi}{2} + \delta\big)X\Big] = R_x\Big(\tfrac{\pi}{2} + \delta\Big),$$

where δ is a small calibration error. Over successive applications, such over- or under-rotations accumulate, resulting in a net unitary deviation

$$\Delta U = U\,U_{\text{ideal}}^{\dagger}.$$

Other leading coherent pulse imperfections include:

- **Crosstalk:** Stray microwave or laser fields intended for qubit i unintentionally drive neighboring qubit j, modeled as a small rotation [46, 47]:

$$U_{\text{xtalk}} = \exp\big[-i\,\epsilon\,X_j\big], \quad \epsilon \ll 1.$$

- **Frequency Drift:** Slow, time-dependent shifts in the qubit resonance frequency [48], $\omega_q \to \omega_q + \delta\omega(t)$, leading to unintended phase rotations

$$U_{\text{drift}} = \exp\Big[-i\int_0^t \delta\omega(t')\,dt'\,Z\Big].$$

- **Leakage:** Population transfer out of the computational subspace (e.g., into $|2\rangle$ in a transmon) which can be captured by non-Pauli components of the pulse unitary [49]. For instance

$$U_{\text{leak}} = \begin{pmatrix} 1 & 0 & 0 \\ & \sqrt{1-\lambda^2} & \lambda \\ & -\lambda & \sqrt{1-\lambda^2} \end{pmatrix}, \quad 0 < \lambda \ll 1.$$

2.2 *Incoherent Noise Channels and Decoherence*

Incoherent errors arise from the unavoidable interaction of qubits with their surrounding environment, resulting in irreversible degradation of quantum information. Representative mechanisms (Fig. 7) include amplitude damping–spontaneous relaxation from $|1\rangle$ to $|0\rangle$; pure dephasing–stochastic phase fluctuations that diminish off-diagonal density-matrix elements; and thermal excitation processes that repopulate higher energy levels. Such effects impose fundamental limits on qubit coherence times and cannot be corrected by unitary operations alone. Mathematically, they are captured by completely positive, trace-preserving (CPTP) maps, either via a discrete Kraus-operator decomposition or through the continuous-time Lindblad master equation.

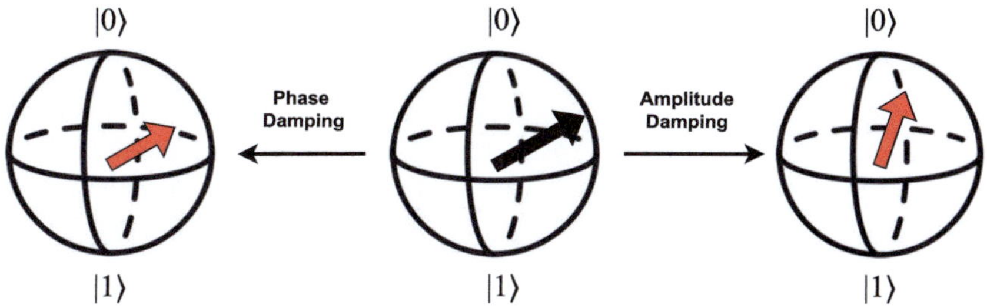

Fig. 7 The effect of amplitude and phase damping on the state of a qubit

2.3 *Measurement Infidelities*

Measurement imperfections arise whenever the physical process of converting a qubit's quantum state into a classical bit produces errors, whether through misassignment, unwanted state changes during measurement, or correlated disturbances among qubits [50]. Accurately characterizing and correcting these imperfections is essential for reliable quantum-circuit verification, as they directly bias the observed outcome statistics and can mask or mimic genuine circuit faults. The primary measurement error sources can be categorized as follows:

2.3.1 Readout Assignment Errors

Quantum readout maps the state of a qubit onto a classical signal, yielding a binary outcome that drives downstream decision logic. In superconducting platforms, this is typically done dispersively by probing a readout resonator whose frequency shifts depending on the qubit state, while trapped-ion systems often use state-dependent fluorescence, and spin qubits employ spin-to-charge conversion with charge sensors. Despite these diverse implementations, all readout schemes suffer from nonzero misassignment probabilities: The detector may report $|0\rangle$ when the qubit is actually in $|1\rangle$, and vice versa [50, 51]. This is illustrated in Fig. 8.

2.3.2 State-Preparation and Measurement Drift Errors

State-Preparation and Measurement (SPAM) errors combine imperfections in preparing the qubit's initial state with drifts in the measurement apparatus, leading to biased outcome statistics even before any circuit action [52]. In practice, an intended $|0\rangle$ may in fact be initialized in a slightly mixed state, and readout thresholds or amplifier gains can slowly shift, causing gradual misassignment. Robust verification protocols therefore must either interleave calibration routines to track SPAM parameters or employ SPAM-resilient techniques such as randomized measurement bases or ref-

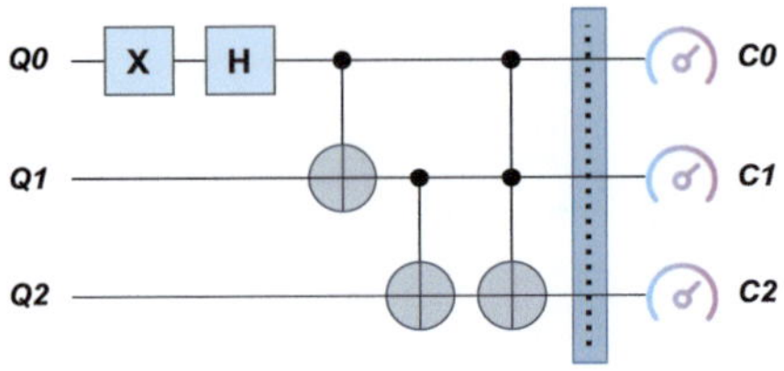

(a) A three-qubit quantum circuit.

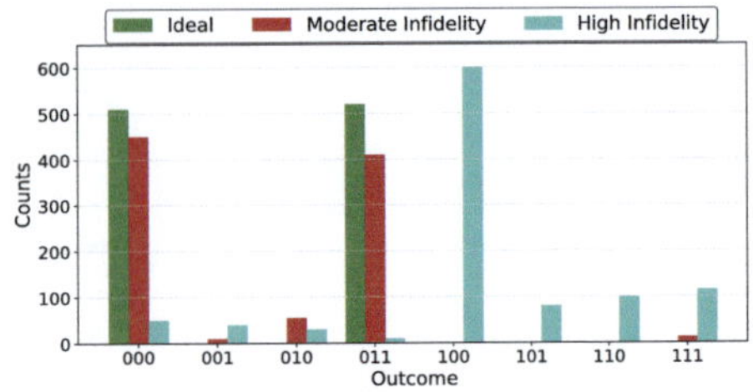

(b) Outcomes.

Fig. 8 Measurement infidelity can cause highly misleading outputs, making computation unusable, as shown in this 3-qubit case

erence state benchmarking–to separate preparation and measurement bias from true circuit errors.

2.3.3 Quantum Nondemolition Violations

Quantum nondemolition (QND) measurements are those designed to preserve the measured observable's eigenstate, allowing repeated reads without perturbing the value [53]. In reality, measurement back-action due to, for example, finite measurement pulse duration or amplifier non-idealities can induce transitions between $|0\rangle$ and $|1\rangle$, violating the QND condition and corrupting subsequent measurements. Such QND violations introduce state-dependent errors that accumulate in multi-shot verification routines, necessitating either shorter measurement pulses, improved amplifier isolation, or explicit modeling of back-action in the verification pipeline.

2.3.4 Measurement Crosstalk

Measurement crosstalk occurs when the readout drive or measurement signal applied to one qubit inadvertently perturbs its neighbors, causing correlated misassignments across multiple channels [54]. In superconducting systems, for example, a strong microwave pulse on qubit A's resonator can shift the frequency or excite qubit B's resonator, leading to false correlations in the observed bit-string. To ensure accurate verification, one must characterize and include cross-assignment matrices or joint readout models that capture these inter-qubit correlations, or otherwise isolate readout lines to suppress crosstalk.

2.3.5 Assignment Matrix and Readout Fidelity

Measurement errors can be encapsulated in the *assignment matrix*

$$M = \begin{pmatrix} P(0\,|\,0) & P(0\,|\,1) \\ P(1\,|\,0) & P(1\,|\,1) \end{pmatrix},$$

where $P(m\,|\,s)$ is the probability of reporting outcome m when the qubit is truly in state s. From M one defines the *readout fidelity*

$$F_{\mathrm{ro}} = \frac{P(0\,|\,0) + P(1\,|\,1)}{2},$$

which for high-performance devices typically exceeds 90%. For instance,

$$M \approx \begin{pmatrix} 0.98 & 0.04 \\ 0.02 & 0.96 \end{pmatrix} \implies F_{\mathrm{ro}} = 0.97.$$

In verification workflows, readout is modeled as a classical post-processing channel by applying the inverse of M to the measured count vector $\mathbf{c}$:

$$\mathbf{c}_{\mathrm{corr}} = M^{-1}\,\mathbf{c}.$$

This correction removes the leading bias from assignment errors. We omit a detailed treatment of statistical propagation through M^{-1} here, as it is peripheral to the core verification framework, and assume that shot counts are large enough for sampling noise to be subdominant compared to systematic readout bias.

2.4 Compilation-Induced Transformations and Routing Errors

High-level quantum circuits must be *transpiled* into hardware-compatible sequences that respect qubit topology, native gates, and timing constraints. These transformations introduce both depth overhead and potential semantic shifts if not verified carefully.

2.4.1 Topology-Driven SWAPs and Qubit Mapping

To satisfy hardware connectivity constraints, logical qubits that interact must often be repositioned using SWAP operations [57]. This necessity arises when the device's coupling graph lacks direct connections between all logical qubit pairs, requiring both

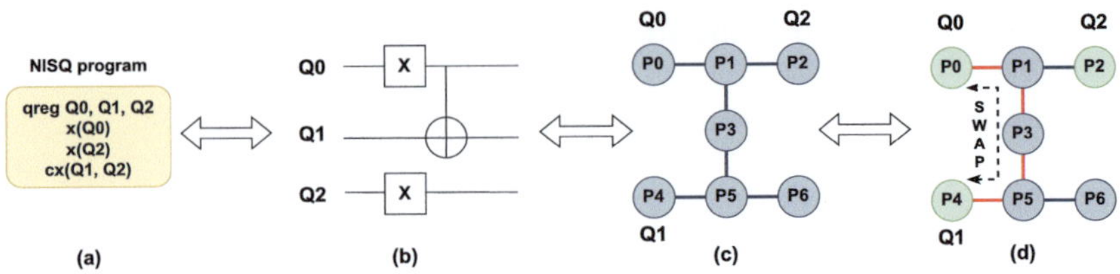

Fig. 9 Naive qubit mapping leading to multiple SWAP gate insertion and high gate overhead

an initial logical-to-physical qubit assignment and dynamic routing throughout the circuit. These transformations are typically guided by heuristics (e.g., SABRE), and often result in SWAP ladders—each composed of three CNOTs—that increase circuit depth and error exposure. The resulting overhead amplifies both decoherence from extended execution time and coherent errors due to imperfect gate calibrations. In addition, overlapping SWAP paths can cause routing contention, and dynamic remapping may introduce subtle misalignments that accumulate over time. An example of such overhead is shown in Fig. 9. A NISQ program utilizing three qubits is designed by a user (Fig. 9a) and converted to its corresponding quantum circuit representation (Fig. 9b). This circuit is then to be mapped onto a seven-qubit quantum computer with a physical layout shown in Fig. 9c. However, in a naive mapping, qubit Q0 is mapped to physical qubit P0, while qubit Q1 is mapped to physical qubit P4, and qubit Q2 is mapped to physical qubit P2 (Fig. 9d). For the CX gate between Q0 and Q1 to be implemented now, at least three SWAP operations will be required, resulting in an additional nine CX gates to be inserted in the resulting quantum circuit before execution. This overhead adds to circuit depth and increases verification complexity significantly.

2.4.2 Gate-Set Translation and Scheduling

Gate-set translation refers to breaking arbitrary multi-qubit gates into the device's supported basis gates, while scheduling arranges commuting operations for parallel execution. This encompasses decomposition of the designed quantum circuit into sequences of CNOTs and single-qubit rotations, and the reordering of gates that commute to reduce overall depth or minimize idle times [56]. An example of this decomposition is shown in Fig. 10, where a three-qubit toffoli gate is decomposed into single and two-qubit gates. Errors introduced include proliferation of two-qubit gates (each with nonzero error rates), accumulation of small over- and under-rotation errors from successive calibrated pulses, interaction with time-dependent noise when gate order changes (e.g., a dephasing spike during a reordered gate), and unintended residual entanglement if pulses overlap or interfere. For instance, decomposing a custom entangler into three CNOTs and six R_z rotations can triple the chance of a two-qubit error and shifts the qubit idle windows into high-noise intervals.

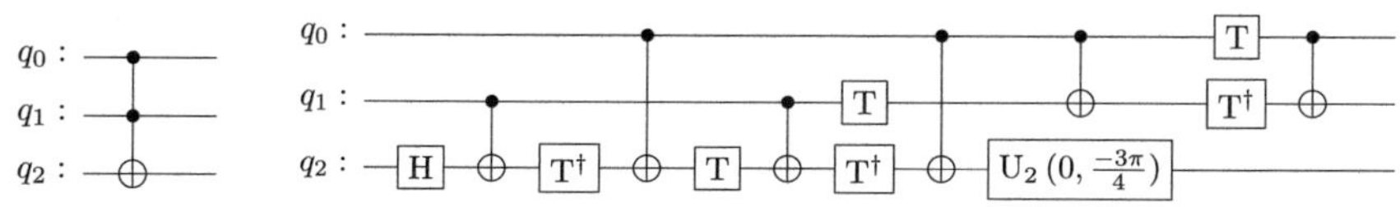

(a) A Toffoli Gate. (b) Decomposed Toffoli Gate.

Fig. 10 Decomposition of a Toffoli gate during the transpilation process

2.4.3 Pulse-Level and Ancilla Effects

Pulse-level effects emerge when logical instructions are converted into real waveforms, and ancilla effects occur when extra helper qubits are introduced [55]. They encompass merging or splitting of calibrated pulses, adjustments to pulse shapes for DRAG correction, and allocation or reset of ancilla qubits for parity checks or intermediate routing. Errors that stem from these include modified crosstalk profiles when pulses are concatenated or overlapped, calibration mismatches if pulse schedules violate assumed timing constraints, leakage into noncomputational levels from improperly shaped pulses, and leftover entanglement or garbage states when ancilla qubits are not fully disentangled and reset before reuse. As a qualitative example, merging two sequential R_x pulses into one longer pulse may introduce additional spectral components that excite neighboring qubits, while failing to uncompute an ancilla used for parity can leave it entangled with the data qubit, corrupting downstream logic.

3 Challenges of Quantum Verification and Validation

Quantum verification faces a set of foundational challenges distinct from those in classical computing. The exponential scaling of quantum state space (2^n for n qubits) renders exhaustive simulation and process-level analysis intractable beyond small systems. Quantum operations are defined by continuous-valued control parameters, making them susceptible to coherent, systematic errors that accumulate over time and evade discrete verification models. Additionally, quantum devices produce inherently probabilistic outputs, necessitating statistical rather than deterministic validation. Toolchain fragmentation—spanning high-level languages, compilers, and hardware control—further complicates consistent specification and validation across abstraction layers. Finally, the lack of a standardized assertion language limits formal specification and inline correctness checking within quantum programs. These challenges collectively constrain the scalability and reliability of current verification methodologies, motivating the development of specialized techniques tailored to the quantum domain.

3.1 Exponential State–Space Explosion

A defining feature of quantum computation is that the state of an n-qubit system is represented by a complex vector in a 2^n-dimensional Hilbert space. This exponential scaling underpins the power of quantum algorithms, enabling quantum systems to encode and manipulate an immense amount of information with relatively few physical qubits. However, it simultaneously presents a fundamental challenge for verification, as the sheer size of the state space renders classical inspection and simulation prohibitively expensive beyond modest values of n.

To quantify this, consider that a pure quantum state over n qubits requires 2^n complex amplitudes to be stored and evolved. A full statevector simulation of just 30 qubits consumes over a gigabyte of memory, while 40 qubits requires tens of terabytes, which is well beyond the limits of standard hardware. When noise is included, representing quantum channels using density matrices or superoperators inflates this requirement to 4^n or even 16^n complex parameters, making even shallow circuits intractable to analyze exactly. The cost grows not only in memory but also in time, as simulating quantum dynamics or computing measurement probabilities generally scales linearly or worse with the size of the state representation.

This exponential blowup limits more than just brute-force simulation [58]. Verification techniques such as quantum process tomography, which aim to reconstruct the behavior of an unknown quantum operation from experimental data, scale as $\mathcal{O}(4^n)$ in the number of experiments and $\mathcal{O}(16^n)$ in the amount of classical post-processing required [59]. Similarly, formal methods based on symbolic representations, such as decision diagrams or ZX-calculus, only offer exponential savings for circuits with particular structure (e.g., low T-count or stabilizer circuits). In general-purpose circuits, especially those that involve high entanglement or randomness, these methods also degrade to exponential complexity.

The practical consequence is that verification approaches requiring full access to the quantum state or operation are inherently restricted to circuits involving a small number of qubits, low depth, or significant structural regularity. As quantum processors scale into regimes beyond 50 or 100 qubits, where even approximate classical simulation becomes infeasible, such methods lose applicability. This exponential barrier thus defines the point at which classical verification breaks down, motivating the need for alternative approaches that do not rely on direct access to the full state space. Subsequent sections will discuss how this limitation influences the design of scalable verification frameworks.

3.2 Lack of Standardized Verification Metrics

Quantum correctness admits multiple, sometimes conflicting, quantitative measures. At the hardware level, one can compute state fidelity to gauge the average overlap between an implemented and ideal state, or employ process fidelity via Choi representations to characterize the full action of a gate [60, 61]. Randomized benchmarking further condenses errors into an average-case decay parameter, while the diamond-norm distance provides a worst-case guarantee—albeit at prohibitive computational cost.

Moving to the circuit and application layers, verification criteria shift again: functional equivalence checks whether two gate sequences realize the same unitary up to numerical tolerance, whereas distributional tests compare the probability distribution of measured outcomes against the ideal, typically via total-variation distance. Composite benchmarks such as quantum volume attempt to unify gate depth, connectivity, and error rates into a single scalar score, but this often obscures algorithm-specific failure modes. In essence, each metric answers a different question. "How accurately does my hardware implement a given operation?" versus "What error rate should I budget for in complex circuits?" versus "Is my compiled circuit logically identical to the specification?", and no single figure of merit spans these concerns. The absence of a coherent framework that maps verification goals to appropriate metrics makes cross-platform comparisons and toolchain integration unreliable. Establishing context-aware standards for metric selection and interpretation is therefore critical to scalable, interoperable quantum verification.

3.3 Fragmentation of Quantum Toolchains and Representations

The quantum software stack is currently fragmented across multiple layers of abstraction, with no universally adopted intermediate representation (IR) or standardization of semantics. High-level frameworks such as Qiskit, Cirq, Braket, Pennylane, and Q# each define their own circuit models, gate sets, and parameter conventions [62–66]. These are compiled down to distinct IRs such as OpenQASM (v2/v3), Quil, or QIR, each supporting different capabilities, ranging from static circuits to dynamic control flow, classical feedforward, and pulse-level instructions. Consequently, the same algorithm can yield structurally and semantically different circuits once transpiled, complicating uniform verification and portability.

Transpilation further exacerbates divergence. Compiler stacks introduce different optimization passes for gate fusion, commutation, mapping to hardware topologies, and decomposition into native gates. Even basic constructs such as controlled operations or phase shifts can be represented differently across backends, affecting circuit layout, depth, and qubit indexing. These discrepancies can alter phase conventions or introduce subtle mismatches that invalidate equivalence checks or interfere with fidelity comparisons, unless manually reconciled.

Noise and calibration data are equally heterogeneous. Qiskit exposes average gate and readout errors via JSON; Rigetti's SDK provides Kraus operators; IonQ offers process matrices; and some platforms report unitary deviations or Pauli error channels. These models vary in both representational format and physical assumptions, Markovian versus non-Markovian, uncorrelated versus correlated errors, time-stationary versus drift-aware, thereby hindering the reuse of benchmarking and verification protocols across toolchains. This fragmentation imposes a significant barrier to scalable verification. Each combination of language, IR, compiler, and hardware backend requires custom pipelines, complicating tool interoperability and obstructing reproducibility. Verification routines must contend with inconsistent representations of circuit structure, gate semantics, and error models, making cross-platform validation brittle and error-prone. As quantum systems grow in complexity, these inconsistencies present a fundamental roadblock to unified, end-to-end verification.

3.4 Modeling and Verifying Under Realistic Noise

In practical quantum hardware, noise is not a single, simple process but a tapestry of overlapping errors that vary in time, space, and origin [67]. Any useful verification framework must begin with a noise model that reflects this complexity–while still remaining simple enough to analyze or simulate. Below, we break down the principal facets of "realistic" noise and explain why each presents a unique challenge.

3.4.1 Combined Coherent and Incoherent Dynamics

In the preceding subsections we have characterized the two principal families of gate errors in quantum processors, namely, coherent control-pulse imperfections (encompassing systematic unitary deviations, frequency drifts *etc.*), and incoherent noise channels (such as amplitude damping and dephasing). Incoherent noise channels are modeled using completely positive, trace-preserving (**CPTP**) maps or the Lindblad master equation with collapse operators $L_{\text{amp}} = \sqrt{1/T_1}\,\sigma^-$ and $L_{\text{deph}} = \sqrt{1/(2T_\phi)}\,Z$ [68, 69]. When these two error classes act simultaneously, the evolution of the density operator ρ is governed by a master equation that intertwines deterministic Hamiltonian drifts and stochastic collapse processes:

$$\frac{d\rho}{dt} = -i\big[H_{\text{ideal}} + H_{\text{err}},\, \rho\big] + \sum_k \Big(L_k\,\rho\,L_k^\dagger - \tfrac{1}{2}\{L_k^\dagger L_k,\, \rho\}\Big),$$

where H_{err} encodes coherent offsets (e.g., small rotation errors δX, detunings $\Delta\omega\, Z$, residual couplings $\chi\, Z \otimes Z$) and the set $\{L_k\}$ captures amplitude and phase-damping channels. Because the unitary part rotates the state in a biased direction while the collapse terms randomly shrink or redirect its Bloch-vector components, their interplay can produce accumulation patterns that neither purely coherent nor purely stochastic models predict. A sequence of small over-rotations can align decoherence "kicks" to amplify infidelity, whereas random phase noise may intermittently reset or aggravate deterministic drift, leading to complex, time-dependent behavior.

These intertwined coherent and incoherent processes impose significant hurdles for reliable verification. Fidelity-estimation protocols that model noise as purely stochastic will inevitably underreport worst-case infidelities, since deterministic over-rotations can align across long gate sequences and yield correlated error patterns that these techniques miss. Developing a complete error model exacerbates the burden: one must decouple unitary calibration offsets (δ, $\Delta\omega$, χ) from non-unitary decay rates ($1/T_1$, $1/T_2$) by executing extensive calibration circuits at varied evolution times and solving high-dimensional semidefinite programs, a resource cost that scales exponentially with the number of qubits. Moreover, both coherent drifts and decoherence rates typically fluctuate on timescales comparable to or shorter than standard verification campaigns, so by the time a lengthy tomography or benchmarking sequence concludes, the device's error characteristics may have shifted, rendering the collected data obsolete. Finally, when coherent perturbations and environmental couplings overlap spectrally, the resulting noise departs from simple Gaussian or Markovian models, defeating verification methods predicated on time-independent CPTP assumptions [71, 72]. Collectively, these challenges underscore the extreme difficulty of attaining scalable, trustworthy performance guarantees on contemporary quantum hardware.

3.4.2 Time-Dependent Drift

Qubit properties are not static. Over the course of hours or even minutes, a qubit's relaxation time T_1 and dephasing time T_2 can fluctuate due to temperature changes, control-electronics drift, or dielectric fluctuations. Gate calibration parameters such as pulse amplitudes, durations, and phases–also shift gradually. We model this by letting

$$T_1(t) = T_1^0\big(1 + \epsilon_1(t)\big), \quad T_2(t) = T_2^0\big(1 + \epsilon_2(t)\big),$$

where $\epsilon_{1,2}(t)$ are slow stochastic processes. Similarly, a gate intended to implement $R_x(\theta)$ may actually perform $R_x\big(\theta + \delta(t)\big)$, with $\delta(t)$ drifting over time. Such time-dependent drift undermines static calibration assumptions, making it necessary for verification procedures to continuously adapt to evolving hardware parameters.

3.4.3 Non-Markovian Memory Effects

Most theoretical treatments assume Markovian noise: the system's future evolution depends only on its current state. However, real devices often exhibit non-Markovian behavior, where the environment retains a memory of past system interactions [73]. Mathematically, the evolution may involve a memory kernel $K(t, s)$:

$$\frac{d\rho(t)}{dt} = \int_0^t K(t, s)\, \rho(s)\, ds.$$

Physically, this arises when qubits couple to slow degrees of freedom, either two-level defects in materials or collective vibration modes in ion traps, that do not immediately relax. Such memory effects can cause error correlations across time: *an error that occurred earlier influences the likelihood of another error later.*

3.4.4 Spatial Correlations and Crosstalk

In multi-qubit devices, noise is rarely independent. Two primary sources of spatial correlation are:

- **Crosstalk**: Control pulses or measurement signals intended for qubit i can inadvertently perturb neighbor j, adding a term $\epsilon_{ij} P_j$ (with $P_j \in \{X, Y, Z\}$) to its Hamiltonian [46].
- **Shared Environmental Noise**: Fluctuations in the cryogenic environment, power supplies, or substrate phonons affect multiple qubits simultaneously, leading to correlated dephasing or relaxation events [74].

Formally, the noise superoperator $\mathcal{E}$ acting on an n-qubit state ρ no longer factorizes into single-qubit channels, but must include multi-qubit collapse operators L_{ij} for correlated jumps, complicating both the model and its verification.

3.4.5 Balancing Fidelity and Tractability

Comprehensively modeling every nuance of realistic noise leads to models with an explosion of parameters far too many for efficient simulation or analytic bounding. Effective verification therefore requires a hierarchical approach:

1. Identify dominant error channels (e.g., primary T_1/T_2 rates, largest crosstalk couplings, principal detunings).
2. Include time-dependent drift terms only when they impact experimental timescales on a significant scale.
3. Capture leading non-Markovian or correlated effects with simplified proxy models (e.g., colored-noise filters, two-point correlation functions).

By striking a balance between *model fidelity* (accurately representing key noise mechanisms) and *tractability* (keeping the model analyzable), one can construct practical noise representations that underpin reliable quantum-circuit validation. Designing and validating these reduced models is itself a central challenge—one addressed by specialized benchmarking and tomography techniques in the following section.

3.5 *Specification Uncertainty*

Specification uncertainty refers to the ambiguities that arise when translating a high-level quantum algorithm into a low-level circuit suitable for hardware execution. In classical design automation, precise, machine-readable specifications and embedded assertions enable automated correctness checks. In quantum computing, however, the absence of such clear specifications undermines systematic verification.

3.5.1 Abstract Algorithm Versus Concrete Circuit

Quantum algorithms are typically specified at the functional level. In practice, developers either hand-craft these kernels or import a vetted library implementation, and then a compiler or transpiler maps them to hardware-native gates, inserts SWAP networks to satisfy connectivity constraints, and schedules operations to respect timing and crosstalk requirements. Each compilation choice can introduce subtle but functionally significant changes (mentioned in Sect. 2.4). Rounding rotation angles during gate decomposition may cause the intended rotation axes to tilt, while SWAP insertion for qubit remapping can alter the circuit depth and entanglement patterns. In the absence of a formal, machine-readable contract—such as an explicit unitary matrix annotation, a domain-specific IR capturing exact semantics, or embedded assertions on measurement outcomes—there is no automated mechanism to detect these semantic deviations. Consequently, verification workflows cannot apply equivalence checking or contract enforcement at compile time, and silent mismatches only surface later as degraded fidelity or incorrect results in end-to-end benchmarks.

3.5.2 Continuous Parameters and Approximation Drift

Quantum routines frequently rely on real-valued parameters like rotation angles θ in single-qubit gates or variational coefficients in ansatz circuits, that are implemented with finite precision and subject to temporal drift. As noted in Sect. 2.1, control-pulse amplitudes and phases slowly vary between calibrations (e.g., systematic drift in pulse amplitude δ or frequency detuning $\Delta\omega$), and in Sect. 2.3.2 we observed analogous shifts in readout thresholds and amplifier gains over time. In practice, each parameter θ is rounded to the nearest representable value and may further drift by an (unknown) offset δ before or during execution, so that the actually implemented gate is

$$R_x(\theta + \delta) \quad \text{instead of} \quad R_x(\theta).$$

Without explicit, machine-readable tolerances on δ, verification routines have no basis for determining whether the compiled circuit remains within acceptable error bounds. These approximations and drifts create four primary verification obstacles. First, propagating parameter uncertainty to bound worst-case fidelity scales exponentially with circuit depth and the number of uncertain parameters. Second, correlated drifts across gates violate independent-error assumptions, forcing static methods (e.g., interval analysis) to track complex multi-parameter dependencies. Third, verification runtimes can exceed drift timescales, rendering calibration data obsolete. Finally, in the absence of formal tolerance contracts (such as interval-annotated gates or embedded assertions), drift-induced deviations remain invisible until end-to-end benchmarks fail.

3.5.3 Probabilistic and Heuristic Outputs

Many quantum applications produce statistical results in the form of expectation values in chemistry simulations or samples from a target distribution in optimization, instead of a single deterministic bitstring. As a result, correctness must be defined in statistical terms (for example, mean-energy error below ϵ, total-variation distance under ε, or an expected approximation ratio above R), and verification becomes a question of whether those criteria hold with high confidence. Yet these performance metrics are rarely codified in executable specifications; instead, teams rely on informal sampling tests and arbitrary pass/fail thresholds.

This lack of formal statistical contracts gives rise to several verification headaches. First, different groups may adopt different error norms or confidence levels, making cross-platform comparisons meaningless. Second, achieving rigorous guarantees (for example, TV-distance $\leq \varepsilon$ with failure probability α) often demands $O(1/\varepsilon^2 \log(1/\alpha))$ measurements, turning verification into a costly Monte Carlo exercise. Third, hardware parameter drift during lengthy sampling campaigns can invalidate early test results, since output distributions may shift before verification completes. Finally, without a domain-specific language for declaring and automatically checking statistical assertions, integrating these tests into automated build pipelines remains a manual, error-prone process. Addressing these challenges—by standardizing statistical metrics, bounding sample complexity, tracking temporal stability, and embedding formal assertions—will be essential for scalable, reliable quantum verification.

4 Quantum Verification and Validation Strategies

To address the challenges of exponential state spaces, analog control errors, probabilistic outputs, fragmented toolchains, and specification uncertainty, a diverse toolkit of verification paradigms has emerged. These range from formal equivalence checking, which uses algebraic and graphical methods to prove circuit identity, to simulation-based testing, which exercises circuits on classical emulators; from randomized benchmarking and gate-set tomography, which extract aggregate and detailed noise characterizations, to state and process tomography, which reconstructs quantum operations at full resolution; and from emerging assertion-based frameworks rooted in quantum Hoare logic to hybrid hardware-in-the-loop approaches (quantum fuzzing) that combine emulation with real-device stress tests. In the following subsections, we survey each strategy's core principles, the specific verification gaps it fills, and its trade-offs in scalability and precision.

4.1 Formal Equivalence Checking

Formal equivalence checking provides a mathematically rigorous means to verify that two quantum circuits—typically an original, high-level specification and its optimized or transpiled counterpart—implement the same unitary transformation. By representing circuits in algebraic or graphical form (e.g., decision diagrams or the ZX-calculus), one can apply rewrite rules and canonicalization procedures to collapse both circuits into a unique normal form, or otherwise prove their equality without exhaustive state enumeration. This approach directly addresses compiler-induced errors and toolchain fragmentation, offering deterministic guarantees of functional correctness up to specified approximations. In the subsections that follow, we describe the principal formalisms, discuss their scalability and heuristics for large circuits, and highlight how they integrate into automated verification pipelines.

4.1.1 ZX-Calculus Reductions

The ZX-calculus is a unified graphical and algebraic formalism that represents quantum circuits as networks of "spider" nodes–each spider corresponding to a simple linear map parameterized by a phase [75]. There are two kinds of spiders: Z-spiders, which enact diagonal phase operations in the computational basis, and X-spiders, which are their Hadamard-rotated counterparts in the X-basis. Concretely, a single Z-spider with phase α implements the map:

$$|0\rangle\langle 0| + e^{i\alpha}|1\rangle\langle 1|,$$

and an X-spider with phase β is given by $H\,Z(\beta)\,H$. Spiders may have multiple inputs and outputs, and wires simply denote tensor composition or contraction of these maps.

The true power of ZX-calculus lies in its local rewrite rules. A small set of graphical transformations—spider fusion, Hadamard-color swapping, the bialgebra law, and a handful of ancillary identities—allows one to reshape and simplify arbitrary spider networks without ever writing down exponentially large matrices. By repeatedly applying these rewrites, one can—often efficiently—collapse a complex circuit diagram toward a normal form that is unique (up to trivial wire permutations and global phases) for any given unitary. Two circuits are therefore equivalent exactly when their ZX normal forms coincide (Fig. 11).

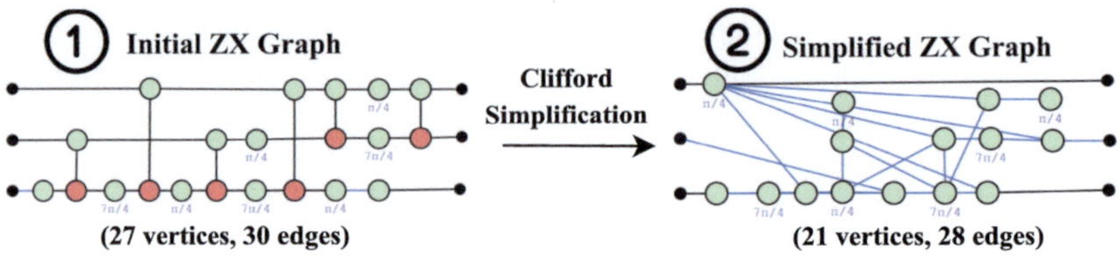

Fig. 11 An example of the simplification of ZX graphs using formal rewrite rules

Why is this critical for equivalence checking? Modern quantum compilers perform numerous non-local optimizations such as gate cancellations, commutation moves, and SWAP insertions to satisfy sparse connectivity, which are difficult to verify by numerical simulation alone once circuit size grows. ZX-calculus reductions offer a symbolic way to absorb and invert these transformations: you translate both the original and the optimized circuits into spider graphs, perform the same rewrite strategy on each, and then check graph isomorphism [32]. If they match, you have a deterministic, machine-checkable proof that the two circuits implement the same unitary—even for medium-sized circuits (tens to low hundreds of qubits) that lie beyond full state-vector simulation.

In practice, tools like PyZX (and research prototypes such as QCEC) implement greedy and heuristic rewrite schedulers to guide the often exponential search [45]. They exploit the fact that Clifford+T circuits, arithmetic blocks, and other structured modules simplify dramatically under spider fusion and bialgebra laws. The rewrite sequence itself can serve as a certificate for audit or debugging: one can replay every spider fusion or color-swap to pinpoint exactly where two circuits diverge.

Overall, ZX-calculus reductions fill a unique niche in quantum verification, as they provide a fully symbolic, language-agnostic, and formally rigorous method for proving circuit equivalence, directly addressing compiler-induced errors and toolchain heterogeneity without resorting to exhaustive simulation.

4.1.2 MPO-Based Equivalence Checking

Matrix Product Operator (MPO) methods represent each quantum circuit as a compressed tensor network, an operator analog of the Matrix Product State, with bond dimensions tuned to the circuit's entanglement structure [44]. Equivalence is then determined by contracting an "intermediary" MPO that encodes the difference between two circuits and evaluating its norm or overlap, avoiding full unitary construction [44]. This approach scales polynomially in the MPO bond dimension rather than exponentially in qubit count, making it well-suited for circuits with limited long-range entanglement or small cut-width. Its strength lies in exploiting locality and low entanglement to verify equivalence of medium-sized circuits that defeat statevector or full-tensor methods. However, performance degrades rapidly if the circuits generate high bond dimensions—such as deep, highly entangling layers—and constructing optimal MPO decompositions can itself be computationally intensive.

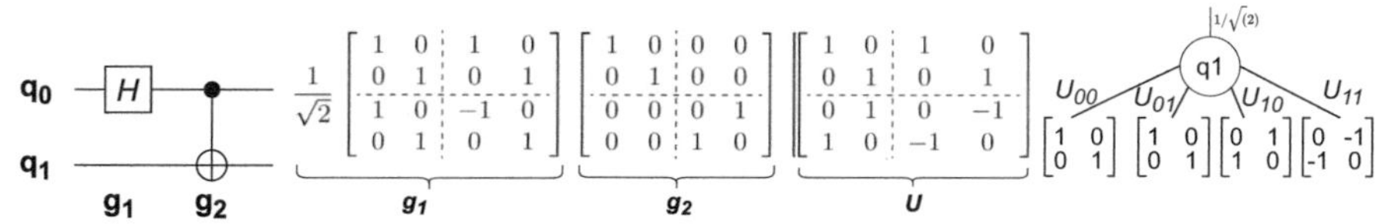

(a) Circuit. (b) Matrix Decomposition. (c) DD.

Fig. 12 Conversion of **a** Quantum circuit to **b** its Matrix representation followed by **c** its Decision diagram Representation

4.1.3 Decision-Diagram Techniques

Decision diagrams (DDs) extend the idea of binary decision diagrams from classical Boolean functions to the quantum realm by representing exponentially large state vectors and operators as directed acyclic graphs whose structure encodes repeated patterns. Instead of storing each of the 2^n amplitudes or $2^n \times 2^n$ matrix entries explicitly, a DD factors the vector or matrix into a hierarchy of smaller pieces—sub-vectors or sub-matrices—arranged according to the qubit decomposition (Fig. 12). Whenever two sub-vectors (or sub-matrices) are identical up to a constant scalar factor, they are merged into a single shared node in the graph, and that scalar is pulled out into the edge weight. This "node sharing" can compress highly regular or structured quantum states (e.g., uniform superpositions, repeated gate patterns) by orders of magnitude [33].

Among DD variants, the *Quantum Multiple-valued Decision Diagram* (QMDD) is the most widely adopted [36]. A QMDD node corresponds to a particular qubit index and has four outgoing edges labeled by the computational basis transitions $(0 \to 0)$, $(0 \to 1)$, $(1 \to 0)$, and $(1 \to 1)$. Each edge carries a complex weight. To recover a matrix element or state amplitude, one multiplies the weights along the unique path from the root to the terminal node. A normalization convention—such as forcing one outgoing edge per node to carry a unit weight and absorbing its magnitude into the parent edge—ensures a canonical, unique representation for each linear map. Building the QMDD for a unitary circuit then involves recursively tensoring and composing smaller DDs corresponding to individual gates, merging identical subgraphs on the fly to maintain compactness.

Quantum Information Decision Diagrams (QuIDDs) take a similar approach but often use binary decomposition: they treat the full $2^n \times 2^n$ matrix as a function of $2n$ Boolean variables (input and output bits) and build a shared BDD-like structure with two edges per node [76]. *Quantum Multiple-valued Decision Diagrams* (QMDDs), in contrast, may decompose by grouping multiple qubits at each level or by using different normalization schemes, trading off sharing efficiency against node-degree complexity. All these variants share the principle of capturing structural redundancies—for example, the repeated submatrix patterns that arise in quantum Fourier transform circuits or block-diagonal structures in many-body Hamiltonians.

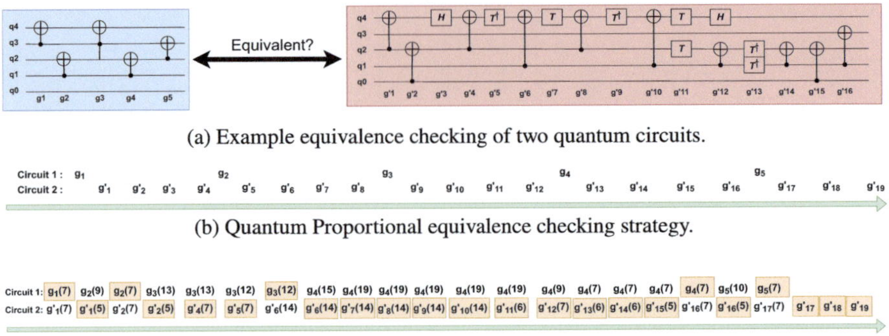

(a) Example equivalence checking of two quantum circuits.

(b) Quantum Proportional equivalence checking strategy.

(c) Quantum Lookahead equivalence checking strategy.

Fig. 13 Example showing quantum equivalence checking using QDD strategies

By translating both the specification and the target circuits into DDs and comparing their canonical forms, one obtains a deterministic proof of unitary equivalence without exhaustive simulation. Although worst-case DD growth remains exponential, the ability to exploit repeated structure makes decision-diagram techniques a cornerstone of modern quantum equivalence checking (Fig. 13).

Diagrammatic decision diagrams (DDs) achieve dramatic compression and instantaneous equivalence checks through four interlocking algebraic strategies. First, *node sharing and canonicalization* merge identical sub-functions by normalizing outgoing edge weights—dividing by a common factor so the first nonzero edge has unit magnitude and absorbing that factor into the incoming edge—ensuring a single canonical node per equivalence class of sub-vectors or sub-matrices. Second, *edge weights as scalar factors* pull repeated constants (for example, the Hadamard amplitude $1/\sqrt{2}$ or T-gate phase $e^{i\pi/4}$) out of node labels onto edges, so subgraphs differing only by a global phase or magnitude collapse into one shared node. Third, *efficient algebraic operations*—such as tensor products realized by constant-time pointer redirections and matrix-vector or matrix-matrix multiplies implemented via memoized recursive routines—scale with the DD size rather than the full 2^n Hilbert-space dimension. Finally, *constant-time equivalence checking* reduces the comparison of two fully reduced DDs to a simple pointer and scalar-factor check at the roots, yielding an $O(1)$ test that confirms whether they represent the same linear map.

For unitary operators, each node has four outgoing edges corresponding to the $(0 \to 0)$, $(0 \to 1)$, $(1 \to 0)$, and $(1 \to 1)$ blocks of the operator on a given qubit. Building DDs for complete circuits and comparing their canonical forms provides a deterministic check of functional equivalence without full statevector simulation. While worst-case DD size remains exponential, practical circuits often exhibit patterns—repeated subcircuits, low T-count regions, arithmetic kernels—that compress dramatically. Advanced techniques include

- *Variable Ordering*: Choosing an optimal qubit order in the DD (via heuristics or profiling) can reduce node count by orders of magnitude.

- *Circuit Partitioning*: Dividing large circuits into smaller subcircuits whose DDs are constructed independently before merging, thus avoiding blow-up in intermediate representations.
- *Garbage Collection and Hash-Consing*: Global node hashing and reference counting maximize sharing and reclaim unused nodes during operations.

Decision-diagram techniques provide a numeric-algebraic, yet symbolic, approach to equivalence checking and operator comparison. They serve as a versatile core for quantum verification tools, capable of handling both state and process descriptions up to several hundred qubits in favorable cases.

4.2 Circuit-Level Simulation and Functional Testing

Before executing circuits on quantum hardware, performing in-depth validation using high-performance classical simulators helps uncover errors early in the workflow [77]. These engines evolve either the full 2^n-amplitude statevector or the $2^n \times 2^n$ density matrix through each gate, yielding exact wavefunction amplitudes or mixed-state statistics for comparison against analytic benchmarks. By enabling user-defined noise channels such as depolarizing, amplitude-damping, or phase-damping maps parameterized by T_1 and T_2, developers can precisely quantify the impact of decoherence and gate infidelity on algorithmic figures of merit. Advanced implementations exploit tensor-network contraction, sparse matrix techniques, and distributed MPI parallelism to extend simulation to 30+ qubits in the statevector regime and 20+ qubits for full density-matrix models. Functional tests can help assert that critical observables (e.g., $\langle Z_i Z_j \rangle$ correlations in quantum chemistry) or complete probability distributions fall within predefined tolerances, flagging issues such as compiler miscompilations, incorrect gate decompositions, or angle-rounding artifacts. Simulators also serve as a sandbox for validating error-mitigation protocols by replaying calibration-derived noise profiles and measuring their corrective effectiveness. While exponential resource requirements ultimately limit classical emulation to small-scale circuits, this simulation-driven validation remains the cornerstone of quantum software quality assurance, enabling confident benchmarking before incurring precious quantum runtime.

4.2.1 Tensor-Network Sampling Verification

Tensor-network contraction methods decompose large quantum circuits into smaller tensors that can be efficiently contracted to estimate output probabilities. The `qFlex` framework implements this approach by slicing and reordering tensor networks to verify sampling-style circuits at scales beyond naive statevector limits [78]. Its memory-efficient slicing and configurable contraction ordering allow verification of random circuits with 40+ qubits under favorable topologies. However, performance degrades sharply for densely entangled or irregular connectivity patterns, and because it targets amplitude sampling rather than full state reconstruction, it cannot directly validate expectation-value-driven algorithms without additional processing.

4.2.2 Parameterized Statevector and Density-Matrix Simulation

High-performance simulators that support both pure and mixed-state evolution enable functional testing of quantum kernels under ideal or noisy conditions. `Qulacs` provides a modular engine for propagating either the 2^n-amplitude statevector or the $2^n \times 2^n$ density matrix, with user-defined noise channels for prototyping error-mitigation strategies [79]. It excels at unit-testing circuits up to 30 qubits in statevector mode and 20 qubits for density-matrix simulations, and its flexible backend supports GPU (Graphics Processing Unit) acceleration or multi-threading. The trade-off is that its general-purpose architecture lags behind specialized tensor-network simulators on very large circuits, and density-matrix mode is limited by quadratic memory scaling.

4.2.3 Distributed Simulation with Noise Injection

Scalable emulation frameworks leverage distributed computing to reproduce both ideal and noisy circuit outputs across HPC resources. `QuEST` uses MPI (Message Passing Interface) and GPU offloading to evolve statevectors or density matrices across multi-node clusters, supporting common noise models for realistic pre-hardware validation [80]. Its near-linear strong scaling makes it suitable for end-to-end debugging of complex algorithms, and built-in noise channels allow assessment of decoherence impacts. On the downside, configuring MPI environments adds operational complexity, and performance gains diminish for poorly partitioned circuits, with exponential scaling ultimately capping practical use to tens of qubits.

4.2.4 Stimuli-Based Input-Output Testing

Stimuli-based testing exercises quantum circuits with targeted input states like computational-basis vectors, GHZ seeds, or randomized Pauli eigenstates, and compares the resulting output distributions against analytic or simulated references [34, 35]. By focusing on corner-case inputs (e.g., high-entanglement seeds or known eigenstates), this method can reveal logic errors, miscompiled primitives, or unexpected noise resonances with minimal overhead. It complements full state vector or sampling verification by reducing the simulation footprint: Only the selected inputs are simulated or executed on hardware, rather than the entire algorithmic space. However, its effectiveness depends on the choice of representative stimuli, as poorly chosen inputs may miss subtle errors, and exhaustive coverage quickly becomes infeasible as the size of the circuit increases.

4.2.5 Local Simulation and Testing Utilities

Integrated SDK (Software Development Kit) simulators provide convenient on-device testing of circuit behavior before hardware submission. Amazon Braket's local testing utilities offer both statevector and shot-based simulators, pre-configured to compare user-specified observables or bit-string distributions against reference data [81]. The seamless integration into the Braket SDK accelerates development by catching miscompilations and logical errors on users' workstations without quantum runtime fees. However, these simulators are limited to the resources of a single node, lack custom tensor-network methods, and depend on cloud-managed SDK versions, which can introduce latency and reproducibility issues.

4.3 Tomographic and Statistical Characterization

Tomographic and statistical characterization techniques form a cornerstone of quantum device assessment by leveraging measurement outcomes to infer underlying error processes. These methods trade off comprehensive insight for resource efficiency, ranging from scalable, aggregate error metrics to detailed reconstructions of quantum states and channels. By harnessing carefully designed experiment sequences and advanced data-processing algorithms, practitioners can quantify both average and worst-case performance across their gate sets. Although inherently limited by sampling complexity and classical post-processing costs, these approaches remain indispensable for calibrating hardware, validating error-mitigation strategies, and benchmarking progress toward fault-tolerant operation.

4.3.1 Randomized Benchmarking and Gate-Set Tomography

To obtain both coarse and fine-grained characterizations of gate errors, practitioners combine scalable statistical benchmarks with comprehensive tomographic reconstructions. Randomized Benchmarking (RB) estimates the average fidelity of native gate operations by applying long sequences of random Clifford gates and fitting the exponential decay of the survival probability as a function of sequence length [82]. Interleaved RB refines this approach by inserting a target gate within each random sequence, thereby isolating its specific error rate without conflating state preparation and measurement (SPAM) imperfections [83]. While RB excels at rapid, SPAM-robust fidelity estimation across multiple qubits, it only yields aggregate error rates and cannot separate coherent miscalibrations from stochastic noise. Gate-Set Tomography (GST) complements RB by performing a self-consistent, high-resolution reconstruction of all preparation, gate, and measurement processes: GST fits a detailed process-matrix model to thousands of carefully chosen circuit fragments, disentangling SPAM errors and extracting coherent offsets (δ, $\Delta\omega$, χ) versus stochastic rates ($1/T_1$, $1/T_2$) [84]. Although GST provides the deepest insight into the structure of errors, its resource demands—both in the number of experiments and the complexity of solving large semidefinite programs—limit its applicability to small qubit subsets. By using RB for rapid system-level benchmarking and GST for targeted diagnostics, verification workflows gain both efficiency and diagnostic depth, enabling informed error-mitigation strategies and reliable performance guarantees.

Beyond these core methods, RB protocols often employ multiple sequence lengths (typically 5–50 Cliffords) and dozens of random seeds to robustly average out sequence-specific fluctuations, yielding a decay parameter p from which the average gate infidelity $r = (d-1)(1-p)/d$ is derived for a d-dimensional system. Modern implementations use cycle benchmarking to isolate error contributions of logical layers or native gate cycles, providing more granular per-layer error budgets. In GST, the choice of fiducial and germ circuits, selected to amplify different error modes, ensures that both low-rank and full-rank errors are detectable, and iterative fitting procedures (such as long-sequence GST) improve the conditioning of the estimator. Toolkits like pyGSTi automate gauge optimization and bootstrap-based confidence region estimation, allowing experimenters to assess the statistical significance of extracted error parameters [85]. When integrated into calibration routines, these methods can track drift and detect calibration regressions by comparing successive RB decay rates or GST parameter sets, closing the loop between characterization and active error suppression.

4.3.2 Quantum State and Process Tomography

Quantum state tomography (QST) and quantum process tomography (QPT) form the gold standard for ground-truth verification of quantum hardware by reconstructing complete density matrices and superoperators from measurement data. In QST, an ensemble of identically prepared states is measured in a tomographically complete basis (e.g., Pauli X, Y, and Z), and the resulting frequencies are used—via linear inversion or maximum-likelihood estimation—to recover the full $2^n \times 2^n$ density matrix, revealing preparation errors, decoherence effects, and leakage [86]. QPT generalizes this procedure to black-box channels by sending a set of informationally complete probe states through the operation under test, performing QST on each output, and assembling the process matrix in Choi or Kraus form to expose coherent miscalibrations, crosstalk, and stochastic noise [87]. By comparing these tomographic reconstructions against ideal models, practitioners can quantify worst-case infidelities, validate calibration routines, and benchmark the effectiveness of error-mitigation strategies. However, the naive implementation of tomography incurs steep experimental and computational costs—$O(4^n)$ measurement settings and $O(16^n)$ classical post-processing—which confines its practical use to small subsystems ($n \leq 4$). Compressed-sensing tomography mitigates this by exploiting low-rank structure: assuming a rank-r state or channel reduces measurements to $O(r2^n)$ and replaces direct inversion with scalable convex optimization [88]. Even so, tomography remains best suited for module-level validation and for generating high-fidelity benchmarks that bootstrap scalable statistical characterization protocols.

4.4 Inline Assertions and Quantum Hoare Logic

Embedding executable specifications directly into quantum programs can catch semantic violations early in the development cycle, bridging the gap between high-level algorithm intent and low-level circuit behavior. Inline assertions let developers declare invariants, parameter bounds, or expected measurement properties alongside circuit definitions, enabling automated checks during simulation or hardware execution. Quantum Hoare logic extends this idea into a formal proof system, allowing rigorous pre- and post-condition reasoning about quantum operations. Together, these approaches promise to transform ad-hoc testing into an auditable, compiler-enforced verification discipline, reducing silent errors and improving reproducibility across toolchains.

4.4.1 Runtime Assertions for Quantum Programs

In classical software development, assertions are inline checks that validate program invariants, pre-conditions, or post-conditions during execution, immediately flagging deviations from expected behavior. They serve as both documentation and automated guardrails, catching bugs early and simplifying debugging by localizing errors at the point of failure. In the quantum domain, where opaque compiler optimizations and hardware noise can silently introduce errors, runtime assertions play an equally critical role in ensuring correctness. Quantum assertions embed executable checks directly into circuits or high-level quantum code, enabling automatic verification of state properties, parameter bounds, and measurement outcomes as the program runs. By coupling the primary computation to ancilla registers or classical control logic, these checks can detect semantic or hardware anomalies without requiring separate end-to-end benchmarking.

Early research introduced projection-based assertions, using projectors Π to verify subspace membership, such as confirming Bell pairs or code-space constraints by measuring only ancillas and leaving the main register largely undisturbed [89]. SWAP-test assertions extend this concept by comparing the on-chip state against a reference via controlled-SWAP circuits, flagging mismatches on a single ancilla measurement without collapsing the entire state [91]. Non-Destructive Discrimination (NDD) circuits further generalize runtime checks by employing phase estimation or entanglement-discrimination protocols to validate complex superposition structures (e.g., GHZ or W states) mid-circuit without perturbing computational qubits [92, 93]. Beyond these foundational techniques, a rich taxonomy of quantum assertions has emerged: parameter-range assertions enforce calibrated bounds on rotation angles or pulse amplitudes; expectation-value assertions ensure observable statistics (such as $\langle Z_i Z_j \rangle$) lie within tolerances; and measurement-outcome assertions verify that critical bit-string patterns occur with sufficient probability [37]. Statistical assertions apply formal hypothesis tests such as confidence intervals or χ^2 criteria, over repeated executions to certify probabilistic guarantees, while approximate assertions adopt Bloom-filter-inspired checks to test membership in a broader codeword set when exact target states are unknown [94]. Additional specialized checks include purity assertions to validate eigenvalue spectra of mixed states, temporal assertions to enforce gate ordering and timing constraints, and resource assertions to monitor ancilla usage and circuit depth against design budgets.

These diverse assertion mechanisms incur only modest overhead—typically a few ancillas and controlled operations—and many designs minimize disturbance through non-destructive or post-selective implementations. When strategically placed during state preparation, mid-circuit computation, or immediately before measurement, runtime assertions provide immediate, localized fault detection, significantly reducing the debugging effort compared to black-box sampling tests. The main challenges lie in selecting assertion types that effectively cover likely failure modes, managing the additional circuit depth, and integrating assertion embedding into existing quantum toolchains. As quantum development environments evolve, a standardized, extensible assertion DSL with built-in compiler support will be essential for automating contract enforcement, rapidly isolating errors, and elevating quantum software quality assurance from ad-hoc tests to a disciplined engineering practice.

4.4.2 Quantum Hoare Logic

Quantum Hoare logic lifts classical Hoare triples $\{P\}\ C\ \{Q\}$ into the quantum setting by interpreting P and Q as projection operators or more general Hermitian predicates on Hilbert space, and C as a completely positive, trace-preserving map acting on density operators. Ying's foundational system defines axioms and inference rules for unitary commands ($\{U^\dagger P U\}\ U\ \{P\}$), measurements ($\{M_k^\dagger P M_k\}\ \texttt{measure}\ \{P\}$), and sequential composition, furnishing a sound and relatively complete proof calculus for partial correctness of purely quantum programs [95]. Feng *et al.* further enrich this framework by introducing rules for classical control flow—if-then-else and while loops—and mixed quantum–classical state representations, enabling verification of hybrid algorithms that interleave measurement with classical post-processing [90].

Building on these advances, Huang et al. develop a domain-specific Hoare logic tailored to stabilizer-based error-correction circuits: they formalize Pauli-predicate assertions, prove soundness and relative completeness in the Coq proof assistant, and implement the Veri-QEC toolchain, which reduces proof obligations to SMT queries and scales to hundreds of qubits under both Pauli and certain non-Pauli noise models [96]. Modern quantum Hoare frameworks also support symbolic execution engines that automatically derive verification conditions from annotated code, and tactic libraries that encapsulate common proof patterns—such as ancilla allocation and deallocation, gate commutation, and code-concatenation lemmas—to dramatically cut down manual proof effort. Integration with interactive development environments (IDEs) can enable real-time proof state visualization, inline suggestion of lemma applications, and seamless transitions between proof and implementation. Finally, a promising hybrid approach couples Hoare logic's high-assurance guarantees for critical kernels with lightweight runtime assertions elsewhere in the code, combining the rigor of formal proofs with the flexibility of inline checks to achieve both soundness and practicality in large-scale quantum software verification.

4.4.3 Hybrid Simulation and Hardware-in-the-Loop Testing (Quantum Fuzzing)

Hybrid simulation and hardware-in-the-loop (HIL) testing combine classical emulation with real-device executions to uncover both software and hardware faults that escape purely offline verification. By generating test circuits through automated metamorphic transformations or symbolic exploration, practitioners alternately run them on simulators and quantum processors to detect inconsistencies caused by compiler bugs, gate miscalibrations, or unmodeled noise. Metamorphic testing frameworks like MorphQ automatically derive structurally related circuit variants whose ideal outputs obey known invariants, flagging any deviations between simulator and hardware results as potential errors [97]. Concolic testing approaches such as Quantum Concolic Testing symbolically explore execution paths, synthesize input states that maximize coverage of branching and measurement outcomes, and then replay these states on hardware to validate actual behavior [98]. In hardware-in-the-loop fuzzing, random or structured perturbations–such as injecting extra gates, varying pulse timings, or shuffling gate order—stress-test both the compilation stack and the device, revealing rare failure modes under realistic operating conditions. These HIL methods excel at catching emergent integration errors across the full software-hardware pipeline, but depend on quantum runtime access, incur queue and execution costs, and require careful alignment of simulated expectations with noisy measurements. When integrated into continuous integration pipelines, hybrid testing provides an essential safety net, continually validating the end-to-end toolchain against evolving hardware and compiler landscapes.

5 Emerging Directions

While the preceding strategies constitute a robust toolkit for ensuring correctness on small to medium-scale quantum devices, the rapid growth of qubit counts and circuit complexity demands new approaches that can keep pace with industrial-scale deployments. Automated specification inference seeks to extract actionable contracts—unitary targets, invariant subspaces, and statistical benchmarks—directly from high-level code or intermediate representations, eliminating much of the manual annotation burden. Fault localization techniques aim to move beyond binary pass/fail outcomes by pinpointing the precise gates, qubits, or transformations responsible for a deviation, using differential analysis, tomographic slicing, and machine-learning-driven hotspot detection. Compiler-assisted circuit repair closes the verification loop by automatically generating corrective rewrites—such as alternative decompositions, dynamic rerouting, or error-mitigation subcircuits—based on actual failure signatures. Finally, integration with machine learning introduces data-driven surrogates for noisy channel emulation, anomaly detection models for real-time fault prediction, and adaptive test generation to prioritize the most error-sensitive regions of a program. Together, these emerging directions promise to transform quantum verification from a largely manual, post-hoc activity into an integrated, self-healing engineering discipline capable of scaling with next-generation quantum hardware.

5.1 Automated Specification Inference

Manual creation of machine-readable contracts (unitary matrices, invariants, statistical goals) is labor-intensive and error-prone. Automated specification inference uses static analysis of high-level code or compiler IR to extract candidate pre/post-conditions, formal invariants, and performance metrics. Techniques such as symbolic execution over domain-specific IRs can mine subcircuit semantics, while pattern mining on annotated examples yields statistical goals (e.g., expected observable ranges) without human intervention. By generating executable specs on the fly, these tools reduce the annotation burden and enable verification frameworks to operate out-of-the-box on new kernels.

5.2 Fault Localization

Detecting a verification failure is only half the battle—pinpointing its origin in a deep, optimized circuit or complex toolchain is critical for efficient debugging. Fault localization methods combine differential analysis (comparing golden versus failing executions), spectral RB variants that identify error hotspots in gate sequences, and hybrid tomographic slicing to isolate malfunctioning subcircuits. Machine-learning approaches can correlate measurement deviations with compiler transformation logs, suggesting candidate fault regions. Rapid localization accelerates repair by focusing developer attention on the minimal set of gates or transformations responsible for the failure.

5.3 Compiler-Assisted Circuit Repair

Closing the loop between verification and synthesis, compiler-assisted repair systems automatically propose corrective transformations when checks fail. Given a counterexample circuit trace, these tools can replace problematic decompositions, reroute logical qubits around noisy hardware regions, or inject corrective subcircuits (e.g., dynamical decoupling or echo pulses). By integrating feedback from RB/GST or runtime assertions, the compiler can iteratively refine gate parameters and layouts to regain compliance with specified fidelity or behavioral contracts. Early prototypes demonstrate automated mitigation of coherent over-rotations and crosstalk by local pulse retuning driven by verification feedback.

5.4 Integration with Machine Learning

Data-driven models offer another avenue for scalable verification by learning compact surrogates of device behavior or by forecasting drift patterns. Neural network emulators can approximate noisy channel maps for rapid fidelity estimation, while anomaly-detection models flag unusual output patterns indicative of hardware faults. Reinforcement-learning agents trained on verification outcomes may guide test generation or circuit repairs, prioritizing checks where failures are most likely. As these ML techniques mature, they will complement formal and statistical methods, enhancing coverage and reducing human effort.

6 Conclusion

Quantum processors are advancing at a breakneck pace, integrating ever-more qubits, heterogeneous control electronics, and sophisticated error-mitigation layers into cohesive systems. As device complexity grows, traditional verification methods, while indispensable, struggle to keep up with the intertwined challenges of compiler optimizations, coherent control drifts, and stochastic noise. In this chapter, we have shown that a multifaceted verification stack, spanning algebraic equivalence checking, statistical benchmarking, tomographic reconstruction, inline assertions, and hardware-in-the-loop fuzzing, is essential to cover the full spectrum of failure modes. Yet, even this arsenal must evolve: automated inference of machine-readable specifications will reduce annotation overhead, precise fault localization will pinpoint the root causes of errors, and compiler-assisted repair loop, enhanced by machine-learning-driven emulators and adaptive test generation will close the verification loop in near real time. Embedding security-grade rigor into quantum software development is no longer optional; it must be woven into every phase of the design, from high-level algorithms to low-level pulse sequences, to preemptively guard against subtle bugs and hardware anomalies. By embracing these emerging directions and unifying them into extensible toolchains, we can forge a resilient foundation for fault-tolerant quantum computing that scales hand-in-hand with hardware innovation.

References

1. Vedral, V., and M. B. Plenio. 1998. Basics of quantum computation. *Progress in Quantum Electronics* 22 (1): 1–39. https://doi.org/10.1016/S0079-6727(98)00004-4.
2. Jiang, Y. Y., C. Q. Deng, H. Fan, B. Y. Li, L. Y. Sun, X. S. Tan, W. T. Wang, G. M. Xue, F. Yan, H. F. Yu, et al. 2025. Advancements in superconducting quantum computing. *National Science Review* 12: nwaf246
3. Bruzewicz, C. D., J. Chiaverini, R. McConnell, and J. M. Sage. 2019. Trapped-ion quantum computing: Progress and challenges. *Applied Physics Reviews* 6(2)
4. Tilly, J., H. Chen, S. Cao, D. Picozzi, K. Setia, Y. Li, E. Grant, L. Wossnig, I. Rungger, G. H. Booth, et al. 2022. The variational quantum eigensolver: A review of methods and best practices. *Physics Reports* 986: 1–128.
5. Choi, J., and J. Kim. 2019. A tutorial on quantum approximate optimization algorithm (QAOA): Fundamentals and applications. In *Proceedings of the 2019 international conference on information and communication technology convergence (ICTC)*, 138–142. IEEE.
6. Wootters, W. K., and W. H. Zurek. 2009. The no-cloning theorem. *Physics Today* 62 (2): 76–77.
7. James, D. F. V., P. G. Kwiat, W. J. Munro, and A. G. White. 2001. Measurement of qubits. *Physical Review A* 64 (5): 052312.
8. Shah, S. M., and M. Irfan. 2005. Embedded hardware/software verification and validation using hardware-in-the-loop simulation. In *Proceedings of the IEEE symposium on emerging technologies*, 494–498. IEEE.
9. Cramer, M., M. B. Plenio, S. T. Flammia, R. Somma, D. Gross, S. D. Bartlett, O. Landon-Cardinal, D. Poulin, and Y.-K. Liu. 2010. Efficient quantum state tomography. *Nature Communications* 1 (1): 149.
10. Knill, E. 2010. Quantum computing. *Nature* 463 (7280): 441–443.
11. Greenbaum, D., and Z. Dutton. 2017. Modeling coherent errors in quantum error correction. *Quantum Science and Technology* 3 (1): 015007.
12. Gutiérrez, M., C. Smith, L. Lulushi, S. Janardan, and K. R. Brown. 2016. Errors and pseudothresholds for incoherent and coherent noise. *Physical Review A* 94 (4): 042338.
13. Holmes, A., S. Johri, G. G. Guerreschi, J. S. Clarke, and A. Y. Matsuura. 2020. Impact of qubit connectivity on quantum algorithm performance. *Quantum Science and Technology* 5 (2): 025009.
14. Burnett, J. J., A. Bengtsson, M. Scigliuzzo, D. Niepce, M. Kudra, P. Delsing, and J. Bylander. 2019. Decoherence benchmarking of superconducting qubits. *npj Quantum Information* 5 (1): 54.
15. Heng, S., D. Kim, S. Heng, and Y. Han. 2022. Decomposition analysis of quantum native gates on various quantum computers. In *Proceedings of the 37th international technical conference on circuits/systems, computers and communications (ITC-CSCC)*, 1–3. IEEE.
16. Wille, R., A. Lye, and R. Drechsler. 2014. Optimal SWAP gate insertion for nearest-neighbor quantum circuits. In *Proceedings of the 2014 19th Asia and South Pacific design automation conference (ASP-DAC)*, 489–494. IEEE.
17. Sanders, Y. R., J. J. Wallman, and B. C. Sanders. 2015. Bounding quantum gate error rate based on reported average fidelity. *New Journal of Physics* 18 (1): 012002.
18. Arute, F., K. Arya, R. Babbush, D. Bacon, J. C. Bardin, R. Barends, R. Biswas, S. Boixo, F. G. S. L. Brandão, D. A. Buell, et al. 2019. Quantum supremacy using a programmable superconducting processor. *Nature* 574 (7779): 505–510.
19. Dalton, K., C. K. Long, Y. S. Yordanov, C. G. Smith, C. H. W. Barnes, N. Mertig, and D. R. M. Arvidsson-Shukur. 2024. Quantifying the effect of gate errors on variational quantum eigensolvers for quantum chemistry. *npj Quantum Information* 10 (1): 18.
20. IBM Quantum: Qiskit Transpiler API. IBM Quantum Documentation. Available via IBM Quantum. https://quantum.cloud.ibm.com/docs/en/api/qiskit/transpiler. Accessed 21 July 2025.
21. Cornelissen, A., J. Bausch, and A. Gilyén. 2021. Scalable benchmarks for gate-based quantum computers. arXiv:2104.10698 [quant-ph].

22. Zhu, D., Z. P. Cian, C. Noel, A. Risinger, D. Biswas, L. Egan, Y. Zhu, A. M. Green, C. H. Alderete, N. H. Nguyen, et al. 2021. Cross-platform comparison of arbitrary quantum computations. arXiv:2107.11387 [quant-ph].
23. Pelofske, E., J. Golden, A. Bärtschi, D. O'Malley, and S. Eidenbenz. 2021. Sampling on NISQ devices: "Who's the fairest one of all?". In *Proceedings of the 2021 IEEE international conference on quantum computing and engineering (QCE)*, 207–217. IEEE.
24. Knill, E., D. Leibfried, R. Reichle, J. Britton, R. B. Blakestad, J. D. Jost, C. Langer, R. Ozeri, S. Seidelin, and D. J. Wineland. 2008. Randomized benchmarking of quantum gates. *Physical Review A* 77 (1): 012307.
25. Zheng, Y.-C., C.-Y. Lai, and T. A. Brun. 2018. Efficient preparation of large-block-code ancilla states for fault-tolerant quantum computation. *Physical Review A* 97 (3): 032331.
26. Nature. 2023. Suppressing quantum errors by scaling a surface code logical qubit. *Nature* 614 (7949): 676–681.
27. Alavi, S. A., S. Ishtiaq, N. Johnson, R. Mishra, D. Oruganti Nagalakshmi, A. Pearl, and J. Snoeijs. 2023. Design verification of the quantum control stack. arXiv:2310.05229 [quant-ph].
28. Quantinuum: Performance validation. Quantinuum documentation. Available via Quantinuum. Accessed 21 July 2025.
29. Liu, J., B. Zhan, S. Wang, S. Ying, T. Liu, Y. Li, M. Ying, and N. Zhan. 2019. Formal verification of quantum algorithms using quantum Hoare logic. In *Proceedings of the international conference on computer aided verification*, 187–207. Springer.
30. Aharonov, D., M. Ben-Or, E. Eban, and U. Mahadev. 2017. Interactive proofs for quantum computations. arXiv:1704.04487 [quant-ph]
31. IBM Quantum: Quantum reliability. IBM Quantum Blog. Available via IBM Quantum. https://www.ibm.com/quantum/blog/quantum-reliability. Accessed 21 July 2025.
32. Peham, T., L. Burgholzer, and R. Wille. 2022. Equivalence checking of quantum circuits with the ZX-calculus. *IEEE Journal on Emerging and Selected Topics in Circuits and Systems* 12 (3): 662–675.
33. Burgholzer, L., and R. Wille. 2020. Improved DD-based equivalence checking of quantum circuits. In *Proceedings of the 2020 25th Asia and South Pacific design automation conference (ASP-DAC)*, 127–132. IEEE.
34. Burgholzer, L., R. Kueng, and R. Wille. 2021. Random stimuli generation for the verification of quantum circuits. In *Proceedings of the 26th Asia and South Pacific design automation conference*, 767–772.
35. Lu, C., N. Choudhury, and K. Basu. 2024. QuaSi: A scalable and reliable quantum simulation-based equivalence checking framework. In *Proceedings of the 2024 IEEE international conference on quantum computing and engineering (QCE)*, vol. 1, 1037–1047. IEEE.
36. Niemann, P., et al. 2015. QMDDs: Efficient quantum function representation and manipulation. *IEEE Transactions on Computer-Aided Design of Integrated Circuits Systems* 35 (1): 86–99.
37. Huang, Y., and M. Martonosi. 2019. Statistical assertions for validating patterns and finding bugs in quantum programs. In *Proceedings of the 46th international symposium on computer architecture*, 541–553
38. Krantz, P., M. Kjaergaard, F. Yan, T. P. Orlando, S. Gustavsson, and W. D. Oliver. 2019. A quantum engineer's guide to superconducting qubits. *Applied Physics Reviews* 6 (2): 021318.
39. Zhao, P., K. Linghu, Z. Li, P. Xu, R. Wang, G. Xue, Y. Jin, and H. Yu. 2022. Quantum crosstalk analysis for simultaneous gate operations on superconducting qubits. *PRX Quantum* 3 (2): 020301.
40. Ficheux, Q., et al. 2018. Dynamics of a qubit while simultaneously monitoring its relaxation and dephasing. *Nature Communications* 9 (1): 1926.
41. Geller, M. R. 2020. Rigorous measurement error correction. *Quantum Science and Technology* 5 (3): 03LT01.
42. Tannu, S. S., and M. K. Qureshi. 2019. Not all qubits are created equal: A case for variability-aware policies for NISQ-era quantum computers. In *Proceedings of the 24th international conference on architectural support for programming languages and operating systems (ASP-LOS '19)*, 1–14. ACM.

43. Das, P., et al. 2021. ADAPT: Mitigating idling errors in qubits via adaptive dynamical decoupling. In *Proceedings of the 54th annual IEEE/ACM international symposium on microarchitecture (MICRO-54)*. IEEE.
44. Sander, A., L. Burgholzer, and R. Wille. 2025. Equivalence checking of quantum circuits via intermediary matrix product operator. *Physics Review Research* 7 (2): 023261.
45. Kissinger, A., and J. van-de-Wetering. 2019. PyZX: Large-scale automated diagrammatic reasoning. arXiv:1904.04735 [quant-ph]
46. Mundada, P., et al. 2019. Suppression of qubit crosstalk in a tunable-coupling superconducting circuit. *Physical Review Applied* 12 (5): 054023.
47. Choudhury, N., et al. 2024. Crosstalk-induced side channel threats in multi-tenant NISQ computers. arXiv:2412.10507 [quant-ph]
48. Zhou, J., et al. 2021. Nonadiabatic geometric quantum gates that are insensitive to qubit-frequency drifts. *Physical Review A* 103 (3): 032609.
49. Chen, Z., et al. 2016. Measuring and suppressing quantum state leakage in a superconducting qubit. *Physical Review Letters* 116 (2): 020501.
50. Tannu, S. S., and M. K. Qureshi. 2019. Mitigating measurement errors in quantum computers by exploiting state-dependent bias. In *Proceedings of the 52nd Annual IEEE/ACM international symposium on microarchitecture*. IEEE/ACM.
51. Hicks, R., et al. 2022. Active readout-error mitigation. *Physical Review A* 105 (1): 012419.
52. Jackson, C., and S. J. Van Enk. 2015. Detecting correlated errors in state-preparation-and-measurement tomography. *Physical Review A* 92 (4): 042312.
53. Braginsky, V. B., Y. I. Vorontsov, and K. S. Thorne. 1980. Quantum nondemolition measurements. *Science* 209 (4456): 547–557.
54. Seo, S., and J. Bae. 2021. Measurement crosstalk errors in cloud-based quantum computing. *IEEE Internet Computing* 26 (1): 26–33.
55. Smith, K. N., et al. 2022. Programming physical quantum systems with pulse-level control. *Frontiers of Physics* 10: 900099.
56. Vartiainen, J.-J., M. Möttönen, and M. M. Salomaa. 2004. Efficient decomposition of quantum gates. *Physical Review Letters* 92 (17): 177902.
57. Barenco, A., C. H. Bennett, R. Cleve, D. P. DiVincenzo, N. Margolus, P. Shor, T. Sleator, J. A. Smolin, and H. Weinfurter. 1995. Elementary gates for quantum computation. *Physical Review A* 52 (5): 3457–3467.
58. Valmari, A. 1996. The state explosion problem. In *Advanced course on petri nets*, vol. 1091, ed. W. Reisig and G. Rozenberg. Lecture Notes in Computer Science, 429–528. Berlin, Heidelberg: Springer.
59. Mohseni, M., A. T. Rezakhani, and D. A. Lidar. 2008. Quantum-process tomography: Resource analysis of different strategies. *Physical Review A* 77 (3): 032322.
60. Nielsen, M. A., and I. L. Chuang. 2000. *Quantum computation and quantum information*. Cambridge: Cambridge University Press.
61. Gilchrist, A., N. K. Langford, and M. A. Nielsen. 2005. Distance measures to compare real and ideal quantum processes. *Physical Review A* 71 (6): 062310.
62. Aleksandrowicz, G., et al. 2019. Qiskit: An open-source framework for quantum computing. Zenodo.
63. Cirq Developers. 2018. Cirq: A Python framework for NISQ quantum circuits. arXiv:1811.00933 [quant-ph]
64. Amazon Web Services: Amazon braket: A fully managed quantum computing service. Available via AWS. https://aws.amazon.com/braket. Accessed 21 July 2025.
65. Bergholm, V., J. Izaac, M. Schuld, C. Gogolin, N. Alam, and P. Rebentrost. 2018. PennyLane: Automatic differentiation of hybrid quantum–classical computations. arXiv:1811.04968 [quant-ph]
66. Svore, K. M., A. W. Cross, C. Chiu, O. deWeck, M. Hauschild, K. S. Kumar, et al. 2019. Qsharp: Enabling scalable quantum computing development with a high-level programming language. In *Proceedings of the 24th ACM international conference on architectural support for programming languages and operating systems (ASPLOS '19)*, 1–12. ACM.

67. Resch, S., and U. R. Karpuzcu. 2021. Benchmarking quantum computers and the impact of quantum noise. *ACM Computing Surveys* 54 (7): 1–35.
68. Morimae, T., and K. Fujii. 2012. Not all physical errors can be linear CPTP maps in a correlation space. *Science and Reports* 2: 508.
69. Manzano, D. 2020. A short introduction to the Lindblad master equation. *AIP Advances* 10 (2): 025106.
70. Serena, P., and A. Bononi. 2015. A time-domain extended Gaussian noise model. *Journal of Lightwave Technology* 33 (7): 1459–1472.
71. Brand, D., I. Sinayskiy, and F. Petruccione. 2024. Markovian noise modelling and parameter extraction framework for quantum devices. *Science and Reports* 14 (1): 4769.
72. Fuliński, A. 1994. Non-Markovian noise. *Physical Review E* 50 (4): 2668.
73. Hou, S. C., S. L. Liang, and X. X. Yi. 2015. Non-Markovianity and memory effects in quantum open systems. *Physical Review A* 91 (1): 012109.
74. Chirolli, L., and G. Burkard. 2008. Decoherence in solid-state qubits. *Advances in Physics* 57 (3): 225–285. https://doi.org/10.1080/00018730802218067.
75. van-de-Wetering, J. 2020. ZX-calculus for the working quantum computer scientist. arXiv:2012.13966 [quant-ph]
76. Viamontes, G. F., I. L. Markov, and J. P. Hayes. 2003. Gate-level simulation of quantum circuits. In *Proceedings of the 2003 Asia and South Pacific design automation conference*, 820–825. IEEE.
77. Doi, J., et al. 2019. Quantum computing simulator on a heterogeneous HPC system. In *Proceedings of the 16th ACM international conference on computing frontiers*. ACM.
78. Villalonga, B., S. Boixo, B. Nelson, C. Henze, E. Rieffel, R. Biswas, and S. Mandrà. 2019. A flexible high-performance simulator for verifying and benchmarking quantum circuits implemented on real hardware. *npj Quantum Information* 5: 86. https://doi.org/10.1038/s41534-019-0196-1
79. Suzuki, Y., Y. Kawase, Y. Masumura, Y. Hiraga, M. Nakadai, et al. 2021. Qulacs: A fast and versatile quantum circuit simulator for research purpose. *Quantum* 5: 559. https://doi.org/10.22331/q-2021-10-06-559.
80. Jones, T., A. Brown, I. Bush, and S. Benjamin. 2019. QuEST and high-performance simulation of quantum computers. *Science and Reports* 9: 10736. https://doi.org/10.1038/s41598-019-47174-9.
81. Amazon Web Services: Testing your quantum tasks with amazon braket. AWS Developer Guide. Available via AWS. https://docs.aws.amazon.com/braket/latest/developerguide/braket-test.html. Accessed 21 July 2025.
82. Magesan, E., J. M. Gambetta, and J. Emerson. 2011. Scalable and robust randomized benchmarking of quantum processes. *Physical Review Letters* 106: 180504.
83. Magesan, E., J. M. Gambetta, B. R. Johnson, et al. 2012. Efficient measurement of quantum gate error by interleaved randomized benchmarking. *Physical Review Letters* 109: 080505. https://doi.org/10.1103/PhysRevLett.109.080505.
84. Nielsen, E., J. K. Gamble, K. Rudinger, T. Scholten, K. Young, and R. Blume-Kohout. 2021. Gate set tomography. *Quantum* 5: 557. https://doi.org/10.22331/q-2021-10-05-557.
85. Nielsen, E., K. Rudinger, T. Proctor, A. Russo, K. Young, and R. Blume-Kohout. 2020. Probing quantum processor performance with pyGSTi. *Quantum Science and Technology* 5 (4): 044002. https://doi.org/10.1088/2058-9565/ab8aa4.
86. James, D. F. V., P. G. Kwiat, W. J. Munro, and A. G. White. 2001. Measurement of qubits. *Physical Review A* 64: 052312. https://doi.org/10.1103/PhysRevA.64.052312.
87. Poyatos, J. F., J. I. Cirac, and P. Zoller. 1997. Complete characterization of a quantum process: The two-bit quantum gate. *Physical Review Letters* 78: 390–393. https://doi.org/10.1103/PhysRevLett.78.390.
88. Gross, D., Y. K. Liu, S. T. Flammia, S. Becker, and J. Eisert. 2010. Quantum state tomography via compressed sensing. *Physics Review Letters* 105: 150401. https://doi.org/10.1103/PhysRevLett.105.150401

89. Li, G., L. Zhou, N. Yu, Y. Ding, and M. Ying. 2020. Projection-based runtime assertions for testing and debugging quantum programs. *Proceedings of the ACM on Programming Languages* 4 (OOPSLA): 217:1–217:29. https://doi.org/10.1145/3428218
90. Feng, Y., and M. Ying. 2021. Quantum Hoare logic with classical variables. *ACM Transactions Quantum Computing* 2 (4). https://doi.org/10.1145/3456877
91. Liu, J., G. T. Byrd, and H. Zhou. 2020. Quantum circuits for dynamic runtime assertions in quantum computation. In *Proceedings of the 25th international conference on architectural support for programming languages and operating systems (ASPLOS '20)*, 1017–1030. ACM, New York, NY, USA. https://doi.org/10.1145/3373376.3378488
92. Gupta, M., A. Pathak, R. Srikanth, and P. K. Panigrahi. 2005. Non-destructive orthonormal state discrimination. In *Proceedings of the 2005 IEEE International Conference on Quantum Computation and Engineering*, 57–60. IEEE.
93. Manu, V. S., and A. Kumar. 2011. Non-destructive discrimination of arbitrary sets of orthogonal quantum states by NMR using quantum phase estimation. arXiv preprint arXiv:1105.2186 [quant-ph]
94. Liu, J., and H. Zhou. 2021. Systematic approaches for precise and approximate quantum state runtime assertion. In *Proceedings of the 2021 IEEE international symposium on high-performance computer architecture (HPCA)*. IEEE.
95. Ying, M. 2012. Floyd–Hoare logic for quantum programs. *ACM Transactions on Programming Languages and Systems 33* (6): 19:1–19:49. https://doi.org/10.1145/2049706.2049708
96. Huang, T., H. Wang, M. Li, F. Zhang, and X. Guo. 2025. Veri-QEC: Automated formal verification of quantum error-correction programs. In *Proceedings of the 46th ACM SIGPLAN conference on programming language design and implementation (PLDI '25)*, 123–136. ACM. https://doi.org/10.1145/3729293
97. Paltenghi, M., and M. Pradel. 2023. MorphQ: Metamorphic testing of the Qiskit quantum computing platform. In *Proceedings of the 45th international conference on software engineering (ICSE)*. ACM. https://doi.org/10.1109/ICSE48619.2023.00202
98. Xia, S., J. Zhao, F. Zhang, and X. Guo. 2025. Quantum concolic testing. arXiv preprint arXiv:2405.04860 [quant-ph]

Quantum Circuit Simulation

Masahiro Fujita and Yusuke Kimura

Abstract As the current quantum computers are still noisy and do not accommodate sufficiently many qubits for practical programs or quantum circuit developments, in order to make sure the correctness and properties of quantum circuits, it makes sense to simulate them on top of classical computers, where no error is expected. There have been proposed varieties of algorithms/methods by which behaviors of given quantum circuits can be generated, i.e., quantum circuit simulation. We review those techniques with discussions of their advantages/disadvantages, and give their evaluations and comparisons by actually running them on benchmark circuits. Finally, we discuss extensions of quantum circuit simulations including symbolic simulations for more robust verification.

1 Introduction

There has been significant progress in implementing and realizing quantum computers for the last several years, and some machines accommodate over 100 or even 1,000 qubits [1]. On the other hand, current quantum computers are noisy and may not be so reliable nor robust, resulting in situations that the computed results are not correct with high probability. In order for a quantum computer today to be reliable, some types of error correction mechanisms are essential, but current error correction methods need more than 1,000 qubits to create a reliable qubit (called "logical" qubit) [2]. Generally speaking, it is not at all easy to develop quantum circuits or quantum software on top of such unreliable machines. Instead, it makes much more sense to simulate the behaviors of quantum circuits and quantum software on top of

M. Fujita (✉)
University of Tokyo, and National Institute of Advanced Industrial Science and Technology (AIST), Tokyo, Japan
e-mail: fujitautokyo@gmail.com

Y. Kimura
Fujitsu Ltd., Kawasaki, Japan
e-mail: yusuke-kimura@fujitsu.com

P. Mishra (ed.), *Design Automation for Quantum Computing*,
https://doi.org/10.1007/978-3-032-09303-5_8

classical machines, where no computation error is expected, though such a simulation may take a long time.

There are several algorithms available for simulating the behavior of quantum circuits on top of classical computers. In this chapter, we review, evaluate, and compare those simulation algorithms and their associated simulator programs using benchmark quantum circuits, such as QASM Bench [3] and QCBM (used in Qulacs [4]). Each quantum circuit simulation algorithm has pros and cons, and works well for some quantum circuits and may not work well for some other quantum circuits. Here we examine and clarify those characteristics by actually running simulators on various benchmark circuits. That knowledge of the performance of quantum circuit simulators on specific types of quantum circuits can be utilized in an effective way to determine which simulation methods to use when developing new quantum algorithms/circuits.

The analysis of the performance of quantum circuit simulation methods is performed not only for one core sequential computation, but also for parallel computing viewpoints. We discuss how much speedup can be expected with the parallel implementation of the quantum circuit simulation methods. In general, parallel computing machines have more memory space globally than single-core machines, which could be useful to simulate large quantum circuits.

This chapter also discusses extensions of quantum circuit simulation so that some types of formal analysis can become feasible. With some types of quantum circuit simulation methods, formal methods that can analyze sets of behaviors of parameterized quantum circuits can be efficiently simulated, which may make a way to utilize existing various formal methods for classical computers in quantum circuit analysis.

This chapter follows the following agenda. We first quickly review the various quantum circuit simulation methods in the next section, followed by the explanation of their usage, experimental evaluations, and analysis of the comparisons in the following sections. Then we discuss extensions of some simulation methods for formal analysis by introducing parameterized representation and abstraction of decision diagrams (DD). The final section gives concluding remarks.

2 Techniques for Quantum Circuit Simulation on Top of Classical Computers

2.1 Quantum States, Quantum Gates, and Quantum Circuits

States of quantum computers can be expressed as quantum states of qubits. A quantum state of N qubits can be expressed as a vector of length 2^N complex numbers, since a state of one qubit is a superposition of value 0 and 1 and the entire state is a combination of states of qubits in quantum computers. Quantum computers are updating quantum states so that the final state gives the intended computational results.

$$I = \begin{bmatrix} 1 & 0 \\ 0 & 1 \end{bmatrix}, X = \begin{bmatrix} 0 & 1 \\ 1 & 0 \end{bmatrix}, H = \frac{1}{\sqrt{2}} \begin{bmatrix} 1 & 1 \\ 1 & -1 \end{bmatrix}, H \otimes I = \frac{1}{\sqrt{2}} \begin{bmatrix} 1 & 0 & 1 & 0 \\ 0 & 1 & 0 & 1 \\ 1 & 0 & -1 & 0 \\ 0 & 1 & 0 & 1 \end{bmatrix}, I \otimes H = \frac{1}{\sqrt{2}} \begin{bmatrix} 1 & 1 & 0 & 0 \\ 1 & -1 & 0 & 0 \\ 0 & 0 & 1 & 1 \\ 0 & 0 & 1 & -1 \end{bmatrix}, CX = \begin{bmatrix} 1 & 0 & 0 & 0 \\ 0 & 1 & 0 & 0 \\ 0 & 0 & 0 & 1 \\ 0 & 0 & 1 & 0 \end{bmatrix}$$

Fig. 1 Basic quantum gates and their matrix representations

Software running on quantum computers can be expressed as a quantum circuit. A quantum circuit consists of quantum gates, each of which evolves quantum states step by step, leading to the final computational results. A quantum gate may operate on only one qubit or multiple qubits together. There are a number of quantum gates that can be relatively easy to implement in quantum computers, and quantum software is developed as a collection of quantum circuits, which are sequences of quantum gates.

Some of the commonly used quantum gates are shown in Fig. 1. I is an identify gate, a special gate which does not change anything. Since most quantum gates are applied only to one or two qubits, behaviors of qubits which have not performed any quantum gates are represented as I gates. X is a complement gate which swaps 0 and 1, and H is called Hadamard gate which turns a state of 0 or 1 into an equal superposition of 0 and 1. CX is a controlled X gate which performs X on the target qubit if the control qubit is 1 and does nothing if it is 0.

If H is applied to the first qubit and I is applied to the second qubit (do nothing on the second qubit), the corresponding transformation can be expressed as $H \otimes I$ where $\otimes$ represents Kronecker product and its matrix representation is shown in the figure. If I is applied to the first qubit and H is applied to the second qubit, the corresponding transformation can be expressed as $I \otimes H$ whose matrix representation is also shown in the figure.

As can be seen above, behaviors of multiple quantum gates can be computed by combining matrices of quantum gates with normal matrix multiplications or Kronecker products. In the case of quantum simulation, we can compute matrices for such overall behaviors first as matrix–matrix operations, and then the resulting matrices are applied to the initial state of quantum circuits. Alternatively, starting with the initial states of the quantum circuit, matrix of each gate in the quantum circuit is applied one by one as matrix–vector operations.

2.2 *Explicit State Vector (SV) Representation of Quantum States*

As stated above, quantum states with N qubits can be precisely represented by a vector of length 2^N complex numbers, and their evolution by a quantum gate can be expressed as matrix–vector multiplication where the matrix precisely represents the transformation given by the quantum gate. So, using length 2^N vectors for quantum states and size $2^N \times 2^N$ matrix for quantum gates, quantum circuit simulation can

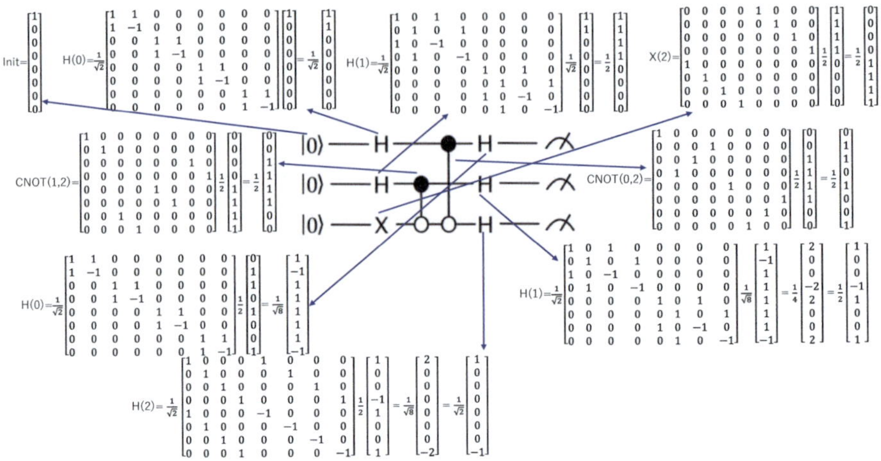

Fig. 2 An example quantum circuit simulation with SV

be performed as a sequence of matrix–vector multiplications for quantum circuits having N qubits. Quantum circuit simulators based on this idea are called state vector-based (SV) simulators in this chapter. An example SV-based quantum circuit simulation is shown in Fig. 2.

In order to keep the accuracy of the computations, double floating points (64-bit presentation of floating point numbers) are commonly used in quantum circuit simulations. Basically, the same memory space is required for SV simulations regardless of the types of quantum circuits once the number of qubits is fixed, and it always increases exponentially with respect to the number of qubits.

With, say, 32GB of memory space, quantum circuits with up to 30 or so qubits can be simulated, and even with supercomputer-type machines where lots of cores+memories are connected, the maximum number of qubits that can be simulated is around 40 to 50.

Please note that regardless of the types of quantum circuits, if the number of qubits is the same, SV-based simulators need the same memories and computation amounts which grow exponentially with respect to the number of qubits. In order to reduce the required memories and computation amounts, there have been proposed several quantum circuit simulation methods other than SV based, which are discussed in the following subsections.

2.3 *Tensor Network (TN)*

Tensors are multi-dimensional arrays of numbers. Intuitively, they can be interpreted as a generalization of scalars, vectors, and matrices. Tensors can be described by their rank, indices, and the dimension of the indices. The rank is the number of indices in

Fig. 3 An example tensor network

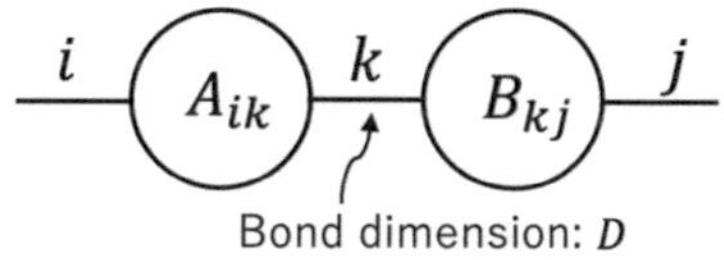

a tensor, i.e., a scalar has rank zero, a vector has rank one, a matrix has rank two, and so on. The dimension of an index is the number of values that index can take. For example, consider a vector with four elements having one index that can take three values. This vector is therefore a rank one tensor, and its index has dimension four, and each index can have three values.

To define tensor networks, it is important to first introduce tensor contraction. Two or more tensors sharing indices can be contracted by summing over repeated indices. In diagrammatic notation, the repeated indices appear as lines connecting tensors, as shown in Fig. 3, which shows an example of a tensor network. We see two tensors of rank two connected by one repeated index, k. The dimension of the repeated index is called the bond dimension.

The contraction of the tensors in Fig. 3 is equivalent to the standard matrix multiplication and can be expressed as $C_{ij} = \sum_k A_{ik} B_{kj}$ where C_{ij} denotes the entry for the i-th row and j-th column of the product $C = AB$.

A tensor network is a collection of tensors where a subset of all indices is contracted. As mentioned above, we can use diagrammatic notation to specify which indices and tensors will be contracted together by connecting individual tensors with lines. Tensor networks can represent complicated operations involving several tensors with many indices contracted in sophisticated ways. The overall behavior of a tensor network can be computed by applying contractions in some ordering until no more contraction is feasible. Since a given quantum circuit can be transformed into the corresponding tensor network, this sequence of contractions corresponds to simulating the quantum circuit.

Figure 4a shows an example quantum circuit where each box represents a quantum gate. Some are single-qubit gates whereas some others are multi(two)-qubit gates. The corresponding quantum tensor network is shown in Fig. 4b. Input and output nodes are added in the tensor network. Input nodes represent the initial states of the quantum circuits, whereas output nodes show the final computed results. By applying contraction to the edges of the tensor network, edges are eliminated. When all edges are eliminated, the resulting tensor network shows the final computed results.

There are alternative ways in which contractions are applied to the tensor network, and depending on the orders of elimination of edges, the intermediate size of the tensor network can be very different. Obtaining optimum ordering is basically NP-complete, but there have been proposed a number of heuristic ways. If the ordering is good for that particular tensor network, the quantum circuit simulation can be speedy. Some similar quantum circuits may share good orderings of contractions. Although

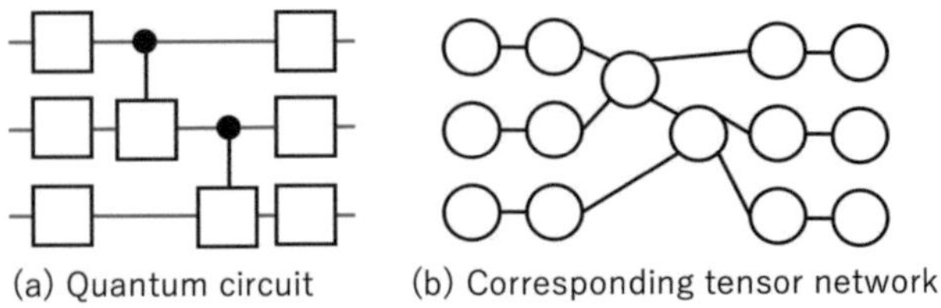

Fig. 4 An example quantum circuit and its corresponding tensor network

there have been proposed good heuristic orderings that work reasonably well for many quantum circuits, depending on the contraction orders used, the performance of the simulation can still vary.

2.4 Matrix Product State (MPS)

Matrix product state (MPS) is a particular case of tensor network, which can be represented as a chain of tensors. Figure 5 presents such a tensor network. To convert a general form tensor into an MPS, a singular value decomposition (SVD) algorithm is employed recursively, starting from the original tensor representing a quantum state, to finally obtain an MPS. The obtained MPS is an exact representation of the original quantum state, and thus the memory footprint of such representation still scales exponentially with the number of qubits. However, by truncating the number of singular values kept in each iteration of the SVD algorithm, the memory footprint can be reduced, though the representation becomes approximated. This number of singular values is the bond dimension D in Fig. 5. Decreasing D reduces the size of the representation with approximation, similar to lossy compression techniques of data. This number can be either set as a fixed number of singular values to keep or by setting a cutoff value below which singular values are eliminated.

SV-based simulation requires 2^N complex values to represent the quantum state of a N qubits system, i.e., an exponential memory usage with respect to the number of qubits. For MPS-based simulation, the bond dimension D is directly linked to the allowed degree of entanglement of the circuit. We can trade off between the size of quantum circuits (or number of qubits) to be simulated and the accuracy of the simulations.

Fig. 5 Tensor network representation of MPS

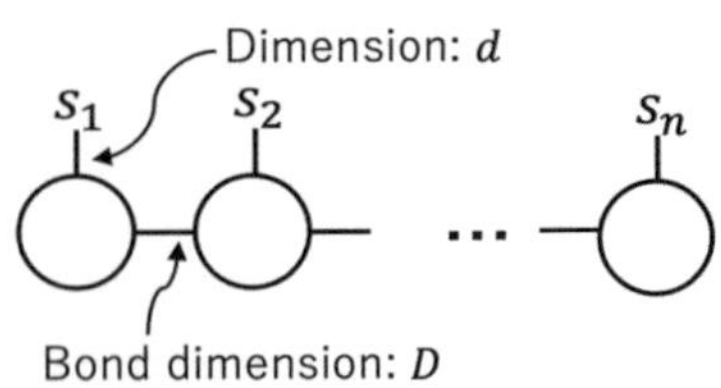

2.5 Decision Diagram (DD)

Decision diagrams (DDs), such as Binary Decision Diagram (BDD) [5], were originally proposed to represent and manipulate logic functions efficiently, but BDD and its multi-valued version Multi-Terminal BDD (MTBDD) have also been used to represent and manipulate matrices and vectors [6]. We can deal with matrices and vectors by introducing index variables to refer to their elements. In the case of vectors, a given vector is partitioned into the top half and the bottom half by the values (0/1) of an index variable. So, if we use N index variables, a length 2^N vector can be represented.

For example, a length 4 vector can be represented with a BDD as shown in Fig. 6. Here, the left edges correspond to 0 and the right edges correspond to 1. Index variable q_0 partitions the vector into the top and bottom halves which are further partitioned by the index variable q_1 as seen from the figure. The values of the four elements in the vector can be retrieved by traversing the BDD from the top to the leaf (constant 1) multiplying all of the edge values. Please note that if the edge value is 1, it is not shown in the diagrams, i.e., only values other than 1 are explicitly associated to the edges. The same vector can be represented in the MTBDD without edge values by having multiple leaves (terminal values). Edge-valued DDs are generally more compact than non-edge-valued DD, so edge-valued DDs are commonly used for the representation and manipulation of matrices and vectors, and this chapter also follows that. Please note that in the edge-valued DDs, since actual values are represented as the products of the edge values, there are some accuracy issues when floating point representations are used. In such cases, multi-terminal DDs may be preferred, which may happen in formal analysis of quantum circuits.

One advantage of DD representation is that the same sub-matrices and sub-vectors can share the sub-graphs, which can drastically reduce the size of the data structures in some cases. As will be seen below, representations of DDs for quantum gates and some quantum circuits can be very compact. Moreover, the time for matrix–matrix and matrix–vector multiplications in DDs is proportional to the size of DDs (products of two DDs) [6]. Therefore, as long as the size of DDs for quantum gates or circuits is manageable, quantum circuit simulations can be efficiently performed.

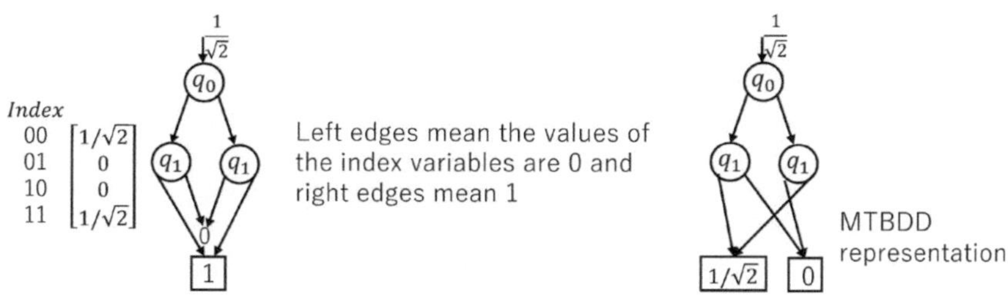

Fig. 6 DD representations of an example vector

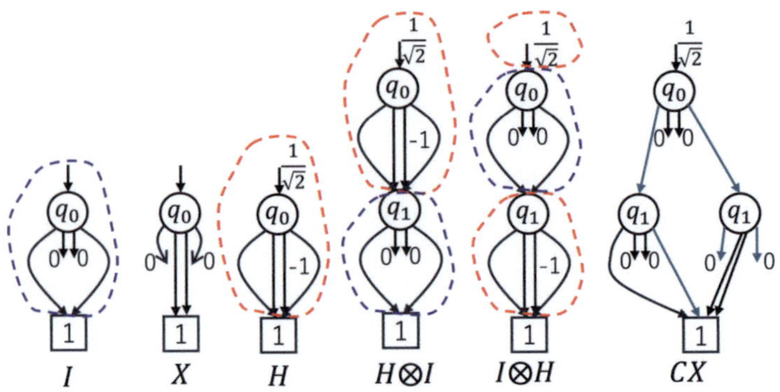

Fig. 7 DD representations of the quantum gates in Fig. 1

This is particularly true for Shor's algorithm as can be seen from the experimental results shown below.

In the case of matrices, a given matrix is partitioned into four sub-matrices, top-left, top-right, bottom-left, and bottom-right. This can be represented either by using two nodes in BDD, or DD with nodes having four edges corresponding to the four sub-matrices. The latter is typically used in quantum circuit simulation and called QMDD [7]. The four edges from a node correspond to four sub-matrices, top-left, top-right, bottom-left, and bottom-right, in the order of left-to-right.

The QMDDs for the quantum gates in Fig. 1 are shown in Fig. 7. In the figure, dotted lines are added to indicate topologically the same sub-graphs. As can be seen in the figure, Kronecker products of the two matrices can be computed by simply connecting the corresponding DDs. This is because in DDs, the same sub-matrices/vectors share the same sub-graphs.

2.6 Parallel Processing of Quantum Circuit Simulation

There are multiple parallelisms in the computation of simulating quantum circuits. As a quantum circuit simulation is a sequence of matrix–vector multiplications starting with the initial state vector, parallel computation algorithms for matrix–vector multiplications can be applied directly [8]. This is rather straightforward in the case of SV-based simulations, but even in the case of DD-based simulations, there are researchers trying to utilize multi-core and multi-node machines [9, 10].

To perform quantum circuit simulation using multiple computing nodes, the quantum state needs to be distributed and stored for memory efficiency. Although there are various ways to distribute, equal partitioning is common in terms of memory efficiency and ease of implementation. When simulating an N-bit quantum circuit with 2^M computing nodes, each node holds $\frac{2^N}{2^M}$ complex numbers. An example of two nodes partitioning of a three-qubit circuit is shown in Fig. 8.

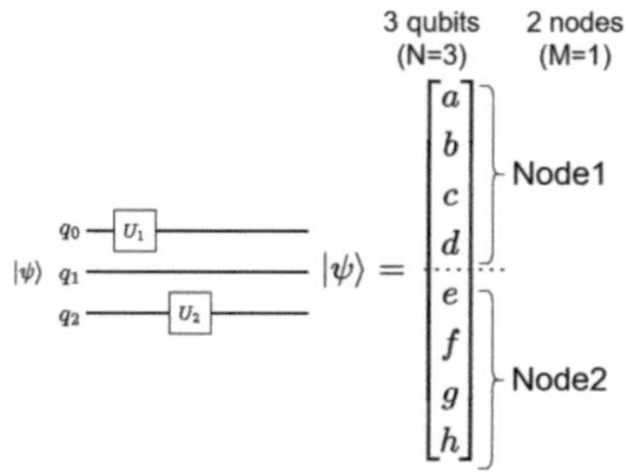

Fig. 8 State vector distribution among multiple nodes

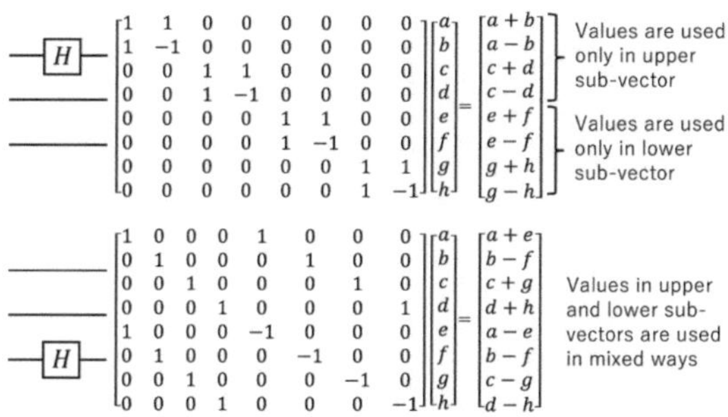

Fig. 9 Qubits are distributed into multiple nod

If H gate is applied to the first qubit and no gate (or I identity gate) is applied to the other two qubits to the circuit shown at the top of Fig. 9, the first four elements of the resulting state vector can be computed only with the first four elements of the current state vector. This can be done only with the elements of the state vector in Node 1. Similarly, the last four elements of the resulting state vector can be computed only with the last four elements of the current state vector, and this can be done only with the elements of the state vector in Node 2. These are shown at the top of Fig. 9.

On the other hand, if H gate is applied to the third qubit and no gate (or I identity gate) is applied to the other two qubits to the circuit shown at the bottom of Fig. 9, in order to compute the first four elements of the resulting state vector, the last four elements of the current state vector, which are stored in Node 2, are required. This results in node–node communications which should be avoided as much as possible for good performance.

When a series of quantum gates is applied, the number of communications can be reduced by inserting SWAP gates as shown in Fig. 10. Although the number of gates is even increased, this SWAP insertion can reduce the overall communication amount and speed up the simulation. Furthermore, while a general SWAP gate moves

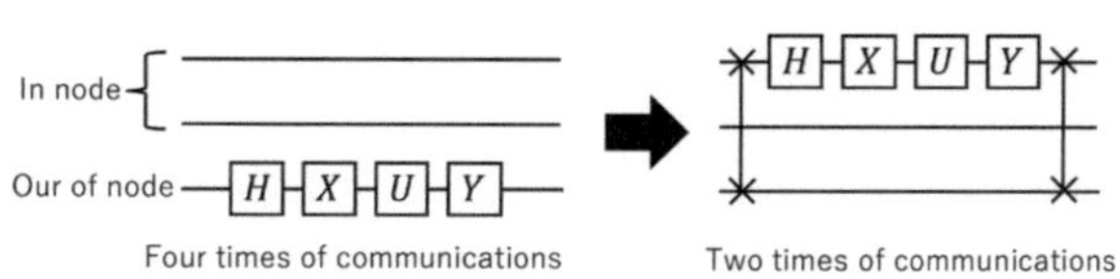

Fig. 10 Reducing communication by SWAP Insertion

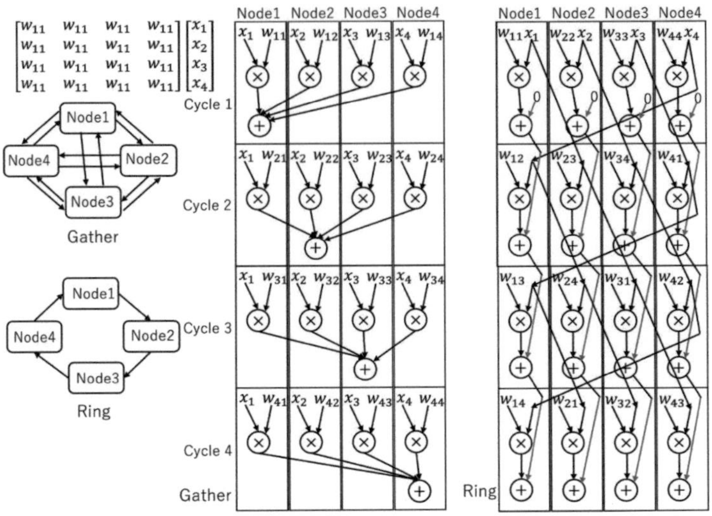

Fig. 11 Difference between gather and ring communication

the positions of only two qubits, a fused SWAP gate, which can move the positions of multiple qubits at the same time, has also been proposed [8]. This helps reduce the overall simulation time.

The communication overhead can also be reduced by changing the communication order. When calculating the next quantum state from top to bottom sequentially, "Gather" instructions are typically used (left-hand side of Fig. 11). However, these instructions tend to have a large communication overhead because one computing node needs to communicate with all other nodes at the same time. The ring-based communication [9] suggests communicating with only neighbor nodes (right-hand side of Fig. 11). The total amount of communication does not change among the two communication styles, but the ring can reduce the communication overhead and help speed up the simulation. The speedup can be more than 30 times [9].

3 Using Quantum Circuit Simulator

3.1 Popular Quantum Circuit Simulator

There are several middleware for designing quantum circuits, e.g., Qiskit [11], Cirq [12], TKET [13], Braket [14], and CUDA-Q [15]. Additionally, PennyLane [16] focuses on quantum machine learning, and Q# [17], Qrisp [18], and Silq [19] are higher-level quantum programming languages. Many middleware and languages have their own SV-based simulators. Some even include multiple types of simulators; for example, NVIDIA's CUDA-Q includes TN- and MPS-based simulators that utilize their own GPUs.

Qiskit has a mechanism called "backend." By switching the backend, Qiskit circuits can be executed on various simulators or real quantum hardware. Qiskit Aer is one of the representative simulator backends, supporting various simulation methods such as SV, MPS, and so on. Furthermore, by supporting Qiskit Backend, third parties can distribute their simulators for Qiskit users. Popular third-party simulators include Qulacs [4] (explicit SV) and DDSIM [20] (DD).

3.2 Quantum Circuit Simulation Using Qiskit

This section introduces how to simulate quantum circuits using Qiskit. Please prepare a Python environment.

3.2.1 Setup

First, you need to install the necessary libraries as in Code 8.1. This example uses two types of simulators: Qiskit Aer [11] and DDSIM [20]. MQT-Bench [21], a library that provides benchmark circuits, is also installed.

Code 8.1 "Preparing Dependencies"

```
$ python -m venv .venv
$ source .venv/bin/activate # In Windows, ".venv\Scripts\
    activate".
$ pip install qiskit~=1.4.0 mqt.ddsim~=1.24.0 mqt.
    bench qiskit-aer
```

3.2.2 Bell State Circuit

This section explains how to design and simulate a Bell state circuit. Lines 4-7 create the circuit `qc`. The `sim` on line 9 is the simulator backend provided by Qiskit Aer. From the execution result on line 12, it can be confirmed that the probabilities of the states 00 and 11 are approximately equal, indicating that the Bell state is correctly simulated.

Code 8.2 "Bell state circuit simulation"

```
from qiskit import QuantumCircuit
from qiskit_aer import AerSimulator

qc = QuantumCircuit(2) # 2 qubits, 2 classic bits
qc.h(0)
qc.cx(0,1)
qc.measure_all()

sim = AerSimulator() # Simulator type is selected
    automatically.
result = sim.run(qc, shots=1024).result()
print("Time taken:", result.time_taken, "sec")
print(result.get_counts())
# Time taken: 0.0026388168334960938 sec
# {'00': 529, '11': 495}
```

3.3 Using Different Types of Quantum Circuit Simulator

In this section, a QFT (quantum Fourier transform) circuit is simulated with various simulators as in Code 8.3. To use the QFT implementation from MQT-Bench [21], we execute `get_benchmark` function as shown on line 5.

Qiskit Aer provides various simulation methods, which can be specified using the option. In line 7, `statevector` is specified in the code, resulting in the SV-based simulation method. Other options like `matrix_product_state` can also be specified.

In line 8, the QFT circuit is transpiled. Quantum circuit simulators do not support all types of quantum gates, and they can handle only a limited set of gates. Such a set of gates is called basis gates. An error occurs if a circuit includes non-basis gates, but the transpiler can convert unsupported gates into combinations of basis gates and optimize the circuit. It is recommended to transpile circuits beforehand, except for those highly optimized manually.

In line 12, another simulator backend `sim_dd` is created. This backend completes the simulation much faster than SV-based simulator. Therefore, it is important to select the appropriate type of simulator for the input circuit.

4 Evaluation of Quantum Circuit Simulators on Benchmark Circuits

In this section, we evaluate and compare various quantum circuit simulators using benchmark circuits. Quantum circuit simulations are performed in the following environment:

- OS: Ubuntu 22.04 (Kernel 6.2.0).
- Software: GCC 11.4.0, Python 3.8.16.

Code 8.3 "QFT circuit simulation with SV based simulator"

```
from qiskit import transpile
from mqt.bench import BenchmarkLevel, get_benchmark
from mqt.ddsim import DDSIMProvider

qc = get_benchmark("qft", BenchmarkLevel.ALG, 22)

sim_sv = AerSimulator(method='statevector')
qc_compiled = transpile(qc, sim_sv)
result = sim_sv.run(qc_compiled).result()
print("SV Time taken:", result.time_taken, "sec")

sim_dd = DDSIMProvider().get_backend("qasm_simulator")
qc_compiled = transpile(qc, sim_dd)
result = sim_dd.run(qc_compiled).result()
print("DD Time taken:", result.time_taken, "sec")
# SV Time taken: 5.489156484603882 sec
# DD Time taken: 14.985747575759888 sec
```

Table 1 Simulators used for our evaluation

Name	Type	Device	#Threads
SV-1CPU	State vector	CPU	1
SV-8CPU	State vector	CPU	8
SV-GPU	State vector	GPU	N/A
MPS-1CPU	MPS	CPU	1
MPS-8CPU	MPS	CPU	8
TN-GPU	Tensor network	GPU	N/A
DD-1CPU	Decision diagram	CPU	1

- CPU: AMD Ryzen 7950X3D (16 cores/32 threads), RAM: 128GB.
- GPU: NVIDIA GeForce RTX 4090.
- Library: Qiskit Aer 0.14.1.

The benchmark circuits used for the evaluations are mostly from QASM Bench [3] and QCBM (used in Qulacs [4]) and additionally some other commonly used quantum circuits.

The quantum circuit simulators used in this section are listed in Table 1. For SV-based simulations, three simulators are used, one runs only with a single core/thread, and another runs on 8 cores/threads, and the last one runs on GPU. For MPS-based simulations, two simulators are used, one runs for single core/tread, and the other runs on 8 cores/threads. As for TN-based simulations, only GPU-based simulator is used. Finally, as for DD-based simulation, the simulator runs only with a single core/thread in the following experiments.

In the following, we give evaluation and comparison results using different benchmark circuits and popular circuits.

4.1 QASM Bench

QASM Bench [3] has quantum circuits based on various algorithms, and classified into small (43 circuits), medium (24 circuits), and large (57 circuits). The large category includes a quantum circuit having 433 qubits, and such very large circuits are impossible to simulate, and so 5 relatively small circuits from the large category having less than 34 qubits are included in the evaluation. Also, we set the timeout for simulation 1,000 seconds.

The experimental results are shown in Figs. 12, 13, and 14 as a scattered plots of the simulation time with SV-1CPU and other simulators. In all of them, the simulation time of SV-1CPU is vertically shown, and the ones of the others are shown horizontally. Circuits that need more than 1,000 seconds are shown on the top-most or right-most line.

In Fig. 12, the three SV-based simulators are compared. For the circuits that need short simulation time, simulation speed is roughly the same regardless of the number of threads, and the simulator that uses GPU is relatively slower. On the other hand,

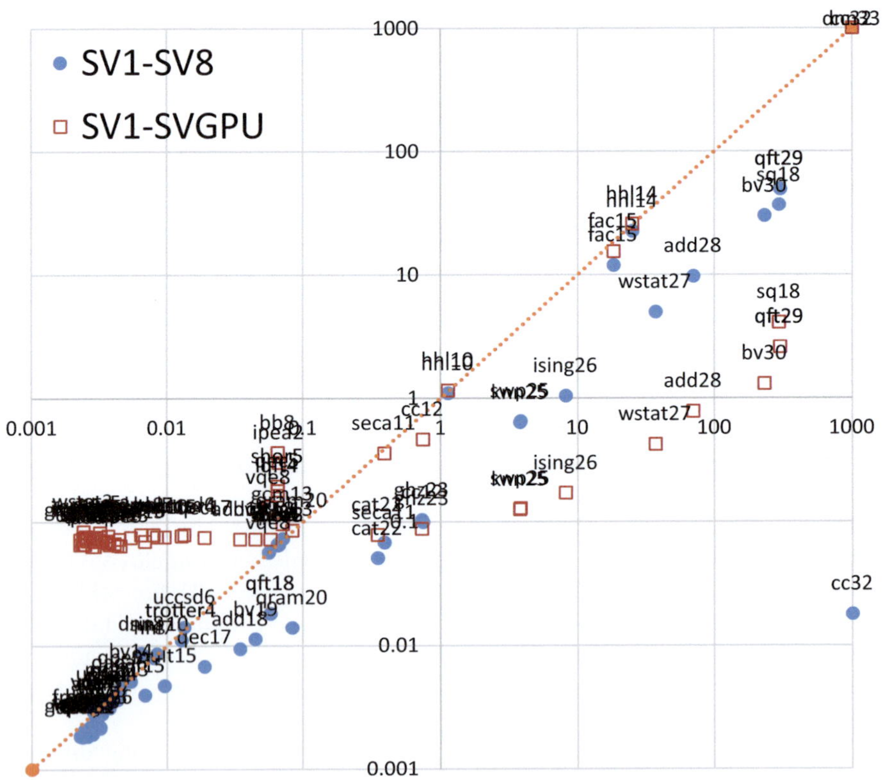

Fig. 12 QASM Bench (sec) (horizontal: SV-1CPU, vertical: SV-8CPU, SV-GPU)

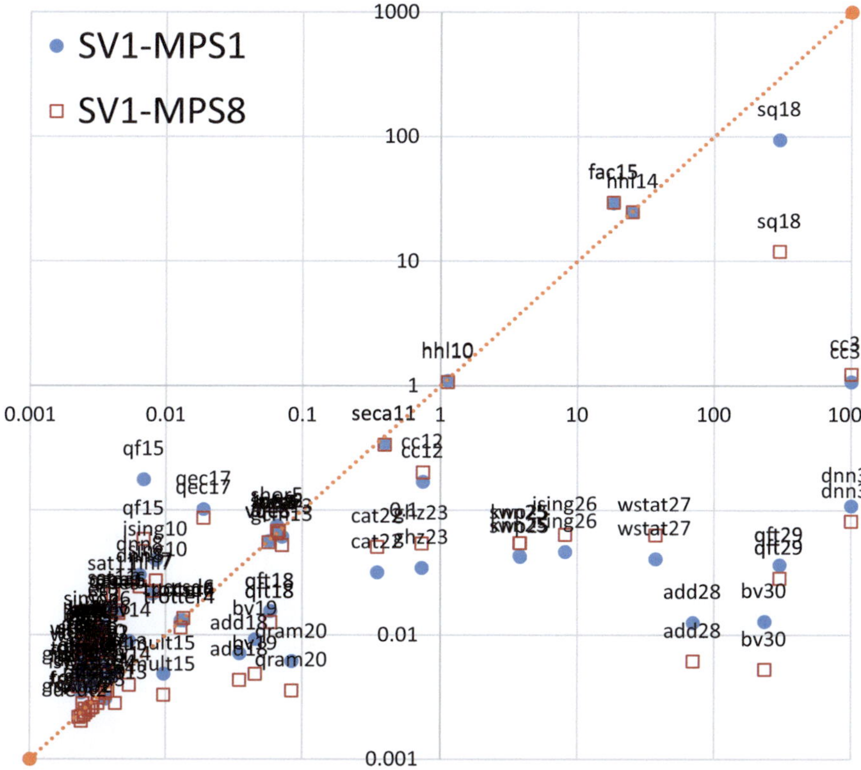

Fig. 13 QASM Bench (sec) (horizontal: SV-1CPU, vertical: MPS-1CPU, MPS-8CPU)

if the number of qubits becomes larger, simulators using multiple threads and GPU are faster. In particular, for large circuits, such as qft29 and bv30, the simulator with GPU becomes more than 10 times faster than the others. This clearly says GPU is very effective for large quantum circuit simulations.

In Fig. 13, the speed of the simulators based on MPS is compared with scattered plots. Except for the circuits that need less than 1 second for their simulations, in many cases, MPS-based simulators are faster than SV-based ones, and MPS-based simulator becomes faster as the number of threads becomes larger.

In Fig. 14, scatter plots on the comparison between SV-based simulator (SV-1CPU) and TN-based/DD-based simulators. TN-based simulators tend to be slower in many circuits, but for some circuits, such as add28 and bv30, they are faster than SV-based simulators. Please note that in TN-based simulations, the order of contraction can influence a lot on the performance. Here we use only the default setting without any further optimization. DD-based simulator is faster in some circuits, such as add28, bv30, and qft29, but for hhl circuit, it becomes slower.

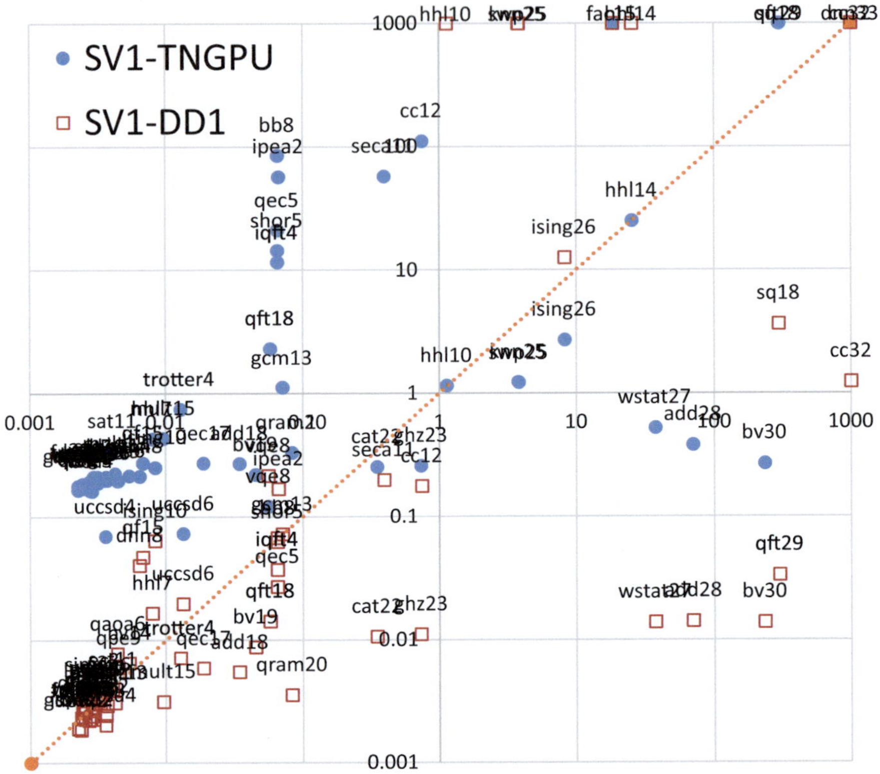

Fig. 14 QASM Bench (sec) (horizontal: SV-1CPU, vertical: TN-GPU, DD-1CPU)

4.2 QCBM

QCBM consists of a kind of random circuits that were used to evaluate Qulacs [4]. It is based on eight iterations of the sub-circuit shown in Fig. 15.

The simulation results are plotted in Fig. 16. Number of qubits are shown horizontally, and the simulation time is shown vertically (log scale) in seconds. From

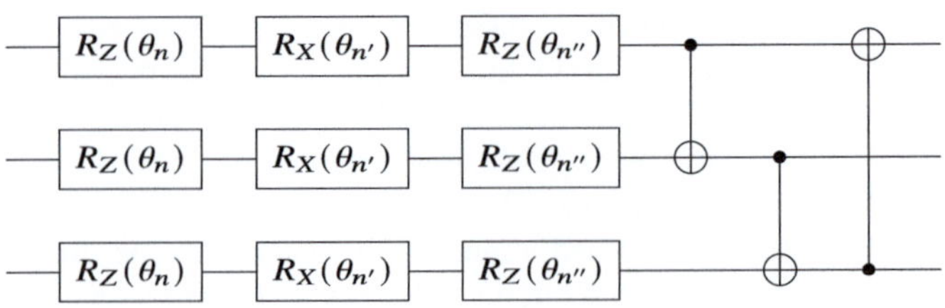

Fig. 15 QCBM (three qubits for repetition)

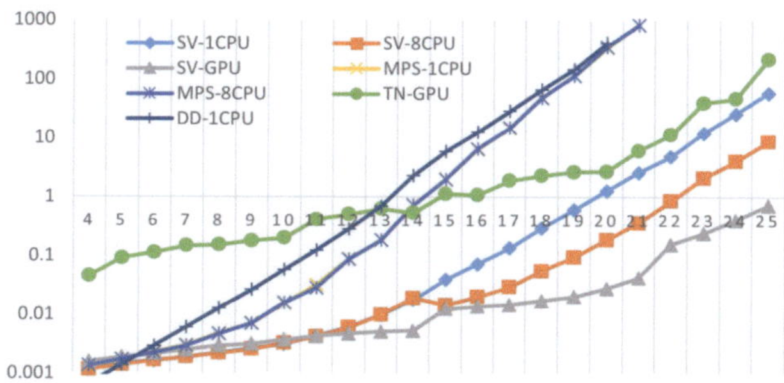

Fig. 16 Results on QCBM (horizontal: number of qubits, vertical: execution time (sec, log10 scale)

the figure, it is clear that SV-based simulation is faster than other methods. Also, with GPU, SV-based simulation is more than 10 times faster than the one with CPU. TN-based simulation is relatively slower when the number of qubits is small, but it becomes competitive to SV-based when the number of qubits is large. On the other hand, MPS-based and DD-based simulations are slower on this benchmark. Due to the randomness of the circuits, there may be more entanglement which is bad for MPS-based simulation, and DD-based simulation has problems in representing many different values in state vectors.

4.3 VQE

VQE (variational quantum eigensolver) [22] repeats classical computing and quantum computing to determine appropriate parameter values in Ansatz. For example, it first obtains the quantum state by using a quantum computer, and then the parameter values are optimized by using a classical computer, which is repeated to get the appropriate parameter values in the quantum circuit. Searching freely in the quantum state space is not practical, so a template circuit, called Ansatz, is given and the parameters in that Ansatz are optimized.

Varieties of Ansatz have been proposed [23, 24], and, in general, there are a number of RX, RY, and RZ gates and their parameters are the optimization targets.

In this experiment, we use a program for Max Cut problem from MQT-Bench [21] provided by the group developing DD-based simulator, DDSIM [20]. This program can generate quantum circuits with the VQE formulation that corresponds to the Max Cut problem for randomly generated graphs. We use SLSQP as the optimizer and RealAmplitude as Ansatz both of which are based on Qiskit implementation [11].

The results are shown in Fig. 17. MPS-based simulations give the best performance, and multi-threads do not improve the results. TN-based simulations are slower than MPS based, and the execution times of both of MPS- and TN-based simulations do not grow significantly as the number of qubits increases. On the other hand, the execution times for SV-based simulations simply grow exponentially with respect

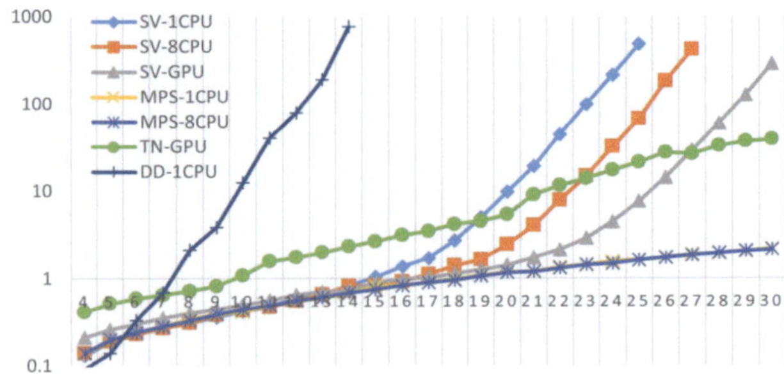

Fig. 17 Results on Maxcut (sec, log10 scale)

to the number of qubits. SV-based simulations can benefit from multi-threads and GPU-based processing. DD-based simulation is very slow as there are many rotation gates with different parameter values in the circuits.

4.4 Shor

Shor's algorithm [25] is used to factor a number. For example, 253 can be factored to 23×11.

In our experiments, quantum circuits using $4n + 2$ qubits are used. Here n is the bit length for the number to be factored. For example, if we like to factor 15 (1111 in binary), $n = 4$ and the number of required qubits becomes 18. There are more compact quantum circuits that use less numbers of qubits, such as the ones using $2n + 3$ [26] and $2n + 2$ [27]. But in our experiments, we use quantum circuits that need $4n + 2$ qubits which are easier for us to access.

Experimental results are shown in Table 2. The timeout is set to be 20 minutes. In the table, simulations that do not finish in 20 minutes are skipped (kept vacant).

Table 2 Results on Shor algorithm (time out = 1,200 s)

N	a	nGates	nQubits	SV-1CPU	SV-8CPU	SV-GPU	MPS-1CPU	MPS-8CPU	TN-GPU	DD-1CPU
15	2	12,600	18	1.95	0.59	0.21	0.45	0.51		0.11
21	2	25,330	22	56.95	13.22	0.66	13.47	13.36		0.65
33	5	44,466	26		664.91	30.8	280.89	276.92		4.01
123	2	118,005	30			1274.39	934.78	933.27		26.55
253	2	204,637	34							1197.18
511	3	400,123	38							818.83

DD-based simulations give the best performance. In the case of factoring 123, DD-based simulation finishes in 30 seconds, whereas the other simulations take 15–20 minutes. TN-based simulation cannot finish for all cases. One of the possible reasons is that the quantum circuits for Shor's algorithm have a large number of gates.

5 Application of Quantum Circuit Simulators for More Robust Verification

5.1 *Introduction of Symbolic Constants*

The basic idea of using symbolic constants in DD for the manipulation of parameterized quantum circuits is shown in Fig. 18. In the figure, initial state vectors are represented with symbolic constants, s_0, s_1, s_2 ($'$ means complement) to cover all possible initial states. So we can reason about the results in terms of changes of the initial state at once.

Figure 19 shows a way to use symbolic constants to represent different circuits uniformly. The quantum circuit has 4 qubits, and if s_{11} is one, the circuit has a controlled-NOT gate between the first and second qubits. If s_{12} is 1, it has a controlled-NOT gate between the second and third qubits, and if s_{13} is 1, between the third and fourth qubits. All of these can be represented with the symbolic constants, s_{11}, s_{12}, s_{13}, and their DD representations are also shown in the figure.

We can combine this parameterized circuit in a multi-level way. When the number of levels increases, DD increases as shown in Fig. 20, and we can say around 10 level

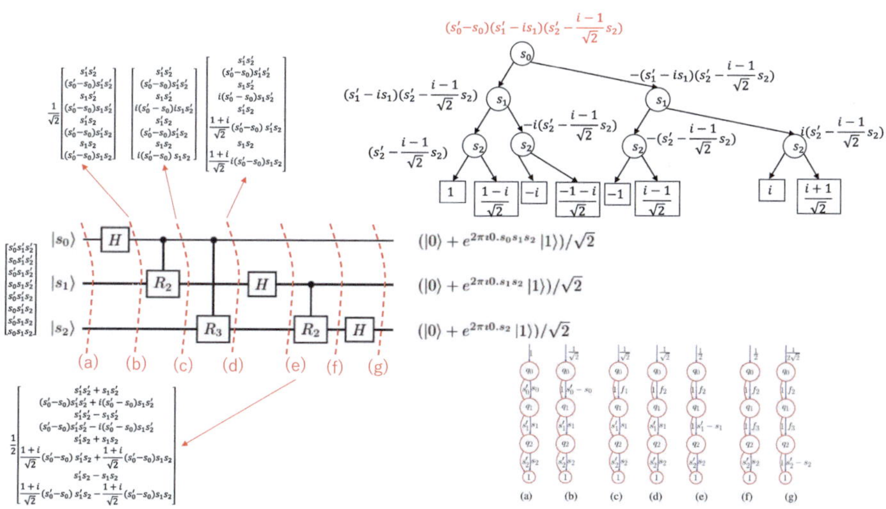

Fig. 18 An example of using symbolic constants to represent parameterized quantum circuits

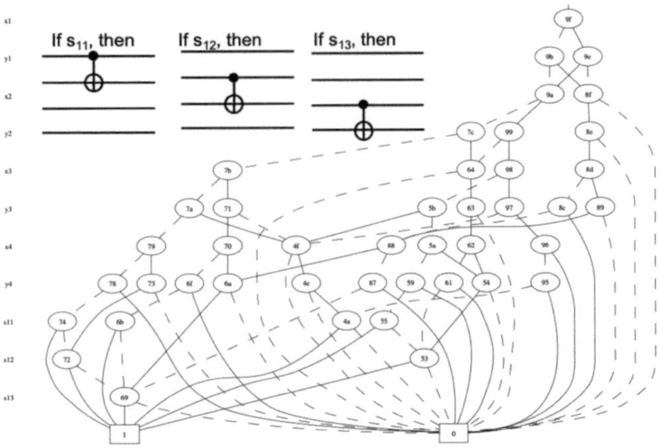

Fig. 19 Parameters determine quantum gates

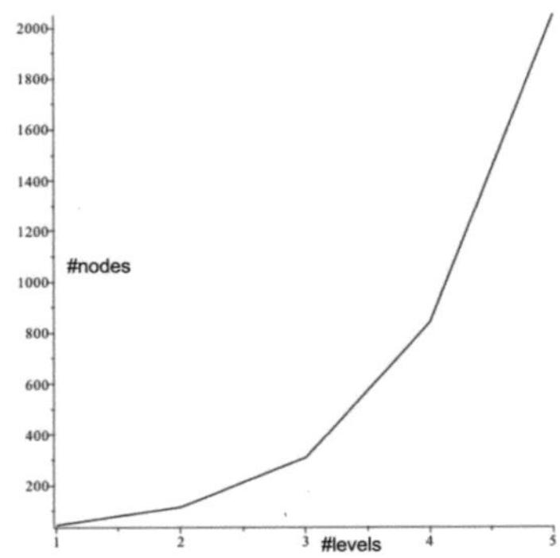

Fig. 20 Multi-level circuits with the parameterized circuit shown in Fig. 19

circuits with 10 qubits can be represented and manipulated as the number of nodes in DDs is expected to be much less than 1 million.

With these methods, a variety of parameterized circuits can be constructed, manipulated, and reasoned about.

5.2 *Approximate Simulation for Larger Quantum Circuits*

In order to deal with larger quantum circuits, one way is to simulate quantum circuits approximately rather than accurately. If the errors in approximation can somehow be well controlled, the simulation results can still be informative. Moreover, the approximation used in such simulations may tell how some quantum circuits can be efficiently simulated on classical computers, whereas some other cannot, which may give us some insights into quantum computing.

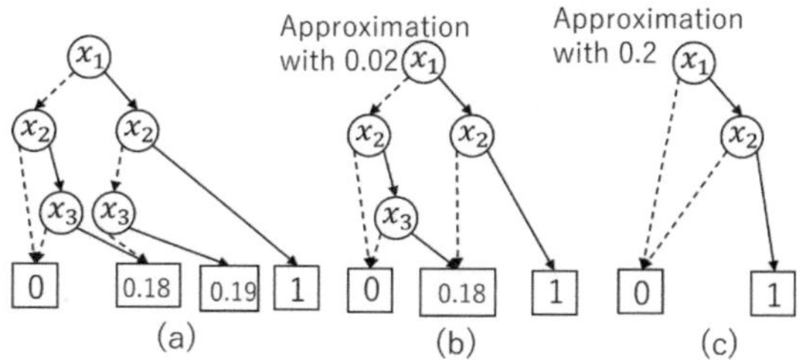

Fig. 21 MTBDD with value approximation

Straightforward ways to reduce the size of DDs using approximation are to approximate the values of terminals so that multiple terminals can be merged as a single terminal. In Fig. 21a, an example MTBDD is shown, and it has four different terminal values, 0, 0.18, 0.19, and 1. If the approximation in terms of error bounds is 0.02, 0.18 and 0.19 can be merged using 0.18 for the representative value as shown in (b) of the figure. If the error bounds is 0.2, all of 0.18, 0.19, and 0 can be merged using 0 for the representative value as shown in (c) of the figure. Here, the issue is how to determine the error bounds. Error bounds can be dynamic, and it may change its values depending on the size of DDs and the target quantum circuits to be analyzed. By experimenting approximated simulations on a number of quantum circuits, good ways for approximation would be naturally suggested.

Also, there have been efforts in using BDD to reason about larger logic circuits approximately or abstractly, such as Abstract BDD [28]. It uses abstraction conditions that determine which nodes in DDs are to be merged and which would be the representative values for them. In the case of (b) and (c) of Fig. 22, abstraction functions are defined as the terminal value differences, but the abstraction function can be more general.

Abstract BDD is an example of such an approximate representation of logic functions and has been shown to be effective in finding good variable orders for BDD [29]. Similar extensions over QDD or DDs for state vectors can be considered and would be similarly effective in finding good variable orderings when representing matrices and vectors in DDs.

Intuitively, an abstract BDD collapses paths in a BDD that have the same abstract value with respect to some abstraction function. The concepts underlying abstract BDDs are most easily explained using Binary Decision Trees (BDTs), but apply to BDDs and other DDs as well. A binary decision tree (BDT) is simply a BDD in which there is no graph sharing.

Here we explain Abstract BDD by an example shown in Fig. 22a shows the BDD and the truth table of the function, and (b) is the corresponding BDT. Suppose the abstraction function is $mod\ x_1 + x_2$ which means as long as the value of $mod\ x_1 + x_2$ is the same, all nodes (functions) are merged together. Nodes P and Q in (b) represent logic functions, x_3 and x_3, respectively, both of which have the same value under $mod\ x_1 + x_2$. So they are merged and if we use P as the representative node, the resulting BDT becomes (c), which is translated into BDD shown in (d).

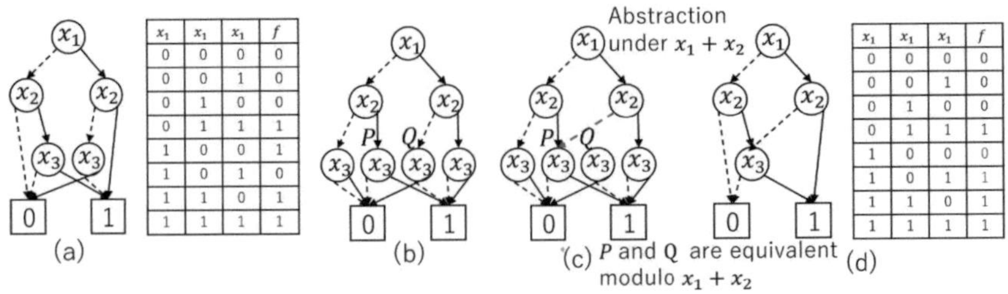

x_1	x_1	x_1	f
0	0	0	0
0	0	1	0
0	1	0	0
0	1	1	1
1	0	0	1
1	0	1	0
1	1	0	1
1	1	1	1

x_1	x_1	x_1	f
0	0	0	0
0	0	1	0
0	1	0	0
0	1	1	1
1	0	0	0
1	0	1	1
1	1	0	1
1	1	1	1

Fig. 22 An example matrix and its decision diagram representation

By defining appropriate abstract functions, useful Abstract BDD or Abstract DD may be able to be defined and utilized. Please note that if two nodes are merged under some abstract function, they are considered to be the same matrix in DD-based quantum circuit simulation. In that sense, the quantum circuit is somehow modified, but larger circuits, e.g., having more qubits, can be simulated and analyzed.

6 Conclusions

In this chapter, we have discussed quantum circuit simulation methods on top of classical computers and their performance on benchmark circuits and various popular circuits. SV-based simulation is most robust in the sense that, regardless of the characteristics of the quantum circuits, the simulation performance is consistent as long as the numbers of the qubits in the quantum circuits are the same.

On the other hand, other quantum circuit simulation methods, such as TN-based, MPS-based, and DD-based simulations, give performance which can vary a lot depending on the types of the quantum circuits. So it is important to use appropriate quantum circuit simulators, once a target quantum circuit is given. In such situations, this chapter gives some ideas on the selection.

We have also discussed a little bit about symbolic constants and abstractions in DD-based analysis. We believe this can be a first step toward formal analysis of quantum circuits by utilizing the various methods available for classic software and hardware.

References

1. IBM. The hardware and software for the era of quantum utility is here. https://www.ibm.com/quantum/blog/quantum-roadmap-2033, 2023.
2. Bravyi, Sergey, Andrew W. Cross, Jay M. Gambetta, Dmitri Maslov, Patrick Rall, and Theodore J. Yoder. March 2024. High-threshold and low-overhead fault-tolerant quantum memory. *Nature* 627 (8005): 778–782.
3. Ang Li, Samuel Stein, Sriram Krishnamoorthy, and James Ang. Qasmbench: A low-level quantum benchmark suite for nisq evaluation and simulation. *ACM Transactions on Quantum Computing*, 4(2), feb 2023.
4. Suzuki, Yasunari, Yoshiaki Kawase, Yuya Masumura, et al. October 2021. Qulacs: a fast and versatile quantum circuit simulator for research purpose. *Quantum* 5: 559.
5. Bryant, Randal E. Aug1986. Graph-based algorithms for boolean function manipulation. *IEEE Trans. Comput.* 35 (8): 677–691.
6. Fujita, M., P. C. McGeer, and J.C.-Y. Yang. Apr1997. Multi-terminal binary decision diagrams: An efficient data structure for matrix representation. *Formal Methods in System Design* 10 (2): 149–169.
7. D.M. Miller and M.A. Thornton. Qmdd: A decision diagram structure for reversible and quantum circuits. In *36th International Symposium on Multiple-Valued Logic (ISMVL'06)*, pages 30–30, 2006.
8. Satoshi Imamura, Masafumi Yamazaki, Takumi Honda, Akihiko Kasagi, Akihiro Tabuchi, Hiroshi Nakao, Naoto Fukumoto, and Kohta Nakashima. mpiqulacs: A distributed quantum computer simulator for a64fx-based cluster systems, 2022.
9. Yusuke Kimura, Shaowen Li, Hiroyuki Sato, and Masahiro Fujita. Accelerating decision diagram-based multi-node quantum simulation with ring communication and automatic swap insertion, 2024.
10. Li, Shaowen, Yusuke Kimura, Hiroyuki Sato, and Masahiro Fujita. 2024. Parallelizing quantum simulation with decision diagrams. *IEEE Transactions on Quantum Engineering* 5: 1–12.
11. Ali Javadi-Abhari, Matthew Treinish, Kevin Krsulich, Christopher J. Wood, Jake Lishman, Julien Gacon, Simon Martiel, Paul D. Nation, Lev S. Bishop, Andrew W. Cross, Blake R. Johnson, and Jay M. Gambetta. Quantum computing with Qiskit, 2024.
12. Cirq Developers. *Cirq*. Zenodo, April 2025.
13. Sivarajah, Seyon, Silas Dilkes, Alexander Cowtan, Will Simmons, Alec Edgington, and Ross Duncan. Nov2020. t-ket⟩: a retargetable compiler for nisq devices. *Quantum Science and Technology* 6 (1): 014003.
14. AWS. Cloud Quantum Computing Service - Amazon Braket - AWS — aws.amazon.com. https://aws.amazon.com/braket/. [Accessed 20-07-2025].
15. NVIDIA. CUDA-Q — developer.nvidia.com. https://developer.nvidia.com/cuda-q. [Accessed 20-07-2025].
16. Ville Bergholm, Josh Izaac, Maria Schuld, et al. Pennylane: Automatic differentiation of hybrid quantum-classical computations, 2022.
17. Krysta Svore, Alan Geller, Matthias Troyer, et al. Q#: Enabling Scalable Quantum Computing and Development with a High-level DSL. In *Proceedings of the Real World Domain Specific Languages Workshop 2018*, RWDSL2018. ACM, February 2018.
18. Seidel, Raphael, Sebastian Bock, René Zander, Matic Petrič, Niklas Steinmann, and Nikolay Tcholtchev. 2024. *and Manfred Hauswirth*. Qrisp: A framework for compilable high-level programming of gate-based quantum computers.
19. Benjamin Bichsel, Maximilian Baader, Timon Gehr, and Martin Vechev. Silq: a high-level quantum language with safe uncomputation and intuitive semantics. In *Proceedings of the 41st ACM SIGPLAN Conference on Programming Language Design and Implementation*, PLDI 2020, page 286–300, New York, NY, USA, 2020. Association for Computing Machinery.
20. Zulehner, Alwin, and Robert Wille. 2019. Advanced simulation of quantum computations. *IEEE Transactions on Computer-Aided Design of Integrated Circuits and Systems* 38 (5): 848–859.

21. Quetschlich, Nils, and Lukas Burgholzer. 2023. *and Robert Wille*. Mqt bench: Benchmarking software and design automation tools for quantum computing. Quantum.
22. Peruzzo, Alberto, Jarrod McClean, Peter Shadbolt, et al. Jul2014. A variational eigenvalue solver on a photonic quantum processor. *Nature Communications* 5 (1): 4213.
23. Kandala, Abhinav, Antonio Mezzacapo, Kristan Temme, et al. Sep2017. Hardware-efficient variational quantum eigensolver for small molecules and quantum magnets. *Nature* 549 (7671): 242–246.
24. Peruzzo, Alberto, Jarrod McClean, Peter Shadbolt, et al. Jul2014. A variational eigenvalue solver on a photonic quantum processor. *Nature Communications* 5 (1): 4213.
25. P.W. Shor. Algorithms for quantum computation: discrete logarithms and factoring. In *Proceedings 35th Annual Symposium on Foundations of Computer Science*, pages 124–134, 1994.
26. Beauregard, Stephane. Mar2003. Circuit for shor's algorithm using 2n+3 qubits. *Quantum Info. Comput.* 3 (2): 175–185.
27. Takahashi, Yasuhiro, and Noboru Kunihiro. Mar2006. A quantum circuit for shor's factoring algorithm using 2n + 2 qubits. *Quantum Info. Comput.* 6 (2): 184–192.
28. Clarke Edmund, Jha Somesh, Lu Yuan, and Wang Dong. Abstract bdds: A technique for using abstraction in model checking. *Lecture Notes in Computer Science*, (1703), 1999.
29. Lu Yuan, Jain Jawahar, Clarke Edmund, and Fujita Masahiro. Improving bdd variable ordering using abstract bdds and sampling. In *Proceedings of Intl. Workshop of High Level Deisgn, Verification and Testing (HLDVT'99)*, 1999.

Quantum Testing and Validation

Aruna Jayasena, Emma Andrews, and Prabhat Mishra

Abstract Quantum computing holds the promise of revolutionizing computation by solving classically intractable problems. However, the inherent complexity, probabilistic behavior, and fragility of quantum systems pose significant challenges to ensuring the correctness and reliability of quantum programs. In this chapter, we present a comprehensive survey of quantum testing techniques aimed at addressing these challenges across various layers of the quantum software and hardware stack. Specifically, we analyze a wide range of testing techniques, including mutation testing, fuzz testing, coverage testing, combinatorial testing, search-based testing, property-based testing, metamorphic testing, fault testing, and concolic testing. We also discuss the applicability and limitations of these testing methods.

1 Introduction

The emergence of quantum computing promises significant computational advantages for a variety of problems in cryptography, optimization, simulation, and machine learning. However, realizing practical quantum applications remains a daunting challenge due to the fragile and error-prone nature of quantum hardware, the probabilistic behavior of quantum programs, and the complexity of quantum software development. As quantum systems grow in scale and complexity, ensuring the correctness and reliability of quantum programs becomes critically important [1–9].

In classical computing, mature testing methodologies have enabled decades of robust software and hardware engineering [10–17]. However, quantum computing

A. Jayasena (✉)
University of Tennessee at Chattanooga, Chattanooga, USA
e-mail: aruna@tennessee.edu

E. Andrews · P. Mishra
University of Florida, Gainesville, USA
e-mail: e.andrews@ufl.edu

P. Mishra
e-mail: prabhat@ufl.edu

P. Mishra (ed.), *Design Automation for Quantum Computing*,
https://doi.org/10.1007/978-3-032-09303-5_9

introduces a fundamentally different paradigm. Qubits can exist in superpositions and become entangled, computations are inherently non-deterministic, and direct observation of the internal state can irreversibly alter it. These features limit the applicability of many classical testing methods. For instance, a traditional test oracle that compares an output to an expected (correct) result becomes difficult or even impossible to define when the output is probabilistic or depends on the collapse of a quantum state. Moreover, the exponential growth of quantum state space with the number of qubits severely limits exhaustive testing or simulation-based validation techniques.

1.1 Quantum Computing and the Need for Testing

As quantum computing transitions from theoretical exploration to practical deployment, rigorous testing becomes essential across the quantum software stack. This stack includes algorithm design with circuit construction, compilation and transpilation, and eventual hardware execution as illustrated in Fig. 1. Each stage introduces distinct risks and testing requirements:

- *Algorithm Design:* Logical and semantic bugs may arise from incorrect circuit structures, improper use of entanglement, or unintended side effects.
- *Transpilation and Compilation:* Optimization routines may modify circuits in ways that subtly change their behavior, especially when reducing gate count or adapting to hardware constraints.

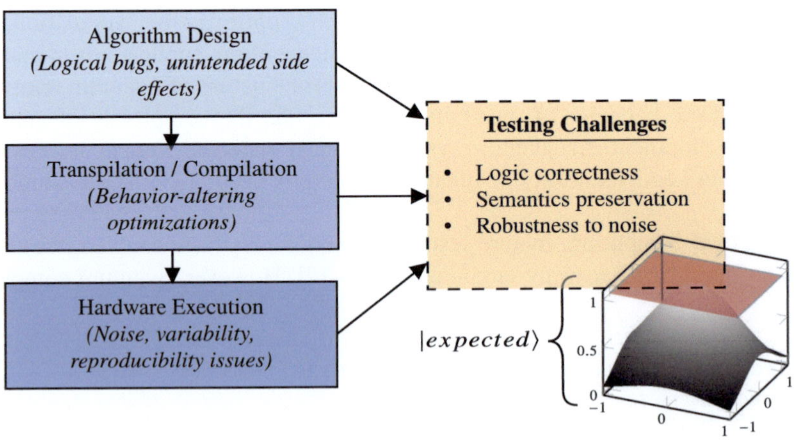

Fig. 1 Testing concerns across the quantum systems and motivation factors for finding efficient quantum verification techniques

- *Hardware Execution:* Quantum hardware is inherently noisy and variable, making test reproducibility difficult and emphasizing the need for robustness testing against realistic noise models.

Testing must therefore be integrated across all development stages in a cross-layer, lifecycle-aware manner to ensure correctness, preserve semantic intent, and validate hardware behavior under practical conditions.

1.2 Overview of this Chapter: From Classical to Quantum Testing

This chapter provides a comparative exploration of foundational testing techniques, starting from their classical origins and extending into the emerging domain of quantum computing. As quantum programs evolve in complexity and scale, systematic testing becomes important to ensure correctness, robustness, and hardware-level alignment. To help transition smoothly from familiar paradigms, we first present classical versions of each technique before introducing their quantum-specific adaptations and challenges. Specifically, we discuss the following topics in detail:

- Coverage testing and test adequacy for systematic exploration of quantum program behavior.
- Assertions and property-based testing for detecting runtime errors and bugs in quantum programs.
- Symbolic execution and formal methods for generation of test inputs for quantum design validation.
- Mutation testing and metamorphic testing for evaluating test suite robustness and generating adversarial test cases for quantum verification.

Throughout the chapter, we emphasize the similarities and departures between classical and quantum testing, providing examples, illustrations, and recent developments to demonstrate that foundational testing techniques must evolve for the quantum era.

The remainder of the chapter is organized as follows. Section 2 provides essential background on classical and quantum computation systems. Section 3 discusses coverage-based testing metrics in both classical and quantum contexts. Section 4 explores assertions and property-based validation techniques across the two paradigms while Sect. 5 discusses validation methods that utilize symbolic execution and formal methods. Section 6 presents a detailed discussion on mutation and metamorphic testing for validating classical and quantum systems. Finally, Sect. 7 concludes the chapter with a summary and reflections on future directions for quantum testing.

2 Computing and Testing Fundamentals

Before we explore the techniques and challenges of testing quantum programs and systems, it is essential to establish a foundational understanding of both classical and quantum computation and how testing principles are traditionally applied. This chapter introduces the fundamental units of classical and quantum information (bits and qubits) and outlines how they are manipulated through logic gates and quantum operations, respectively. We begin by reviewing classical computing principles, such as deterministic control flow and Boolean logic, which create the foundation for conventional testing methodologies. We then introduce core quantum concepts, including superposition, entanglement, and measurement, which give rise to fundamentally different system behaviors. Building on this, we examine classical testing strategies that rely on predictable and reproducible execution, and contrast them with the new constraints introduced by quantum mechanics, such as the no-cloning theorem, probabilistic outcomes, and noisy hardware, which necessitate a rethinking of test design, evaluation, and automation.

2.1 Classical Computing Basics

Classical computing is built on the deterministic manipulation of binary information. At its core are bits, logic gates, and structured control flow that define how computation proceeds step by step.

Bits and Logic Gates: In classical systems, information is represented using *bits*, which can take on one of two values: 0 or 1. Logical operations are performed on these bits using *logic gates*. These gates implement basic Boolean functions, which are the building blocks of classical digital circuits.

Figure 2 illustrates a simple combinational logic circuit that computes the Boolean expression: $Z = (A \wedge B) \vee (\neg A \wedge C)$. In classical digital systems, such circuits are fully observable: we can probe internal wires and logic gates to monitor signal values at various stages of computation. This fine-grained observability enables step-by-step debugging, waveform tracing, and precise fault localization, features that are foundational to classical hardware testing. The ability to directly inspect intermediate

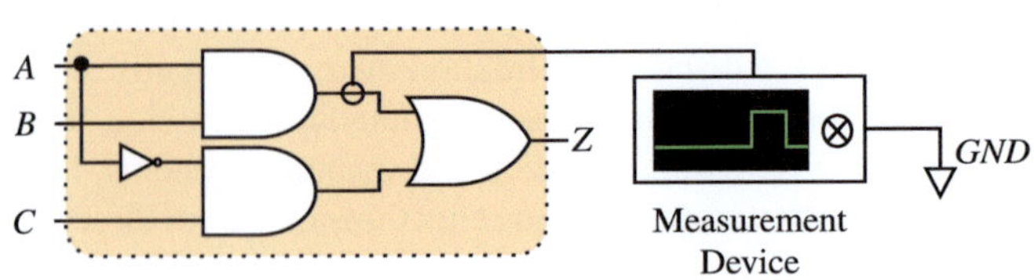

Fig. 2 Example logic circuit computing $Z = (A \wedge B) \vee (\neg A \wedge C)$

values without altering the circuit behavior simplifies the verification process and supports the development of tools for simulation, formal analysis, and test coverage measurement.

Control Flow in Classical Programs: Control flow governs the execution path through a classical program on a classical computing device. It is determined by conditionals (e.g., `if-else`) and loops (e.g., `for`, `while`). This structured flow enables predictable, repeatable behavior, which is a fundamental requirement for classical testing and debugging. These fundamental components (bits, logic gates, and control structures) form the predictable substrate upon which classical computation and testing are built. This deterministic behavior simplifies both test generation and debugging by enabling the same input to always produce the same output.

Figure 3a illustrates a typical function written in **C** language, representing a software-level implementation of control logic. Figure 3b shows the corresponding Control Flow Graph (CFG), where each basic block represents a straight-line sequence of instructions, and edges represent possible transitions between them. Note that branch B_3 introduces two outgoing edges, capturing the conditional execution paths in the source code. From a hardware perspective, such control constructs are implemented using multiplexers, comparators, and program counters in digital logic. The execution path is determined deterministically by evaluating the conditions and driving control signals. In classical hardware systems, this behavior is both observable and reproducible: signal transitions through logic gates and state changes in flip-flops can be monitored using probes or simulation waveforms. This dual view of software control structures and their hardware realization forms the basis for comprehensive testing and verification in classical systems.

```
#include <stdio.h>
int write_reg(int x) { ← [B1]
 volatile int *data_reg =
            (int *)0x40000000; ← [B2]
 if (x == 0) { ← [B3]
  *data_reg = 1; ← [B4]
 } else {
  *data_reg = 0; ← [B5]
 }
 return 0; ← [B6]
}
```

(a) Simple function

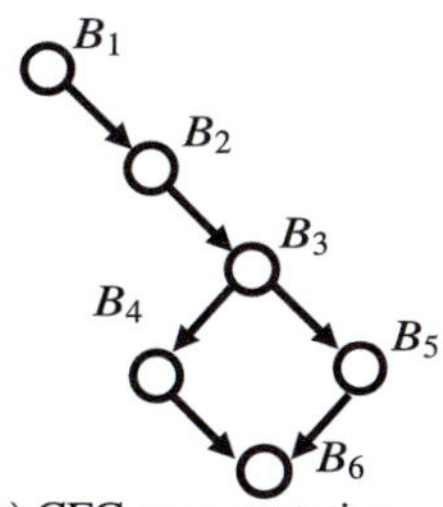

(b) CFG representation

Fig. 3 An illustrative example of a function and corresponding directed control-flow graph (CFG). Here, B_i represents the ith basic block in the function

2.2 Quantum Computing Basics

Quantum computing departs from classical computing by leveraging the principles of quantum mechanics to represent and manipulate information. The basic unit of quantum information is the *qubit*, which, unlike a classical bit, can exist in a superposition of states.

Qubits and Superposition:: A qubit is a two-level quantum system described by a linear combination of the computational basis states:

$$|\psi\rangle = \alpha\,|0\rangle + \beta\,|1\rangle\,, \quad \text{where } \alpha, \beta \in \mathbb{C},\ |\alpha|^2 + |\beta|^2 = 1.$$

Unlike a classical bit, which can be either 0 or 1, a qubit can exist in a superposition of both states simultaneously. The complex amplitudes α and β determine the probability of observing each basis state upon measurement: the qubit collapses to $|0\rangle$ with probability $|\alpha|^2$, and to $|1\rangle$ with probability $|\beta|^2$.

This state can be geometrically represented on the surface of the Bloch sphere, as shown in Fig. 4a. In this visualization, the north and south poles correspond to the classical states $|0\rangle$ and $|1\rangle$, while all other pure states lie on the surface of the sphere. For example, the state $|+\rangle = \frac{1}{\sqrt{2}}(|0\rangle + |1\rangle)$ lies on the equator of the sphere, representing an equal superposition of $|0\rangle$ and $|1\rangle$. The Bloch sphere thus provides an intuitive way to understand the continuous and complex-valued nature of qubit states, as well as the phase differences that have no classical counterpart. This geometric representation is particularly useful in quantum testing and debugging, where visualizing state evolution can guide circuit design and validation.

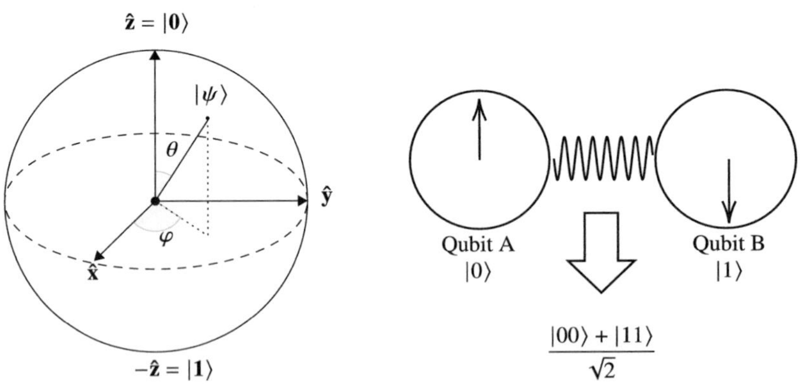

(a) Bloch sphere projection (b) Quantum entanglement

Fig. 4 An overview of superposition and entanglement: **a** visualization of superposition using Bloch sphere projection: a qubit state $|\psi\rangle = \alpha\,|0\rangle + \beta\,|1\rangle$ lies on the surface of the unit sphere, and **b** the illustrative effect of quantum entanglement

Qudits: While qubits are the fundamental unit of information in most quantum computing systems, quantum devices are not strictly limited to two-level systems. More generally, quantum hardware can operate on *qudits*, quantum units with d discrete levels. A qudit state can be represented as a linear combination of basis states:

$$|\psi\rangle = \sum_{i=0}^{d-1} \alpha_i |i\rangle, \tag{1}$$

where the complex coefficients α_i satisfy the normalization condition $\sum |\alpha_i|^2 = 1$. When $d = 2$, the system corresponds to a standard qubit; when $d = 3$, it is referred to as a qutrit. Qudits can offer increased information density and improved error resilience, and are actively explored in certain quantum platforms such as trapped ions and photonic systems.

Entanglement: Quantum entanglement is a uniquely non-classical phenomenon in which the quantum states of two or more qubits become deeply correlated such that the state of each qubit cannot be described independently of the others, even when they are spatially separated. In an entangled system, the global state carries information that is not present in any individual qubit alone, as illustrated in Fig. 4b. A canonical example of entanglement is the two-qubit Bell state that can be illustrated as follows:

$$|\Phi^+\rangle = \frac{1}{\sqrt{2}}(|00\rangle + |11\rangle),$$

which represents a superposition where both qubits are simultaneously in the $|00\rangle$ and $|11\rangle$ states. The crucial property of this entangled state is that the outcome of a measurement on one qubit is perfectly correlated with the outcome of a measurement on the other. For instance, if the first qubit is measured and found to be in the state $|0\rangle$, the second qubit will also collapse to $|0\rangle$. Similarly, if the first qubit collapses to $|1\rangle$, so will the second, regardless of the physical distance between them. This form of non-local correlation defies classical intuition and has implications for quantum communication, teleportation, and some cryptographic protocols. It also introduces complexity into quantum testing, as the behavior of an entangled qubit cannot be validated in isolation, and the full system state must be considered as a whole.

Quantum Circuits and Measurement: Quantum algorithms are constructed as sequences of operations on qubits or qudits, where each operation is typically represented by a unitary transformation (gate) acting on one or more quantum states. These transformations manipulate the quantum state coherently, preserving the overall probability amplitude. Unlike classical circuits, which use irreversible logic gates, quantum gates are reversible and linear, enabling interference and entanglement to play a computational role.

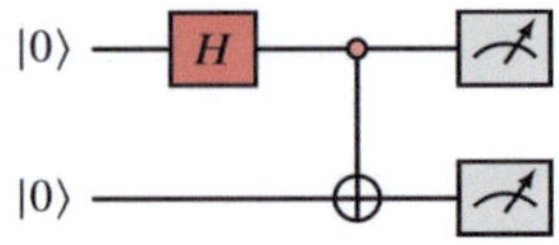

Fig. 5 Quantum circuit creating the Bell state $\left|\Phi^+\right\rangle = \frac{1}{\sqrt{2}}(|00\rangle + |11\rangle)$

Figure 5 shows a simple quantum circuit that prepares the entangled Bell state $\left|\Phi^+\right\rangle = \frac{1}{\sqrt{2}}(|00\rangle + |11\rangle)$. The circuit begins with two qubits initialized in the $|0\rangle$ state. The first gate applied is a Hadamard gate (H) on the top qubit, which transforms it into an equal superposition state:

$$H\,|0\rangle = \frac{1}{\sqrt{2}}(|0\rangle + |1\rangle).$$

Next, a CNOT (controlled-NOT) gate is applied, with the top qubit acting as the control and the bottom qubit as the target. This entangles the two qubits by flipping the target qubit only when the control qubit is in the $|1\rangle$ state. The resulting joint state is the Bell state $\left|\Phi^+\right\rangle$, an entangled state that cannot be separated into individual qubit descriptions. The final stage of the circuit involves measurement, indicated by the measurement symbols in the diagram. When measured in the computational basis, the qubits collapse to either $|00\rangle$ or $|11\rangle$, each with 50% probability. Note that the measurement outcomes are correlated due to the entangled nature of the state: observing $|0\rangle$ on the first qubit guarantees $|0\rangle$ on the second, and likewise for $|1\rangle$.

This probabilistic and non-local behavior makes testing and debugging quantum programs substantially more complex than testing classical systems. A single run may not reveal the correct behavior; instead, multiple executions and statistical analysis are often required to validate quantum program correctness. Additionally, measurement irreversibly collapses the quantum state, meaning intermediate inspection of a circuit (common in classical debugging) is typically not possible in quantum systems.

Beyond standard quantum circuits, advanced models have emerged to support more flexible and expressive computation. Parameterized quantum circuits incorporate gates with tunable parameters, which can be fixed or optimized during training, especially in hybrid quantum-classical machine learning. Another powerful abstraction is the dynamic quantum circuit, which introduces classical control flow, such as "if" or "for" constructs, within a quantum program. These circuits enable adaptive behaviors based on measurement outcomes during execution. An example of a dynamic circuit with an if-else branching structure is shown in Fig. 6. The figure illustrates a simple example of a dynamic quantum circuit with classical control flow based on measurement outcomes. The first qubit q_0 is initialized with a Hadamard gate and then measured. Its measurement result is stored in a 2-bit classical register c_1. This classical information is used to dynamically control operations on the second qubit q_1 depending on the value in c_1, the circuit conditionally applies a

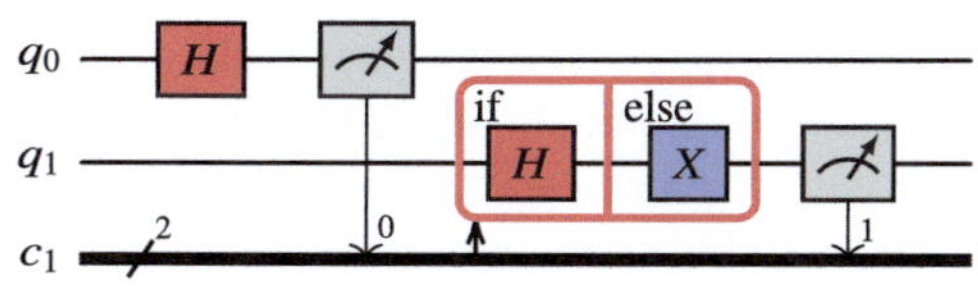

Fig. 6 A dynamic quantum circuit with if-else branching based on measurement outcome. Depending on the value in c_1, the circuit conditionally applies a Hadamard gate or an X gate on q_1

Hadamard gate or an X gate on q_1. The control structure, indicated by a box labeled `"if"...`"else"`, shows how classical computation is integrated within the quantum circuit to influence its behavior adaptively. Finally, q_1 is measured.

2.3 *Testing in Classical Systems*

In classical computing, testing is a fundamental practice to ensure that systems behave according to specifications [11]. Testing encompasses both software and hardware domains, and relies on deterministic, reproducible behavior to detect deviations from expected and unexpected outcomes. Figure 7 illustrates the overall flow that is used for testing the classical systems. It has several components, including a test generation algorithm, an oracle that produces the correct expected output from the system, and the evaluator, which compares the outputs of the system that is being tested and the oracle.

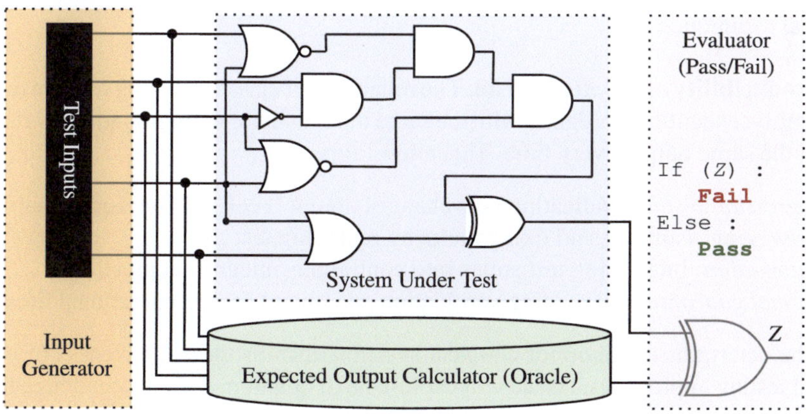

Fig. 7 Basic flow of classical system testing using an oracle for expected output

Testing Goals and Methodologies: The primary goals of classical system testing span a wide range of quality assurance concerns, from functional correctness to security assurance. These goals guide the development of test strategies, automation frameworks, and validation tools across both software and hardware systems:

- *Correctness*: Ensuring that the system produces the correct and expected output for all specified inputs under normal operating conditions. This typically involves validating the system against its specification, using unit tests, assertions, and formal verification tools [12].
- *Robustness*: Verifying that the system behaves predictably and safely under stress conditions, corner cases, and invalid inputs. Robustness testing is essential for identifying unhandled exceptions, buffer overflows, or undefined behavior in low-level systems [18].
- *Coverage*: Measuring how thoroughly the test suite exercises the system under test. Coverage metrics like statement coverage, branch coverage, and path coverage provide quantitative feedback to guide test case generation. Higher coverage often correlates with a greater likelihood of detecting latent bugs.
- *Security*: Testing the system for vulnerabilities that could be exploited by adversaries. This includes fault injection, side-channel analysis, privilege escalation testing, and other forms of security validation to ensure confidentiality, integrity, and availability [18–29]. Security testing often involves both white-box (design-aware) and black-box (behavioral) approaches.

Role of Test Oracles: A test oracle defines the expected output for a given input. In classical systems, oracles are straightforward because of the deterministic nature of computation, where the same input will always produce the same output. This makes it feasible to define exact oracles either manually [30, 31], through specifications [18], or automatically [12], through reference models [32–34] in the case of formal techniques.

Reproducibility and Determinism: The reliability of classical testing rests on reproducibility: executing a program with the same input under the same conditions should yield the same output every time. This allows for:

- *Regression testing*: Validating new changes do not break existing functionality.
- *Debugging*: Isolating and fixing faults by replaying test cases.
- *Automation*: Integrating test suites into continuous integration pipelines.
- *Formal guarantees*: Providing verification guarantees for the operational lifespan.

The deterministic nature of classical systems sustains the effectiveness of traditional testing approaches. As we will explore next, quantum systems challenge these assumptions through inherent uncertainty and non-cloning constraints.

2.4 Constraints in Quantum Testing

While classical testing benefits from determinism and reproducibility, quantum systems introduce fundamental physical and operational constraints that challenge these assumptions. Quantum testing must contend with the unique behaviors of qubits and quantum mechanics: notably, the no-cloning theorem, probabilistic measurement, and system noise.

No-Cloning and Limited Observability: A foundational principle in quantum mechanics is the *no-cloning theorem* [36], which asserts that it is impossible to create an identical copy of an arbitrary unknown quantum state. While it may be possible to construct a machine that perfectly clones known basis states such as $|0\rangle$ and $|1\rangle$, such a machine would necessarily fail when applied to a superposition state like $|0\rangle + |1\rangle$, producing incorrect or inconsistent results due to the linearity of quantum operations as illustrated in Fig. 8. This fundamental constraint arises from the structure of quantum mechanics itself and implies that no universal quantum cloning device can exist. The no-cloning theorem has significant implications for quantum testing and debugging, as it restricts the ability to probe or duplicate quantum states during execution without disturbing them. This phenomenon poses a significant challenge for testing:

- It is impossible to clone a qubit to preserve its state while measuring it.
- The inability to duplicate state limits the use of classical test strategies such as snapshot-based testing or branching test cases from a common state.

This constraint necessitates test strategies that compensate for the inability to directly observe quantum states, typically by executing the same quantum circuit multiple times and analyzing the distribution of measurement outcomes. Statistical sampling across repeated runs allows testers to infer probabilistic behavior and detect deviations from the expected output distribution, rather than relying on single-run determinism like classical systems.

Probabilistic Outputs and Oracle Challenges: Measurement in quantum computing is inherently probabilistic: a qubit collapses into $|0\rangle$ or $|1\rangle$ based on the squared

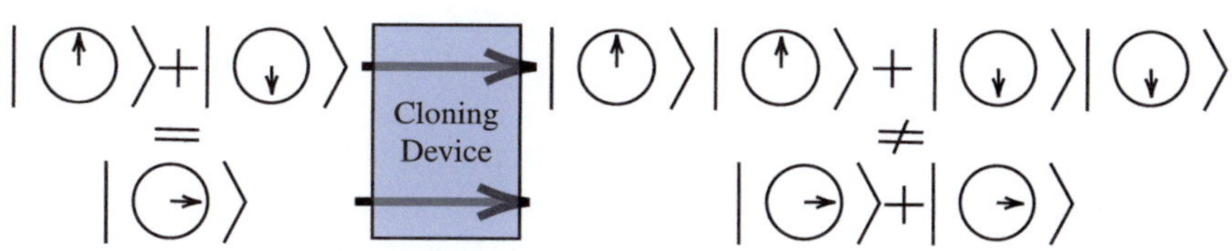

Fig. 8 Illustration of the no-cloning theorem. A hypothetical cloning machine that can perfectly copy the basis states $|0\rangle$ and $|1\rangle$ would fail when applied to a superposition state like $|0\rangle + |1\rangle$, producing incorrect or inconsistent results. This fundamental limitation implies that no universal quantum cloning machine can exist [35]

amplitude of its state components. This introduces uncertainty even when running the same quantum program repeatedly with the same input.

- A test may yield different outputs across executions due to quantum randomness.
- It is not always possible to define a single expected output; test oracles must now specify distributions or expected probabilities.

This calls for statistical test evaluation, where the correctness of a quantum program is determined by comparing the observed distribution of measurement outcomes (collected over many repeated executions) with the theoretically expected distribution. Rather than expecting a single deterministic result, test success is defined in terms of how closely the empirical frequencies align with the target probabilities, often using metrics such as fidelity, total variation distance, or hypothesis testing.

Hardware Noise and Decoherence: Quantum hardware is still in its infancy and remains highly susceptible to various sources of noise, which can introduce errors at every stage of computation. These errors affect the fidelity of quantum operations and ultimately distort the output distribution, making reliable testing and debugging more complex.

- *Decoherence*: This refers to the loss of quantum information due to interactions between the qubits and their surrounding environment. Decoherence causes superposition and entanglement to degrade over time, limiting the depth and duration of quantum computations. It is a major challenge in preserving quantum coherence during execution.
- *Gate errors and crosstalk*: Quantum gates, though mathematically idealized as unitary operations, are physically implemented using analog pulses that are prone to imprecision. Gate errors arise from inaccurate pulse shaping, calibration drift, and qubit–qubit interactions. Crosstalk between adjacent qubits can introduce unintended operations, particularly in multi-qubit gates such as CNOT.
- *Readout errors*: Even after an ideal computation, the final measurement step can produce incorrect classical outcomes. These readout errors are often asymmetric (e.g., $|0\rangle$ being misread as $|1\rangle$ more frequently than vice versa) and may vary across qubits.

Figure 9 illustrates how these errors can manifest at the circuit level. In this example, a clean quantum circuit (Fig. 9a) is compared against a noisy version (Fig. 9b) where gates have been accidentally inserted, deleted, or replaced—mimicking the effects of gate infidelity, crosstalk, and decoherence. Such noise-induced alterations impact the distribution of measurement results and complicate the evaluation of correctness. As a result, a test failure may not always indicate a logical error in the quantum program itself, but could instead reflect a transient hardware fault or instability in the underlying device. Quantum testing frameworks must therefore disentangle true algorithmic bugs from stochastic physical effects, often relying on repeated executions, statistical validation, and noise-aware modeling to make meaningful assessments of correctness.

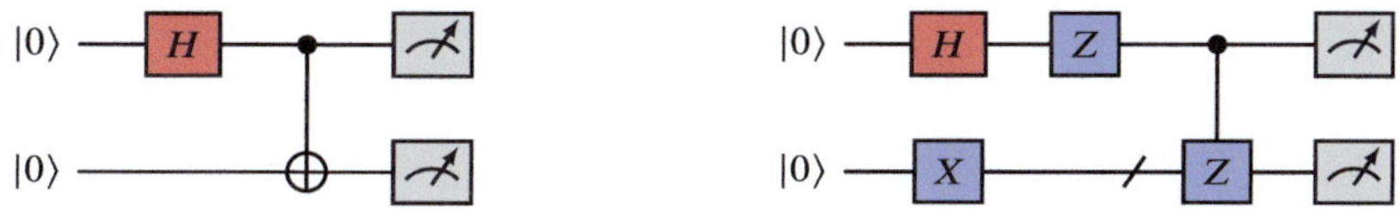

(a) Desired circuit ($|\psi_{\text{desired}}\rangle$)

(b) Actual noisy circuit ($|\psi_{\text{noisy}}\rangle$)

Fig. 9 Illustration of how noise can affect a quantum circuit. A useful way to conceptualize this is that a noisy quantum computer may alter the intended circuit by inserting, deleting, or replacing gates, ultimately changing the distribution of measurement outcomes

3 Coverage Metrics and Test Adequacy

Coverage metrics are essential for assessing how thoroughly a test suite exercises a program. Classical software relies on well-defined metrics such as statement coverage, branch coverage, and path coverage to guide test generation and evaluate adequacy. However, quantum programs introduce unique challenges, such as superposition, entanglement, and probabilistic behavior, that render many classical metrics insufficient. This section first reviews classical coverage foundations and then introduces quantum-specific approaches, including gate-level metrics and advanced techniques like combinatorial and search-based testing, to address the distinct nature of quantum computation.

3.1 Classical Coverage Metrics and Their Foundations

Classical software and hardware testing have long relied on well-defined coverage metrics to assess test adequacy. Statement coverage ensures that each executable instruction in the program is visited by at least one test case. Branch coverage, also known as decision coverage, ensures that every possible outcome (true and false) of each conditional statement is exercised. Consider the following example in Fig. 10a. To achieve full statement coverage in this example code, one test input (e.g., `x = 1`) is sufficient. However, to satisfy branch coverage, both branches must be executed, requiring inputs like `x = 1` and `x = -1`. Thus, branch coverage is stricter and often a better indicator of control-flow exploration.

Beyond these basic metrics, condition coverage requires that each Boolean subexpression within a compound condition be evaluated both to true and false. Decision coverage is similar but considers the overall truth value of a decision. Path coverage aims to cover all possible execution paths in the program, and loop coverage focuses on executing loops zero, one, and multiple times. Consider the example in Fig. 10b. Here, full loop coverage requires tests with `n` $=\emptyset$, `n = 1`, and `n > 1` to cover the no-iteration, single-iteration, and multiple-iteration behaviors, respectively. These classical metrics form the foundation of adequacy assessment in conventional software and inspire corresponding metrics for quantum programs.

```c
int foo(int x) {
    if (x > 0) {
        return 1;
    } else {
        return -1;
    }
}
```

(a) Simple conditional program

```c
void loop_example(int n) {
    for (int i = 0; i < n; i++) {
        printf("%d\n", i);
    }
}
```

(b) Loop-based example

Fig. 10 Conditional and loop-based program examples

In hardware testing, similar coverage metrics are used to ensure adequate stimulation of circuit elements. Gate-level coverage refers to ensuring that all logic gates (e.g., AND, OR, NOT, etc.) are activated during simulation. Toggle coverage measures whether each signal bit in the circuit transitions from 0 to 1 and from 1 to 0 during testing, capturing dynamic behavior. These metrics are essential for evaluating the effectiveness of testbenches in hardware verification workflows and are typically collected using hardware simulation tools.

3.2 *Quantum Coverage Metrics*

As classical coverage metrics fall short in capturing the probabilistic and entangled nature of quantum programs, researchers have proposed quantum-specific alternatives. These include metrics defined at the quantum gate level, such as coverage over controlled operations like CNOT, as well as high-level test strategies such as combinatorial and search-based testing. Unlike classical branches, quantum control flow may emerge from entanglement, superposition, or dynamic circuit behavior, requiring fundamentally different notions of test adequacy. This section discusses recent work on gate-level metrics, combinatorial coverage strategies, and search-based test generation techniques, highlighting how these approaches aim to explore quantum state space and validate circuit correctness under quantum-specific constraints.

Input/Output Level Metrics: Coverage can be measured in different ways, each considering a different facet of the quantum circuit under test. The most popular coverage metrics are input, output, input–output, and gate branch [37, 38]. Input coverage is defined as all possible state values for the number of qubits of the circuit being tested as input. Conversely, output coverage is the output being in all possible state values for the number of qubits of the circuit. Input–output coverage is thus that for every input–output pair tested, all possible states within the pair are covered.

Quito is an automated framework for generating tests to achieve high coverage in input, output, and input–output coverage metrics [39]. Tests are generated iteratively,

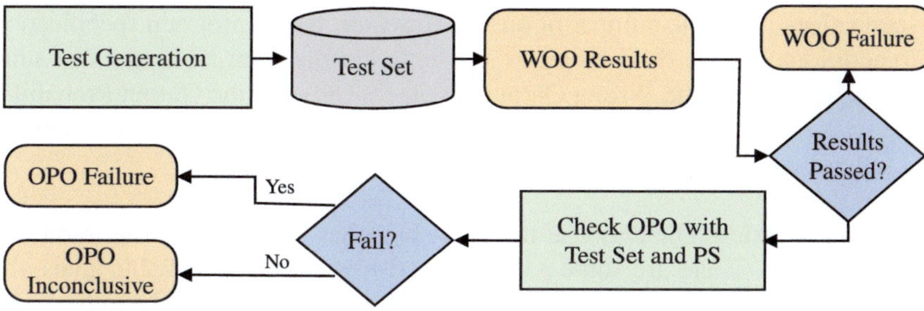

Fig. 11 Test generation and assessment in Quito [39]. If the program specification (PS) is available, Quito runs two oracles to assess the results from running a test suite: Wrong Output Oracle (WOO) and Output Probability Oracle (OPO)

with the previous iteration of tests and their coverage used as feedback to guide the generation of the next tests to improve overall coverage, as illustrated in Fig. 11. Quito operates on a quantum program, and optionally, a program specification (PS) of the quantum program. This program specification details the expected output distributions for specific inputs and is expressed as

$$\mathrm{PS}(i) = \left\{(o_1, p_1), \ldots, (o_{|O|}, p_{|O|})\right\}, \tag{2}$$

where $\mathrm{PS}(i)$ is the program specification under input i, $o_j \in O$ is the jth output from the set of possible output values O of the circuit as determined by the number of qubits, and p_j is the probability that output o_j occurred. The probabilities create a distribution, and thus $\sum_{j=1}^{|O|} p_j = 1$.

As quantum circuits are non-deterministic and can result in different outputs across different runs of the quantum circuit, it is imperative that any test suite and coverage metric can determine when the generated outputs are a failure, indicating an issue with the quantum circuit. Several test generation techniques define two different failure cases, as follows [39–43]:

- *Incorrect Output State:* Every output generated by running the test suite is compared to the expected outputs in the program specification. For every output generated with the test suite, it must also occur in the program specification. In other words, if there is an output generated by running the test suite that is not listed in the program specification, the test suite encountered a failure and contains a fault or other bug in the quantum circuit.
- *Incorrect Output Distribution:* It examines every output generated with the test suite and compares it to the expected outputs in the program specification, but with a focus on the distribution of the outputs in both. The generated output distribution must be within a margin of error of the expected output distribution for a given input.

With no program specification, the three coverages (input, output, and input–output) are measured according to the encountered state values out of the total possible state values given the number of qubits. However, with a program specification, Quito additionally runs two oracles to assess the results from running a test suite. These two oracles are the Wrong Output Oracle (WOO) and the Output Probability Oracle (OPO). The WOO follows the incorrect output state failure case, while the OPO follows the incorrect output distribution.

Gate-Level Metrics: The classical notion of branches does not exist in traditional quantum circuits. There are some gates, collectively called controllable gates, that exhibit conditional behavior by activating a gate based on the value of another qubit. For example, CNOT, or controllable NOT, is a two-qubit gate that uses one qubit as the control signal for whether the second qubit has the NOT gate processed on its current state. Therefore, there are different definitions of branch coverage in quantum domain compared to classical domain.

Gate branch coverage is a branch coverage metric defined specifically for quantum circuits [38]. It measures the different paths taken from controllable gates, ensuring that the one or more control qubits of the controllable gate did (or did not) activate the gate on the other qubit. Specifically, the gate branch coverage (GBC) can be expressed as

$$\mathrm{GBC}(G) = \frac{\sum_{g \in G} e(g)}{\sum_{g \in G} 1 + c(g)}, \tag{3}$$

where gate g in the set of controllable gates G of the quantum circuit has $e(g)$ exercised branches and $c(g)$ control qubits.

Figure 12 illustrates an example setup for getting the gate branch coverage of the execution paths for the quantum circuit under test. In this example, the quantum

Fig. 12 Gate branch coverage **b** of a basic quantum circuit with a Hadamard gate and a CNOT gate **a**

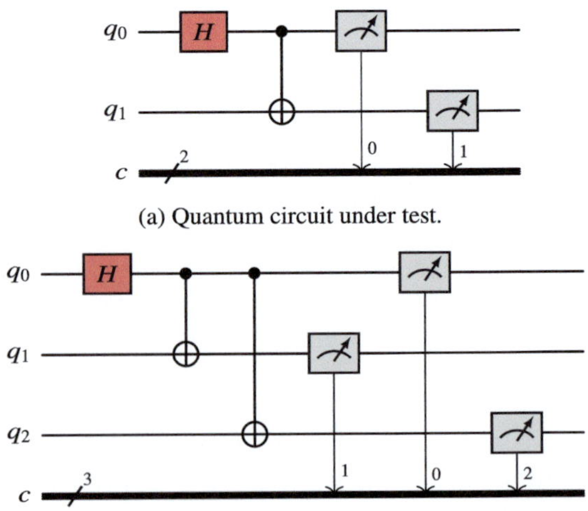

(a) Quantum circuit under test.

(b) Quantum circuit under test with gate branch coverage measurements embedded in it.

circuit being tested is a two-qubit circuit as shown in Fig. 12a, with a Hadamard gate acting on qubit q_0, followed by a CNOT gate on q_0 and q_1 such that q_0 is the control qubit for the gate. To get the execution path without using the measurement operator and collapsing the state of the qubit, the quantum circuit can be redefined such that an extra qubit, such as q_2 in Fig. 12b, is inserted. Connecting a CNOT gate between q_0 and q_2 with q_0 as the control bit will cause the NOT gate to process on q_2 if q_0 is $|1\rangle$. In other words, if the measured value of q_2 after the CNOT gate indicates that the NOT gate was processed, then it is clear that the previous CNOT gate between q_0 and q_1 was also taken. This CNOT gate followed by a measurement with the extra qubit can provide an effective way to get the execution path of a circuit without changing its functionality at the cost of an extra qubit during simulation.

Coverage-guided test generation can also be extended to specific quantum applications, such as quantum machine learning. Quantum neural networks, the quantum equivalent of classical neural networks, consist of parameterized quantum circuits. These circuits thus have a defined structure with a learned output. Ensuring that these quantum neural networks are functioning as intended is crucial for verifying that their learned outputs are functionally correct. Shao and Zhao present a coverage-guided test generation technique for quantum neural networks that can sufficiently generate test cases with high coverage [44].

Combinatorial Testing: Combinatorial testing uses a combination of inputs to test a given quantum circuit. These input combinations can be tailored for specific purposes, such as finding faults or bugs within the quantum circuit under test [40, 43]. One automated combinatorial testing framework is QuCAT [43], which generates k-value schemas of a specified k strength to test as inputs. A k-value schema is defined as the combination of input values for the k input qubits to the circuit. Note that the input qubits can be a subset of the total qubits of the circuit under test. k also determines the type of combinatorial test suite to use, such as pairwise for $k = 2$, 3-wise for $k = 3$, and so on. Additionally, the test suites can be created at either a defined strength or increasing strength. For the increasing strength test suites, they are executed and increased in strength until a failure is encountered. This failure indicates that there is a fault within the quantum circuit under test that needs to be rectified.

Failures are indicated with incorrect output state and incorrect output distribution. The program specification of the quantum circuit under test is thus required, as defined in Equation 2. The program specification is utilized to determine the validity of the outputs generated by running the circuit many times with the test suite against the expected outputs, measuring failures with incorrect output state and incorrect output distribution, as defined earlier. Both strength strategies are effective at creating robust test suites for the quantum circuit under test.

Search-Based Testing: In search-based testing, specific inputs are generated with their corresponding expected outputs as determined by a program specification. QuSBT is a genetic algorithm to generate test suites with a maximum number of failing test cases [41, 42]. As quantum circuits are non-deterministic, a quantum circuit must be run many times to get the distribution of resulting output states. For each test case generated, there is a set of input values v also generated. These input values are each executed n times to get the output distribution. The outputs and the distribution of the outputs are checked for failures according to the incorrect output state and the incorrect output distribution. A failure f is added to the set of encountered failures F for the test case.

The fitness function of the search-based genetic algorithm is designed to maximize the number of failures that test cases can produce. By prioritizing test cases that reveal faults, the algorithm can evolve toward more effective fault-detection capabilities. The fitness function is formally defined as

$$\text{fitness}(v) = \left|\{f_j \in F | f_j = \text{true}\}\right|, \tag{4}$$

where v represents the input values for a test case and F is the set of failures encountered when executing the test case. This fitness-driven approach enables the genetic algorithm to systematically generate diverse test cases that can effectively expose faults across different quantum algorithms and circuit implementations.

4 Assertion and Property-Based Testing

Assertions and property-based testing are foundational techniques in classical software and hardware verification, used to specify, check, and enforce expected behavior during program or circuit execution. Assertions are declarative conditions embedded in the system under test that must hold at specific execution points or across time. In the context of quantum computing, these techniques must be revisited to account for inherent quantum phenomena such as measurement-induced collapse, non-determinism, and the no-cloning theorem. This section begins by reviewing assertion mechanisms in classical software and hardware systems and then explores their quantum counterparts, building a discussion of how property-based testing principles are adapted to the quantum domain.

4.1 Classical Assertions

Assertions in classical computing, which span across both software and hardware domains, are critical tools for validating the correctness of computations, enforcing specifications, and detecting anomalies during testing. Assertions are typically Boolean conditions inserted into programs or hardware designs that must evaluate to `true` during execution. If an assertion fails, it signals a deviation from the expected behavior [16]. These assertions often measure the value of a variable in use during the program, such as variable `b` in Fig. 13a.

```c
int divide(int a, int b) {
    assert(b != 0);
    return a / b;
}
```

(a) Inline software assertion in *C*

```verilog
assert property (@(posedge clk)
    state == S_LOCKED
        |-> key_valid == 1);
```

(b) Hardware assertion in *Verilog*

Fig. 13 Examples of assertions used in classical systems

Inline Software Assertions: Inline assertions are Boolean expressions embedded within the source code that validate program conditions dynamically during execution. They are primarily used to enforce preconditions (assumptions about inputs before a function executes), postconditions (guarantees about outputs after execution), and invariants (conditions that must remain true throughout a program's operation, such as within loops or object lifecycles). For example, in Fig. 13a, the `assert()` macro evaluates an expression and terminates the program if the expression is false. These assertions serve as executable documentation and help expose unexpected behavior or logic flaws at runtime. While not a substitute for thorough testing, they act as safeguards during development and debugging by immediately flagging violations of intended behavior, thus reducing the likelihood of silent failures or undefined behavior propagating through the system.

Assertions in Hardware Design: In hardware design and verification, assertions serve as formal, embedded properties that specify the expected behavior of signals and modules over time [16]. Unlike software assertions, which typically check conditions at a single point in time, hardware assertions often express temporal relationships, defining how signals should evolve across clock cycles. These assertions are commonly written using hardware assertion languages, such as SystemVerilog Assertions (SVA) or Property Specification Language (PSL), and are integrated into simulation or formal verification flows [16, 45]. Hardware assertions can capture a wide range of properties, including protocol compliance, state machine transitions, data validity windows, and safety conditions [45]. For example, the SystemVerilog assertion in Fig. 13b expresses that if the system enters the `S_LOCKED` state, the signal `key_valid` must be asserted (i.e., logic high) in the following clock cycle. This is a temporal implication: it states that on the rising edge of the clock, if state equals `S_LOCKED`, then `key_valid` must hold true in the next cycle. Such assertions are essential in verifying control logic and ensuring compliance with hardware design specifications, especially in safety-critical and security-sensitive systems.

Assertion Placement in Classical Test Flow: In both hardware and software testing workflows, assertions serve as checkpoints to validate system correctness at critical phases of execution. While their underlying implementations may differ, their structural roles when guarding inputs, monitoring intermediate behavior, and verifying outputs are fundamentally aligned. Assertions can be systematically positioned before the input (stimulus) is applied (to ensure valid initial conditions), during exe-

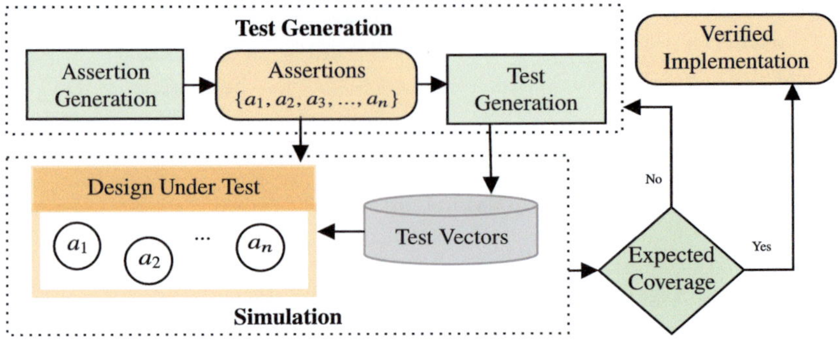

Fig. 14 Assertion checkpoints in designs under test are monitored during simulation using test vectors

cution (to monitor ongoing behavior), and after response observation (to confirm the correctness of outcomes). Figure 14 highlights the overall classical assertion-based validation, specifically, assertions are generated based on the properties that should hold by the system and then they are inserted into the system under test. The system is simulated with test vectors to activate the assertions while monitoring their behavior and coverage. This assertion-based validation flow applies uniformly to unit tests in software as well as testbenches in hardware verification. Each assertion point plays a distinct role in isolating the source of failures and supporting traceable, modular verification.

4.2 *Quantum Assertions*

Due to the unique constraints of quantum computing, including non-deterministic outcomes, the no-cloning theorem, and measurement-induced collapse, traditional assertions used in classical programs are not directly applicable to quantum programs. Instead, the inability to clone or observe quantum states without altering them introduces significant challenges for defining test oracles and writing assertions. To address these challenges, researchers have developed a variety of assertion-based validation techniques adapted to quantum contexts.

Statistical Assertions: Statistical assertions perform statistical tests on the measured quantum states as the assertion condition [46]. Specifically, the statistical tests of the chi-square test and contingency table analysis are performed on three different types of assertions related to the measured quantum state, as follows:

1. *Classical assertions* test whether the quantum state is a classical value, meaning a bit value of either 0 or 1, at a specific measurement.
2. *Superposition assertions* check the probability distribution of the superposition of a quantum state at a specific measurement.

3. *Entanglement assertions* examine the entanglement value of two or more entangled quantum states at a specific measurement.

Runtime Assertions: As statistical assertions require measuring the quantum states to check if the values are as expected, runtime assertions try to avoid using the measurement operation directly [47]. This is done to avoid collapsing the state of the qubit being measured, which can affect any circuit computations that are processed after the measurement for assertion purposes. Within runtime assertions, there are different types of assertions, each testing different aspects of the quantum circuit:

1. *Dynamic assertions* allow for creating assertions around the value of qubits during runtime [47]. These can be used to measure the three types of statistical assertions [46] through the use of an ancilla qubit [47]. This ancilla qubit uses CNOT gates to measure the current value of the circuit qubits. Measuring the ancilla qubit and thus collapsing its current state value does not impact the execution of the remainder of the circuit. This is similar to the strategy used to measure the gate branch coverage as described in Sect. 3.2. These types of dynamic assertions, which utilize the ancilla qubit to measure statistical assertions, have also been demonstrated to work in qutrit systems [48].
2. *Projection-based assertions* establish the assertions with projections based on Birkhoff–von Neumann logic [49]. The quantum state values to be analyzed are measured using projection measurements, which can then be compared through the assertion's projection. By doing so, the resulting projection-based assertions can provide more insight into the quantum circuit under test, as they do not solely rely on classical logic that lacks complete equivalents or parallels to quantum.
3. *SWAP assertions* adjust how the ancilla qubit is measured to allow for a broader range of states to be observed [50]. To do so, instead of using gates like CNOT to determine the value of a qubit in the circuit, a SWAP-based circuit is used. This SWAP-based circuit consists of a unitary gate U on the ancilla qubits, the inverse of the unitary gate U^{-1} on the real qubits, and, finally, a SWAP gate between the real qubits and the ancilla qubits. This allows for the state in circuit under test to remain a known value as $U^{-1}U = I$ and the ancillas to be measured directly, containing a more precise state value.
4. *Approximate quantum state assertions* check if the set of states measured via the ancilla qubits is part of the set of expected state values for the circuit at that measurement point within the circuit execution [50]. This assertion is beneficial as it enables developers to approximate the results instead of having specific knowledge of the exact value the state needs to be.
5. *Concurrent assertions* run in a separate circuit from the actual circuit under test when run on a real quantum device with noise present [51]. As both circuits are executing, a SWAP test is performed, similar to SWAP assertions, that checks if the states of both circuits are equivalent in the presence of noise.

Assertion Generation: Generating assertions manually is a time-consuming process that also requires expert knowledge of the circuit under test. quAssert is a framework for automatically generating statistical assertions for a quantum circuit, allowing

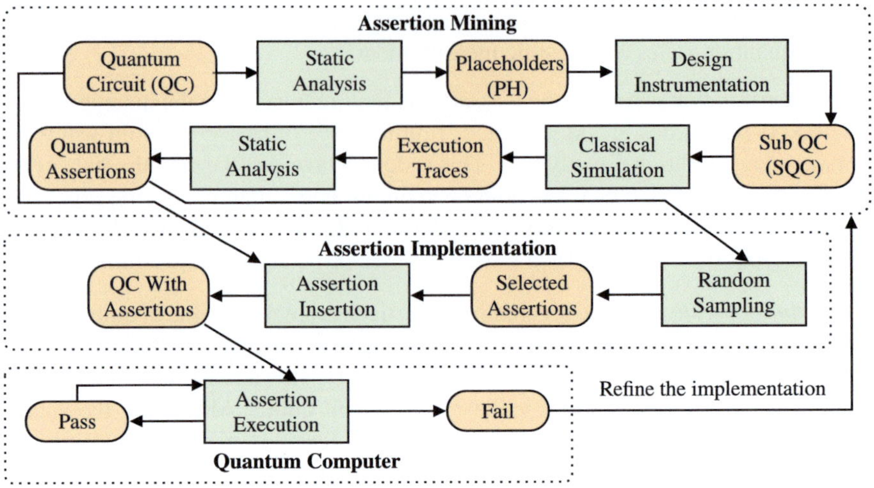

Fig. 15 An overview of the quAssert framework for automatically generating quantum assertions [52]

for fast and correct insertion of assertions into the quantum circuit under test [52]. Assertions are generated through static analysis and then randomly sampled to insert into the quantum circuit at specified locations. This resulting assertion circuit is then run to determine the condition of the inserted assertions (Fig. 15).

The location placement of the assertions is of key importance to find bugs and faults within the designed quantum circuit. Rovara et al. proposed an automated assertion generation framework for statistical assertions that moves the locations of assertions across subsequent iterations to find the location of bugs [53]. This is done by moving current assertions up the circuit, adjusting the statements as needed through commutation rules, and inserting new assertions further down the circuit.

Property Testing: Another automated testing strategy is property testing, which uses a model to generate test cases based on properties of the circuit under test. QSharpCheck adapts QuickCheck from classical to perform property testing on Q# programs [54]. It defines a test language for these programs, which includes definitions of the properties to test. Each property contains specific values as input and output to satisfy the property that are used in the generation of a test case for the property. These properties are often derived from the program specification. With the test cases generated from the properties, each are executed, gathering the output results from each test. Statistical analysis is performed on the resulting test output to determine the success of each test to locate potential issues. QuCheck is a similar tool that has been proposed for use with Qiskit programs [55].

Recent efforts expand on property testing to additionally include delta debugging for regression testing of the test cases generated with property testing [56]. When a quantum circuit is developed, it may be tested many times at different points within the development cycle. The test cases are generated based on the properties as defined by the program specification for the final circuit. As development proceeds, these test cases are run. If a test that succeeded in the past is now failing, there was a regression in the quantum circuit, and it needs to be debugged to find the error. Delta debugging is useful in quantum computing, as it is a lightweight solution in pinpointing the potential cause of the regression [56]. It does so by having an oracle examine the deltas between the passing and failing circuits to find potential changes, namely, the failure that is causing the regression.

The adaptation of assertion and property-based testing to quantum computing represents a fundamental shift from classical verification paradigms. Quantum assertions must navigate unique challenges, including measurement-induced collapse, probabilistic outcomes, and the no-cloning theorem through innovative techniques such as statistical assertions and runtime assertions using ancilla qubits. Automated frameworks discussed in this section demonstrate promising progress toward systematic quantum verification, though significant challenges remain in assertion placement, measurement trade-offs, and scalability to complex quantum systems.

5 Symbolic Execution and Concolic Testing

Symbolic and formal testing techniques play a central role in validating both classical and quantum programs by enabling systematic reasoning about program behavior. In classical systems, symbolic execution explores program paths using symbolic inputs and constraint solvers to uncover bugs and increase test coverage, while formal methods such as model checking and theorem proving provide strong guarantees about system correctness. As quantum programs grow in complexity, these techniques are being reimagined to account for the unique properties of quantum computation, such as superposition, entanglement, and probabilistic measurement. This section presents a comparative view of symbolic and formal testing in classical and quantum domains, beginning with classical foundations and then detailing the adaptations required for quantum systems, including path amplitude reasoning, state vector constraints, and hybrid concolic techniques suited for quantum behaviors.

5.1 Classical Symbolic Execution

Symbolic execution is a powerful program analysis technique used to explore multiple execution paths in a program by treating inputs as symbolic variables rather than concrete values [12, 18, 33]. It systematically generates path constraints and checks their feasibility using constraint solvers, such as SMT (satisfiability modulo theories) solvers [57]. The core idea is to reason about all possible execution paths that depend on symbolic inputs to uncover errors such as assertion violations, crashes, or security vulnerabilities.

```c
int foo(int x, int y) {
  int z = 0;
  if (x > 0) {
    z = x + y;
  } else {
    z = x - y;
  }
  assert(z != 42);
}
```

(a) Solving branching conditions

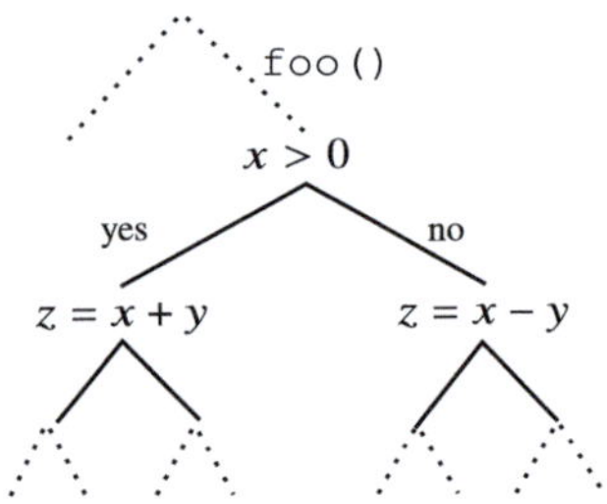

(b) Symbolic execution tree.

Fig. 16 Illustration for solving branch condition with symbolic execution tree

Path Exploration: In classical symbolic execution, a program is executed using symbolic inputs (e.g., x, y), and at each branch, the symbolic executor forks the execution into separate paths depending on the condition (e.g., $x > 0$ vs. $x \leq 0$). Each path accumulates a *path condition*—a set of logical constraints that must hold for that path to be taken. Consider the simple C program in Fig. 16a. Using symbolic execution, we treat `x` and `y` as symbolic variables. There are two main execution paths:

- *Path 1:* `x > 0` $\Rightarrow$ `z = x + y`; path condition: $\phi_1 = (x > 0)$.
- *Path 2:* `x` $\leq$ `0` $\Rightarrow$ `z = x - y`; path condition: $\phi_2 = (x \leq 0)$.

The symbolic executor then checks whether each path can lead to the assertion being violated (i.e., whether $z = 42$ is satisfiable under the path condition). For example, for Path 1, the expression $\phi_1 \wedge (x + y = 42)$ is sent to the SMT solver. If satisfiable, the tool reports a counterexample such as $x = 22$, $y = 20$.

Concolic Testing: Concolic testing provides a synergistic combination of *concrete* simulation and *symbolic* execution. In other words, it enhances symbolic execution by combining it with concrete program execution [13, 58]. During a test run, the program is executed with concrete input values, and symbolic expressions are simultaneously tracked along the execution path. After the execution, the collected symbolic path constraints are negated one at a time to generate new input values that force the program execution through previously unexplored paths. One of the primary motivations for concolic testing is to address the problem of *state space explosion*, which is a phenomenon where the number of possible execution paths grows exponentially with the number of branches in the program. Pure symbolic execution, while powerful, can quickly become intractable as the number of feasible paths exceeds the capacity of both memory and solver time [12, 18]. Concolic testing mitigates this issue by focusing on one path at a time. By leveraging concrete executions to guide symbolic reasoning, it avoids simultaneous branching at each decision point. Instead of forking at every conditional, it records the current execution path and uses constraint solvers selectively to steer future tests toward alternate paths. This targeted exploration strategy significantly reduces the number of paths that must be analyzed at once, enabling deeper exploration of complex programs while still maintaining reasonable computational cost.

5.2 *Quantum Symbolic Execution*

Similar to its classical counterpart, quantum equivalence checking involves comparing the behavior of two quantum circuits to determine whether they produce equivalent or identical outcomes. As quantum circuits are executed either on simulators or real quantum hardware, they often undergo compilation and transpilation into intermediate representations tailored to the target backend. This transformation process makes equivalence checking essential, not only to verify the correctness of compiler optimizations and hardware mappings, but also to ensure that the high-level quantum logic is preserved across different execution platforms.

Equivalence Checking: Burgholzer and Wille [59] have designed an equivalence checking framework that compares the original quantum circuit with its compiled equivalent to ensure that both circuits produce the same result. Figure 17 illustrates the overall architecture of the proposed framework. Quantum circuits have the reversibility property where the matrix associated with a gate, such as a unitary gate U, has an inverse U^{-1}. Therefore, multiplying the two gates together results in the identity matrix I, such that $UU^{-1} = I$. As it is the gates that are reversible, this means the entire circuit can also be reversed resulting in the identity function $\mathbb{I}$, $GG^{-1} = \mathbb{I}$, where G is the original circuit and G^{-1} is its inverse. This idea is extended to equivalence checking by using the reversibility property.

With two quantum circuits G and G' to test, G' can invert all of its gates to result in $\mathbb{I}$ when applied with the gates of G. This needs to be done in a specific manner, and is aided by decision diagrams of the gates. Therefore, the decision diagram is constructed such that $G \Rightarrow \mathbb{I} \Leftarrow G'$, or the gates from G are applied normally and the gates from G' are applied as their inverses, resulting in $\mathbb{I}$. If applying the gates does not result in $\mathbb{I}$, then G and G' are not equivalent. Figure 18 contains a sample decision

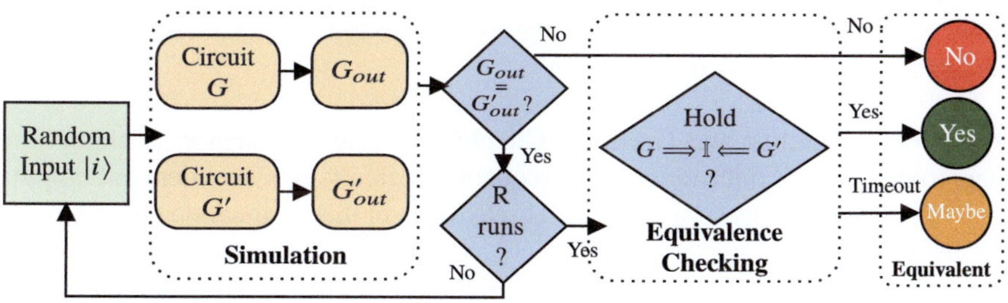

Fig. 17 Equivalence checking framework proposed by Burgholzer and Wille [59]

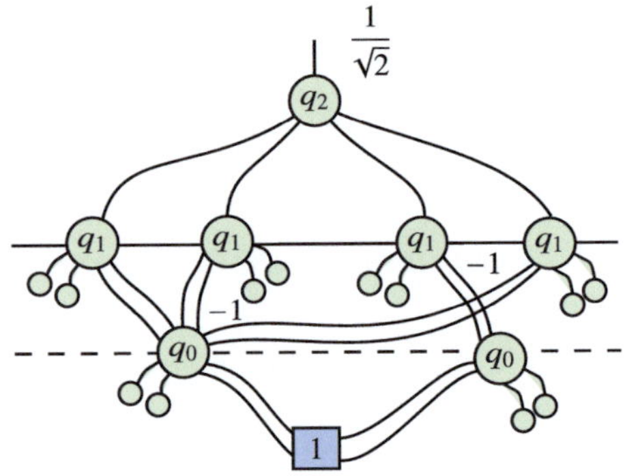

Fig. 18 Sample decision diagram for unitary gate U

diagram for a unitary gate U. Specifically, the figure illustrates how the authors have recursively defined the different decompositions of the unitary gate. With the recursive definition resulting in different levels, the decision diagram establishes complete functionality of the unitary gate while maintaining a low number of nodes.

In addition to basic quantum circuits, equivalence checking frameworks for parameterized and dynamic quantum circuits have also been proposed [60]. For parameterized quantum circuits, ZX-calculus [61] is used to check the equivalence between a parameterized quantum circuit and its compiled version. Unlike with traditional quantum circuits, where the equivalence checking can be elementwise for each gate in the circuit, parameterized gates must be checked for all possible parameter values θ. The parameterized circuits are instead represented in ZX-diagrams where ZX-calculus can be performed on them to simplify the resulting circuits, primarily the parameterized assignments. With the assignments in place, the two circuits can thus be checked for equivalence through traditional equivalence checking methods.

For dynamic circuits, the classical condition statements must be checked for equality. However, traditional approaches using decision diagrams are not sufficient to represent all aspects of the condition statements [62]. The dynamic quantum circuit is instead represented in its tensor network form, with the decision diagram becoming a tensor decision diagram based on the tensor network. Tensor networks are a reduced dimensionality view of the tensors and matrices that make up the gate operations of the quantum circuit on the qubit states. The condition statements can be easily represented with tensor networks, and its associated tensor decision diagram allows for easy equivalence checking between two dynamic quantum circuits and the associated tensor decision diagrams.

Equivalence, Identity, and Unitarity Checking: While equivalence checking often refers to the task of confirming that two quantum circuits produce identical outputs, Long et al. [63] provide a more nuanced breakdown of this verification task into three core problems:

1. *Equivalence Checking:* A common approach for equivalence checking between two quantum programs P and P' is the *Swap Test*, a standard quantum subroutine that estimates the overlap (fidelity) between their output states. The programs are applied to identical copies of input states sampled from the Pauli basis, chosen for its completeness, since any quantum state can be expressed as a linear combination of Pauli eigenstates. If the output states are nearly identical (within a small tolerance ϵ) across a representative set of such inputs, the two programs are considered equivalent.
2. *Identity Checking:* Verifies whether a program P performs the identity operation (i.e., $P = I$). This fidelity-based approach extends naturally to identity checking, where the goal is to determine whether a program P acts as the identity operation. After applying P to a randomly selected Pauli basis state $|\psi\rangle$, the result is passed through the inverse of the state preparation (mapping it back to $|0\rangle^{\otimes n}$), and the outcome is measured. Consistent measurement of $|0\rangle^{\otimes n}$ across trials indicates functional equivalence to the identity.
3. *Unitarity Checking:* Confirms whether P is a valid unitary operation, preserving orthogonality and state purity. To check whether a quantum program is unitary, a fundamental requirement for most quantum operations, two key properties are tested: preservation of orthogonality and consistency under superposition. This involves preparing pairs of orthogonal basis states (e.g., $|i\rangle, |j\rangle$) and superposition states (e.g., $\frac{1}{\sqrt{2}}(|i\rangle \pm |j\rangle)$), applying the program to them, and then using the Swap Test to verify that the output states remain orthogonal. A failure in either test suggests a violation of unitarity. Collectively, these checks ensure the program preserves linearity, orthogonality, and purity, which are the core principles of valid quantum transformations.

Unlike circuit optimization-based equivalence checkers (e.g., in Qiskit or Tket), the proposed algorithms assume black-box access to the quantum programs. This black-box model is highly relevant for post-compilation testing or when dealing with proprietary or obfuscated quantum circuits.

Concolic Testing: Quantum programs have been defined for use with concolic testing through the definition of quantum constraints, including generation techniques [64]. The overall workflow of quantum concolic testing is similar to classical with several extra steps to accommodate quantum logic, and the steps are summarized below and in Fig. 19:

1. Generation of initial test case, including inputs.
2. Creation of the symbolic objects.
3. Execution of the quantum circuit using the input values specified by the test case. The execution path is saved.
4. Generate a quantum path constraint using the quantum constraint generator on the execution path.

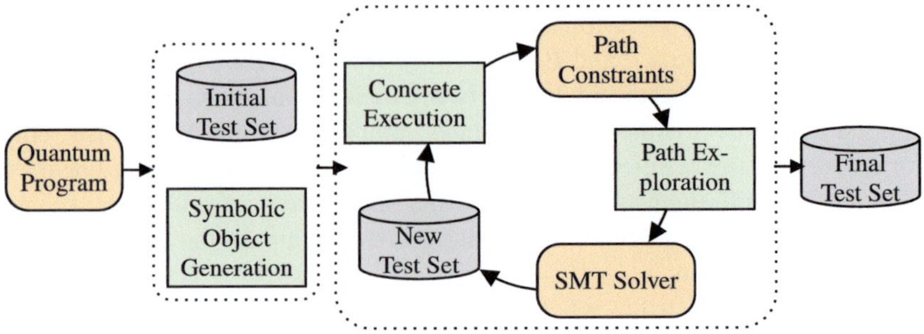

Fig. 19 Overview of quantum concolic testing [64]

5. Determine a new path constraint through modification of the saved one.
6. Generate a test case by solving the new path constraint with an SMT solver.
7. Repeat steps (3)–(6), checking that the desired path was executed, until coverage is satisfactory.

Let us define a few terms relevant for discussing symbolic execution in the context of quantum concolic testing.

- *Quantum control statement* is the measurement of a specified set of qubits in the circuit and their resulting output.
- *Quantum conditions* are the set of desired results from the quantum control statements.
- *Quantum symbolic object* represents the entire quantum circuit, recording the execution path and all operations along it.
- *Quantum constraint* determines if the output state from the quantum control statement satisfies all quantum conditions. It is defined by

$$\bigwedge_{a \in S \setminus S_r} (\langle a|\phi\rangle = 0) = \text{true}, \tag{5}$$

where $|\phi\rangle$ is the output state of quantum control statement, S is the set of all possible observable outputs as determined by the number of qubits, and S_r is the set of output states from the quantum control statement that satisfy the quantum condition.

Quantum constraints must be generated specifically to account for the non-deterministic nature of quantum circuits. As one input can produce several different outputs, just checking the resulting output states alone is not sufficient to determine correctness or obeying a defined quantum condition. Therefore, three different constraint generation methods are used, as follows:

- *State equality* checks if the distribution of output states measured is equivalent within a margin of error to the expected output distribution, and is formalized as

$$\bigwedge_{a \in S} \left(\left| |\langle a|\phi\rangle|^2 - D_a \right| < \delta \right) = \text{true}, \tag{6}$$

where δ is the margin of error and D_a is the expected probability distribution for a.
- *State lower limit* checks if the probability that a specific output occurred is greater than a defined lower limit within a margin of error, and is formalized as

$$\bigwedge_{s \in P \wedge s=(a_s, p_s)} \left(|\langle a_s|\phi\rangle|^2 > p_s - \delta \right) = \text{true}, \tag{7}$$

where δ is the margin of error and p_s is the lower limit for state pair s in the set of expected state pairs P.
- *State upper limit* checks if the probability that a specific output occurred is less than a defined upper limit within a margin of error, and is formalized as

$$\bigwedge_{s \in P \wedge s=(a_s, p_s)} \left(|\langle a_s|\phi\rangle|^2 \leq p_s + \delta \right) = \text{true}, \tag{8}$$

where δ is the margin of error and p_s is the lower limit for state pair s in the set of expected state pairs P.

In summary, symbolic execution and formal testing techniques are evolving from classical foundations to address quantum computing's unique challenges. While classical approaches rely on deterministic constraint solving, quantum adaptations must handle probabilistic measurements, superposition states, and unitary operations through novel methods like quantum constraint generation, equivalence checking via decision diagrams, and specialized concolic testing frameworks. These formal methodologies will be essential for ensuring quantum program correctness as quantum applications mature.

6 Mutation and Metamorphic Testing

Metamorphic and mutation testing have proven effective across both classical hardware and software testing, serving as powerful techniques to assess test adequacy and uncover faults in systems where traditional test oracles may be insufficient. Mutation testing involves introducing small, systematic changes (mutants) to a program or hardware description to evaluate the ability of test cases to detect these faults [65, 66]. Metamorphic testing, on the other hand, leverages known relationships among inputs and outputs to validate correctness even when expected outcomes are unknown or difficult to compute. In this section, we examine the application of these techniques in both classical and quantum contexts. We first discuss their use in classical software and hardware systems, including digital circuit mutation and input–output relation modeling. We then explore the emerging adaptations of mutation and metamorphic testing in quantum computing, where quantum gates, entangled states, and probabilistic execution introduce new challenges for test design and fault detection across both quantum software and quantum hardware architectures.

6.1 Classical Mutation Testing

Classical mutation testing is a fault-based testing technique used to evaluate the quality of a test suite. It operates by systematically introducing small changes called as *mutants* into a program's source code or a hardware description, with the goal of assessing whether the existing test cases can detect the injected faults. If a test case can distinguish the mutant from the original version (i.e., cause different outputs), the mutant is said to be *killed*; otherwise, it is considered *alive*, possibly indicating a weakness in the test suite.

Software Mutation: Consider the simple example function Fig. 20a from a classical software program. A mutation operator might replace the relational operator > with >= to generate a mutant for the statement `if (a >= b)`. A good test suite should include inputs such as `a = 5` and `b = 5`, which would cause the original version to return `b`, but the mutant to return `a`, thus "killing" the mutant.

Hardware Mutation: Mutation testing can also be applied to hardware, typically at the Register Transfer Level (RTL) description. Consider a Verilog module implementing a 1-bit full adder in Fig. 20b. A mutant might modify the carry-out expression by changing a bitwise AND to OR, making it `assign cout = (a | b) | (b & cin) | (a & cin)`. This mutation alters the carry logic. A test vector such as `a=1, b=1, cin=0` should produce different carry-out values for the original and mutant designs, allowing the testbench to detect the fault.

```
int max(int a, int b) {
    if (a > b) // change to (a>=b)
        return a;
    else
        return b;
}
```

(a) Software example

```
module full_adder(input a, b, cin,
                  output sum, cout);
  assign sum = a ^ b ^ cin;
  assign cout = (a & b) | (b & cin)
    | (a & cin); // change to (a | b)
endmodule
```

(b) Hardware example.

Fig. 20 Classical mutation testing examples

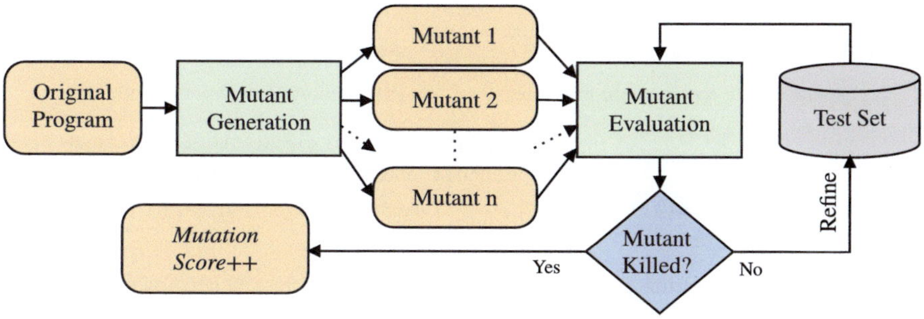

Fig. 21 Workflow of classical mutation testing: each mutant is derived from the original, tested, and evaluated for behavioral deviations

Mutation Testing Workflow: The classical mutation testing workflow involves a systematic process to assess the effectiveness of a test suite. The key steps are illustrated in Fig. 21 and described below:

1. *Mutant Generation:* Mutation operators are applied to the original source code or hardware description to generate a set of mutants. Each mutant contains a small syntactic change, such as replacing an arithmetic or logical operator, inverting a condition, or altering a signal in a hardware netlist. The goal is to mimic common developer mistakes or design flaws.
2. *Test Execution:* The existing test suite is executed against each mutant, typically in an automated fashion. For software programs, this involves compiling and running each mutant with the test inputs. For hardware designs, this may require simulation or synthesis runs with testbenches.
3. *Mutant Evaluation:* After execution, the output of the mutant is compared with the output of the original version. If at least one test case causes a behavioral deviation, the mutant is considered *killed*. Otherwise, it is said to be *alive*. Mutants that survive may point to inadequacies in the test suite or may be functionally equivalent to the original (and thus undetectable).
4. *Score Computation:* The effectiveness of the test suite is quantified using a *mutation score*, calculated as the ratio of killed mutants to the total number of non-equivalent mutants. A higher mutation score indicates a more thorough and fault-sensitive test suite.

6.2 *Quantum Mutation Testing*

Like in classical mutation testing, quantum mutation testing involves defining several mutation operations that can change the functionality of a quantum circuit minimally, with the hope of having the test suite catch the buggy execution of the resulting mutated quantum circuit. Muskit [67] and QMutPy [68] define three basic mutation

```
- circuit.x(qubits[0])
+ circuit.h(qubits[0])
```

(a) Quantum gate replacement mutation.

```
- circuit.x(qubits[0])
+ pass
```

(b) Quantum gate deletion mutation.

```
  circuit.x(qubits[0])
+ circuit.y(qubits[0])
```

(c) Quantum gate insertion mutation.

```
  up_cqreg=ClassicalRegister(2*self._n ...
  circuit.add_register(up_cqreg)
- circuit.measure(self._up_qreg, up_cqreg)
+ pass
```

(d) Quantum measurement deletion mutation.

Fig. 22 Quantum mutation testing examples

operations on quantum circuits, centered around the gates. QMutPy additionally defines two other mutation operations on the measurements within the quantum circuit. Some examples of these mutations are shown in Fig. 22.

1. *Quantum gate replacement* replaces one gate in the quantum circuit with another. The replacement gate must be syntactically equivalent, meaning it uses the same number of operands and produces the same number of results as the original gate. Figure 22a shows an example of quantum gate replacement mutation.
2. *Quantum gate deletion* removes a gate from the quantum circuit. Figure 22b shows an example of quantum gate deletion mutation.
3. *Quantum gate insertion* adds a syntactically equivalent gate to the gate before it. Figure 22c shows an example of quantum gate insertion mutation.
4. *Quantum measurement insertion* adds a measurement operator on a qubit.
5. *Quantum measurement deletion* removes a measurement operator on a qubit. Figure 22d shows an example of quantum measurement deletion mutation.

To check how well the introduced bugs into the mutants were able to be identified by the tests, a mutation score is calculated [68]. The mutation score measures the rate of how many mutants were killed against all mutants tested. Formally, this is expressed as

$$\sum_{o \in O} \frac{\frac{|K_o|}{|M_o|-|I_o|}}{|O|} \times 100\%, \tag{9}$$

where o is a mutation operator in the set of all mutation operators O, $|M_o|$ is the number of mutants o created, $|I_o|$ is the number of mutants that are incompetent meaning they did not contain changes, $|K_o|$ is the number of mutants created by o that were killed. Note that in the case where the number of incompetent mutants is equal to the number of total mutants, the mutation score is 0% (Fig. 23).

Fuzz Testing: Mutations can also be used to generate new inputs to test in a testing strategy called fuzz testing. An initial input can be generated following some criteria, such as the program specification, and several additional test inputs can be

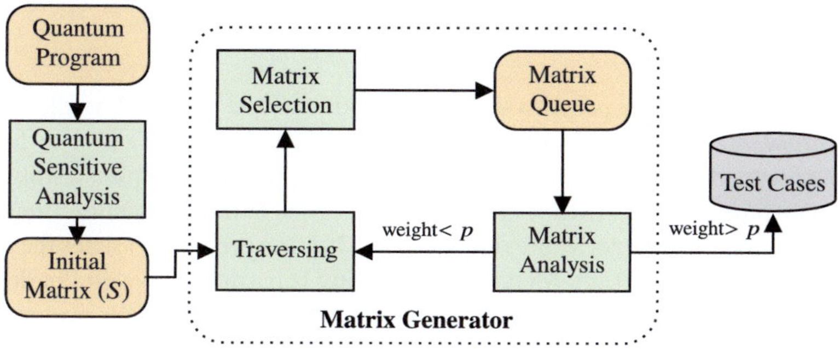

Fig. 23 Overview of QuanFuzz [69] quantum program fuzzing framework

generated with minor modifications via mutation. QuanFuzz is a fuzz testing framework that utilizes a search-based input generator to generate inputs to achieve high coverage [69]. The inputs are taken as their density matrices. The inputs along with their resulting measurement outputs are analyzed for any sensitive branches. This is an execution path within the program, and the search-based framework mutates the input matrix to generate a new test case. The aim of this mutation is to activate the non-taken paths, until all possible paths are taken.

One drawback of traditional input generation techniques is that they use basis states, such as $|0\rangle$ or $|1\rangle$, but quantum programs can exist in superposition and entanglement with other qubits. QuraTest adds an additional circuit in front of the quantum circuit under test to transform the original basis state input into input with superposition and entanglement [70]. These input changing circuits are generated as UCNOT using unitary and CNOT gates, IQFT using unitary gates and inverse quantum Fourier transform, and random using randomly selected gates from a specified list. Figure 24 shows an overview of the framework. The additional input states that are produced with the insertion of the additional circuit enable dynamic testing of quantum circuits, aligning closer to realistic scenarios on real quantum devices.

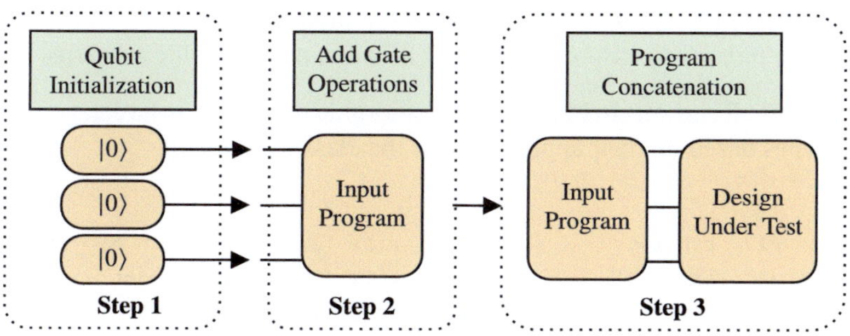

Fig. 24 Three-step process of generating a test case in QuraTest framework [70]

6.3 Classical Metamorphic Testing

Metamorphic testing is a property-based testing technique designed to alleviate the *oracle problem*, where the correct output of a program is unknown or difficult to compute. Instead of requiring a precise expected output, metamorphic testing relies on *metamorphic relations* (MRs), necessary properties or transformations that relate multiple inputs and their corresponding outputs [71]. If a program violates these relations, a fault is likely present. Metamorphic testing is particularly useful in domains such as machine learning, simulation, scientific computing, and numerical analysis, where establishing a ground truth output for arbitrary inputs is often infeasible.

Software Example: Consider testing an implementation of the sine function. We may not know the exact output of `sin(x)` for arbitrary inputs, but we do know a mathematical property, $\sin(-x) = -\sin(x)$. This identity serves as a metamorphic relation. We can test with the assertion statement in Fig. 25a. If the assertion fails, then the implementation violates the known metamorphic property, indicating a bug.

Hardware Example: In hardware design, metamorphic testing can be applied to verify arithmetic units such as floating-point square root modules. Consider the identity, $\sqrt{a^2} = |a|$. Given a hardware module `sqrt_unit`, a metamorphic test could be the assertion statement in Fig. 25b. This relation holds across a wide range of input values and allows the detection of subtle bugs in arithmetic pipeline logic, rounding modes, or handling of special cases (e.g., denormals).

Metamorphic Testing Workflow: The workflow of metamorphic testing involves generating follow-up test cases from source test cases by applying known metamorphic relations. The results are then checked for consistency according to the expected transformation of outputs.

1. *Identify metamorphic relations* relevant to the target function or circuit.
2. *Generate source test cases* and record their outputs.
3. *Transform inputs* using metamorphic relations to produce follow-up test cases.
4. *Execute the program or circuit* on follow-up inputs.
5. *Compare outputs* to ensure consistency with the metamorphic relations.

Metamorphic Relations in Practice: Metamorphic relations can be categorized into various types based on their applications in the classical domain.

```
double y1 = sin(x);
double y2 = sin(-x);
assert(y2 == -y1);
```

(a) Software example

```
wire [31:0] sq = a * a;
wire [31:0] res = sqrt_unit(sq);
assert(res == abs(a));
```

(b) Hardware example.

Fig. 25 Classical mutation testing examples

- *Algebraic relations:* e.g., `f(x + y) = f(x) + f(y)` for linear systems.
- *Geometric transformations:* e.g., rotation or scaling in graphics applications.
- *Permutations:* e.g., sorting output should be invariant to input order.
- *Data scaling or normalization:* e.g., normalizing input vectors should not affect cosine similarity.

Metamorphic testing complements mutation testing by targeting different classes of faults, particularly those that arise due to algorithmic inconsistencies or floating-point instability.

6.4 Quantum Metamorphic Testing

When a direct oracle is unavailable, which is the case when testing most of the quantum circuits, quantum metamorphic testing offers a promising alternative by exploiting known relationships between inputs and outputs rather than requiring exact expected values. Abreu et al. [72] defined a formal structure for specifying metamorphic relations applicable to quantum circuits under test. The metamorphic rules are Boolean functions defining a relationship between the input and output of the quantum circuits. The implementation of the metamorphic rules in quantum circuits must be reversible. The metamorphic rules are thus crafted using unitary gates to ensure effective reversability, such that $UU^\dagger = I$. These metamorphic rules can replace measurement operations within the quantum circuit to avoid collapsing the state while gaining important information about the execution of the quantum circuit under test. The resulting quantum circuit with the metamorphic rules inserted is referred to as an oracle quantum program.

MorphQ is an automated metamorphic testing framework that can establish metamorphic rules in quantum circuits under test while also being able to automatically generate the quantum circuits to test [73]. The framework begins by generating a valid quantum circuit in Qiskit to test, ensuring that it is syntactically correct and exhibits valid gate placements, as illustrated in Fig. 26. Next, the metamorphic transformations are inserted into the generated quantum circuit. Three categories of potential transformations can be inserted:

1. *Circuit transformations* change the circuit itself, modifying qubits and gates.
2. *Representation transformations* adjust the intermediate representation of the quantum circuit, in this case, the QASM assembly language equivalent of the Qiskit circuit.
3. *Execution transformations* change transpilation and simulation settings within Qiskit for simulating the quantum circuit.

Once the transformations are inserted into the quantum circuit, both the original circuit and the transformed circuit are simulated, gathering their execution results. If there are any deviations in the execution results, then there is a bug or other fault within the test.

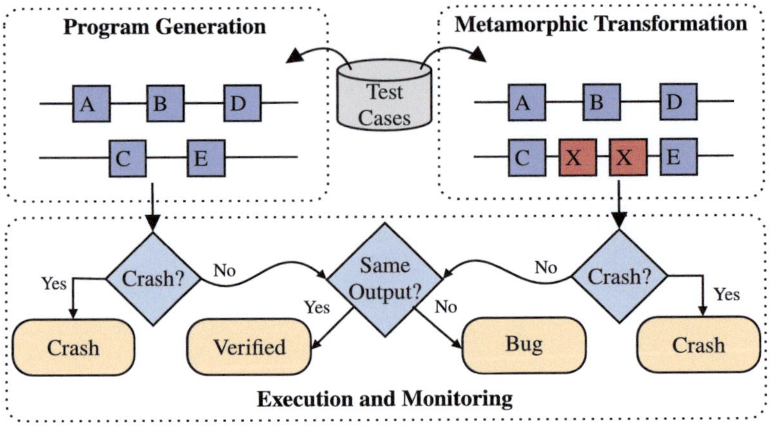

Fig. 26 Overview of MorphQ quantum metamorphic testing framework [73]

Mutation and metamorphic testing techniques have successfully transitioned from classical to quantum domains, adapting to address the unique challenges of quantum computation. Classical mutation testing evaluates test suite quality by systematically introducing faults into software and hardware, while metamorphic testing leverages input–output relationships to overcome oracle problems. In quantum systems, these approaches evolve to handle quantum-specific constructs: quantum mutation testing introduces gate replacements, deletions, and insertions alongside measurement modifications, while quantum metamorphic testing exploits unitary transformations and quantum circuit properties to establish testable relationships without collapsing quantum states. These methodologies provide essential tools for validating quantum programs and circuits where traditional testing approaches may be insufficient.

7 Summary

Classical computing has benefited from decades of mature testing practices, including well-defined fault models, deterministic execution, and extensive tool support for test generation, mutation analysis, and formal verification. These techniques enabled reliable validation even in large-scale systems, aided by benchmarks, coverage metrics, and automated oracles. As quantum computing emerges as a new computational paradigm, it inherits this rich legacy but also diverges significantly due to its probabilistic behavior, fragile hardware, and fundamentally different models of computation and observation. The evolution of quantum testing methodologies over recent years, as illustrated in Fig. 27, demonstrates the rapid development of specialized techniques to address these unique challenges. While many classical techniques, such as mutation testing, symbolic execution, and metamorphic testing, can be adapted to quantum systems, they face substantial challenges when applied to

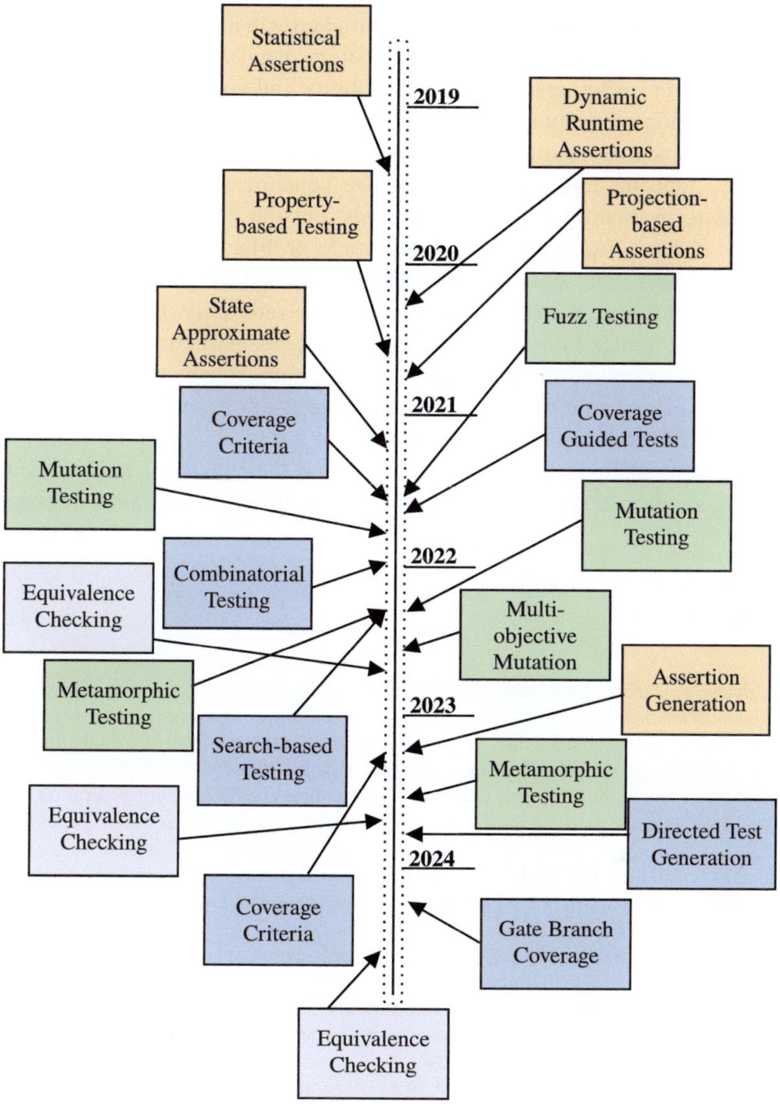

Fig. 27 Timeline of different quantum testing studies discussed in this chapter

circuits involving ten or more qubits. Exponential growth in the state space introduces scalability concerns, and the no-cloning theorem, measurement-induced collapse, and entanglement further restrict test observability and fault isolation.

The maturation of quantum testing as a discipline will require addressing several critical gaps that distinguish it from classical approaches. The absence of standardized fault models and comprehensive benchmark suites currently hinders the systematic evaluation of testing methods and the establishment of universal adequacy criteria. Most significantly, the oracle problem—trivial in deterministic classical systems—represents a fundamental challenge in quantum testing due to inherent non-determinism and limited ground truth availability. The path forward lies in developing automated oracle generation techniques, particularly through formal equivalence checking and metamorphic relations, alongside hardware-aware testing strategies that incorporate device-specific characteristics such as noise models, gate fidelity, and qubit connectivity. These cross-layer approaches will ultimately bridge the gap between abstract quantum algorithms and their physical implementations, establishing quantum testing as a robust foundation for reliable quantum computing.

References

1. Utt, Zachery, Daniel Volya, and Prabhat Mishra. 2024. Quantum measurement classification using statistical learning. *ACM Transactions on Quantum Computing (TQC)* 5 (2): 1–16.
2. Daniel Volya and Prabhat Mishra. Quantum spectral clustering of mixed graphs. In *2021 58th ACM/IEEE design automation conference (DAC)*, pages 463–468. IEEE, 2021.
3. Daniel Volya and Prabhat Mishra. Quantum data compression for efficient generation of control pulses. In *Asia and South Pacific Design Automation Conference (ASPDAC)*, pages 216–221, 2023.
4. Daniel Volya and Prabhat Mishra. Quantum steering of surface error correcting codes. In *2023 IEEE International Conference on Quantum Computing and Engineering (QCE)*, volume 1, pages 1394–1400. IEEE, 2023.
5. Daniel Volya and Prabhat Mishra. Qudcom: Towards quantum compilation for qudit systems. arXiv preprint arXiv:2311.07015, 2023.
6. Daniel Volya and Prabhat Mishra. Quantum benchmarking via random dynamical quantum maps. In *2024 IEEE International Conference on Quantum Computing and Engineering (QCE)*, volume 1, pages 1277–1283. IEEE, 2024.
7. Volya, Daniel, and Prabhat Mishra. 2024. State preparation on quantum computers via quantum steering. *IEEE Transactions on Quantum Engineering (TQE)* 5: 1–14.
8. Daniel Volya, Andrey Nikitin, and Prabhat Mishra. Fast quantum characterization via riemannian optimization. In *2024 IEEE International Conference on Quantum Computing and Engineering (QCE)*, volume 1, pages 1270–1276. IEEE, 2024.
9. Daniel Volya, Tao Zhang, Nashmin Alam, Mark Tehranipoor, and Prabhat Mishra. Towards secure classical-quantum systems. In *2023 IEEE International Symposium on Hardware Oriented Security and Trust (HOST)*, pages 283–292. IEEE, 2023.
10. Farahmandi, Farimah, and Prabhat Mishra. 2018. Automated test generation for debugging multiple bugs in arithmetic circuits. *IEEE Transactions on Computers* 68 (2): 182–197.
11. Jayasena, Aruna, and Prabhat Mishra. 2024. Directed test generation for hardware validation: A survey. *ACM Computing Surveys* 56 (5): 1–36.

12. Jayasena, Aruna, and Prabhat Mishra. 2024. Hive: Scalable hardware-firmware co-verification using scenario-based decomposition and automated hint extraction. *IEEE Transactions on Computer-Aided Design of Integrated Circuits and Systems* 43 (10): 3278–3291.
13. Lyu, Yangdi, and Prabhat Mishra. 2020. Scalable concolic testing of rtl models. *IEEE Transactions on Computers* 70 (7): 979–991.
14. Lyu, Yangdi, Xiaoke Qin, Mingsong Chen, and Prabhat Mishra. 2018. Directed test generation for validation of cache coherence protocols. *IEEE Transactions on Computer-Aided Design of Integrated Circuits and Systems* 38 (1): 163–176.
15. Prabhat Mishra and Farimah Farahmandi. *Post-Silicon Validation and Debug*. Springer, 2019.
16. Witharana, Hasini, Yangdi Lyu, Subodha Charles, and Prabhat Mishra. 2022. A survey on assertion-based hardware verification. *ACM Computing Surveys (CSUR)* 54 (11s): 1–33.
17. Witharana, Hasini, Hansika Weerasena, and Prabhat Mishra. 2024. Formal verification of virtualization-based trusted execution environments. *IEEE Transactions on Computer-Aided Design of Integrated Circuits and Systems* 43 (11): 4262–4273.
18. Aruna Jayasena and Prabhat Mishra. Firmwall: Directed symbolic execution of firmware binaries for defending against unauthorized system calls. *IEEE Transactions on Information Forensics and Security (TIFS)*, 2025.
19. Farahmandi, Farimah, and Yuanwen Huang. 2020. *and Prabhat Mishra*. System-on-Chip Security: Validation and Verification. Springer.
20. Aruna Jayasena, Emma Andrews, and Prabhat Mishra. Tvla*: Test vector leakage assessment on hardware implementations of asymmetric cryptography algorithms. *IEEE Transactions on Very Large Scale Integration (VLSI) Systems*, 31(9):1269–1279, 2023.
21. Aruna Jayasena, Richard Bachmann, and Prabhat Mishra. Evilcs: An evaluation of information leakage through context switching on security enclaves. In *2024 Design, Automation & Test in Europe Conference & Exhibition (DATE)*, pages 1–6. IEEE, 2024.
22. Jayasena, Aruna, Richard Bachmann, and Prabhat Mishra. 2025. Ciseleaks: Information leakage assessment of cryptographic instruction set extension prototypes. *IEEE Transactions on Information Forensics and Security (TIFS)* 20: 1551–1565.
23. Jayasena, Aruna, and Prabhat Mishra. 2023. Scalable detection of hardware trojans using atpg-based activation of rare events. *IEEE Transactions on Computer-Aided Design of Integrated Circuits and Systems* 42 (12): 4450–4462.
24. Aruna Jayasena, Sai Suprabhanu Nallapaneni, and Prabhat Mishra. Fuss: Coverage-directed hardware fuzzing with selective symbolic execution. *ACM Transactions on Embedded Computing Systems (TECS)*, 2025.
25. Prabhat Mishra, Swarup Bhunia, and Mark Tehranipoor. *Hardware IP security and trust*. Springer, 2017.
26. Pan, Zhixin, and Prabhat Mishra. 2022. A survey on hardware vulnerability analysis using machine learning. *IEEE Access* 10: 49508–49527.
27. Zhixin Pan and Prabhat Mishra. *Explainable AI for Cybersecurity*. Springer, 2023.
28. Pan, Zhixin, Jennifer Sheldon, and Prabhat Mishra. 2022. Hardware-assisted malware detection and localization using explainable machine learning. *IEEE Transactions on Computers* 71 (12): 3308–3321.
29. Sahan Sanjaya, Aruna Jayasena, and Prabhat Mishra. Information leakage through physical layer supply voltage coupling vulnerability. *IEEE Transactions on Very Large Scale Integration (VLSI) Systems*, 33(6):1715–1728, 2025.
30. Bo-Yuan Huang, Hongce Zhang, Aarti Gupta, and Sharad Malik. Ilang: a modeling and verification platform for socs using instruction-level abstractions. In *International Conference on Tools and Algorithms for the Construction and Analysis of Systems*, pages 351–357. Springer, 2019.
31. Hongce Zhang, Caroline Trippel, Yatin A Manerkar, Aarti Gupta, Margaret Martonosi, and Sharad Malik. Ila-mcm: integrating memory consistency models with instruction-level abstractions for heterogeneous system-on-chip verification. In *2018 Formal Methods in Computer Aided Design (FMCAD)*, pages 1–10. IEEE, 2018.

32. Anish Athalye, Adam Belay, M Frans Kaashoek, Robert Morris, and Nickolai Zeldovich. Notary: A device for secure transaction approval. In *Proceedings of the 27th ACM Symposium on Operating Systems Principles*, pages 97–113, 2019.
33. Anish Athalye, M Frans Kaashoek, and Nickolai Zeldovich. Verifying hardware security modules with {Information-Preserving} refinement. In *USENIX Symposium on Operating Systems Design and Implementation*, pages 503–519, 2022.
34. Noah Moroze, Anish Athalye, M Frans Kaashoek, and Nickolai Zeldovich. RTLV: push-button verification of software on hardware. In *Workshop on RISC-V for Computer Architecture Research (CARRV)*, 2021.
35. Wolfgang Dür and Stefan Heusler. What we can learn about quantum physics from a single qubit. arXiv preprint arXiv:1312.1463, 2013.
36. Bužek, Vladimir, and Mark Hillery. 1996. Quantum copying: Beyond the no-cloning theorem. *Physical Review A* 54 (3): 1844.
37. Shaukat Ali, Paolo Arcaini, Xinyi Wang, and Tao Yue. Assessing the Effectiveness of Input and Output Coverage Criteria for Testing Quantum Programs. In *2021 14th IEEE Conference on Software Testing, Verification and Validation (ICST)*, pages 13–23, April 2021.
38. Daniel Fortunato, José Campos, and Rui Abreu. Gate Branch Coverage: A Metric for Quantum Software Testing. In *Proceedings of the 1st ACM International Workshop on Quantum Software Engineering: The Next Evolution*, QSE-NE 2024, pages 15–18, New York, NY, USA, July 2024. Association for Computing Machinery.
39. Xinyi Wang, Paolo Arcaini, Tao Yue, and Shaukat Ali. Quito: A Coverage-Guided Test Generator for Quantum Programs. In *2021 36th IEEE/ACM International Conference on Automated Software Engineering (ASE)*, pages 1237–1241, November 2021.
40. Xinyi Wang, Paolo Arcaini, Tao Yue, and Shaukat Ali. Application of Combinatorial Testing to Quantum Programs. In *2021 IEEE 21st International Conference on Software Quality, Reliability and Security (QRS)*, pages 179–188, December 2021.
41. Xinyi Wang, Paolo Arcaini, Tao Yue, and Shaukat Ali. Generating Failing Test Suites for Quantum Programs With Search. In *Search-Based Software Engineering*, pages 9–25. Springer International Publishing, 2021.
42. Xinyi Wang, Paolo Arcaini, Tao Yue, and Shaukat Ali. QuSBT: Search-based testing of quantum programs. In *Proceedings of the ACM/IEEE 44th International Conference on Software Engineering: Companion Proceedings*, ICSE '22, pages 173–177, New York, NY, USA, October 2022. Association for Computing Machinery.
43. Xinyi Wang, Paolo Arcaini, Tao Yue, and Shaukat Ali. QuCAT: A Combinatorial Testing Tool for Quantum Software. In *2023 38th IEEE/ACM International Conference on Automated Software Engineering (ASE)*, pages 2066–2069, September 2023.
44. Minqi Shao and Jianjun Zhao. A coverage-guided testing framework for quantum neural networks. arXiv preprint arXiv:2411.02450, 2024.
45. Witharana, Hasini, Aruna Jayasena, Andrew Whigham, and Prabhat Mishra. 2023. Automated generation of security assertions for rtl models. *ACM Journal on Emerging Technologies in Computing Systems* 19 (1): 1–27.
46. Yipeng Huang and Margaret Martonosi. Statistical assertions for validating patterns and finding bugs in quantum programs. In *Proceedings of the 46th International Symposium on Computer Architecture*, ISCA '19, pages 541–553, New York, NY, USA, June 2019. Association for Computing Machinery.
47. Ji Liu, Gregory T. Byrd, and Huiyang Zhou. Quantum Circuits for Dynamic Runtime Assertions in Quantum Computation. In *Proceedings of the Twenty-Fifth International Conference on Architectural Support for Programming Languages and Operating Systems*, ASPLOS '20, pages 1017–1030, New York, NY, USA, March 2020. Association for Computing Machinery.
48. Ehsan Faghih and Huiyang Zhou. Dynamic Runtime Assertions in Quantum Ternary Systems. In *2024 IEEE International Conference on Quantum Computing and Engineering (QCE)*, volume 01, pages 42–49, September 2024.
49. Gushu Li, Li Zhou, Nengkun Yu, Yufei Ding, Mingsheng Ying, and Yuan Xie. Projection-based runtime assertions for testing and debugging Quantum programs. *Proc. ACM Program. Lang.*, 4(OOPSLA):150:1–150:29, November 2020.

50. Ji Liu and Huiyang Zhou. Systematic Approaches for Precise and Approximate Quantum State Runtime Assertion. In *2021 IEEE International Symposium on High-Performance Computer Architecture (HPCA)*, pages 179–193, February 2021.
51. Hasini Witharana, Daniel Volya, and Prabhat Mishra. QcAssert: Quantum Device Testing with Concurrent Assertions. In *2024 29th Asia and South Pacific Design Automation Conference (ASP-DAC)*, pages 491–496, January 2024.
52. Hasini Witharana, Daniel Volya, and Prabhat Mishra. quassert: Automatic generation of quantum assertions. arXiv preprint arXiv:2303.01487, 2023.
53. Damian Rovara, Lukas Burgholzer, and Robert Wille. Automatically refining assertions for efficient debugging of quantum programs. arXiv preprint arXiv:2412.14252, 2024.
54. Shahin Honarvar, Mohammad Reza Mousavi, and Rajagopal Nagarajan. Property-based Testing of Quantum Programs in Q#. In *Proceedings of the IEEE/ACM 42nd International Conference on Software Engineering Workshops*, ICSEW'20, pages 430–435, New York, NY, USA, September 2020. Association for Computing Machinery.
55. Pontolillo, Gabriel. March 2025. *Mohammad Reza Mousavi, and Marek Grzesiuk*. QuCheck: A Property-based Testing Framework for Quantum Programs in Qiskit.
56. Gabriel Joseph Pontolillo and Mohammad Reza Mousavi. Delta Debugging for Property-Based Regression Testing of Quantum Programs. In *Proceedings of the 5th ACM/IEEE International Workshop on Quantum Software Engineering*, Q-SE 2024, pages 1–8, New York, NY, USA, August 2024. Association for Computing Machinery.
57. Vanegue, Julien, Sean Heelan, and Rolf Rolles. 2012. Smt solvers in software security. *USENIX WOOT Conference on Offensive Technologies* 12: 9–22.
58. Witharana, Hasini, Aruna Jayasena, and Prabhat Mishra. 2024. Incremental concolic testing of register-transfer level designs. *ACM Transactions on Design Automation of Electronic Systems* 29 (5): 1–23.
59. Burgholzer, Lukas, and Robert Wille. September 2021. Advanced Equivalence Checking for Quantum Circuits. *IEEE Transactions on Computer-Aided Design of Integrated Circuits and Systems* 40 (9): 1810–1824.
60. Tom Peham, Lukas Burgholzer, and Robert Wille. Equivalence Checking of Parameterized Quantum Circuits: Verifying the Compilation of Variational Quantum Algorithms. In *Proceedings of the 28th Asia and South Pacific Design Automation Conference*, ASPDAC '23, pages 702–708, New York, NY, USA, January 2023. Association for Computing Machinery.
61. John van de Wetering. Zx-calculus for the working quantum computer scientist. arXiv preprint arXiv:2012.13966, 2020.
62. Xin Hong, Yuan Feng, Sanjiang Li, and Mingsheng Ying. Equivalence Checking of Dynamic Quantum Circuits. In *Proceedings of the 41st IEEE/ACM International Conference on Computer-Aided Design*, ICCAD '22, pages 1–8, New York, NY, USA, December 2022. Association for Computing Machinery.
63. Long, Peixun, and Jianjun Zhao. 2024. Equivalence, identity, and unitarity checking in black-box testing of quantum programs. *Journal of Systems and Software* 211: 112000.
64. Shangzhou Xia, Jianjun Zhao, Fuyuan Zhang, and Xiaoyu Guo. Quantum concolic testing. *Proceedings of the ACM on Software Engineering*, 2(ISSTA):1146–1166, 2025.
65. Jia, Yue, and Mark Harman. 2010. An analysis and survey of the development of mutation testing. *IEEE transactions on software engineering* 37 (5): 649–678.
66. Mike Papadakis, Marinos Kintis, Jie Zhang, Yue Jia, Yves Le Traon, and Mark Harman. Mutation testing advances: an analysis and survey. In *Advances in computers*, volume 112, pages 275–378. Elsevier, 2019.
67. Eñaut Mendiluze, Shaukat Ali, Paolo Arcaini, and Tao Yue. Muskit: A Mutation Analysis Tool for Quantum Software Testing. In *2021 36th IEEE/ACM International Conference on Automated Software Engineering (ASE)*, pages 1266–1270, November 2021.
68. Daniel Fortunato, JOSÉ CAMPOS, and RUI ABREU. Mutation Testing of Quantum Programs: A Case Study With Qiskit. *IEEE Transactions on Quantum Engineering*, 3:1–17, 2022.
69. Jiyuan Wang, Ming Gao, Yu Jiang, Jianguang Lou, Yue Gao, Dongmei Zhang, and Jiaguang Sun. Quanfuzz: Fuzz testing of quantum program. arXiv preprint arXiv:1810.10310, 2018.

70. Jiaming Ye, Shangzhou Xia, Fuyuan Zhang, Paolo Arcaini, Lei Ma, Jianjun Zhao, and Fuyuki Ishikawa. QuraTest: Integrating Quantum Specific Features in Quantum Program Testing. In *2023 38th IEEE/ACM International Conference on Automated Software Engineering (ASE)*, pages 1149–1161, September 2023.
71. Tsong Yueh Chen, Fei-Ching Kuo, Huai Liu, Pak-Lok Poon, Dave Towey, TH Tse, and Zhi Quan Zhou. Metamorphic testing: A review of challenges and opportunities. *ACM Computing Surveys (CSUR)*, 51(1):1–27, 2018.
72. Rui Abreu, João Paulo Fernandes, Luis Llana, and Guilherme Tavares. Metamorphic testing of oracle quantum programs. In *Proceedings of the 3rd International Workshop on Quantum Software Engineering*, Q-SE '22, pages 16–23, New York, NY, USA, February 2023. Association for Computing Machinery.
73. Matteo Paltenghi and Michael Pradel. MorphQ: Metamorphic Testing of the Qiskit Quantum Computing Platform. In *2023 IEEE/ACM 45th International Conference on Software Engineering (ICSE)*, pages 2413–2424, May 2023.

Assertion-Based Debugging of Quantum Circuits: Techniques, Effectiveness, and Challenges

Abhoy Kole, Anton Maidl, and Rolf Drechsler

Abstract The rapid advancements in quantum hardware and software have heightened the need for reliable, error-free quantum program development. Unlike classical computing, where decades of research have led to mature debugging techniques, directly applying these methods to quantum software remains challenging due to the fundamental differences rooted in quantum mechanics. Moreover, the inherent complexity of quantum programs makes manual debugging impractical, even for small circuits. In this context, this work provides a comprehensive review of existing techniques developed to support quantum program correctness, with a particular focus on assertion-based debugging approaches. The paper also highlights key research challenges and open issues in this emerging area.

1 Introduction

Quantum computing is advancing rapidly, with the potential to transform the landscape of information processing. A wide range of quantum algorithms has been developed to address critical problems in domains such as finance [1, 2], materials science [3, 4], and cryptography [5, 6]. Ongoing efforts from both industry and academia have led to the emergence of practical, moderately scaled quantum computers. Notable hardware platforms include superconducting quantum computers from IBM and Google [7, 8], trapped-ion systems by IonQ and Quantinuum [9, 10], and topological quantum computing initiatives led by Microsoft and QuTech [11, 12].

A. Kole (✉)
DFKI, Bremen, Germany
e-mail: abhoy.kole@dfki.de

A. Maidl
German Aerospace Center, Bremerhaven, Germany
e-mail: anton.maidl@dlr.de

R. Drechsler
University of Bremen/DFKI, Bremen, Germany
e-mail: drechsler@uni-bremen.de

P. Mishra (ed.), *Design Automation for Quantum Computing*,
https://doi.org/10.1007/978-3-032-09303-5_10

In parallel, a rich ecosystem of quantum software frameworks has been developed to support the design, simulation, and execution of quantum algorithms [13]. These include IBM's Qiskit [14], Google's Cirq [15], Microsoft's Q# [16], Xanadu's PennyLane [17], and Quantinuum's TKET [18]. Similar to classical software systems, the development of large-scale quantum applications targeting real-world industry challenges requires robust and efficient strategies throughout key stages of the software lifecycle, including systematic testing, effective debugging, and long-term maintainability [19, 20].

As quantum programming languages continue to evolve, the challenges associated with testing and debugging quantum programs are receiving growing attention. Despite notable progress [21–23], the development and maintenance of quantum software remain intrinsically difficult. The fundamental principles of quantum mechanics, such as superposition, entanglement, no-cloning, and interference—introduce unique complexities that significantly hinder the identification and resolution of bugs in quantum programs [24].

Beyond formal verification [25–27] and equivalence checking [28, 29], a variety of approaches have been proposed for testing and debugging quantum software, many of which are inspired by established practices in classical software engineering. Among these, several studies have introduced mechanisms for implementing static and dynamic quantum assertions to validate program behavior [21, 30, 31]. In addition, automated tools have been developed to support fault localization, assertion evaluation, and bug detection in quantum programs [22, 32, 33].

Complementary efforts have also emphasized the development of systematic approaches for test generation and coverage in quantum software. Unit test case generation driven by system properties is discussed in [34, 35], while the role of white-box and black-box testing, along with the design of test suites aligned with program specifications, is emphasized in [36]. Design of input and output coverage criteria and corresponding test case generation techniques are explored in [37], and benchmark repositories to assess test suite effectiveness are provided in [23, 38, 39]. Furthermore, debugging techniques for parameterized quantum algorithms, including parameter tuning, noise mitigation, and classical component optimization, are examined in [40].

Software bugs can significantly hinder computational reliability, compromise the integrity of results, and ultimately impact economic stability, system security, and the quality of life of end users. Debugging and repair tasks consume substantial resources, making effective bug detection crucial. Many software engineering tasks, including analysis, testing, and debugging, focus on developing efficient tools for identifying and correcting defects. This chapter surveys the current landscape of testing and debugging techniques for quantum software and highlights key challenges and potential directions for future improvement.

The remainder of the paper is structured as follows: Sect. 2 presents the necessary background on quantum computing, along with an overview of testing and debugging within the quantum software development lifecycle. Section 3 explores various assertion-based debugging approaches for quantum circuits, with a focus on the interpretation and evaluation of assertions. Finally, Sect. 4 concludes the paper by discussing the strengths and limitations of assertion-based debugging methods, and highlighting potential directions for future improvement.

2 Background

2.1 Quantum Program

A quantum program operates on quantum bits (qubits), as opposed to classical bits used in conventional digital computing. In classical systems, the state of a program is represented by a sequence of 0s and 1s. In contrast, the state of a quantum program is described using Dirac notation [41], capturing the probabilistic and complex nature of quantum information. For instance, the state of an n-qubit system can be written as

$$|\phi\rangle = \alpha_0 |0\ldots00\rangle + \alpha_1 |0\ldots01\rangle + \cdots + \alpha_{2^n-1} |1\ldots11\rangle . \quad (1)$$

Here, each $|\cdot\rangle$ represents a basis state, and the complex coefficients α_i (for $i = 0, \ldots, 2^n - 1$) are called *amplitudes*. These amplitudes encode both magnitude and phase, and their squared magnitudes, $|\alpha_i|^2$, give the probabilities of observing the corresponding states upon measurement. Due to normalization, the total probability satisfies: $\sum_{i=0}^{2^n-1} |\alpha_i|^2 = 1$. This representation implies that a quantum program can exist in a *superposition* of multiple basis states simultaneously, as illustrated in Eq. 1, a fundamental departure from classical computing.

The most widely recognized and actively researched approach to quantum computing is the gate-based model [42]. In this model, quantum programs are constructed by applying a sequence of quantum gates to qubits. These gates act on one or more qubits and modify their quantum states during the execution of the program on a quantum computer. Table 1 lists some of the most commonly used quantum gates. For a more detailed discussion of quantum gates, readers may refer to [43].

Quantum programs typically consist of gate operations applied in sequence to specific qubits. Although several high-level quantum programming languages exist, such as Qiskit [14], Cirq [15], Q# [16], and PennyLane [17], this work uses OpenQASM [44] as the primary representation. OpenQASM is a widely adopted low-level standard for describing quantum programs, and many high-level quantum languages compile down to it before execution on actual quantum hardware.

Figure 1a presents the OpenQASM code for generating one of the Bell states [45]:

$$|\phi^+\rangle = \frac{1}{\sqrt{2}} (|00\rangle + |11\rangle) . \quad (2)$$

Table 1 Summary of some basic quantum gates, showing how they transform qubit states

Gate	Action	Description
Hadamard (H)	h q_i: $\|0\rangle \rightarrow \frac{1}{\sqrt{2}}(\|0\rangle + \|1\rangle)$	Creates equal superposition
Pauli-X	x q_i: $\|0\rangle \rightarrow \|1\rangle$	Bit-flip
Pauli-Z	z q_i: $\|1\rangle \rightarrow -\|1\rangle$	Phase-flip
Phase (P)	p(θ) q_i: $\|1\rangle \rightarrow e^{i\theta}\|1\rangle$	Phase-shift by angle θ
Rotation-X	$r_x(\theta)$ q_i: $\|0\rangle \rightarrow \cos(\frac{\theta}{2})\|0\rangle - i\sin(\frac{\theta}{2})\|1\rangle$	Rotates around X-axis by angle θ
CNOT	cx q_i, q_j: $\|1\rangle \otimes \|0\rangle \rightarrow \|1\rangle \otimes \|1\rangle$	Flips target q_j if control q_i is $\|1\rangle$
The $\otimes$ symbol, called the tensor product, is used to combine the states of multiple qubits		
However, it's often left out for simplicity, e.g., $\|0\rangle \otimes \|1\rangle$ is usually written as just $\|01\rangle$		

```
OPENQASM 3.0;
include "stdgates.inc";

qubit[2] q;
bit[2] c;

h q[0];
cx q[0], q[1];
c[0] = measure q[0];
c[1] = measure q[1];
```

(a)

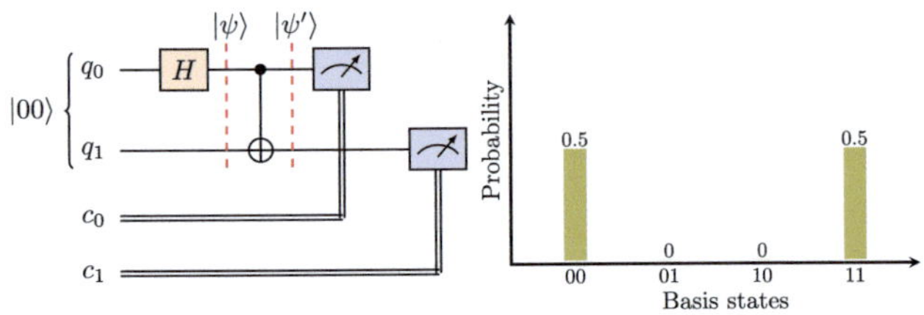

(b) (c)

Fig. 1 Quantum program for generating the Bell state $|\phi^+\rangle$: **a** OpenQASM code representation, **b** Quantum circuit diagram, **c** Measurement outcome probability distribution

This is an example of an *entangled* state, where the two qubits cannot be described independently. Measuring one qubit immediately determines the state of the other. Figure 1b illustrates the corresponding quantum circuit.

Starting from the initial state $|00\rangle$ of qubits q_0 and q_1, applying an H gate to q_0 transforms the program state to

$$|\psi\rangle = H|0\rangle \otimes |0\rangle = \frac{1}{\sqrt{2}}(|0\rangle + |1\rangle) \otimes |0\rangle . \tag{3}$$

Subsequently, applying a CX gate with q_0 as control and q_1 as target results in the state:

$$|\psi'\rangle = CX\left(\frac{1}{\sqrt{2}}(|0\rangle + |1\rangle) \otimes |0\rangle\right) = \frac{1}{\sqrt{2}}(|00\rangle + |11\rangle) \tag{4}$$

Under ideal conditions, repeated execution of this quantum program yields the measurement probability distribution shown in Fig. 1c. This also highlights a key distinction: quantum computers produce probabilistic outcomes upon measurement, in contrast to the deterministic behavior of classical computers. Therefore, ensuring the final program state $|\psi'\rangle$ in Eq. 4 to match the target Bell state $|\phi^+\rangle$ in Eq. 2 is crucial for obtaining the expected distribution of outcomes.

2.2 *Quality Management for Quantum Applications*

Advancements in quantum hardware and software, combined with the potential to revolutionize a wide range of computationally intensive problems across societal and industrial domains, are giving rise to new challenges. Developing practical, real-world quantum applications requires not only the implementation of quantum algorithms in software [46], but also adherence to a structured software development life cycle (SDLC) [19, 20]. Supporting this effort further necessitates a comprehensive quantum software stack that includes programming languages, compilers, and operating systems [47].

Inspired by classical software engineering practices, the development of reliable quantum software involves multiple phases, such as requirements engineering, system architecture and design, implementation, testing, debugging, and maintenance. While the growing availability of quantum programming languages [48] provides flexibility in writing quantum programs, programming itself represents only one aspect of the broader SDLC. To ensure the correctness, robustness, and longevity of quantum software systems, it is essential to address other critical phases, particularly verification, validation, and long-term maintainability.

To deliver high-quality software, both verification and validation are essential and complementary processes that work together to ensure not only the correctness of the build but also its practical utility. Verification focuses on ensuring the software meets its specified requirements through internal consistency checks and correctness assessments. In contrast, validation ensures the software fulfills its intended purpose and aligns with real user needs [36].

Verification spans the entire development lifecycle, encompassing activities such as static analysis, code reviews, unit testing, integration testing, and system testing. On the other hand, validation typically occurs later in the cycle and involves acceptance testing to confirm the software's suitability for deployment. While testing aims to identify defects and inconsistencies within quantum software, debugging is a critical companion activity at each testing stage, aimed at diagnosing and resolving those issues. Figure 2 illustrates the testing and debugging workflow in quantum software development.

In general, test strategies comprise three core components: test oracles, test data, and test cases that are designed by considering certain fault types. Upon observing failures during quantum program testing, they must be diagnosed through debugging, which involves isolating and correcting the source of the issue [46]. The debugging process typically follows two key steps:

1. Bug Analysis and Isolation: Analyze the code to pinpoint the precise location and root cause of the defect.

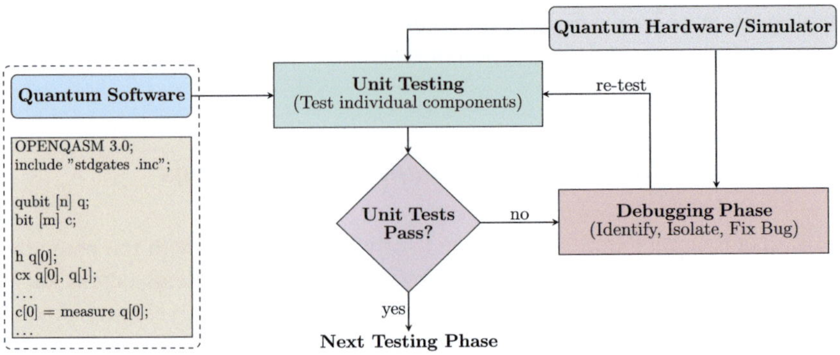

Fig. 2 Process flow for testing and debugging quantum software as an integral component of quality assurance and management

2. Bug Fixing and Re-testing: Apply the necessary fix and immediately verify it by re-running the relevant tests to ensure the issue has been resolved.

Quantum software presents unique challenges due to its black-box nature. Intermediate states of a quantum program cannot be observed during execution without altering the quantum state, since measuring a qubit collapses its superposition [42]. As a result, conventional debugging techniques must be adapted or reimagined for quantum systems. Several approaches have been proposed in the literature, including:

1. Assertion-Based Debugging: Embedding assertions—conditions expected to hold true at specific execution points—into the code. A failed assertion signals an unexpected state and helps locate the defect.
2. Program Slicing: Extracting only the relevant code segments that influence a specific output, thereby narrowing the scope of analysis.

A brief discussion of these techniques, along with potential research opportunities, is presented in the following section.

3 Debugging Quantum Circuits

3.1 Statistical Assertions

The primary objective of quantum assertions is to facilitate debugging and validate the functional correctness of quantum programs. Statistical quantum assertions address this challenge by exploiting the probabilistic characteristics of quantum mechanics. They work by performing repeated measurements at specific points in a quantum program and applying statistical analysis to the resulting distribution of outcomes.

This enables inference about the intermediate quantum state and detection of deviations from the expected behavior [16, 30]. The foundational concepts, along with the practical utility of these assertions, are elaborated in the following subsections.

3.1.1 Assertion Types

Implementing quantum assertions faces unique challenges compared to classical assertions, mainly due to destructive nature of quantum measurement, no-cloning theorem, and entanglement [42]. To gain insight into the internal state of a quantum program, statistical analysis techniques, such as state tomography [42], are employed on the distribution of measurement outcomes. For instance, a two-qubit quantum program can exist in a superposition of four basis states, leading to measurement outcomes that can generally be categorized into three types: classical, superposition, and entangled, as illustrated in Fig. 3.

By analyzing the distribution of measurement outcomes, one can derive state-specific assertions corresponding to classical, superposition, and entangled states, as demonstrated in [16, 30]:

1. **Classical State:** The measurement distribution is typically *unimodal*, with a sharp peak at the deterministic outcome defined by the computational basis (e.g., $|00\rangle$). This behavior reflects the absence of quantum interference or entanglement (see Fig. 3a).
2. **Superposition State:** For an n-qubit uniform superposition state,

$$\frac{1}{2^n}\left(|0\ldots00\rangle + |0\ldots01\rangle + \cdots + |1\ldots11\rangle\right), \tag{5}$$

   the measurement outcomes are distributed approximately *uniformly* across all 2^n computational basis states. This reflects the equal probability amplitude across all basis vectors (e.g., for a two-qubit system, see Fig. 3b).
3. **Entangled State:** Measurement outcomes exhibit *non-classical correlations* across qubits. For example, in the Bell state

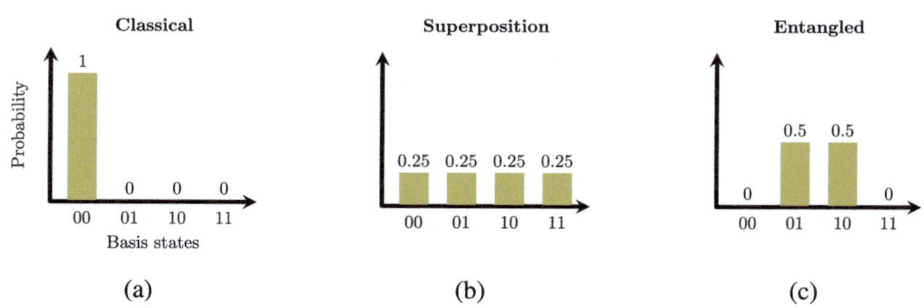

Fig. 3 Measurement outcome distribution for a two-qubit quantum program: **a** Classical state, **b** Separable superposition state, **c** Entangled state

$$\left|\psi^{+}\right\rangle = \frac{1}{\sqrt{2}}\left(|01\rangle + |10\rangle\right), \tag{6}$$

measuring one qubit instantaneously determines the outcome of the other, regardless of spatial separation, illustrating quantum entanglement (see Fig. 3c).

These distribution-based assertions act as quantum breakpoints, enabling inspection and verification of the quantum program's state at specific points during execution. This facilitates systematic debugging by providing insight into the internal state evolution of the program.

3.1.2 Assertion Evaluation

To evaluate whether the observed measurements align with the expected quantum state, statistical hypothesis testing [49] is applied to ensembles of repeated measurements. In particular, for an n-qubit quantum program, the chi-square (χ^2) goodness-of-fit test is used to compare observed outcome frequencies (O_i) with expected counts (E_i). The test statistic is computed as

$$\chi^2 = \sum_{i=0}^{2^n-1} \frac{(O_i - E_i)^2}{E_i}. \tag{7}$$

The null hypothesis is defined as the assumption that the observed outcome distribution is consistent with the expected distribution. To evaluate this hypothesis, a significance level α (e.g., 5%) is selected as the threshold for statistical confidence. The corresponding p-value is computed by evaluating the cumulative distribution function (CDF) of the chi-square distribution for the calculated χ^2 statistic with the appropriate degrees of freedom (df):

$$p = 1 - \mathrm{CDF}(\chi^2, df) \tag{8}$$

If the resulting p-value is less than or equal to the chosen significance level (i.e., $p \leq \alpha$), the null hypothesis is rejected. This indicates that the observed outcomes deviate significantly from the expected distribution, and the assertion is considered to have failed. Conversely, if the p-value exceeds α, the observed measurements are deemed statistically consistent with the expected behavior, and the assertion is considered to pass.

As an illustrative example, consider the Bell state $\left|\phi^{+}\right\rangle$, which ideally results in measurement outcomes in the $|00\rangle$ and $|11\rangle$ basis states, each with 50% probability. An assertion corresponding to this entangled program state is represented as OpenQASM code in Fig. 4a and depicted as a quantum circuit diagram in Fig. 4b. When this assertion is executed, the associated code segment is run multiple times to collect a representative distribution of measurement outcomes. The resulting contingency table, shown in Fig. 4c, summarizes both the observed program state (O) and the

```
OPENQASM 3.0;
include "stdgates.inc";

qubit[2] q;

...
cx q[0], q[1];
assert_E q[0,1] = |ϕ+⟩;
...
```

(a)

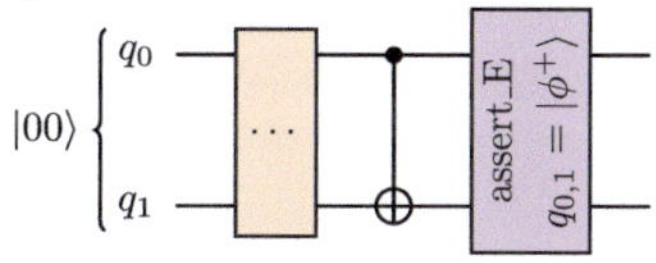

(b)

Contingency Table:

$\lvert . \rangle$	$\lvert 00\rangle$	$\lvert 01\rangle$	$\lvert 10\rangle$	$\lvert 11\rangle$
O_i	485	25	20	470

O_i: 1000 times code run

(c)

```
Compute χ²;
Compute p-value;
if p ≤ α then
    "ASSERTION ERROR";
```

(d)

Fig. 4 Statistical assertion for verifying the Bell state $\lvert\phi^+\rangle$ in a two-qubit quantum program: **a** OpenQASM code representation, **b** Corresponding quantum circuit diagram, **c** Contingency table of measured outcomes, **d** Assertion verification based on statistical hypothesis testing

expected frequencies for the $\lvert\phi^+\rangle$ state (E). Based on these observations, a χ^2 test is performed to statistically evaluate the agreement between the measured outcomes and the ideal distribution. The hypothesis testing process, including the rejection of the null hypothesis, is illustrated in Fig. 4d.

With a sufficient number of measurements, statistical tests can determine whether a quantum assertion holds, thereby revealing potential bugs in the program. These assertions guide the compiler to insert quantum breakpoints at specific points in the quantum program where execution halts and qubit states are measured. Each compiled variant of the program includes execution up to a designated breakpoint, followed by an early measurement and an evaluation of assertions against expected quantum state properties. The tool then aggregates measurement outcomes across multiple simulations and applies statistical hypothesis testing to assess whether the observed distributions are consistent with expected quantum state classifications (classical, superposition, or entangled states). If the statistical test rejects the null hypothesis, this indicates a violation of the assertion, which may either signal a bug in the quantum program or an incorrectly specified assertion.

3.2 Dynamic Assertions

As described in [30], quantum programs can incorporate three types of assertions: classical, superposition, and entanglement. Inspired by the principles of quantum error correction, the dynamic evaluation of these assertions facilitates debugging by capturing runtime information critical for identifying program errors. In the absence of such errors, the program proceeds with its normal execution flow. Unlike direct measurement of data qubits—which may induce undesired side effects by collapsing portions of the quantum state—these approaches aim to preserve the system's coherence by employing ancilla qubits and projective measurements [21]. This preservation, however, comes at the cost of increased circuit complexity [50]. In [31], quantum circuits for all three types of assertions are presented and demonstrated on IBM quantum computers. The concepts behind these assertions, along with their practical effectiveness, are discussed in the subsections below.

3.2.1 Classical Assertion

This type of assertion checks whether a set of qubits $q_0, q_1, \ldots, q_{n-1}$is correctly initialized or satisfies specific intermediate classical conditions, e.g.,

$$|\psi\rangle = |N\rangle, \qquad \text{for } 0 \leq N \leq 2^n - 1. \tag{9}$$

For example, consider a two-qubit quantum program where, after the execution of gate U, a classical assertion is inserted to verify the expected classical state of the program, as shown in Fig. 5a. The corresponding quantum circuit diagram is presented in Fig. 5b. To perform this verification, n ancilla qubits $a_0, a_1, \ldots, a_{n-1}$ are introduced. Each ancilla qubit is entangled with its corresponding data qubit using a controlled-NOT (CX) gate, $\mathrm{CX}(q_i, a_i), i = 0, 1, \ldots, n-1$. The OpenQASM rep-

```
OPENQASM 3.0;
include "stdgates.inc";

qubit[2] q;
...
U q[0], q[1];
assert_C q[0,1] = |N⟩;
V q[0], q[1];
...
```

(a)

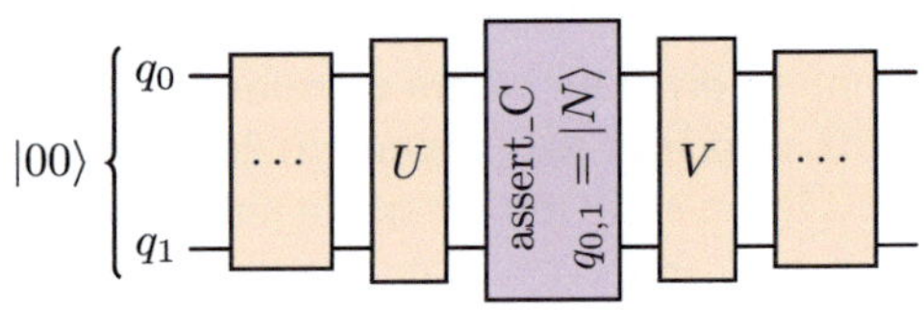

(b)

```
assert q[0,1] = |N⟩:

  qubit a[2];
  bit c[2];

  cx q[0], a[0];
  cx q[1], a[1];
  c[0] = measure a[0];
  c[1] = measure a[1];

if c[0,1]! = N then
  "ASSERTION ERROR";
```

(c)

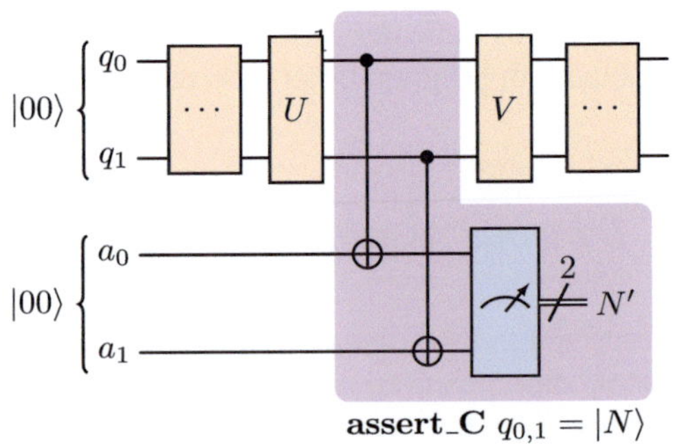

(d)

Fig. 5 Illustration of a dynamic classical assertion $q_{0,1} = |N\rangle$ for $0 \leq N \leq 3$: OpenQASM representation of the quantum program with the assertion inserted; **b** corresponding quantum circuit showing the assertion placed between unitary blocks U and V; **c** OpenQASM semantics interpreting the assertion logic; **d** quantum circuit after expanding the assertion. The assertion verifies at runtime that the system is in state $|N\rangle$ after U, ensuring correct execution of the subsequent block V

resentation of the assertion for the example two-qubit program (refer to Fig. 5a) is provided in Fig. 5c. After the entanglement via CX gates, the ancilla qubits are measured to determine whether the assertion holds—specifically, whether the measured value N' matches the expected value N, as depicted in Fig. 5d. This runtime check ensures the correctness of the quantum state $|\psi\rangle = U|0\rangle$, and in particular, that the subsequent computation $V|\psi\rangle$ operates on a valid intermediate state. The validity is confirmed if the measurement results of the ancilla qubits a_i (for $i = 0, 1, \ldots, n-1$) in the computational basis match the expected classical state.

If $|\psi\rangle$ is a classical basis state, i.e., $|\psi\rangle = |N_i\rangle$ for some $0 \leq N_i \leq 2^n - 1$, then the entanglement behaves as follows:

$$CX^{\otimes n}(|\psi\rangle \otimes |0\rangle) = |\psi\rangle \otimes |\psi\rangle\,. \tag{10}$$

Measuring the ancilla qubits in the computational basis yields a classical value N'. If $N' = N$, the assertion passes (i.e., $|\psi\rangle = |N\rangle$); otherwise, $N' \neq N$ indicates a violation (i.e., a bug), and the result of $V|\psi\rangle$ must be discarded.

If, instead, $|\psi\rangle$ is a superposition state due to an error in U or another runtime fault:

$$|\psi\rangle = \sum_{i\in\{0,1,\ldots,2^n-1\}} \alpha_i\,|N_i\rangle\,, \qquad \text{where} \sum_i |\alpha_i|^2 = 1 \tag{11}$$

then the entanglement operation yields

$$CX^{\otimes n}(|\psi\rangle \otimes |0\rangle) = \sum_{i\in\{0,1,\ldots,2^n-1\}} \alpha_i\,|N_i\rangle \otimes |N_i\rangle\,. \tag{12}$$

Upon measuring the ancilla qubits in the computational basis, if the result N' matches the target value N, it implies the data qubits have collapsed to the desired classical state $|\psi\rangle = |N\rangle$. If $N' \neq N$, it means the state has been projected onto an incorrect basis state, and the result of $V|\psi\rangle$ is invalid and must be disregarded.

3.2.2 Assertions for Superposition

Another category of dynamic runtime assertions focuses on detecting superposition within a quantum circuit. The most basic scenario involves asserting a uniform superposition on a single qubit. Two distinct methods for this have been proposed in [31]. Both approaches operate on a target quantum state $|\psi\rangle$ and utilize an ancillary qubit to record the outcome of the assertion, as illustrated in Fig. 6.

The key distinction between the two lies in their gate complexity and error resilience. While the first method is generally preferred due to the lower error rates associated with avoiding the CX gate, the second, though more gate-intensive, has the advantage of deterministically returning the tested qubit to the $|+\rangle$ state. This not

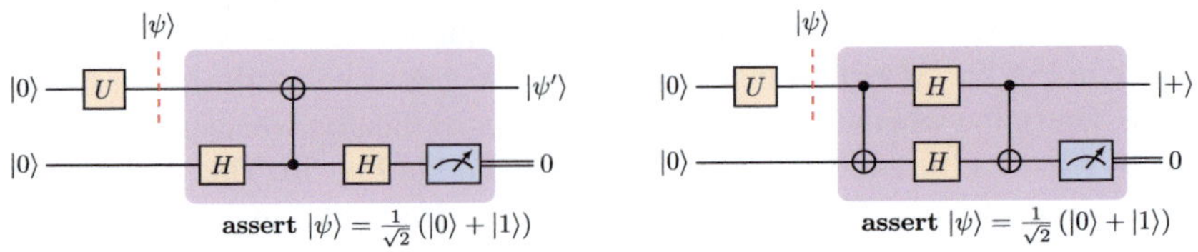

(a) (b)

Fig. 6 Alternative implementations of a single-qubit assertion for the uniform superposition state $|\psi\rangle = \frac{1}{\sqrt{2}}(|0\rangle + |1\rangle)$: **a** **assertion_S**$_1$—a lightweight implementation using a single CX gate to check for superposition; **b** **assertion_S**$_2$—a more gate-intensive approach employing two CX gates, which also projects the state of $|\psi\rangle$ into $|+\rangle$ regardless of its original state

only preserves the qubit for continued use but also enables fault correction, all while registering any deviation through the ancilla.

In more detail, consider the initial single-qubit state:

$$|\psi\rangle = \alpha_0 |0\rangle + \alpha_1 |1\rangle . \tag{13}$$

Applying the first assertion circuit, S_1, to the composite state $|\psi\rangle \otimes |0\rangle$ results in

$$\begin{aligned} S_1 : I \otimes H \cdot CX \cdot I \otimes H \,(|\psi\rangle \otimes |0\rangle) = & \frac{(\alpha_0 + \alpha_1)|00\rangle}{2} + \frac{(\alpha_0 - \alpha_1)|01\rangle}{2} \\ & + \frac{(\alpha_0 + \alpha_1)|10\rangle}{2} + \frac{(-\alpha_0 + \alpha_1)|11\rangle}{2} . \end{aligned} \tag{14}$$

In the fault-free case where $|\psi\rangle = |+\rangle$ (i.e., $\alpha_0 = \alpha_1 = \frac{1}{\sqrt{2}}$), the ancilla qubit will be measured in the $|0\rangle$ state, indicating that the target qubit is in uniform superposition. Conversely, if $|\psi\rangle = |-\rangle$, the ancilla will be in the $|1\rangle$ state after assertion. If a fault occurs in the circuit, resulting in a deviation from the expected superposition, the ancilla qubit will instead be found in a superposition of $|0\rangle$ and $|1\rangle$, with the measurement statistics reflecting the extent of the deviation.

Similarly, applying the second assertion, S_2, to the composite state $|\psi\rangle \otimes |0\rangle$ yields

$$\begin{aligned} & S_2 : CX \cdot H \otimes H \cdot CX \,(|\psi\rangle \otimes |0\rangle) \\ & \quad = \frac{1}{\sqrt{2}} \left(|+\rangle \otimes [(\alpha_0 + \alpha_1)|0\rangle + (\alpha_0 - \alpha_1)|1\rangle]\right) . \end{aligned} \tag{15}$$

As a result, the qubit under test is always projected into the $|+\rangle$ state, regardless of its original state $|\psi\rangle$. The ancilla qubit, however, carries information about the input state: in the ideal case where $|\psi\rangle = |+\rangle$, the ancilla ends up in the $|0\rangle$ state, while for $|\psi\rangle = |-\rangle$, the ancilla is in the $|1\rangle$ state. In the presence of faults, where $|\psi\rangle$ deviates from a known superposition, the ancilla qubit becomes entangled with that deviation, and its measurement outcome reflects a superposition, thereby indicating the presence of an error.

In addition to assertions for uniform superposition, [31] also introduces a method to verify whether a qubit state $|\psi\rangle$ matches a specific arbitrary superposition. To assert that the test qubit $|\psi\rangle$ corresponds to the target state:

$$|\phi\rangle = \cos\left(\frac{\theta}{2}\right)|0\rangle + e^{i\varphi}\sin\left(\frac{\theta}{2}\right)|1\rangle, \tag{16}$$

a custom assertion circuit can be employed, as illustrated in Fig. 7. This circuit uses a specially constructed gate G, designed such that $|\phi\rangle$ is an eigenvector of G with eigenvalue 1. In the fault-free case, where $|\psi\rangle = |\phi\rangle$, the application of G leaves the qubit undisturbed, thereby confirming the validity of the asserted state.

Starting from the initial state $|\psi\rangle \otimes |0\rangle$, the first Hadamard gate applied to the ancilla transforms the system into

$$\frac{1}{\sqrt{2}}(|\psi\rangle \otimes |0\rangle + |\psi\rangle \otimes |1\rangle). \tag{17}$$

Next, the application of the controlled-G gate results in the state:

$$CG\left(\frac{1}{\sqrt{2}}(|\psi\rangle \otimes |0\rangle + |\psi\rangle \otimes |1\rangle)\right) = \frac{1}{\sqrt{2}}(|\psi\rangle \otimes |0\rangle + G|\psi\rangle \otimes |1\rangle). \tag{18}$$

If $|\psi\rangle = |\phi\rangle$ (the target superposition state) then, by construction, $G|\psi\rangle = |\psi\rangle$, since $|\phi\rangle$ is an eigenvector of G with eigenvalue 1. Applying the second Hadamard gate to the ancilla then yields the final state $|\psi\rangle \otimes |0\rangle$, leaving the test qubit unchanged and producing a measurement result of 0 on the ancilla, confirming the assertion. Conversely, if $|\psi\rangle$ is orthogonal to $|\phi\rangle$, it corresponds to another eigenvector of G with eigenvalue -1. In this case, the final state becomes $|\psi\rangle \otimes |1\rangle$, again preserving the qubit under test while indicating orthogonality via a measurement outcome of 1 on the ancilla.

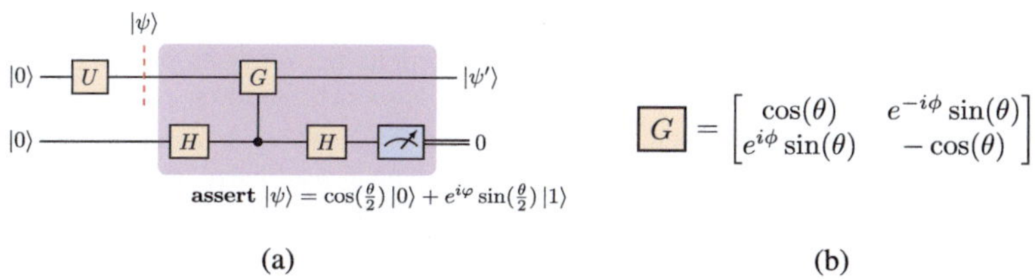

Fig. 7 Implementation of a single-qubit assertion for an arbitrary superposition $|\psi\rangle = \cos\left(\frac{\theta}{2}\right)|0\rangle + e^{i\varphi}\sin\left(\frac{\theta}{2}\right)|1\rangle$: **a** circuit implementation using a controlled-G gate constructed based on the expected quantum state; **b** matrix representation of the G gate as a single-qubit unitary operator

The assertion depicted in Fig. 6a can be interpreted as a special case of the more general assertion framework shown in Fig. 7. Specifically, when the target state (defined in Eq. 16) is the uniform superposition, i.e., $|\phi\rangle = \frac{1}{\sqrt{2}}(|0\rangle + |1\rangle)$, it corresponds to the Bloch sphere angles $\theta = \frac{\pi}{2}$ and $\varphi = 0$. Under these parameters, the custom gate G simplifies to the Pauli-X gate, and the CG operation reduces to a standard CX gate, as used in Fig. 6a.

3.2.3 Assertions for Entanglement

To verify that a set of n program qubits is entangled in the form

$$|\psi\rangle = \frac{1}{\sqrt{2}}(|0\ldots00\rangle + |1\ldots11\rangle), \tag{19}$$

a parity measurement-based assertion scheme can be employed [31]. The structure of the parity-checking circuit depends on both the number of qubits (even or odd) and the specific nature of the entangled state. As an example, Fig. 8a illustrates the parity-checking circuit for asserting the Bell state $|\phi^+\rangle$.

Consider the program state

$$|\psi\rangle = \alpha_0|00\rangle + \alpha_1|11\rangle. \tag{20}$$

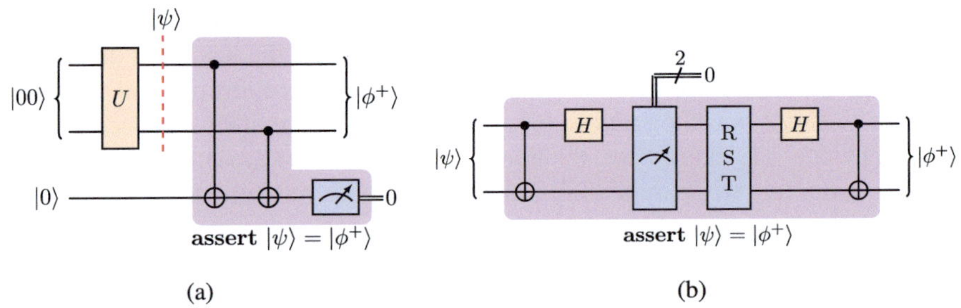

Fig. 8 Implementations of dynamic entanglement assertions for verifying the Bell state $|\psi\rangle = \frac{1}{\sqrt{2}}(|00\rangle + |11\rangle)$: **a assertion_E$_1$**—a resource-efficient approach based on parity measurement, employing an ancilla qubit and implemented using two CX gates [31]; **b assertion_E$_2$**—a more resource-intensive, ancilla-free variant requiring intermediate measurement and active reset (RST) of the data qubits, in addition to Bell state ($|\phi^+\rangle$) preparation and its inverse [21]

Applying the first CX gate of the assertion circuit, with control on the first program qubit and target on the ancilla initialized to $|0\rangle$, extends the entanglement to include the ancilla:

$$E_1 : CX^1 \left((\alpha_0 |00\rangle + \alpha_1 |11\rangle) \otimes |0\rangle \right) = \alpha_0 |000\rangle + \alpha_1 |111\rangle . \tag{21}$$

Subsequently, applying the second CX gate, controlled by the second program qubit, disentangles the ancilla and restores it to its original state:

$$E_1 : CX^2 \left(\alpha_0 |000\rangle + \alpha_1 |111\rangle \right) = (\alpha_0 |00\rangle + \alpha_1 |11\rangle) \otimes |0\rangle . \tag{22}$$

Measuring the ancilla in the $|0\rangle$ state confirms the expected entangled state. However, this assertion provides only partial information, as the same ancilla outcome can result from classical states such as $|00\rangle$ and $|11\rangle$, which are not entangled. Therefore, while effective in detecting certain fault conditions, the parity-based assertion does not fully distinguish entangled from non-entangled states.

An alternative and potentially more efficient technique for verifying entangled states is the projection-based measurement approach [21]. This assertion method consists of four main stages: (1) projecting the program state into the computational basis, (2) measuring the resulting state, (3) resetting the measured program qubits, and (4) restoring the original state. These stages are illustrated in Fig. 8b for the case of asserting the Bell state $|\phi^+\rangle$.

1. **Projection:** The program state $|\psi\rangle$ (see Eq. 20) is first projected into the computational basis using a standard Bell-state inverse circuit (i.e., CX followed by $H \otimes I$):

$$E_2 : H \otimes I \cdot CX \left(\alpha_0 |00\rangle + \alpha_1 |11\rangle \right) = \left[(\alpha_0 + \alpha_1) |0\rangle + (\alpha_0 - \alpha_1) |1\rangle \right] \otimes |0\rangle . \tag{23}$$

2. **Measurement:** If the original state is indeed the Bell state $|\phi^+\rangle$, where $\alpha_0 = \alpha_1 = \frac{1}{\sqrt{2}}$, the measurement outcome of the projected state yields $|00\rangle$ with certainty.
3. **Reset:** After measurement, both program qubits are reset to the state $|00\rangle$.
4. **Restoration:** Finally, the inverse of the projection circuit is applied to restore the program qubits to their original Bell state $|\phi^+\rangle$.

While this approach is more efficient than the parity-based method and can accommodate a broader class of quantum states, including classical and superposition states, it shares a key limitation with statistical assertions: it requires repeated measurement of the program qubits. Although the original state is restored before resuming execution, the act of measurement introduces additional overhead and potential complications. To address this, the following section presents an ancilla-based assertion scheme that verifies the quantum state without directly measuring the program qubits.

3.2.4 Quantum Assertion

Unlike the previous three assertions, which are generally designed to verify specific types of quantum states, such as classical, superposition, or entangled states, quantum assertions more generally aim to verify whether a set of n data qubits corresponds to a specific quantum state $|\phi\rangle$ at a given stage of the computation. The desired quantum state is given by the following expression:

$$|\phi\rangle = \sum_{i \in \{0,1,\ldots,2^n-1} \alpha_i |N_i\rangle , \qquad \text{where} \sum_i |\alpha_i|^2 = 1 \text{ and } 0 \leq N_i \leq 2^n - 1 \quad (24)$$

and can be generated through the application of an appropriate unitary transformation $\widehat{U}$ on a set of n quits representing the ground state $|0\rangle$, i.e., $|\phi\rangle = \widehat{U}\,|0\rangle$. To assert that a program-generated state $|\psi\rangle = U\,|0\rangle$ matches the expected state $|\phi\rangle$, the following procedure is stated in [50]:

1. Construct the n-qubit unitary such that $|\phi\rangle = \widehat{U}\,|0\rangle$.
2. Apply $\widehat{U}$ to a set of n ancilla qubits, preparing them in the state $|\phi\rangle$.
3. Simultaneously, apply the inverse operation $\widehat{U}^\dagger$ to the data qubits that hold the program state $|\psi\rangle$.
4. Apply SWAP gates between corresponding pairs of data (q) and ancilla (a) qubits: SWAP(q_i, a_i) for all $i = 0, 1, \ldots, n-1$.
5. Measure the ancilla qubits to verify correctness.

Figure 9a shows the quantum circuit representation of this assertion for a two-qubit program state. The verification ensures that the state output by block U is indeed $|\phi\rangle$, ensuring subsequent operations on a validated quantum state.

The SWAP operation exchanges the states of the data and ancilla qubits, resulting in the following transformation due to the assertion:

$$Q_1 : SWAP\left(\widehat{U}^\dagger\,|\psi\rangle \otimes \widehat{U}\,|0\rangle\right) = \widehat{U}\,|0\rangle \otimes \widehat{U}^\dagger\,|\psi\rangle\,. \quad (25)$$

The assertion $|\psi\rangle = |\phi\rangle$ holds if and only if:

$$\widehat{U}^\dagger\,|\psi\rangle = |0\rangle\,, \quad (26)$$

since $\widehat{U}\widehat{U}^\dagger = I$ and $|\phi\rangle = \widehat{U}\,|0\rangle$. In this case, the ancilla qubits—now holding $\widehat{U}^\dagger\,|\psi\rangle$—collapse to $|0\rangle$ upon measurement, resulting in an all-zero classical outcome. If the assertion fails, the measurement of ancilla qubits yields a non-zero classical value, indicating a mismatch between $|\psi\rangle$ and $|\phi\rangle$. This signals a bug, and further execution of the quantum program should be aborted or disregarded.

As an alternative to the SWAP-based scheme, which requires n ancilla qubits, a quantum assertion can also be implemented using a multi-controlled OR gate with only a single ancilla qubit [50], as illustrated in Fig. 9a. In this method, the unitary $\widehat{U}^\dagger$ is applied to the data qubits with the goal of transforming their state to $|0\rangle$. This state

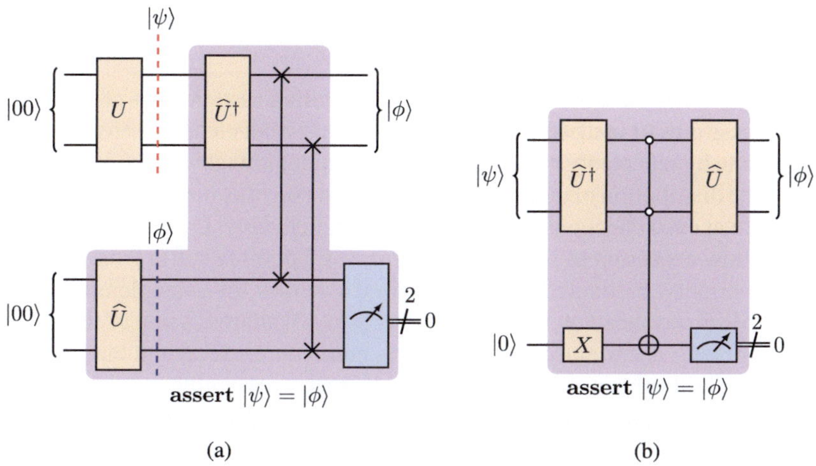

Fig. 9 Implementations of two-qubit dynamic quantum assertions for verifying $|\psi\rangle = |\phi\rangle$: **a assertion_Q**$_1$—a resource-efficient approach using two ancilla qubits, implemented with two SWAP gates (equivalent to 6 CX gate [42]); **b assertion_Q**$_2$—a more resource-intensive variant using only one ancilla qubit, requiring a logical OR gate (realized with 6 CX, 7 P, and 2 H gates [51]), in addition to $\widehat{U}$ and $\widehat{U}^\dagger$ gates. Both implementations enable runtime verification of arbitrary quantum program states, including classical, superposition, and entangled states

then serves as the control for an OR gate (i.e., negatively controlled version of the multiple-controlled-X gate, denoted $\widehat{\mathrm{C}}^n\mathrm{X}$ where n indicates the number of control qubits operating on a flipped target qubit [51]), which targets the single ancilla qubit leading to the assertion transformation:

$$Q_2 : \widehat{U} \otimes I \cdot \widehat{\mathrm{C}}^n\mathrm{X}(\widehat{U}^\dagger |\psi_n\rangle \otimes X |0_1\rangle) = |\phi_n\rangle \otimes |0_1\rangle \, , \tag{27}$$

provided that $\widehat{U}^\dagger |\psi_n\rangle = |0_n\rangle$.[1]

In this case, the ancilla qubit flips from $|1\rangle$ to $|0\rangle$, and a measurement result of 0 confirms that the assertion holds. If the ancilla is measured in the $|1\rangle$ state, a mismatch is detected, indicating the presence of a bug. If the assertion succeeds, the original quantum state is recovered by reapplying the unitary operation $\widehat{U}$, allowing the quantum program to proceed with its normal execution. While this method requires only a single ancilla qubit, it comes at the cost of increased circuit depth and gate complexity. This is because both $\widehat{U}^\dagger$ and $\widehat{U}$ must act on the data qubits, and the realization of the $\mathrm{C}^n\mathrm{X}$ gate is significantly more resource-intensive than performing n SWAP operations [52].

[1] The subscripts in ψ_n and 0_1 explicitly denote the presence of n program qubits and one ancilla qubit, respectively.

4 Conclusion

Several approaches have been proposed for interpreting assertions as checkpoints in quantum programs to verify their intermediate states. Statistical assertions verify the program state by repeatedly measuring it in the computational basis and applying the χ^2 test to the distribution of outcomes [16, 30]. However, this method requires halting the program at each checkpoint, which limits its practicality. Dynamic assertions, in contrast, allow execution to proceed uninterrupted, but often suffer from limitations such as ambiguity, particularly in parity-based schemes for entanglement verification [31]. More advanced quantum assertions [21, 50] address some of these limitations by reducing uncertainty and supporting continuous execution, but often incur significant gate overhead. Given that assertion-based debugging is mainly applicable on quantum hardware, since simulators are restricted by program size, statistical assertions may remain more feasible on today's noisy quantum devices. Additionally, although several automated slicing and bug-detection frameworks have been proposed [22, 32, 33], the probabilistic nature of quantum computation necessitates optimization. Efficient assertion strategies and slicing techniques can significantly reduce the number of program executions required during debugging, thereby minimizing resource overhead and improving practicality.

References

1. Egger, Daniel J., Claudio Gambella, Jakub Marecek, Scott McFaddin, et al. 2020. Quantum computing for finance: State-of-the-art and future prospects. *IEEE Transactions on Quantum Engineering* 1: 1–24.
2. Herman, Dylan, Cody Googin, Xiaoyuan Liu, Yue Sun, Alexey Galda, et al. 2023. Quantum computing for finance. *Nature Reviews Physics* 5 (8): 450–465.
3. Cao, Yudong, Jonathan Romero, Jonathan P. Olson, Matthias Degroote, et al. 2019. Quantum chemistry in the age of quantum computing. *Chemical Reviews* 119 (19): 10856–10915.
4. Bauer, Bela, Sergey Bravyi, Mario Motta, and Garnet Kin-Lic Chan. 2020. Quantum algorithms for quantum chemistry and quantum materials science. *Chemical Reviews* 120 (22): 12685–12717.
5. Pirandola, S., U. L. Andersen, L. Banchi, M. Berta, D. Bunandar, et al. 2020. Advances in quantum cryptography. *Advances in Optics and Photonics* 12 (4): 1012–1236.
6. Gisin, Nicolas, Grégoire Ribordy, Wolfgang Tittel, and Hugo Zbinden. 2002. Quantum cryptography. *Reviews of Modern Physics* 74: 145–195.
7. Krantz, P., M. Kjaergaard, F. Yan, T. P. Orlando, S. Gustavsson, and W. D. Oliver. 2019. A quantum engineer's guide to superconducting qubits. *April* 6 (2): 021318.
8. Siddiqi, Irfan. 2021. Engineering high-coherence superconducting qubits. *Nature Reviews Materials* 6 (10): 875–891.
9. Bruzewicz, Colin D., John Chiaverini, Robert McConnell, and Jeremy M. Sage. 2019. Trapped-ion quantum computing: Progress and challenges. *Applied Physics Reviews* 6 (2): 021314.
10. Häffner, H., C. F. Roos, and R. Blatt. 2008. Quantum computing with trapped ions. *Physics Reports* 469 (4): 155–203.
11. Das Sarma, Sankar, Michael Freedman, and Chetan Nayak. 2005. Topologically protected qubits from a possible non-abelian fractional quantum hall state. *Physical Review Letters* 94: 166802.

12. Aguado, Ramón, and Leo P. Kouwenhoven. 2020. Majorana qubits for topological quantum computing. *Physics Today* 73 (6): 44–50.
13. Montanaro, Ashley. 2016. Quantum algorithms: An overview. *npj Quantum Information* 2 (1): 15023.
14. Javadi-Abhari, Ali, Matthew Treinish, Kevin Krsulich, Christopher J Wood, et al. 2024. Quantum computing with qiskit. arXiv preprint arXiv:2405.08810 [quant-ph].
15. Omole, Victory, Akhilesh Tyagi, Calista Carey, AJ Hanus, Andrew Hancock, et al. 2023. Cirq: A python library for writing, manipulating, optimizing and running quantum circuits.
16. Svore, Krysta, Alan Geller, Matthias Troyer, John Azariah, Christopher Granade, et al. 2018. Q#: Enabling scalable quantum computing and development with a high-level dsl. In *Proceedings of the Real World Domain Specific Languages Workshop 2018*, RWDSL2018, New York, NY, USA. Association for Computing Machinery.
17. Bergholm, Ville, Josh Izaac, Maria Schuld, Christian Gogolin, Shahnawaz Ahmed, et al. 2018. Pennylane: Automatic differentiation of hybrid quantum-classical computations. arXiv preprint arXiv:1811.04968 [quant-ph].
18. Sivarajah, S., S. Dilkes, A. Cowtan, W. Simmons, A. Edgington, and R. Duncan. 2020. t|ket>: A retargetable compiler for NISQ devices. *Quantum Science and Technology* 6 (1): 014003.
19. Weder, Benjamin, Johanna Barzen, Frank Leymann, Marie Salm, and Daniel Vietz. 2020. The quantum software lifecycle. In *Proceedings of the 1st ACM SIGSOFT International Workshop on Architectures and Paradigms for Engineering Quantum Software*, APEQS 2020, 2–9, New York, NY, USA. Association for Computing Machinery.
20. Metwalli, Sara Ayman, and Rodney Van Meter. 2024. Testing and debugging quantum circuits. *IEEE Transactions on Quantum Engineering* 5: 1–15.
21. Li, Gushu, Li Zhou, Nengkun Yu, Yufei Ding, Mingsheng Ying, and Yuan Xie. 2020. Proq: Projection-based runtime assertions for debugging on a quantum computer. arXiv:1911.12855v2 [cs.PL].
22. Metwalli, Sara Ayman, and Rodney Van Meter. 2022. A tool for debugging quantum circuits. In *2022 IEEE International Conference on Quantum Computing and Engineering (QCE)*, 624–634.
23. Mendiluze, Eñaut, Shaukat Ali, Paolo Arcaini, and Tao Yue. 2021. Muskit: A mutation analysis tool for quantum software testing. In *2021 36th IEEE/ACM International Conference on Automated Software Engineering (ASE)*, 1266–1270.
24. Miranskyy, Andriy, Lei Zhang, and Javad Doliskani. 2020. Is your quantum program bug-free? arXiv preprint arXiv:2001.10870 [cs.SE].
25. Liu, Junyi, Bohua Zhan, Shuling Wang, Shenggang Ying, Tao Liu, et al. 2019. Formal verification of quantum algorithms using quantum hoare logic. In *Computer Aided Verification*, ed. Isil Dillig and Serdar Tasiran, 187–207. Cham: Springer.
26. Davidson, Timothy AS, Simon J Gay, Hynek Mlnarik, Rajagopal Nagarajan, and Nick Papanikolaou. Model checking for communicating quantum processes. *The International Journal of Unconventional Computing* 8 (1): 73–98 (2012).
27. Sun, Xin, and Feifei He. 2020. A first step to the categorical logic of quantum programs. *Entropy* 22 (2).
28. Ardeshir-Larijani, Ebrahim, Simon J. Gay, and Rajagopal Nagarajan. 2018. Automated equivalence checking of concurrent quantum systems. *ACM Transactions on Computational Logic* 19 (4).
29. Burgholzer, Lukas, and Robert Wille. 2021. Advanced equivalence checking for quantum circuits. *IEEE Transactions on Computer-Aided Design of Integrated Circuits and Systems* 40 (9): 1810–1824.
30. Huang, Yipeng, and Margaret Martonosi. 2019. Statistical assertions for validating patterns and finding bugs in quantum programs. In *Proceedings of the 46th International Symposium on Computer Architecture*, ISCA '19, 541–553, New York, NY, USA. Association for Computing Machinery.

31. Liu, Ji, Gregory T. Byrd, and Huiyang Zhou. 2020. Quantum circuits for dynamic runtime assertions in quantum computation. In *Proceedings of the Twenty-Fifth International Conference on Architectural Support for Programming Languages and Operating Systems*, ASPLOS '20, 1017–1030, New York, NY, USA. Association for Computing Machinery.
32. Maidl, Anton, Abhoy Kole, Kamalika Datta, Jannis Stoppe, and Rolf Drechsler. 2025. Towards an automated debugging approach for fault identification in quantum circuits. In *2025 IEEE 28th International Symposium on Design and Diagnostics of Electronic Circuits and Systems (DDECS)*, 19–24.
33. Rovara, Damian, Lukas Burgholzer, and Robert Wille. 2025. A framework for the efficient evaluation of runtime assertions on quantum computers. arXiv preprint arXiv:2505.03885 [quant-ph].
34. Honarvar, Shahin, Mohammad Reza Mousavi, and Rajagopal Nagarajan. 2020. Property-based testing of quantum programs in q#. In *Proceedings of the IEEE/ACM 42nd International Conference on Software Engineering Workshops*, ICSEW'20, 430–435, New York, NY, USA. Association for Computing Machinery.
35. Abreu, Rui, João Paulo Fernandes, Luis Llana, and Guilherme Tavares. 2023. Metamorphic testing of oracle quantum programs. In *Proceedings of the 3rd International Workshop on Quantum Software Engineering*, Q-SE '22, 16–23, New York, NY, USA. Association for Computing Machinery.
36. Miranskyy, Andriy, and Lei Zhang. 2019. On testing quantum programs. In *2019 IEEE/ACM 41st International Conference on Software Engineering: New Ideas and Emerging Results (ICSE-NIER)*, 57–60.
37. Ali, Shaukat, Paolo Arcaini, Xinyi Wang, and Tao Yue. 2021. Assessing the effectiveness of input and output coverage criteria for testing quantum programs. In *2021 14th IEEE Conference on Software Testing, Verification and Validation (ICST)*, 13–23.
38. Zhao, Pengzhan, Zhongtao Miao, Shuhan Lan, and Jianjun Zhao. 2023. Bugs4q: A benchmark of existing bugs to enable controlled testing and debugging studies for quantum programs. *Journal of Systems and Software* 205: 111805.
39. Campos, José, and André Souto. 2021. Qbugs: A collection of reproducible bugs in quantum algorithms and a supporting infrastructure to enable controlled quantum software testing and debugging experiments. In *2021 IEEE/ACM 2nd International Workshop on Quantum Software Engineering (Q-SE)*, 28–32.
40. Hao, Tianyi, Kun Liu, and Swamit Tannu. 2023. Enabling high performance debugging for variational quantum algorithms using compressed sensing. In *Proceedings of the 50th Annual International Symposium on Computer Architecture*, ISCA '23, New York, NY, USA. Association for Computing Machinery.
41. Dirac, Paul Adrien Maurice. 1939. A new notation for quantum mechanics. In *Mathematical Proceedings of the Cambridge Philosophical Society*, vol. 35, 416–418. Cambridge University Press.
42. Nielsen, M., and I. Chuang. 2000. *Quantum Computation and Quantum Information*. Cambridge University Press.
43. Williams, Colin P. 2011. *Quantum Gates*, 51–122. London: Springer.
44. Cross, Andrew, Ali Javadi-Abhari, Thomas Alexander, Niel De Beaudrap, et al. 2022. OpenQASM 3: A broader and deeper quantum assembly language. *ACM Transactions on Quantum Computing* 3 (3).
45. Bell, J. S. 1964. On the Einstein Podolsky Rosen paradox. *Physics Physique Fizika* 1: 195–200.
46. Ali, Shaukat, Tao Yue, and Rui Abreu. 2022. When software engineering meets quantum computing. *Communications of the ACM* 65 (4): 84–88.
47. Bertels, K, A Sarkar, A Krol, R Budhrani, J Samadi, E Geoffroy, et al. 2021. Quantum accelerator stack: A research roadmap. arXiv preprint arXiv:2102.02035 [quant-ph].
48. Heim, Bettina, Mathias Soeken, Sarah Marshall, Chris Granade, et al. 2020. Quantum programming languages. *Nature Reviews Physics* 2 (12): 709–722.
49. Press, William H. 2007. *Numerical Recipes 3rd Edition: The Art of Scientific Computing*. Cambridge University Press.

50. Liu, Ji, and Huiyang Zhou. 2021. Systematic approaches for precise and approximate quantum state runtime assertion. In *2021 IEEE International Symposium on High-Performance Computer Architecture (HPCA)*, 179–193.
51. Amy, Matthew, Dmitri Maslov, Michele Mosca, and Martin Roetteler. 2013. A meet-in-the-middle algorithm for fast synthesis of depth-optimal quantum circuits. *IEEE Transactions on Computer-Aided Design of Integrated Circuits and Systems* 32 (6): 818–830.
52. Maslov, Dmitri. 2016. Advantages of using relative-phase toffoli gates with an application to multiple control toffoli optimization. *Physical Review A* 93: 022311.

Quantum Security and Post-quantum Cryptography

Security Attacks Abusing Pulse-Level Quantum Circuits

Chuanqi Xu and Jakub Szefer

Abstract This work presents the first thorough exploration of the attacks on the interface between gate-level and pulse-level quantum circuits and pulse-level quantum circuits themselves. Typically, quantum circuits and programs that execute on quantum computers, are defined using gate-level primitives. However, to improve the expressivity of quantum circuits and to allow better optimization, pulse-level circuits are now often used. The attacks presented in this work leverage the inconsistency between the gate-level description of the custom gate, and the actual, low-level pulse implementation of this gate. By manipulating the custom gate specification, this work proposes numerous attacks: qubit plunder, qubit block, qubit reorder, timing mismatch, frequency mismatch, phase mismatch, and waveform mismatch. This work demonstrates these attacks on the real quantum computer and simulator, and shows that most current software development kits are vulnerable to these new types of attacks. In the end, this work proposes a defense framework. The exploration of security and privacy issues of the rising pulse-level quantum circuits provides insight into the future development of secure quantum software development kits and quantum computer systems.

1 Introduction

Quantum computing is advancing rapidly with more and bigger quantum computers coming online every year: from quantum computers with one or two qubits two decades ago, to computers with 1121 qubits presently available; and projections for quantum computers with 200-qubit error-corrected modularity capable of running 100 million gates by the current decade's end [24]. These existing quantum computers are categorized as Noisy Intermediate-Scale Quantum (NISQ) devices [51] since

C. Xu (✉) · J. Szefer
Yale University, New Haven, CA, USA
e-mail: chuanqi.xu@yale.edu

J. Szefer
e-mail: jakub.szefer@northwestern.edu

P. Mishra (ed.), *Design Automation for Quantum Computing*,
https://doi.org/10.1007/978-3-032-09303-5_11

they do not provide error correction [15] and have noisy qubits and operations. Nevertheless, they show promise in applications like optimization, natural sciences, artificial intelligence, and finance [8, 18, 23, 28, 32, 45, 50].

Today, quantum computers are easily accessible through cloud-based services such as IBM Quantum [25], Amazon Braket [1], or Microsoft Azure [47]. These services are not just for academics, but actual pay-as-you-go cloud services that anybody can use. Many startups and companies do not have their own quantum computers, and they depend on cloud-based quantum computers to run their, often proprietary, quantum circuits. For example, J.P. Morgan Chase leverages cloud-based Quantinuum H-series to develop algorithms for solving linear systems on quantum hardware [65]. As the utilization of quantum computers is on the rise, the importance of securing quantum circuits and programs that execute on the quantum computers, also becomes increasingly evident.

Quantum circuits serve as the foundational building blocks for quantum computation, orchestrating the manipulation and processing of quantum bits. These circuits encode quantum algorithms and enable the execution of complex computations, offering the potential to outperform classical computers in specific tasks. Typical quantum circuits are usually built using quantum gates, which are abstract units that define specific operations on qubits.

However, quantum circuits are not restricted to be defined only using quantum gates. In fact, quantum gates are only one abstract layer, and they need to be compiled into actual operations on qubits by quantum software development kits (SDKs). For example, quantum gates on superconducting quantum computers are realized by electromagnetic pulses. Recent advancements highlight the substantial advantages of deconstructing gate-level circuits into these lower layers. Defining circuits using low-level control pulses proves advantageous for tasks such as quantum circuit decomposition, compilation, and optimization [4, 5, 19, 36, 40, 54], as well as enhancing the expressivity and efficiency of quantum circuits [34, 35, 37, 38]. Despite the many benefits of specifying quantum circuits using lower-level control pulses, security and privacy have seldom been touched on. One recent research work introduced higher-energy state attacks, exploiting control pulses to excite qubits into higher-energy levels, and then abusing the properties of higher-energy states to circumvent standard operations in quantum computers [60].

To address research gaps in this field, this paper undertakes the first comprehensive exploration of security and privacy aspects concerning quantum circuits built partly or wholly from low-level pulses rather than purely from gates. In particular, we introduce a new class of attacks, which leverage the inconsistency between the definition of the custom quantum gate, and the low-level implementation of the pulses that realize the gate operations. For the attacks, we assume attackers can clandestinely implant or modify the low-level pulse implementation such that it is not revealed to the high-level gate definition. The proposed attacks are particularly dangerous for two reasons: Firstly, most current quantum software development kits are deficient in the interface between gate-level and pulse-level, and pulse configurations specifying pulse-level gates are in the analog form. These make our attacks less understandable and difficult to verify. Secondly, pulse configuration is determined by the hardware properties and environment. Due to reasons such as variations in manufacturing and the continuously changing environment within and near the quantum computer, pulse configuration is distinct and unique for a specific quantum hardware and needs frequent updates to maintain proper functionality. This, in turn, means that pulse specifications are individualized and time-restricted. It is almost impossible to verify pulses by setting an overall standard for pulses for all quantum computers, or by comparing them to some prior, correct pulses, since even correct pulses will have changed with time. The pulse data is updated as often as the quantum computers are calibrated, which can be every day, or even every few hours, providing space for adversaries to install attacks. Both factors result in a unique new attack surface in the quantum software supply chain, which is unique and different from classical computing, making the attacks challenging to detect and prevent.

The high-level workflow schematic of our proposed attacks is shown in Fig. 1. Our attacks can primarily manifest themselves in two places in the quantum software development stack. The first location is the interface between gate-level and pulse-level circuits, referred to as *channel attacks*. These attacks, encompassing *qubit spoilage*, *qubit reorder*, and *qubit block*, focus more on software development kit aspects. We scrutinized two widely used quantum SDKs: Qiskit and Amazon Braket Python SDK, and identified the possibility of exploiting *channel attacks* in both SDKs. The second location is in the pulse configuration, which directly specifies the pulse details. We categorize this as *pulse attacks*. These attacks include *timing mismatch*, *frequency mismatch*, *phase mismatch*, and *waveform mismatch*, each exploiting specific features of pulses to manipulate qubits. We demonstrate these attacks on the real quantum computer and simulator, focusing on extensively

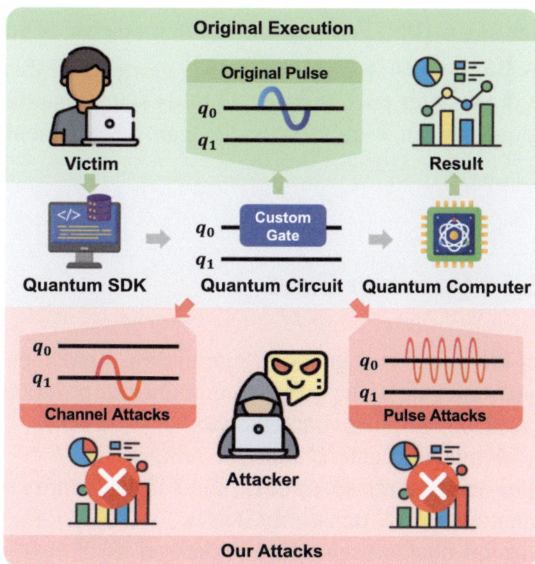

Fig. 1 The workflow of a quantum computer, demonstrating where the channel and pulse attacks proposed in this paper can occur. The quantum software development kit (Quantum SDK) is used to develop Quantum Circuits which are then executed on a Quantum Computer. The proposed attacks target the Quantum Circuit specification and can be deployed, among others, through supply chain attacks

utilized quantum algorithms: quantum teleportation, Grover's search, and quantum neural networks.

To counteract these attacks, we present a defense framework that checks each type of the discussed attacks, as well as propose a standard for the reuse of pulse-level circuits. Our defense strategy encompasses a multifaceted approach, leveraging both hardware and software components to fortify quantum circuits against the attacks.

2 Background

This section aims to offer background on how data is manipulated in quantum computers and a typical workflow for running quantum computing programs.

2.1 *Qubits and Quantum States*

In the realm of quantum computing, the fundamental unit of information is the quantum bit, or qubit. Conceptually akin to the classical bit in conventional computing, a qubit possesses two basis states, symbolized in bra-ket notation as $|0\rangle$ and $|1\rangle$. However, unlike classical bits which are constrained to values of either 0 or 1, a qubit can exist in a superposition of these states: $|\psi\rangle = \alpha |0\rangle + \beta |1\rangle$, where α, β are complex and $|\alpha|^2 + |\beta|^2 = 1$.

Qubits are often represented using vector representation. For instance, the single-qubit basis states can be denoted as two-dimensional column vectors: $|0\rangle = \begin{bmatrix}1, & 0\end{bmatrix}^T$ and $|1\rangle = \begin{bmatrix}0, & 1\end{bmatrix}^T$, where T is the transpose. Thus, the state $|\psi\rangle$ can be expressed as $|\psi\rangle = \alpha |0\rangle + \beta |1\rangle = \begin{bmatrix}\alpha, & \beta\end{bmatrix}^T$. Extending this notion to n qubits, the space of n-qubit states encompasses 2^n basis states, ranging from $|0\ldots0\rangle$ to $|1\ldots1\rangle$. Consequently, an n-qubit state $|\phi\rangle$ can be represented as $|\phi\rangle = \sum_{i=0}^{2^n-1} a_i |i\rangle$, where $\sum_{i=0}^{2^n-1} |a_i|^2 = 1$.

2.2 *Quantum Gates*

In the domain of quantum computing, fundamental operations are represented as quantum gates. The quantum circuits and programs execute a sequence of quantum gates to manipulate qubits toward desired states.

A quantum gate U must be unitary, i.e., $UU^\dagger = U^\dagger U = I$, where $U^\dagger$ signifies the conjugate transpose of U, and I is the identity matrix. Operating on a qubit $|\psi\rangle$, a quantum gate U transforms it as $|\psi\rangle \rightarrow U |\psi\rangle$. Employing the matrix representation, n-qubit quantum gates are expressed as $2^n \times 2^n$ matrices. Quantum gates can be classified as single-qubit gates or multi-qubit gates. U3 gate is one general single-qubit gate that is parameterized by three Euler angles. Some other single-qubit gates include Pauli-X gate: a single-qubit gate similar to the classical NOT gate, flips $|0\rangle$ to $|1\rangle$ and vice versa; RZ gate: a phase shift between $|0\rangle$ and $|1\rangle$. Unlike classical computing, multi-qubit gates can create quantum entanglement. One notable example is the CNOT gate, also known as the CX gate, a two-qubit gate that applies a Pauli-X gate to the target qubit if the control qubit is in state $|1\rangle$, otherwise, it remains unchanged. Another example is the control U3 gate, which also uses one bit to control the application of the U3 gate.

Matrix representations of some gates are presented below, adhering to the qubit order specified by Qiskit [53]:

$$\mathtt{U3}(\theta, \phi, \lambda) = \begin{bmatrix} \cos(\frac{\theta}{2}) & -e^{i\lambda}\sin(\frac{\theta}{2}) \\ e^{i\phi}\sin(\frac{\theta}{2}) & e^{i(\phi+\lambda)}\cos(\frac{\theta}{2}) \end{bmatrix}$$

$$\mathtt{RZ}(\theta) = \begin{bmatrix} e^{-i\frac{\theta}{2}} & 0 \\ 0 & e^{i\frac{\theta}{2}} \end{bmatrix}, \mathtt{CX} = \begin{bmatrix} 1 & 0 & 0 & 0 \\ 0 & 0 & 0 & 1 \\ 0 & 0 & 1 & 0 \\ 0 & 1 & 0 & 0 \end{bmatrix}$$

Research proved that any unitary quantum gate can be approximated with minor error using a small set of quantum gates [13]. Consequently, contemporary quantum computers typically support a limited set of basis gates. By amalgamating these basis gates, they can emulate other quantum gates. These fundamental gates, termed native gates, serve as pivotal configurations for quantum processors.

2.3 Control Pulses

Microwave pulses usually serve as the primary means to control qubits. A control pulse is analog and specified by a set of parameters, such as the envelope, frequency, and phase. The envelope delineates the shape of the signal generated by the arbitrary waveform generator (AWG), a standard laboratory instrument; frequency specifies a periodic signal utilized to modulate the envelope signal; phase manipulates the frame phase of the equipment. Together, these components constitute the output signal directed toward the qubit.

Envelopes are often discretized into time steps, with each value denoting the amplitude at a specific time step. An alternative approach involves parameterized pulses determined by predefined shapes. These parameters typically encompass the duration (pulse length), amplitude (relative pulse strength), and other factors dictating pulse shape.

Pulses for all native gates on current quantum computers are often predefined, but their parameters are updated periodically. The parameters of pulses are determined by many factors such as qubit properties and environment, and thus they are not fixed over time and are unique for a specific quantum computer: Due to the constantly changing environment, the physical properties of quantum computers also continuously vary. As a result, the parameters of pulses are regularly updated over time via automated measurements and calibrations to maintain high fidelity. Because of this, gate-level circuits need to be recompiled before they are executed so that the most recent pulses representing the gates are used. Apart from changing continuously, due to reasons such as variations in manufacture and environment, the parameters of pulses are also distinct for qubits. Parameters of pulses for different qubits on the same quantum computer, qubits on different quantum computers with the same type and model, have different parameters.

2.4 Pulse-Level Quantum Circuit

Quantum circuits described by quantum gates are called gate-level circuits. The gate-level circuits are transformed into pulse-level circuits by the compiler so that they can be executed on the quantum computer, or the user can directly write pulse-level circuits themselves. To realize quantum circuits, the control pulses must be properly arranged, such as the starting time of the pulses, and to which channels pulses should be applied. All of this information constitutes the pulse-level quantum circuit. Pulse-level circuits hold significant value for users. For example, pulse-level circuits are actively investigated for optimization and applications as mentioned in Sect. 1, improving both the expressivity and efficiency.

How to incorporate pulses into quantum circuits is the design choice of quantum software development kits. One widely used method is to define the custom gates, and specify what effect these custom gates should have on qubits. Functioning similarly to normal quantum gates, custom gates require the specification of the number of qubits and the target qubits for application within a quantum circuit. The detailed operations of custom gates are either independently specified as a list of quantum gates or linked to specifications of pulse-level controls. Custom gates can be recursively defined to encompass another custom gate.

Pulse-level circuits are necessary because they specify the actual operations on quantum computers. While they are not public on all quantum computers due to complexity, an increase in exposing them has been seen, including IBM Qiskit [52], Amazon Braket [27], and PennyLane [30]. Though the advantages of pulse-level circuits are widely explored, security and privacy aspects have not been researched, which are the focus of this paper.

2.5 From Gate-Level to Pulse-Level

Typically, quantum circuits and programs are defined using gate-level primitives, with quantum software development kits such as Qiskit [53], Amazon Braket Python

SDK [1], Q# [47], Cirq [14]. Similar to compilation in classical computing, gate-level quantum circuits often need to be modified so that they satisfy the requirements of specific hardware. This process involves decomposing non-native quantum gates into native gates, optimizing gates by grouping and eliminating redundant gates, mapping logical qubits to physical qubits, routing circuits within the constrained topology, etc.

The next step is to transform gate-level circuits to lower-level specifications, usually pulse-level circuits. This process is like assembling in classical computing. Leveraging pre-calibrated data for native gates on individual qubits or qubit pairs, microwave pulse sequences are generated and tailored for specific hardware. Besides native gates, arbitrary pulses can also be included with custom gates. Consequently, pulse-level circuits can be processed by quantum computer controllers to generate the actual operations manipulating qubits on the target quantum computer.

3 Threat Model

In this paper, we introduce attacks targeting quantum circuits that incorporate pulse-level controls. Typically, pulse-level controls are integrated into gate-level circuits using user-defined custom gates.

3.1 Assumptions

The target of the attacks is quantum circuits that contain custom pulse-level controls. We assume that attackers are able to modify the specification of the control pulses on victims' quantum circuits. We assume that after the malicious modifications have been performed unknowingly to victims, victims execute the circuit, thus triggering the operation of the modified pulse gates.

Pulse attacks can be performed in two distinct ways. The first way involves the supply chain attack, where attackers are assumed to perform malicious modifications at various points in the software development process. This could include directly altering local code, tampering with public repositories from which victims download code, or using malicious toolchains, such as compromised compilers, to change pulse information covertly. Importantly, the supply chain attack only requires attackers to be able to modify user-level programs secretly. It does not depend on the ability to alter downstream components in the workflow, nor the ability to interfere with the infrastructure that executes quantum programs. This type of attack will be the main focus of this paper.

The second way involves more powerful attackers who are capable of altering the infrastructure itself. Instead of manipulating user-level quantum programs, these attackers can modify the infrastructure so that the pulse specifications do not produce the expected results. For instance, an attacker might manipulate the environment so that the pulse specifications in the program no longer accurately correspond to the actual conditions in the hardware. This will be left for future research.

While it is conceivable for attackers to modify other parts of quantum circuits, and this could be combined with our attacks, we assume attackers only alter pulse configurations for quantum circuits. This assumption is rooted in the clandestine nature of the attack. We argue that detecting the installation of the attack becomes more challenging for victims when only pulse configurations are modified, as opposed to modifications on other parts. For instance, gate-level modifications can be easily detected through methods performing exact match checks of previous and current quantum circuits, such as quantum computer antivirus [11, 12] or maybe simply the hash check. Conversely, modifications on pulse configurations remain more secret. To the best of our knowledge, no current tool exists that could verify pulse-level quantum circuits, and our work raises the necessity of verifying quantum circuits at the lower level.

3.2 Attack Stealthiness

All victims are assumed to understand gate-level circuits and any higher levels above gate-level circuits, but their capabilities are distinguished by how much they can verify the pulse-level circuits. The verification of pulse-level circuits requires the two-fold explainability of pulses:

1. Syntax: relations between pulses and logic operations, more specifically, the effect of pulses on qubits. E.g., the pulse rotates a qubit 90 degrees along the X axis.
2. Semantics: relations between pulses and meanings or motivations. E.g., applying a Hadamard gate and a CNOT gate typically creates an entanglement state.

Syntax verification is heavily dependent on translating pulses into gates or operations, which is specific to the hardware but applicable to all quantum programs executed on that hardware. Conversely, semantics verification is contingent on the specific quantum algorithms and the underlying logic of their designs, which is general across different hardware platforms but presupposes that the pulses are syntactically correct.

The verification process at both levels poses significant challenges, thereby enhancing the stealthiness of our attacks introduced in this paper.

3.2.1 Challenges of Syntax Verification

The pulse-level controls are similar to embedded assembly or machine code in classical computing, enabling direct low-level hardware control. However, unlike the fixed and limited number of instruction set architectures in classical computing, pulses are defined by analog data, such as the frequency and amplitude of electromagnetic waves. We present three major challenges for the syntax verification:

- **Interpretability**: Due to analog characteristics, pulse data is not easily interpretable by humans. Moreover, since this data is hardware-dependent, different qubits possess distinct calibration data. The same analog data does not necessarily have the same functionality on two qubits or even the same qubits under varying environmental conditions. This complexity hinders victims from understanding the actual impact of the pulses.
- **Volatility**: The qubit properties are frequently changing [64], and regular machine calibrations on cloud platforms accentuate the timeliness of pulse data. This means that pulse data may quickly become outdated, necessitating timely re-calibrations. Although changes in a short time frame may be minor, they can accumulate if not properly calibrated, prompting victims to seek updated pulse data. The time interval for updates depends on the hardware and properties, ranging from days to months, which is not too short to invalidate the attack quickly nor too long to free circuits from updates.
- **Cost**: Calibration experiments are resource-intensive and costly. In addition, victims may find these experiments complex, leading them to rely on quantum circuit providers for detailed experiment information. Even if victims conduct calibrations themselves, the intricacy of the process may still drive them to seek assistance from quantum circuit providers, and attackers may install attacks in these places rather than the pulse data directly.

3.2.2 Challenges of Semantics Verification

A key distinction between quantum programs and classical programs that complicates semantics verification is the nature of operations in quantum programs, which focuses on controlling the evolution of qubits. This results in control flows that are more abstract and challenging to comprehend. While this complexity is significant, it is not the only hurdle. Another critical issue is that not all quantum algorithms are inherently interpretable. For instance, parametric algorithms such as quantum deep learning or quantum approximate optimization algorithms have parameters trained to fit specific data sets. These parameters do not possess an explainable meaning, adding another layer of difficulty.

3.3 Victim Classification

Book Chapters 2025—Security Attacks Abusing Pulse-level Quantum Circuits

Based on the victim's capabilities to verify the pulse-level circuits at the syntactical and semantic level, victims are classified into five levels as shown in Table 1:

Table 1 Victim classification based on victim's capabilities to verify gate-level and pulse-level circuits, and attacks that can be effectively applied. Checkmark means the victim has the corresponding capability to verify the code at that level

Victim level	Victim verification capability				Vulnerability to our attacks	
	Gate-level	Gate-pulse interface	Pulse-level syntax	Pulse-level semantics	Channel attack	Pulse attack
Level 1					Can attack	Can attack
Level 2	✓				Can attack	Can attack
Level 3	✓	✓			–	Can attack
Level 4	✓	✓	✓		–	Algorithm-dependent
Level 5	✓	✓	✓	✓	–	–

1. Level 1: the weakest victims that cannot verify either gate-level circuits or pulse-level circuits.
2. Level 2: victims that can only verify gate-level circuits. In Sect. 4, we propose the *channel attacks* which leverage the fact that the gate-level circuits are all the same but the underlying pulse-level circuits can be arbitrarily modified. Therefore, for level 1 victims, they cannot detect the channel attacks.
3. Level 3: victims that can verify the gate-level circuits and check pulse-level circuits, but they cannot verify the pulse-level circuits at the syntactical level. Notice that victims are assumed to understand the quantum algorithms at the logic level in this case, but they still cannot verify the pulse-level circuits at the semantic level because verification at the syntactical level is the precondition. In Sect. 4, we propose the *pulse attacks* which modify the underlying pulse configuration.
4. Level 4: victims that can verify the gate-level circuits and pulse-level circuits at the syntactical level, but they cannot verify the pulse-level circuits at the semantic level. The pulse attacks may apply here depending on the algorithm's interpretability.
5. Level 5: the strongest victims that can verify the gate-level circuits and pulse-level circuits at both the syntactical level and the semantic level.

Level 1 victims are too weak, and thus they are not the focus of this paper. On the contrary, level 5 victims are too strong. Even though they can defend themselves from the attacks proposed in this paper, we claim that the assumption of the capability to verify the pulse-level semantics for all quantum algorithms is too demanding and may not be possible, such as for quantum neural networks. Therefore, we mainly discuss the level 2–4 victims in this paper.

4 Proposed Attacks

In this section, we present a classification of our attacks. The pulse-level controls are typically integrated into quantum circuits through user-defined custom gates according to most current quantum software development kits. The custom gate, a gate-level operation similar to a standard quantum gate, can encapsulate pulse-level controls within its structure. Consequently, the attacks can be inserted at two primary places: the interface between the gate-level and pulse-level description, which we refer to as the *channel attacks*, and the detailed configuration of the pulse-level controls, which we refer to as the *pulse attacks*. Besides the distinction at the definition level, our attacks can also be performed together with themselves or with other attacks, or be abused as the approach to realize other attacks, which we refer to as the *compound attacks*. The illustration of our new attacks is shown in Fig. 2.

4.1 Channel Attacks

The channel attacks stem from inconsistencies at the interface between the gate-level quantum gate and the pulse-level pulses. We have identified a prevalent flaw in most current quantum software development kits, where the matching between the qubits of the quantum gate and the channels of pulses is not adequately verified and sufficiently enforced. This defect leads to two types of attacks, *qubit plunder* and *qubit block*. In addition, the problem that the quantum gate and the pulses are considerably loosely combined entails the third type of attack, *qubit reorder*:

- **Qubit Plunder**: Qubit plunder arises directly from the misalignment between the qubits of the quantum gate and the channels of the underlying pulses. In most quantum software development kits, custom gates can be applied to any qubit set

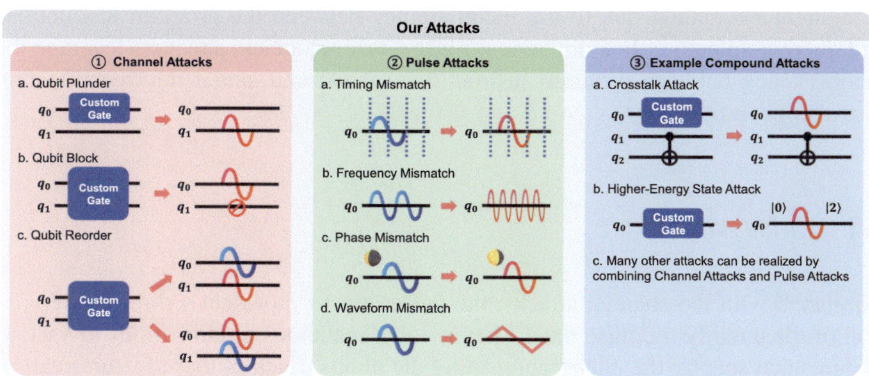

Fig. 2 Illustration of our attacks

without corresponding checks on the channels to which the pulses are applied. For instance, a custom gate may be designated for qubits 0 and 1, while the underlying pulses could be intended for channels corresponding to qubits 2 and 3. Victims of this attack suffer when only gate-level verification is performed, as channels can be arbitrarily selected. Besides, this attack can be performed on any type of channel. With qubit plunder, a single-qubit quantum gate could maliciously affect control or readout channels, leading to unwanted entanglement or measurements. This attack may be detected when the plundered channels are already occupied with pulses, but the pulse overlap error does not clearly point to the root cause of the problem.

- **Qubit Block**: In contrast to qubit plunder, qubit block involves claiming more qubits at the gate level than the number of channels to which pulses are applied. While theoretically inconsequential in perfect quantum computers or with error correction, this attack prolongs execution time. On current NISQ (Noisy Intermediate-Scale Quantum) computers, it can induce qubit decoherence, causing unidirectional decay to $|0\rangle$. For instance, one can specify an empty custom gate with only a delay inside, and attackers can control the decay rate and hence the results by manipulating the delay time of the custom gate.
- **Qubit Reorder**: Apart from misalignments between gate-level qubits and pulse-level channels, the loose definition of custom gates relative to pulses allows the reordering of the encapsulated pulses. Custom gates merely serve as containers to store pulses and lack mechanisms to enforce the arrangement of pulses underneath. For instance, the direction of a custom CNOT gate can be exchanged by exchanging channels (typically also need to add or remove some pulses together). Consequently, channels associated with multi-qubit gates can be arbitrarily permuted.

The inconsistency between gate-level and pulse-level in most SDKs has not been explored before and thus can lead to security attacks. We claim the attacks stem from the inconsistency between the high-level and low-level abstractions, which appear to be problems in many fields. For instance, Spectre [29] and Meltdown [39] in classical computers are mainly due to the inconsistency between the program instructions and microarchitecture. In comparison, for the channel attack, attackers leverage the inconsistency between program instructions (gates) and actual microarchitecture instructions (pulses).

4.2 Pulse Attacks

Compared with the channel attacks, the pulse attacks represent a direct manipulation of the underlying pulse data. They involve malicious modifications to various features that specify the pulses controlling the qubits. Each of these features can be exploited by attackers to execute their attacks. For a comprehensive understanding of the low-level pulse controls on qubits, readers are directed to [31]. The attacks are characterized by the mismatch between the original pulse data and the maliciously altered pulse data:

- **Timing Mismatch**: This type of attack introduces synchronization issues by altering the timing of pulses. One type of timing attack is to change the duration of the pulses and thus change the influence of the pulses, which can cause the operation to be different or also cause the decoherence similar to the qubit block. However, due to the duration of the pulse being part of the waveform of the pulses, we classify this type of attack as waveform mismatch described in the following. The precise timing of pulses and delays is crucial in the pulse definition, complicated further by constraints such as sampling time intervals, acquire alignment, pulse alignment, and granularity of the pulse. For instance, the starting time of a pulse must be the minimum common multiple of acquire alignment and pulse alignment, and the duration of a pulse must be multiples of the pulse granularity. Otherwise, the synchronization problem will cause the qubit to oscillate between $|0\rangle$ and $|1\rangle$. Attackers can exploit these constraints by introducing slight delays to displace pulses from the allowed increments, thereby affecting results. Fortunately, while earlier versions of Qiskit and IBM machines overlooked timing checks, subsequent updates have addressed this vulnerability. We list it here to notify this type of attack when designing the systems.
- **Frequency Mismatch**: Quantum states $|0\rangle$ and $|1\rangle$ are encoded using energy levels. The frequency of a pulse must correspond to the energy difference between these energy levels. The fundamental principle in quantum mechanics dictates that only pulses with frequencies matching the energy difference between specific energy levels of the qubit can effectively control it. Therefore, attackers can manipulate the influence of a pulse by modifying its frequency. For instance, altering the frequency of a pulse to the forbidden range so that the pulse can be disabled, or to match the unwanted energy levels could drive qubits to another state, similar to higher-energy state attack [60], compromising the integrity of the quantum computation.
- **Phase Mismatch**: In pulse-based quantum computing, the phase of a pulse is a crucial parameter that determines the rotation applied to the qubit. By altering the phase, attackers can effectively induce rotations akin to specific quantum gates, such as the rotational-Z gate commonly used for manipulating qubit states. This manipulation exploits the fact that phase changes can rapidly update the frame tracking the qubit state, allowing for instantaneous and nearly error-free rotations. Attackers may strategically introduce phase shifts at specific points in the pulse sequence to disrupt the intended quantum computation, leading to erroneous results. One advantage of phase mismatch is that it can realize any gate operation: any unitary $U \in \mathrm{SU}(2)$ can be decomposed as [44]:

$$U(\theta, \phi, \lambda) = Z_{\phi-\pi/2} X_{\pi/2} Z_{\pi-\theta} X_{\pi/2} Z_{\lambda-\pi/2} \tag{1}$$

where Z_θ is the rotational-Z gate with the rotational angle θ, which is the same as the phase of the pulse on many quantum computers, and $X_{\pi/2}$ is the rotational-X gate with rotational angle $\pi/2$, or SX gate with a global phase.
- **Waveform Mismatch**: The waveform of a pulse, representing the envelope function of its shape, influences the dynamics of qubit manipulation. A detailed understanding of pulse waveforms is essential for controlling qubit rotations accurately. A simple model mentioned in [31] (Eq. 91) gives the Hamiltonian: $\bar{H}_d = -\frac{\Omega}{2} V_0 s(t)(I\sigma_x + Q\sigma_y)$, where the I-component is the "in-phase" pulse that corresponds to rotations around the x-axis, and the Q-component is the "out-of-phase" pulse that corresponds to the rotations about the y-axis. The qubit satisfies the Schrödinnger equation $H\,|\psi(t)\rangle = i\hbar\frac{\partial}{\partial t}\,|\psi(t)\rangle$. Simply speaking, the amplitude of the I and Q components correspond to the rotational angle along the x- and y-axis, respectively. Attackers can exploit this by tampering with the waveform, thereby controlling the gates applied to qubits. For instance, by modifying the amplitude or shape of the waveform, attackers can induce unintended rotations or interfere with the coherence of qubit states.

4.3 Compound Attacks

The channel attacks and pulse attacks can be used as building blocks for compound attacks. In compound attacks, individual channel attacks or pulse attacks are used as a building block to achieve higher-level attack goals. We list several interesting compound attacks that can be implemented by leveraging our attacks:

- *Qubit Flip*: This attack aims to flip the state of a qubit, thereby altering its computational output.
- *Entanglement Injection*: Attackers may seek to inject unwanted entanglement between qubits, disrupting the coherence of the quantum system.
- *Decoherence Amplification*: Decoherence amplification attacks aim to accelerate the decoherence process in qubits, leading to faster loss of quantum information and degradation of computational fidelity.
- *Trojan Attack*: The control qubit can be used as the trojan qubit to trigger the gates.
- *Measurement Manipulation*: manipulate the end-of-circuit or mid-circuit measurement results.
- *Multi-tenant Attack*: entanglement can also be created in the multi-tenant environment. Though multi-tenant quantum computers are not supported yet, this paradigm is actively researched [41, 48, 56, 58], and our attacks can be applied in terms of multi-tenant use cases, especially with the channel attacks.

Moreover, our attacks can be combined with other attack vectors or leveraged as a means to facilitate other attacks. Below is an incomplete list based on related works:

- *Crosstalk Attack*: Injecting pulses with significant crosstalk effects or modifying pulse frequencies to exacerbate crosstalk effect can lead to crosstalk attacks [2, 3, 11, 12, 16, 43], where the parallel execution of some gates leads to larger errors.
- *Higher-Energy State Attack*: Driving qubits to a higher-energy state outside the two-level computational system $|0\rangle$ and $|1\rangle$, which inherently requires pulse operations, enables attackers to execute higher-energy state attacks [60] that can invalidate normal quantum gates.
- *"Horizontal" Information Leakage*: Modifying the measurement pulse can lead to "horizontal" information leakage [6, 46, 55, 63] in quantum circuits that exists through executions and interferes the input states of the next execution.
- *Power Attack*: Pulses are related to power traces of quantum circuits [17, 62], and thus our attacks can influence the equipment power features.
- *Fault-Injection*: Changing the pulses at the target locations can lead to fault-injection [49, 61]. This can also be done by reasoning the relations between the pulses and control electronics [9].

By leveraging the flexibility and intricacies of pulse-based quantum computing, attackers can orchestrate compound attacks that exploit multiple vulnerabilities in quantum systems simultaneously, posing significant challenges to quantum security protocols.

5 Attack Demonstration

In this section, we present attack implementations. Our attacks are general, so there are infinite implementations. We first propose one straightforward example of qubit flip. Then we demonstrate several algorithm-specific examples.

5.1 Toy Attack Example: Qubit Flip Attack

Given the extensive control over quantum circuits afforded by the attacks, one generic approach involves the attacker utilizing an extra layer of custom gates on all qubits either in the end to manipulate final results or at the beginning to control input states. This requires the modification of gate-level circuits, which requires attackers to be more powerful and is beyond the main threat model of this paper. This example serves to show one most straightforward way to perform the attack. Attackers may claim that the custom gate layer is for better fidelity, such as optimized readout pulses for lower measurement error or higher fidelity state preparation. Notice that level 1 victims have the lowest verification ability, and attackers can directly substitute the whole program. Here the goal is to present one simplest use case of the attack, and victims can check anything except the custom gate layer.

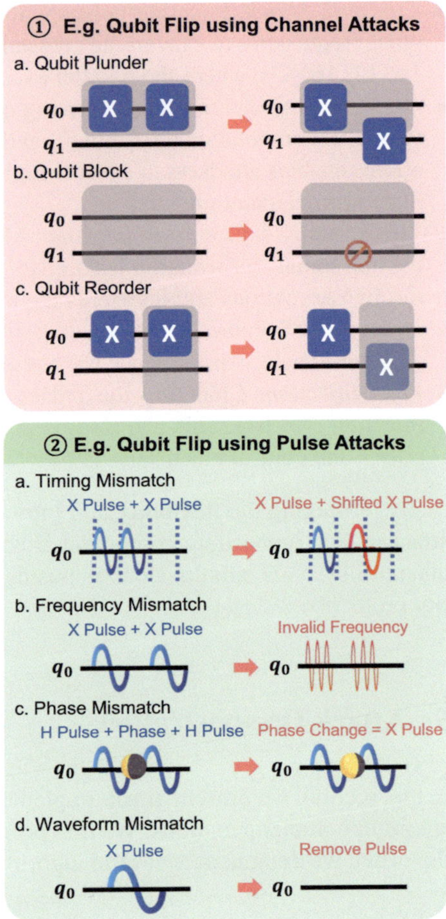

Fig. 3 Implementations of qubit flip with channel attacks and pulse attacks. The gray block in the channel attack section represents the custom gate that is manipulated in the attacks

We present some straightforward implementations of the qubit flip attack. The implementations are shown in Fig. 3, but the attack clearly extends beyond mere qubit flip.

5.1.1 Level 1 Victim: Qubit Flip Attack Using Channel Attacks

The power of channel attacks diminishes in the following order due to their control abilities:

- **Qubit Plunder:** If an additional channel is provided, qubit flip through qubit plunder can be realized by manipulating pulse movement. A direct configuration involves adding a layer of two Pauli-X pulses on each qubit. Two Pauli-X gates are the same as the identity gate, leaving the qubit unchanged. The qubit flip occurs when one of the Pauli-X pulses is moved to an unused channel, altering the probability between $|0\rangle$ and $|1\rangle$ on that qubit.

- **Qubit Reorder:** Similarly, results can be altered by reordering pulses. A straightforward scheme entails adding two additional Pauli-X gates on one qubit, incorporating one of the Pauli-X gates in the custom gate. Thus, keeping the pulse does nothing, while reordering the pulses flips two qubits simultaneously. This is a weaker attack than the qubit plunder in that it can only flip an even number of qubits simultaneously, while qubit plunder can flip any qubit independently under the condition of extra channels.
- **Qubit Block:** Qubit block is the least potent, incapable of arbitrary result modification. If the qubit is perfect with no decoherence or can be error-corrected, it cannot modify the results. One way as described previously is to add a long delay before the measurement, and when accounting for decoherence, qubits will decay to $|0\rangle$.

5.1.2 Level 1 Victim: Qubit Flip Attack Using Pulse Attacks

Because pulse attacks can directly control the pulses, they are more powerful than channel attacks:

- **Timing Mismatch**: In supported platforms, synchronization leads to qubit to oscillate between $|0\rangle$ and $|1\rangle$. Therefore, the results can be easily controlled by manipulating the number of units for the delay, though the ability is limited by the granularity of the machine. How the number of units for the delay corresponds to the state change can be easily measured with simple experiments.
- **Frequency Mismatch**: Only frequencies correlated to energy between two energy levels can lead to energy level transitions, which means the operation of one gate can be easily disabled by tuning the frequency into the forbidden range. The scheme can be the same as qubit plunder by replacing the moving of pulse to changing of frequency, i.e., adding two Pauli-X pulses and sandwiching one frequency setting between them, and setting the frequency to the forbidden range when a qubit flip is desired.
- **Phase Mismatch**: Qubit flip through phase change can be implemented with the identity $X = H \cdot RZ(\pi) \cdot H$. The custom gate is composed of three gates, the Hadamard gate, one Rotational-Z gate, and another Hadamard gate. When setting the phase, i.e., setting the rotational angle of the Rotational-Z gate, to be 0, the gate is identical to the identity gate. However, when setting the phase to be π, the gate is identical to the Pauli-X gate.
- **Waveform Mismatch**: Waveform attack is the most flexible attack. The gate can be disabled by setting the amplitude to 0, and the rotational angle can be tuned by easily setting the waveform. The qubit flip can be easily realized by adding one empty placeholder pulse and changing the waveform of the placeholder pulse to the Pauli-X pulse when the qubit flip is wanted.

5.2 Algorithm-Specific Attack Examples

While the aforementioned universal implementations are simple, straightforward, and applicable to all kinds of quantum circuits, they are easily detectable through in-depth circuit examination, even by humans. For example, custom gates with only two Pauli-X pulses on each qubit before measurement usually make no sense within the quantum circuit. Though shifting the custom gates to the middle may be feasible and make detection less obvious, it typically remains implausible for arbitrarily targeted attacks. Given that the configuration of custom gates hinges on the specific quantum algorithm, such as decisions on pulse utilization and placement, a universal scheme for installing attacks to control results may not exist. Nonetheless, mapping between pulses and results may be acquired through methods such as brute force or trial and error.

A more clandestine approach involves implementing the quantum algorithm as a pulse-optimized or parametric quantum circuit, as discussed in Sect. 1. For instance, pulses employed for quantum circuit decomposition, compilation, and optimization exhibit complexity and modifiability. These pulses are typically optimized for specific quantum gates, such as Toffoli gates; or tailored for particular objectives, such as the multiply-controlled gate in Grover's search. Although these pulses serve fixed goals, they necessitate periodic updates due to changes in the physical properties of quantum computers, as explained in Sect. 3.2.1. Another instance is pulse learning for variational quantum algorithms like quantum machine learning, where the parametric nature involves learning features of pulses during the process. These parameters, akin to weights in neural networks, are challenging to interpret individually, and may be totally different when training again depending on the initial values, making them susceptible to our attacks.

In this section, we provide some examples to demonstrate our attacks with algorithm-specific implementations. Because the stealthiness of the attack is highly dependent on the capability of the victims as discussed in Sect. 3.3, we explore the attacks on level 2–4 victims: channel attack on quantum teleportation for level 2 victim; pulse attack on Grover's search for level 3 victim; pulse attack on quantum neural networks for level 4 victim. The experiments for level 2 and 3 victims were done on `ibm_osaka` on IBM Quantum, which is a 127-qubit quantum computer, while the experiments for level 3 victims were done on simulators.

5.2.1 Level 2 Victim: Channel Attack on Quantum Teleportation

This demonstrates the channel attack to maliciously couple and decouple qubits.

One of the most important features that differentiate quantum computing from classical computing is quantum entanglement. Qubits can be coupled together and the measurement result will depend on how they are connected, rather than their quantum states independently. Quantum entanglement is the feature that is used in almost every quantum algorithm. One example of using quantum entanglement is quantum teleportation [7], which is a technique for transferring quantum information from a sender at one location to a receiver some distance away. The quantum circuit for the demonstration is shown in Fig. 4a. Alice prepares her quantum state on q_2, and this quantum state can be transmitted to Bob through the circuit when the coupling gate in the middle is the CNOT gate. However, assuming that the coupling gate is provided as a service in the form of a custom gate with pulses that are provided by Eve, then Eve can perform many types of attacks to influence the results. For instance, we show that Eve can secretly couple or decouple qubits.

In our demonstration, the gate-level circuit is the same for all cases, so the channel attack is hidden from the gate-level verification. Specifically, Eve can perform *Coupling* q_2 and q_1 by applying the CNOT pulse even though the coupling gate is applied on q_2 and q_3, *Decoupling* by simply doing nothing, or *Del H* by applying the pulse of Hadamard gate to cancel Hadamard gate after the coupling gate.

In the experiment, we use a rotational-X gate with angle θ and Pauli-Z gate to prepare the state for transmission: $|\psi\rangle = \cos\frac{\theta}{2}|0\rangle + \sin\frac{\theta}{2}|1\rangle$. The probabilities of measuring $|1\rangle$ with the above pulse configurations are shown in Fig. 4b. "Theory" shows the theoretical results with no errors, and "Benchmark" shows the results with the original quantum teleportation circuit.

"Coupling: Eve" shows Eve's measurement result when Eve performed the "Coupling" gate, and it shows that Eve can secretly copy the quantum teleportation circuit to steal the quantum state in transmission, while "Coupling: Bob" shows that Bob cannot receive any state because the coupling of q_2 and q_3 is secretly occupied by Eve. Similarly, "Decoupling" does not couple q_2 and q_3, so Bob cannot receive the quantum state. Although "Del H" shows the same probability as the benchmark, the correct state should be a pure state $|\psi\rangle = \alpha|0\rangle + \beta|1\rangle$, but Bob actually receives a mixed state of $|0\rangle$ with probability $|\alpha|^2$ and $|1\rangle$ with probability $|\beta|^2$. This attack can be successful even though we assume Bob can do simple checks on the measurement results which behave the same as the correct results.

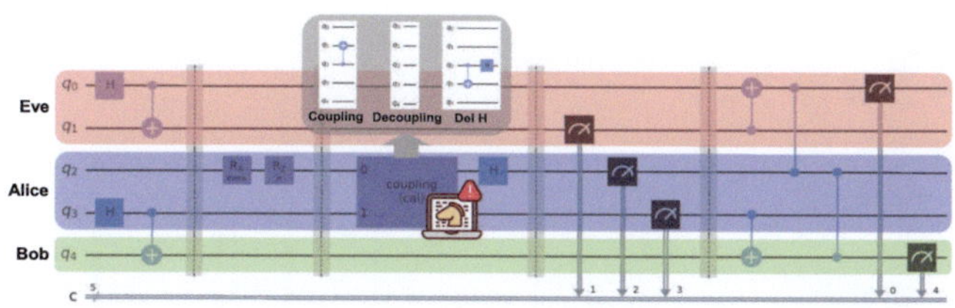

(a) Quantum circuit of quantum teleportation. Alice (blue block) wants to transport her quantum state to Bob (green block), with the coupling gate provided by Eve (red block). Eve can tamper the pulses to achieve many goals, some of which are listed in the gray block.

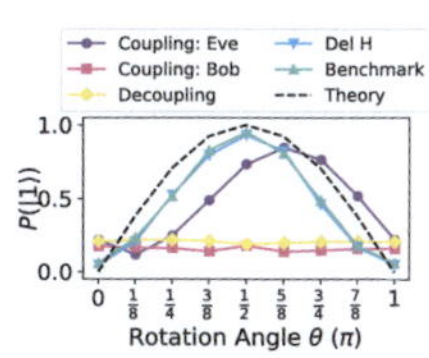

(b) Bob's $P(|1\rangle)$ ("Coupling: Eve" is Eve's) with different pulse configurations of the coupling gate.

Fig. 4 Example demonstration of the algorithm-specific attack on quantum teleportation

5.2.2 Level 3 Victim: Pulse Attack on Grover's Search

This demonstrates the pulse attack to manipulate results.

In the experiment, we implement 2-qubit Grover's search [20] as the parametric quantum circuits. The whole Grover operator is provided as one custom gate. Grover's search is a quantum algorithm for unstructured search that finds with high probability the unique input to a black box function that produces a particular output value using just $O(\sqrt{N})$ evaluations of the function. The parameters we used are the quantum state to search in the computational basis, i.e., $|00\rangle$, $|01\rangle$, $|10\rangle$, $|11\rangle$, and we assume that attackers provide this parametric Grover operator.

For different search states, the Grover operator is different, as shown in Fig. 5a. However, only several gates are different, and all of them are rotation-Z gates. The percentages of different gates for search states in the computational basis are shown in Fig. 5b. It shows that only a small portion of the quantum circuit is different, and the percentage is even smaller when counting pulses rather than gates. Therefore, attackers can only modify a small portion of the Grover operator by phase mismatch to change the search states. The results are shown in Fig. 5c. Inside the whole custom gate for the Grover operator, only the phases of four gates (eight underlying pulses) are modified, and the output is successfully redirected to the desired attack states.

5.2.3 Level 4 Victim: Pulse Attack on Quantum Neural Networks

Quantum algorithms are also often implemented with parameters to control the quantum circuits, such as the rotational angles for quantum gates. The parametric quantum circuits are widely used in quantum algorithms that require training or variational operations like quantum machine learning and quantum approximate optimization algorithm [34, 35, 37, 38]. The parameters can also be parameters for pulses, as

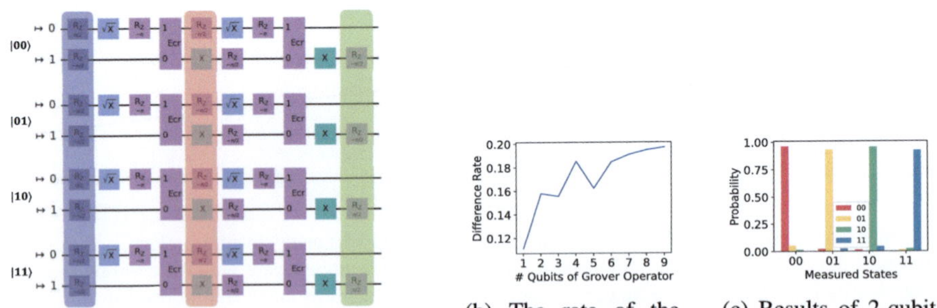

(a) Part of the quantum circuit of 2-qubit Grover's operator. Translucent blocks show the difference for search states from $|00\rangle$ to $|11\rangle$.

(b) The rate of the number of different gates to the number of total gates in Grover operator

(c) Results of 2-qubit Grover's search with targeted states modified by phase mismatch.

Fig. 5 Demonstration on 2-qubit Grover's search

they can be directly learned and the quantum circuits can be directly optimized at the pulse level. Similar to classical neural networks, quantum neural networks include many trainable weights, and the final trained weights are not necessarily explainable. In the following, we implement one quantum neural network for image classification and evaluate the pulse attack on it.

We implement the quanvolutional neural networks proposed in [22], which is similar to the convolutional neural networks (CNN) in classical computing [33]. More specifically, we implement the example quanvolutional neural network provided by TorchQuantum [59]. The architecture is shown in Fig. 6a. We use MNIST [10] as the dataset, which is a database of handwritten digits from 0 to 9. Each image has 28×28 pixels. The quanvolutional layer is a 4-qubit circuit that includes three parts: (1) Encoder: one rotational Y gate on each qubit where each rotation angle is one image pixel value. This is used to encode the classical data into quantum circuits. (2) Kernel: one U3 gate on each qubit and one control U3 gate on each adjacent qubit pair. Each U3 gate includes the three Euler angles to rotate the qubit, and the control U3 gate is the controlled version of the U3 gate. All three angles in each U3 and control U3 gate are trainable parameters to be learned during the training process. The kernel is applied 5 times in a row. (3) Measurement: the quantum circuit results are measured, and the expected values on each qubit will be the output of the quanvolutional layer. Other parts are the same as classical neural networks, including the average pooling layer before the quanvolutional layer and the flatten

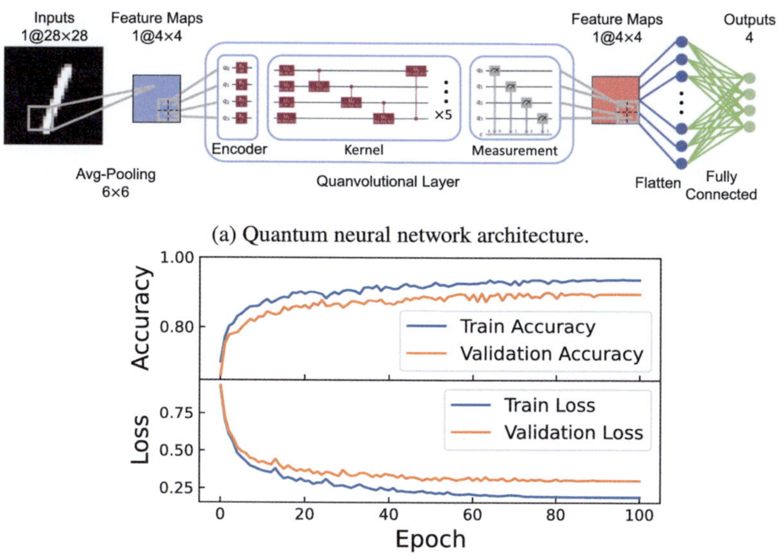

(a) Quantum neural network architecture.

(b) Accuracy and loss in training.

Fig. 6 The quanvolutional neural network (QNN) architecture and training data in the experiment. The quanvolutional layer is computed on quantum computers, while others are the same and computed in classical computers

and fully connected layers after it. Notice that for the training efficiency of QNN, many operations are done to decrease the computational complexity, including the large 6×6 average pooling layer, images with only four labels from 0 to 3, and in total 1000 images are used for training and validation. The experiments are done in the simulators. The training process is shown in Fig. 6b. With 100 epochs, the model reaches an accuracy of 93.7% on the training dataset and 89.6% on the validation dataset.

First, the influence of weight changes on the quanvolutional layer on the accuracy, and the corresponding norm differences, are experimented with. The norm difference is computed as: view three parameters in a U3 gate as a three-dimensional vector, compute the norm of the change divided by the norm of the original vector. We evaluate four ways to change weights: (1) All Weights: all weights are added by the ratio multiplied by their gradients; (2) All Weights with Cutoff: if one U3 gate norm difference is larger than a threshold, then its weights will not be changed. In the experiment, the threshold is set to be 0.05. (3) One Gate: only the first U3 gate is changed. (4) One Weight: only the first rotation angle of the first layer is changed, and the ratio in the figure is the absolute value of the change. The gradients are computed by forwarding only the first epoch of the training dataset.

The results are shown in Fig. 7, which demonstrates that with a small change, the model results can be influenced much. With a maximum of around 10% difference and a mean of 3% difference on all parameters, the accuracy drops from 89.6 to

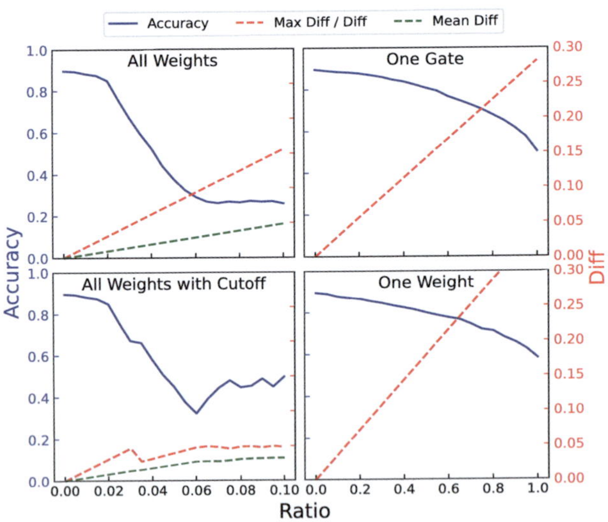

Fig. 7 The accuracy drops with weight changes in the quanvolutional layer. Diff is the norm difference. The weights are added by the ratio multiplied by their gradients, except for "One Weight" where the weight is directed added by ratio. "All Weights" changes weights in all gates. "All Weights with Cutoff" changes weights in all gates when the norm difference is smaller than 0.05. "One Gate" changes only the weights in the first U3 gate. "One Weight" changes the first weight in the first U3 gate

26.8%. Moreover, with the 5% cutoff threshold to limit the maximum difference, only a mean of 2.7% change, the lowest point in the figure shows that the accuracy can be decreased to 32.3%. If changing most weights of the layers is difficult, the accuracy decrease to 60% can also be realized with around 25% change for the first rotational gate or 35% change for only one weight in the first layer. Note that we did not do the experiments to find the best weight to change, but arbitrarily chose the first layer. With careful explorations, a smaller change in the weight may be found and thus the attack can be more covert.

More secret attacks can be implemented. Rather than directly changing the weight and destroying the model, some attacks such as the backdoor attack [21, 42] can insert trojan into the model, leading to malicious results only when trojan is triggered. We implement a simple backdoor attack on the previously trained model. We mixed 20% backdoor samples into the training dataset and fine-tuned the model. The trojan pattern is a 3×3 white pattern on the upper left of the figure. The trojan can be other patterns. This pattern is arbitrarily chosen, and it can be smaller if the average pooling layer is smaller or removed. Images with the pattern are arbitrarily assigned to target 0.

Figure 8 shows the fine-tuning results. Without fine-tuning, the model has an accuracy of 21.6% on the backdoor validation data. When training around 40 epochs, the model is well-backdoored. The accuracy of the clean training data and validation data is 90.9% and 87.1%, with 2.8% and a 2.5% drop from the original model, while the accuracy of the backdoor training data and validation data is 87.0% and 89.0%, achieving similar accuracy as the clean data. The changes to realize the backdoor

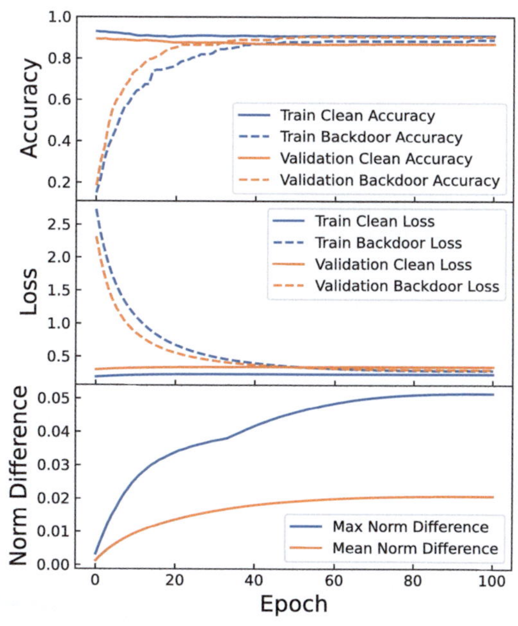

Fig. 8 The training accuracy, loss, and norm difference in the fine-tuning process of the backdoor attack

attack are small, with a maximum difference of 4.1% and a mean difference of 1.8%. The figure shows that further training will not change the weights and accuracy much, i.e., with around 2% change, the model can be inserted with a trojan.

6 Defense Framework

In this section, we propose one framework to defend from the attacks. The verification of quantum programs can be inspired by the verification of classical programs, though with some peculiarities. Many traditional methods can be used and may be helpful to some degree. For example, the quantum program can be checked and distributed with associated hash codes such as SHA256. This is a common way of distributing classical software. This could be helpful when the properties of quantum computers do not have a big change, similar to classical computing. However, as we explained in Sect. 3.2.1, the volatility of quantum computers may make this method fruitless because the physical properties of the quantum hardware may undergo big changes, and thus the same pulse configuration may not have the same functionality. On the other hand, instead of resorting to the provided pulse data, users may calibrate by themselves to get the up-to-date pulse data. Nevertheless, the cost issue described in Sect. 3.2.1 may discourage users from doing this, and users may also need to acquire the programs for calibration from attackers and these programs may also be tampered with by attackers.

In fact, not all of the proposed attacks are due to the same reasons. The cause for the attacks can be divided into:

- **Software flaw**: This is the cause of the qubit plunder and the qubit block. These two attacks arise from the fact that current quantum software development kits overlook the correspondence between gate-level qubits and pulse-level channels. As a result, they can be detected and fixed at the software level.
- **Interface defect**: This is the cause of the qubit reorder. The current software design for adding pulse-level controls is deficient as we explained before. The custom gate is just a container used to store the pulse-level controls. Even though we can enforce the correspondence of gate-level qubits and pulse-level channels, we still cannot guarantee the functionality of pulses inside. This attack must be detected if the function of the custom gate is specified beforehand. The loose interface design in most current SDKs may be due to the consideration of convenience in development. The custom gate with the abstract functionality can be applied to any qubits, while the underlying pulse configuration is qubit-dependent. In some sense, they are contradictory and cannot be easily automatically modified, otherwise, the pulse configuration can also be easily updated since the automatic modification requires the pulse syntax verification. The straightforward way to implement the custom gate with the same functionality on all qubits will either require the current design where the custom gate is flexible and pulses are selected for the specific

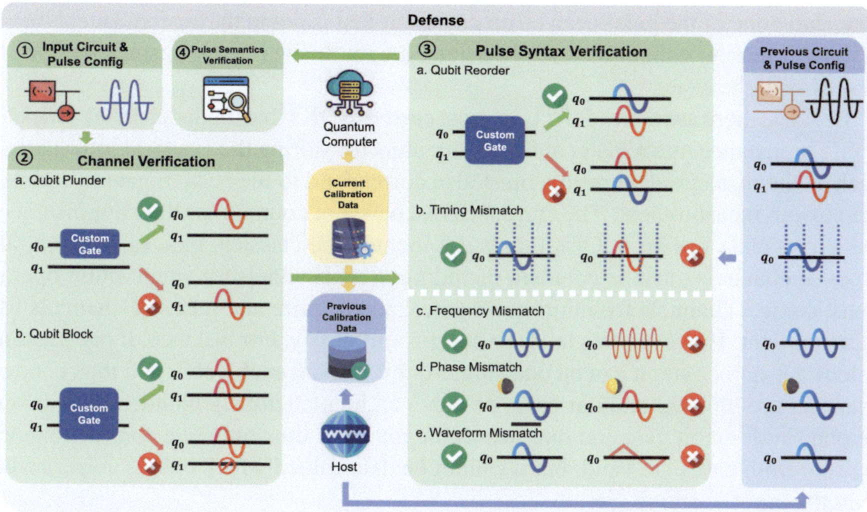

Fig. 9 Illustration of the defense. With the quantum circuit (and pulse configuration included) as the input, the channel verification can detect the qubit plunder and the qubit block attacks by checking the exact match of the gate-level qubits and the pulse-level channels. For the qubit reorder and the timing mismatch attack, we propose that the circuits must be associated with the previous pulse configuration for instructions comparison. For the remaining attacks, the previous and current calibration data are also needed for parameter comparison

qubits, or the custom gate is inflexible and qubit-dependent. Both of them entail inflexibility and complexity in software development.

- **Challenges for the pulse-level verification**: This is the cause of other attacks. Again, the pulse configuration is only a temporary representation of the functionality of the custom gate. The challenges are discussed in Sect. 3.2.

In the remainder of this section, we provide one straightforward framework for detecting the attacks through *channel verification*, *pulse syntax verification*, and *pulse semantics verification*. Figure 9 shows the workflow.

6.1 Channel Verification

The qubit plunder and the qubit block can be directly detected with the software given a quantum circuit. For each custom gate in the quantum circuit, the qubits to which the custom gate is applied are provided, and the pulse configuration for the custom gate is also included in the quantum circuit. Therefore, with the mapping between gate-level qubits and pulse-level channels, the qubits and the channels must be the exact match, i.e., there must be at least one pulse on each qubit of the custom gate, and none of the pulses can be on any qubit that is not in the custom gate. Notice that qubit reorder cannot be directly detected with this verification, since the match may already be exact.

Besides, there are many defects in most current SDKs that hinder channel verification. For instance, the register and memory slots for storing the classical information such as the measurement results must also correspond to the custom gate to address the custom measurement. The current SDKs have not counted for this. For instance, the `Gate` class design in Qiskit does not include the classical bits. In addition, all types of channels must be counted, including drive channels for single-qubit operations, control channels for multi-qubit gates, and acquire and measure channels for measurement. However, this leads to design complexity. For instance, if one custom gate is a 3-qubit gate, it can be one single-qubit gate on each qubit, and thus control channels are blocked and not used, or they can be maliciously plundered; or it can be one single-qubit gate and one two-qubit gate, and thus only one control channel is used. Both cases are valid, but it cannot be determined which case is correct with only the input quantum circuit.

6.2 Pulse Syntax Verification

The other attacks cannot be directly detected with only the quantum circuit as the input. As explained, in order to detect these attacks, the syntax and semantics of each pulse in the custom gate must be specified. This is reasonable for custom gates with clear goals, such as the pulse-optimized Toffoli gate. Nevertheless, it is not possible for many other gates, such as the end-to-end quantum neural network.

The pulse syntax verification is to verify that each pulse in the input pulse configuration functions correctly as desired. The pulse syntax verification may be the most difficult part of the verification for quantum programs. It is a complicated problem that if two calibration data are given, how to verify if these two pulse configurations have similar functionality. The related problem is quantum tomography [57]: given a black box (the custom gate), how to figure out its functionality? One reason is that there are too many intertwined parameters to specify one pulse. For example, if the frequency of the pulse is the calibrated qubit frequency, then the new frequency may also need to be the new calibrated qubit frequency. However, maybe the original data contains errors, and thus all parameters were measured with bias. In this case, the verification will become very difficult. Similarly, different pulse waveforms may result in the same effect, this is again hard to deal with.

We propose to associate the quantum circuit with both the pulse configuration and the calibration data at the time the pulse configuration is calibrated, or identically the date on which the pulse configuration is calibrated. This associated data should be stored at some trusted authorities. Before users execute the quantum circuit or before the circuit publisher updates the pulse configuration, the input pulse configuration will be compared with the trusted pulse configuration, together with the analysis of the current calibration data from the quantum computer vendors or user custom experiments with the trusted calibration data.

For the pulse configurations that only include the native pulses provided by the quantum computer vendors, the verification can be the exact checking. However, in most cases, the parameters of pulses are specified. Since the pulse data is analog and the calibration process is noisy and erroneous, it is difficult to provide exactly the same functionality check. The tolerance may need to be provided for the verification.

6.3 Pulse Semantics Verification

Once the pulse syntax is verified, the semantics verification may be straightforward. Gate-level circuits can be described as the directed acyclic graph (DAG) [26], and pulse-level circuits may also be described using the same language. Qubits in the gate-level DAGs correspond to channels in the pulse-level DAGs, and the verification problem can be reduced to the graph problems.

Besides, once the syntax of each pulse is determined, the analog operations can be described similarly to the "instructions" in classical computing, and thus many verification methods for classical programs may be easily applied to quantum programs.

6.4 Evaluation

The defense framework was tested against all the attacks described in Sect. 5, and successfully detected all of them. However, the current implementation of the defense operates using an exact matching approach, allowing for variations in pulse amplitude within a specified tolerance range, as mentioned earlier. This design results in low precision but high recall for detecting malicious modifications: it effectively detects circuits where pulses or even gates have been tampered with, but it struggles to handle circuits that vary but generate identical results as the original circuits. For instance, a pulse intended for an X gate could be split into two pulses for an SX gate, and the current implementation would incorrectly flag this as a malicious circuit.

The trade-off between precision and recall can be adjusted a bit by modifying the tolerance range—higher tolerance increases precision while lowering recall. Usually, as circuits only need to be calibrated by simply adjusting the amplitude and the frequency to follow the current conditions of the environment, this exact matching approach proves useful. Nevertheless, a more advanced implementation that can accommodate variations in other pulse specifications is worth exploring and remains an area for future research.

7 Discussion and Future Work

One impactful future work is to improve the verification. While pulse semantics verification may resemble the verification for classical programs, pulse syntax verification raises a big challenge. One way to define the syntax may be to return to the matrix representation since all quantum gates should be unitary. Then the pulse check can be computing the similarity between two unitary matrices. However, again, this problem is similar to tomography, where we must find a way to efficiently compute the unitary matrix given the pulses.

Another important topic is how errors are correlated with tolerance in the defense. This can depend on the size of the custom gate, the type of the parameter, etc. The current defense requires input tolerance, where the relation between it and the results is vague. If users can give the largest error threshold and then the tolerance can be computed, the defense will be more useful and powerful.

In addition, if the pulse parameters can be automatically updated given the calibration data from the cloud provider, then the attacks can be mitigated to a great extent. Superficially, if giving the pulses and checking their functionality is like the P problem, then giving the functionality and providing the pulses is like the NP problem. However, if we fix all pulses and only change some of the parameters, this problem may be greatly simplified.

Lastly, if we think from the other direction from the qubit properties to the pulse data, because pulse data needs to be updated when qubit properties change, maliciously changing the qubit properties can be a new type of attack. This can be one type of supply chain attack that happens in the manufacturing process, or by directly altering the environment in the data center of the cloud platforms.

8 Conclusion

This work presented the first thorough exploration of the attacks on the interface between the gate-level and pulse-level quantum circuits and pulse-level quantum circuits. The attacks presented in this work leverage the mismatch between the gate-level definition and description of the custom gate, and the actual, low-level pulse implementation of this gate. By manipulating the pulse definition numerous attacks are possible as this work proposed: qubit plunder, qubit block, qubit reorder, timing mismatch, frequency mismatch, phase mismatch, and waveform mismatch. The attacks are in part possible because there is a lack of sufficient verification in the current quantum software development kits. This work thus also proposed a defense framework to protect from these attacks. In summary, this work provides insight into the future development of secure quantum software development kits and quantum computer systems.

Acknowledgements This work was supported in part by NSF grant 2332406.

References

1. Amazon Web Services. Amazon Braket. 2024.
2. Ash-Saki, Abdullah, Mahabubul Alam, and Swaroop Ghosh. 2020. Analysis of crosstalk in nisq devices and security implications in multi-programming regime. In *Proceedings of the ACM/IEEE international symposium on low power electronics and design*, ISLPED '20, 25–30, New York, NY, USA. Association for Computing Machinery.
3. Ash-Saki, Abdullah, Mahabubul Alam, and Swaroop Ghosh. 2020. Experimental characterization, modeling, and analysis of crosstalk in a quantum computer. *IEEE Transactions on Quantum Engineering* 1: 1–6.
4. Baker, Jonathan M., Casey Duckering, and Frederic T. Chong. 2020. Efficient quantum circuit decompositions via intermediate qudits. In *2020 IEEE 50th international symposium on multiple-valued logic (ISMVL)*, 303–308.
5. Jonathan M. Baker, Casey Duckering, Pranav Gokhale, Natalie C. Brown, Kenneth R. Brown, and Frederic T. Chong. 2020. Improved quantum circuits via intermediate qutrits. *ACM Transactions on Quantum Computing* 1(1).
6. Bell, Brennan, and Andreas Trügler. 2022. Reconstructing quantum circuits through side-channel information on cloud-based superconducting quantum computers. In *IEEE international conference on quantum computing and engineering (QCE)*, 259–264.
7. Bennett, Charles H., Gilles Brassard, Claude Crépeau, Richard Jozsa, Asher Peres, and William K. Wootters. 1993. Teleporting an unknown quantum state via dual classical and Einstein-Podolsky-Rosen channels. *Physics Review Letters* 70: 1895–1899.
8. Biamonte, Jacob, Peter Wittek, Nicola Pancotti, Patrick Rebentrost, Nathan Wiebe, and Seth Lloyd. 2017. Quantum machine learning. *Nature* 549 (7671): 195–202.
9. Das, Subrata, Avimita Chatterjee, and Swaroop Ghosh. 2024. Investigating impact of bit-flip errors in control electronics on quantum computation.
10. Deng, Li. 2012. The mnist database of handwritten digit images for machine learning research [best of the web]. *IEEE Signal Processing Magazine* 29 (6): 141–142.
11. Deshpande, Sanjay, Chuanqi Xu, Theodoros Trochatos, Yongshan Ding, and Jakub Szefer. 2022. Towards an antivirus for quantum computers. In *IEEE international symposium on hardware oriented security and trust (HOST)*, 37–40.
12. Deshpande, Sanjay, Chuanqi Xu, Theodoros Trochatos, Hanrui Wang, Ferhat Erata, Song Han, Yongshan Ding, and Jakub Szefer. 2023. Design of quantum computer antivirus. In *IEEE international symposium on hardware oriented security and trust (HOST)*, 260–270.
13. Deutsch, David Elieser, Adriano Barenco, and Artur Ekert. 1995. Universality in quantum computation. *Proceedings of the Royal Society of London. Series A: Mathematical and Physical Sciences* 449 (1937): 669–677.
14. Cirq Developers. Cirq. 2022. See full list of authors on Github: https://github.com/quantumlib/Cirq/graphs/contributors
15. Devitt, Simon J., William J. Munro, and Kae Nemoto. 2013. Quantum error correction for beginners. *Reports on Progress in Physics* 76(7): 076001.
16. Ding, Yongshan, Pranav Gokhale, Sophia Fuhui Lin, Richard Rines, Thomas Propson, and Frederic T. Chong. 2020. Systematic crosstalk mitigation for superconducting qubits via frequency-aware compilation. In *53rd Annual IEEE/ACM international symposium on microarchitecture (MICRO)*, 201–214.
17. Erata, Ferhat, Chuanqi Xu, Ruzica Piskac, and Jakub Szefer. 2024. Quantum circuit reconstruction from power side-channel attacks on quantum computer controllers.
18. Farhi, Edward, Jeffrey Goldstone, and Sam Gutmann. 2014. A quantum approximate optimization algorithm. arXiv preprint arXiv:1411.4028.
19. Gokhale, Pranav, Jonathan M. Baker, Casey Duckering, Natalie C. Brown, Kenneth R. Brown, and Frederic T. Chong. 2019. Asymptotic improvements to quantum circuits via qutrits. In *Proceedings of the 46th international symposium on computer architecture*, ISCA '19, 554–566, New York, NY, USA. Association for Computing Machinery.

20. Grover, Lov K. 1996. A fast quantum mechanical algorithm for database search. In *Proceedings of the twenty-eighth annual ACM symposium on theory of computing*, STOC '96, 212–219, New York, NY, USA. Association for Computing Machinery.
21. Gu, Tianyu, Brendan Dolan-Gavitt, and Siddharth Garg. 2019. Badnets: Identifying vulnerabilities in the machine learning model supply chain.
22. Henderson, Maxwell, Samriddhi Shakya, Shashindra Pradhan, and Tristan Cook. 2019. Quanvolutional neural networks: Powering image recognition with quantum circuits.
23. Herman, Dylan, Cody Googin, Xiaoyuan Liu, Yue Sun, Alexey Galda, Ilya Safro, Marco Pistoia, and Yuri Alexeev. 2023. Quantum computing for finance. *Nature Reviews Physics* 5 (8): 450–465.
24. IBM Quantum. 2023. The hardware and software for the era of quantum utility is here. https://research.ibm.com/blog/quantum-roadmap-2033.
25. IBM Quantum. 2024. https://www.ibm.com/quantum.
26. Iten, Raban, Romain Moyard, Tony Metger, David Sutter, and Stefan Woerner. 2022. Exact and practical pattern matching for quantum circuit optimization. *ACM Transactions on Quantum Computing* 3 (1): 1–41.
27. Jaskula, Jean-Christophe, Kshitij Chhabra, Peter Karalekas, and Stefan Natu. 2022. Amazon braket launches braket pulse to develop quantum programs at the pulse level.
28. Jones, Jonathan A., Michele Mosca, and Rasmus H. Hansen. 1998. Implementation of a quantum search algorithm on a quantum computer. *Nature* 393 (6683): 344–346.
29. Kocher, Paul, Jann Horn, Anders Fogh, Daniel Genkin, Daniel Gruss, Werner Haas, Mike Hamburg, Moritz Lipp, Stefan Mangard, Thomas Prescher, Michael Schwarz, and Yuval Yarom. 2019. Spectre attacks: Exploiting speculative execution. In *IEEE symposium on security and privacy (SP)*, 1–19.
30. Kottmann, Korbinian. 2023. Differentiable pulse programming with qubits in PennyLane.
31. Krantz, P., M. Kjaergaard, F. Yan, T. P. Orlando, S. Gustavsson, and W. D. Oliver. 2019. A quantum engineer's guide to superconducting qubits. *Applied Physics Reviews* 6 (2): 021318.
32. Lanyon, Benjamin P., James D. Whitfield, Geoff G. Gillett, Michael E. Goggin, Marcelo P. Almeida, Ivan Kassal, Jacob D. Biamonte, Masoud Mohseni, Ben J. Powell, Marco Barbieri, et al. 2010. Towards quantum chemistry on a quantum computer. *Nature Chemistry* 2 (2): 106–111.
33. LeCun, Y., B. Boser, J. S. Denker, D. Henderson, R. E. Howard, W. Hubbard, and L. D. Jackel. 1989. Backpropagation applied to handwritten zip code recognition. *Neural Computation* 1 (4): 541–551.
34. Liang, Zhiding, Jinglei Cheng, Hang Ren, Hanrui Wang, Fei Hua, Zhixin Song, Yongshan Ding, Fred Chong, Song Han, Xuehai Qian, and Yiyu Shi. 2024. Napa: Intermediate-level variational native-pulse ansatz for variational quantum algorithms.
35. Liang, Zhiding, Jinglei Cheng, Zhixin Song, Hang Ren, Rui Yang, Hanrui Wang, Kecheng Liu, Peter Kogge, Tongyang Li, Yongshan Ding, and Yiyu Shi. 2023. Towards advantages of parameterized quantum pulses.
36. Liang, Zhiding, Zhixin Song, Jinglei Cheng, Zichang He, Ji Liu, Hanrui Wang, Ruiyang Qin, Yiru Wang, Song Han, Xuehai Qian, and Yiyu Shi. 2023. Hybrid gate-pulse model for variational quantum algorithms. In *60th ACM/IEEE design automation conference (DAC)*, 1–6.
37. Liang, Zhiding, Zhixin Song, Jinglei Cheng, Hang Ren, Tianyi Hao, Rui Yang, Yiyu Shi, and Tongyang Li. 2023. Spacepulse: Combining parameterized pulses and contextual subspace for more practical VQE.
38. Liang, Zhiding, Hanrui Wang, Jinglei Cheng, Yongshan Ding, Hang Ren, Zhengqi Gao, Zhirui Hu, Duane S. Boning, Xuehai Qian, Song Han, Weiwen Jiang, and Yiyu Shi. 2022. Variational quantum pulse learning. In *IEEE international conference on quantum computing and engineering (QCE)*, 556–565.
39. Lipp, Moritz, Michael Schwarz, Daniel Gruss, Thomas Prescher, Werner Haas, Anders Fogh, Jann Horn, Stefan Mangard, Paul Kocher, Daniel Genkin, Yuval Yarom, and Mike Hamburg. 2018. Meltdown: Reading kernel memory from user space. In *Proceedings of the 27th USENIX conference on security symposium*, SEC'18, 973–990, USA. USENIX Association.

40. Litteken, Andrew, Jonathan M. Baker, and Frederic T. Chong. 2022. Communication trade offs in intermediate qudit circuits. In *IEEE 52nd international symposium on multiple-valued logic (ISMVL)*, 43–49.
41. Liu, Lei, and Xinglei Dou. 2021. Qucloud: A new qubit mapping mechanism for multi-programming quantum computing in cloud environment. In *IEEE international symposium on high-performance computer architecture (HPCA)*, 167–178.
42. Liu, Yunfei, Xingjun Ma, James Bailey, and Feng Lu. 2020. Reflection backdoor: A natural backdoor attack on deep neural networks. In *Computer vision–ECCV 2020: 16th European conference*, Glasgow, UK, Aug 23–28, Proceedings, Part X 16, 182–199. Springer.
43. Maurya, Satvik, Chaithanya Naik Mude, Benjamin Lienhard, and Swamit Tannu. 2024. Understanding side-channel vulnerabilities in superconducting qubit readout architectures.
44. McKay, David C., Christopher J. Wood, Sarah Sheldon, Jerry M. Chow, and Jay M. Gambetta. 2017. Efficient z gates for quantum computing. *Physics Review A* 96: 022330.
45. Mermin, N. David. 2007. *Quantum computer science: An introduction*. Cambridge University Press.
46. Mi, Allen, Shuwen Deng, and Jakub Szefer. 2022. Securing reset operations in nisq quantum computers. In *Proceedings of the 2022 ACM SIGSAC conference on computer and communications security*, CCS '22, 2279–2293, New York, NY, USA. Association for Computing Machinery.
47. Microsoft Azure. 2024. Azure quantum.
48. Niu, Siyuan, and Aida Todri-Sanial. 2023. Enabling multi-programming mechanism for quantum computing in the NISQ era. *Quantum* 7: 925.
49. Oliveira, Daniel, Edoardo Giusto, Emanuele Dri, Nadir Casciola, Betis Baheri, Qiang Guan, Bartolomeo Montrucchio, and Paolo Rech. 2022. Qufi: A quantum fault injector to measure the reliability of qubits and quantum circuits. In *2022 52nd Annual IEEE/IFIP international conference on dependable systems and networks (DSN)*, 137–149. IEEE.
50. Orús, Román, Samuel Mugel, and Enrique Lizaso. 2019. Quantum computing for finance: Overview and prospects. *Reviews in Physics* 4: 100028.
51. Preskill, John. 2018. Quantum computing in the NISQ era and beyond. *Quantum* 2: 79.
52. Qiskit. Qiskit Pulse. 2024.
53. Qiskit contributors. 2023. Qiskit: An open-source framework for quantum computing.
54. Seifert, Lennart Maximilian, Jason Chadwick, Andrew Litteken, Frederic T. Chong, and Jonathan M. Baker. 2022. Time-efficient qudit gates through incremental pulse re-seeding. In *IEEE international conference on quantum computing and engineering (QCE)*, 304–313.
55. Tan, Jerry, Chuanqi Xu, Theodoros Trochatos, and Jakub Szefer. 2023. Extending and defending attacks on reset operations in quantum computers.
56. Tannu, Swamit S., Poulami Das, Michael L. Lewis, Robert Krick, Douglas M. Carmean, and Moinuddin K. Qureshi. 2019. A case for superconducting accelerators. In *Proceedings of the 16th ACM international conference on computing frontiers*, CF '19, 67–75, New York, NY, USA. Association for Computing Machinery.
57. Tilly, Jules, Hongxiang Chen, Shuxiang Cao, Dario Picozzi, Kanav Setia, Ying Li, Edward Grant, Leonard Wossnig, Ivan Rungger, George H. Booth, et al. 2022. The variational quantum eigensolver: A review of methods and best practices. *Physics Reports* 986: 1–128.
58. Upadhyay, Suryansh, and Swaroop Ghosh. 2023. Stealthy swaps: Adversarial swap injection in multi-tenant quantum computing.
59. Wang, Hanrui, Yongshan Ding, Jiaqi Gu, Zirui Li, Yujun Lin, David Z. Pan, Frederic T. Chong, and Song Han. 2022. Quantumnas: Noise-adaptive search for robust quantum circuits. In *The 28th IEEE international symposium on high-performance computer architecture (HPCA-28)*.
60. Xu, Chuanqi, Jessie Chen, Allen Mi, and Jakub Szefer. 2023. Securing nisq quantum computer reset operations against higher energy state attacks. In *Proceedings of the 2023 ACM SIGSAC conference on computer and communications security*, CCS '23, 594–607, New York, NY, USA. Association for Computing Machinery.
61. Xu, Chuanqi, Ferhat Erata, and Jakub Szefer. 2023. Classification of quantum computer fault injection attacks.

62. Xu, Chuanqi, Ferhat Erata, and Jakub Szefer. 2023. Exploration of power side-channel vulnerabilities in quantum computer controllers. In *Proceedings of the 2023 ACM SIGSAC conference on computer and communications security*, CCS '23, 579–593, New York, NY, USA. Association for Computing Machinery.
63. Xu, Chuanqi, Jamie Sikora, and Jakub Szefer. 2024. A thorough study of state leakage mitigation in quantum computing with one-time pad.
64. Xu, Chuanqi, and Jakub Szefer. 2023. Long-term analysis of the dependability of cloud-based nisq quantum computers. In *Proceedings of the 18th international conference on availability, reliability and security*, ARES '23, New York, NY, USA. Association for Computing Machinery.
65. Yalovetzky, Romina, Pierre Minssen, Dylan Herman, and Marco Pistoia. 2024. Solving linear systems on quantum hardware with hybrid hhl++. *Scientific Reports* 14 (1): 20610.

Design of Quantum Computer Antivirus

Sanjay Deshpande, Chuanqi Xu, Theodoros Trochatos, Hanrui Wang, Ferhat Erata, Song Han, Yongshan Ding, and Jakub Szefer

Abstract The rapid advancement of quantum computing in recent years has been marked by the construction of increasingly larger machines and their integration into the internet as cloud-based quantum computing services. With this growing accessibility, there arises a risk that malicious users may attempt to execute code on these systems to leak sensitive information from other users, disrupt or manipulate computational outcomes, or reverse engineer the underlying architecture and its associated intellectual property. To address these emerging security threats in cloud-based quantum environments, this chapter introduces and examines various forms of quantum computer viruses. The study demonstrates that such malicious circuits can degrade the performance and correctness of quantum applications, including Grover's search algorithm and quantum machine learning classification tasks. As a defense, this chapter presents a first-of-its-kind quantum computer antivirus designed to safeguard expensive and fragile quantum hardware from such threats.

S. Deshpande (✉) · C. Xu · T. Trochatos · F. Erata · Y. Ding · J. Szefer
Yale University, New Haven, CT, USA
e-mail: sanjay.deshpande1@northwestern.edu

C. Xu
e-mail: chuanqi.xu@yale.edu

T. Trochatos
e-mail: theodoros.trochatos@yale.edu

F. Erata
e-mail: ferhat.erata@yale.edu

Y. Ding
e-mail: yongshan.ding@yale.edu

J. Szefer
e-mail: jakub.szefer@northwestern.edu

H. Wang · S. Han
Massachusetts Institute of Technology, Cambridge, MA, USA
e-mail: hanrui@mit.edu

S. Han
e-mail: songhan@mit.edu

P. Mishra (ed.), *Design Automation for Quantum Computing*,
https://doi.org/10.1007/978-3-032-09303-5_12

The proposed antivirus inspects quantum programs, also referred to as circuits, prior to execution, enabling the detection of potentially harmful code. Implemented as a compile-time measure, the approach ensures zero additional overhead during the actual runtime of the quantum computer.

1 Introduction

Quantum computing represents a transformative paradigm in computation, offering the capability to address problems that are intractable for classical digital computers, such as integer factorization via Shor's algorithm [9]. Rapid advances in hardware have fueled a global race to build increasingly powerful quantum processors. Systems with 133 qubits are already commercially available, and IBM has projected the development of 2,000 logical qubits by 2033 [16]. These machines are now widely accessible through cloud-based platforms, including IBM Quantum, Amazon Braket, and Microsoft Azure.

While much of the research community remains focused on improving the functional performance, scalability, and error-correction capabilities of quantum computers, comparatively less attention has been devoted to their security. Unlike classical systems, quantum computers are inherently vulnerable to low-level attacks for two main reasons. First, most current architectures expose fine-grained hardware controls, granting users direct access to physical qubits and their manipulation through gates or pulses. Such access can be exploited by adversarial users to disrupt computation or damage hardware. Second, quantum devices are intrinsically sensitive to environmental disturbances and control noise, making them susceptible to even subtle adversarial interference.

This chapter investigates these security concerns by examining how malicious quantum circuits, referred to here as virus circuits, can degrade the performance of victim computations. The evaluation demonstrates that targeted attacks can significantly reduce the success probability of critical workloads, including Grover's search algorithm [14], quantum machine learning classifiers [12], and variational quantum eigensolver (VQE) applications for molecular energy estimation [18], potentially leading to incorrect or misleading results.

To mitigate such threats, this chapter introduces the first *quantum computer antivirus*, a framework designed to detect and block virus circuits before execution. The antivirus operates at the compilation stage, analyzing circuit source code against a curated database of known malicious patterns. This database is designed for extensibility, allowing the incorporation of new virus definitions as attack methodologies evolve.

Recognizing that a local attacker could attempt to bypass antivirus protections, for example, by altering compilation software, the proposed design also supports secure execution within a Trusted Execution Environment (TEE), specifically Intel SGX [1]. A TEE provides hardware-enforced isolation [13], safeguarding the antivirus from tampering by other software, the operating system, or the hypervisor. Circuits verified inside SGX are digitally signed, enabling quantum cloud providers to authenticate that the program has been validated. At the same time, this architecture preserves user privacy, as the circuit source code can be verified locally without disclosure to the cloud provider, protecting intellectual property.

2 Principles of Quantum Computers

The bit is the fundamental unit of classical computation, while quantum computing builds on an analogous concept: the quantum bit, or qubit. Unlike a classical bit that can be either 0 or 1, a qubit exists in a superposition of two basis states, denoted by $|0\rangle$ and $|1\rangle$. A general qubit state $|\psi\rangle$ is a linear combination of these basis states:

$$|\psi\rangle = \alpha|0\rangle + \beta|1\rangle,$$

where α and β are complex coefficients such that $|\alpha|^2 + |\beta|^2 = 1$.

Quantum states can also be represented using two-dimensional vectors. For instance, the basis states can be written as $|0\rangle = [1\ 0]^T$ and $|1\rangle = [0\ 1]^T$, which allows a general qubit state to be expressed as $|\psi\rangle = [\alpha\ \beta]^T$. For systems with more qubits, the dimensionality of the representation grows exponentially. A two-qubit system has four basis states: $|00\rangle$, $|01\rangle$, $|10\rangle$, and $|11\rangle$, and is described using four-dimensional vectors. In general, an n-qubit quantum state requires a 2^n-dimensional vector to represent its state.

Quantum computation is performed by applying a sequence of quantum gates to the qubits. These gates are unitary operations, meaning they preserve the total probability and map one quantum state to another: $|\psi\rangle \rightarrow U|\psi\rangle$, where U is a unitary matrix. The most common gates operate on one or two qubits, as any n-qubit gate can be decomposed into a combination of one- and two-qubit gates. One-qubit gates are represented by 2×2 matrices. Two-qubit gates are represented by 4×4 matrices.

A common one-qubit gate is the Hadamard gate (H), which transforms the $|0\rangle$ state into $|+\rangle = (|0\rangle + |1\rangle)/\sqrt{2}$ and the $|1\rangle$ state into $|-\rangle = (|0\rangle - |1\rangle)/\sqrt{2}$. And a frequently used two-qubit gate is the Controlled-NOT or CX gate. When the most significant qubit serves as the control and the least significant as the target, the CX gate flips the target qubit if the control qubit is in the $|1\rangle$ state and leaves it unchanged otherwise. That is, $|0x\rangle \rightarrow |0x\rangle$ and $|1x\rangle \rightarrow |1\bar{x}\rangle$ (where $x \in 0, 1$ and $\bar{0} = 1, \bar{1} = 0$).

The matrix representations of the (H) and CX gates are:

$$\mathrm{H} = \frac{1}{\sqrt{2}}\begin{bmatrix}1 & 1\\ 1 & -1\end{bmatrix}, \quad \mathrm{CX} = \begin{bmatrix}1&0&0&0\\0&1&0&0\\0&0&0&1\\0&0&1&0\end{bmatrix}$$

Publicly accessible quantum hardware, such as IBM's quantum computers, typically supports only a limited set of native gates. For example, IBM systems currently support: Four one-qubit gates: `I`, R_z, $\sqrt{X}$, and `X` and one two-qubit gate: `CX` Any other gate, such as the Hadamard gate, must be decomposed into a sequence of these native gates for execution on the hardware.

2.1 Device Topologies

One of the fundamental constraints in quantum hardware design is qubit connectivity, that is, which pairs of qubits can directly interact with each other. In superconducting quantum computing architectures, connectivity is established through physical couplers (such as capacitors) that link pairs of qubits via circuit wiring. These couplers enable two-qubit gates, which are essential for creating entanglement.

However, not all qubit pairs in a device are physically connected. Two-qubit gates can only be applied between directly coupled qubits. When a gate operation is required between two non-adjacent qubits, their quantum states must be moved closer together through a series of logical SWAP operations, introducing additional overhead and potential noise.

The overall layout of physical qubit connections is referred to as the device's topology. In theory, an all-to-all topology where every qubit is connected to every other qubit would offer the most flexibility for algorithm design. But in practice, this is extremely difficult to realize in superconducting systems due to engineering complexity, fabrication limitations, and increased noise.

As a result, most superconducting quantum processors adopt sparse topologies, such as two-dimensional meshes or lattice graphs. While these provide fewer direct connections, they offer benefits like improved control precision, reduced crosstalk, and better scalability. Such trade-offs make sparse topologies a practical and widely used choice in real-world quantum devices. Figure 1 shows examples of the topologies used in some IBM quantum computers.

2.2 Noise and Crosstalk in Quantum Computers

Noise in quantum processors is typically classified into two broad categories: *idle errors* and *operational errors*.

Idle errors occur spontaneously when qubits are not actively manipulated. These include:

- **Qubit decoherence errors**, which arise from interactions between a qubit and its surrounding environment, leading to a gradual loss of quantum information.

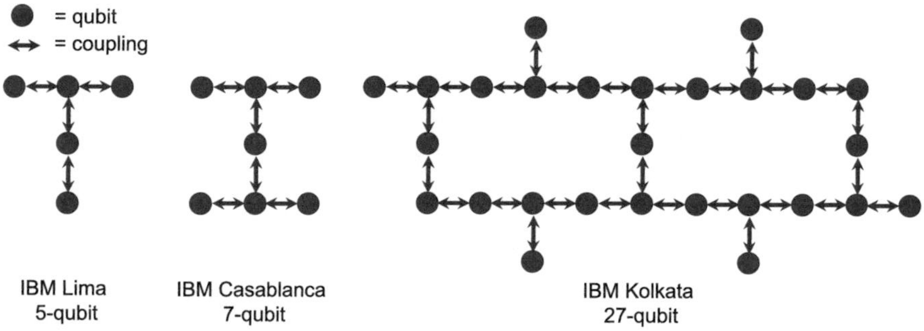

Fig. 1 Sample topologies of IBM quantum computers. Observe that only certain qubits are coupled together. Lima and Casablanca are examples of computers freely available to researchers, while Kolkata is a bigger chip only available with special access. Note that Kolkata and all the larger IBM quantum computers today follow a similar qubit connectivity pattern by repeating the basic pattern of arranging the qubits in the rectangle shape, which can be tiled to form even bigger machines

- **Idle crosstalk**, resulting from always-on, unwanted ZZ-type interactions between qubits due to residual resonances in the hardware. Such effects can cause correlated errors even when the qubits are not part of the computation.

In contrast, *operational errors* occur during active quantum operations, measurements, or other control processes due to imperfect implementation. These include:

- **Gate errors**, caused by inaccuracies in the control pulses used to implement quantum gates, such as the Hadamard or `CX` gates.
- **Readout errors**, which arise during measurement and distort the probability distribution of the recorded outcomes.
- **Operational crosstalk**, where the application of gates to one or more target qubits unintentionally perturbs the states of nearby qubits. This can significantly degrade both qubit state fidelity and the accuracy of gates applied to affected qubits.

Both idle and operational errors represent key challenges in quantum computing, directly influencing algorithmic fidelity, error mitigation strategies, and the scalability of quantum devices.

2.3 Multi-tenancy in Quantum Computers

A key trend in quantum computing research and development is the rapid scaling of the number of qubits available in hardware platforms. Current state-of-the-art superconducting quantum processors provide up to 133 qubits, and systems with more than 200 qubits are expected to become commercially available by next year [26]. IBM has announced roadmaps projecting the development of 2000 logical qubits in the near future [16]. Such large-scale quantum systems are anticipated to enable

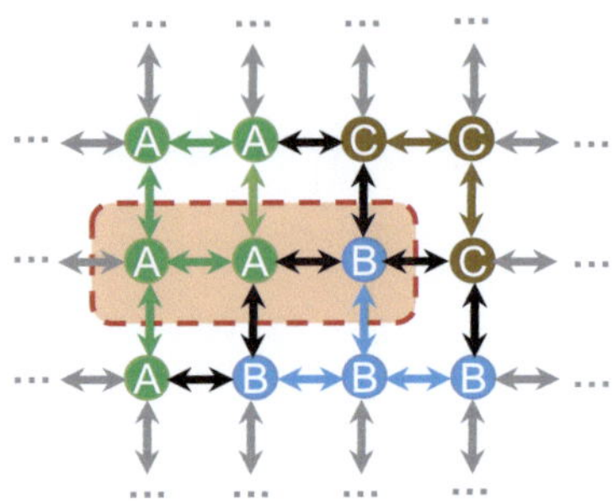

Fig. 2 Example diagram of three users, A, B, and C, being assigned a disjoint set of qubits within a quantum computer. Each user is assigned qubits and couplings between the qubits. Couplings between different users are inactive but physically present, as they cannot be dynamically reconfigured in current superconducting quantum devices. Red dashed lines highlight a potential scenario where 2 qubits from user A can cause crosstalk errors in a qubit of user B

the execution of more complex algorithms, solve larger problem instances, and, in a cloud-based model, support multiple concurrent workloads.

This motivates the concept of *multi-tenant quantum computing*, in which multiple users share the same quantum hardware resources. Early proposals for multi-tenancy in quantum systems have been discussed in the literature [8]. The approach is conceptually similar to resource sharing in classical cloud computing, where workloads from multiple users may run simultaneously on the same physical infrastructure [4]. In classical systems, this sharing may occur through the allocation of separate computational blocks to different users or via *time multiplexing*, where the same hardware is sequentially shared.

IBM's current public quantum cloud services already support time-multiplexed access to quantum computers, but do not implement true multi-tenancy. In a fully realized multi-tenant quantum architecture, both *time sharing* and *spatial sharing* would be possible, allowing two or more quantum programs to execute concurrently on disjoint subsets of qubits within the same processor (Fig. 2). However, spatial sharing introduces new challenges, most notably *crosstalk* between qubits assigned to different programs. In such scenarios, unintended interactions between adjacent qubits can degrade the fidelity of one or more programs, making crosstalk mitigation a critical design consideration for multi-tenant quantum systems.

3 Threat Model

The threat model considered in this chapter assumes that users have *gate-level* or *pulse-level* access to a quantum computer and are able to submit quantum programs or circuits for execution on a target device. All submitted programs are assumed to be compiled by a *trusted compiler* prior to execution. For instance, in current IBM quantum systems, user submissions are compiled using the Qiskit framework. In future scenarios where users may not fully trust the quantum service provider, compilation could instead be performed locally within a trusted execution environment (e.g., Intel SGX), producing both the compiled circuit and an attestation verifying that it was generated using a provider-approved compiler.

The model further assumes that certain users may be *malicious*, with potential objectives including, (1) leaking information from other users' computations, (2) interfering with or manipulating the results of other users' programs, (3) reverse engineering the underlying quantum hardware architecture or extracting proprietary information.

Malicious functionality may be introduced explicitly by the user or indirectly through the use of compromised third-party libraries. In either case, the circuits are assumed to be expressed at the gate or pulse level, consistent with current practice.

Two deployment settings are considered:

1. **Single-tenant setting**: This corresponds to present-day commercial systems, in which a single user occupies the quantum processor at any given time. In this case, a malicious circuit could be embedded within a library or executed deliberately as a *virus circuit*.
2. **Multi-tenant (multi-programming) setting**: In this scenario, multiple users execute quantum programs concurrently on disjoint subsets of qubits. Although not yet supported in commercial systems, initial proposals for such architectures have been presented in the literature [8]. In this setting, a malicious user's circuit may run simultaneously with a victim circuit on a separate set of qubits.

In both settings, the primary attack vector under consideration is *crosstalk*. A malicious circuit could exploit crosstalk to (1) induce incorrect results in the victim circuit, (2) extract information about the victim's state by observing the impact of the victim's operations on the malicious qubits, and (3) perform probing operations, such as frequency sweeps, to infer hardware characteristics. These threats arise because crosstalk effects can propagate between physically adjacent qubits, even when they belong to logically separate computations.

4 Attack Evaluation

One of the principal weaknesses of quantum computers is *crosstalk*, which is therefore considered in this work as a primary attack vector. To evaluate the impact of crosstalk, the study employs the experimental setup illustrated in Fig. 3.

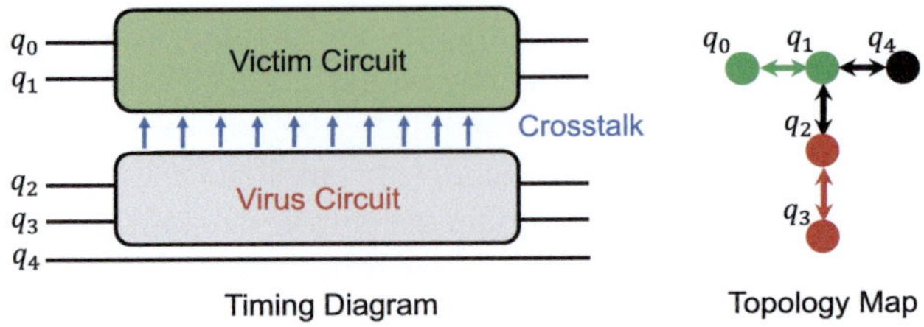

Fig. 3 Topology map of IBM Lima machine and timing diagram of a crosstalk attack example. The example shows a 2-qubit victim circuit on the green qubits and a 2-qubit attacker circuit on the red qubits. Although the couplings between victim and attacker qubits are physically present, they are not used to emulate a multi-tenant setup and isolation of the two users. Note the fifth qubit is idle in this example as only four qubits total are needed by the victim and attacker

4.1 Benchmark Victim Circuits

This section presents a set of representative quantum algorithms, illustrated in Fig. 4, which are employed as benchmark victim circuits for evaluating the impact of attacker circuits on their output distributions.

4.1.1 Grover's Algorithm

Grover's algorithm exemplifies the capability of achieving a quadratic speed-up for unstructured search problems. An *unstructured search* can be defined as follows: given a list of $\mathtt{N} = 2^n$ items (where n is the number of bits) arranged without any specific order, the objective is to locate a specific target item $\mathtt{X}$. In classical computation, identifying $\mathtt{X}$ requires, on average, $\mathtt{N}/2$ queries, and up to $\mathtt{N}$ queries in the worst case. Grover's algorithm reduces this search complexity to $\sqrt{\mathtt{N}}$ queries. A detailed description of the algorithm is provided in [14].

4.1.2 Deutsch–Jozsa Algorithm

The Deutsch–Jozsa algorithm [10] addresses the problem of determining whether an unknown Boolean function $f : \{0, 1\}^n \rightarrow \{0, 1\}$ is *constant* or *balanced*. A constant function f produces the same output (either always 0 or always 1) for all possible inputs, whereas a balanced function returns 0 for exactly half of the inputs and 1 for the other half. The objective is to distinguish between these two cases.

4.1.3 Bernstein–Vazirani Algorithm

The Bernstein–Vazirani algorithm [5] can be viewed as an extension of the Deutsch–Jozsa algorithm. Instead of determining the type of the function, this algorithm returns

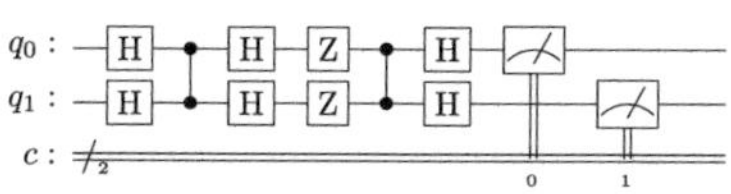

(a) Circuit for 2-qubit Grover's algorithm.

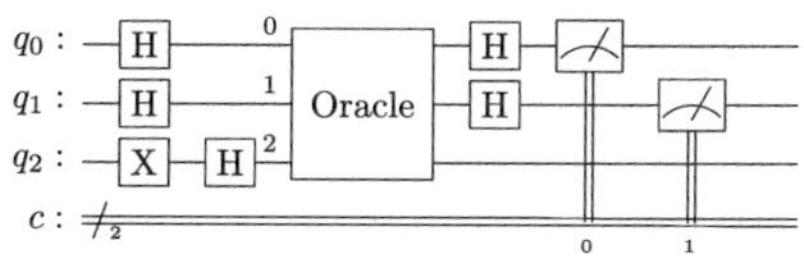

(b) Circuit for 2-qubit Deutsch-Jozsa algorithm, details of the Oracle are not shown due to limited space.

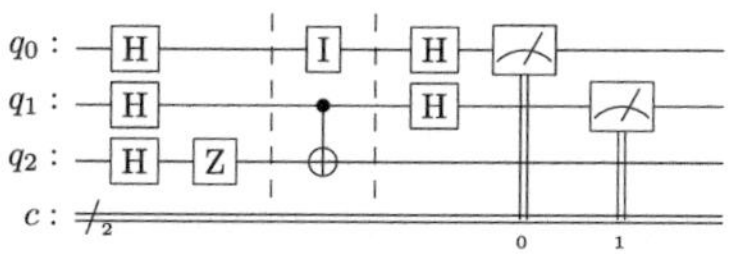

(c) Circuit for 2-qubit Bernstein-Vazirani algorithm.

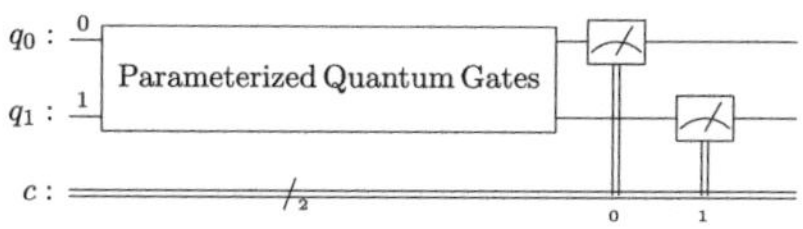

(d) Circuit for 2-qubit VQE algorithm, details of the Parametrized Quantum Gates are not shown due to limited space.

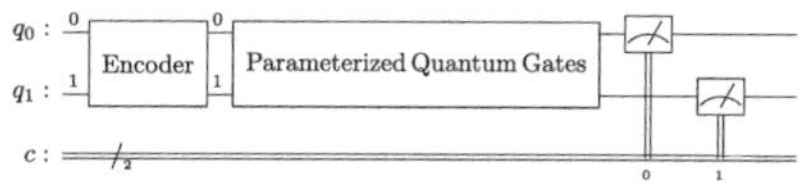

(e) Circuit for 2-qubit QNN algorithm, details of the Encoder and Parametrized Quantum Gates are not shown due to limited space.

Fig. 4 Timing diagrams of the benchmark victim circuits, later used to demonstrate the effectiveness of different virus circuits' ability to affect output of these victim circuits through crosstalk. The horizontal lines labeled q represent qubits, and the lines labeled c represent the output classical bits read out from the qubits

a bitwise product of the n-bit input with some n-bit binary string "s" (where "s" is unknown). Classically, recovering s requires n function evaluations, whereas the quantum version of the algorithm retrieves s with a single query, demonstrating a clear separation in query complexity.

4.1.4 VQE

The Variational Quantum Eigensolver (VQE) [21] is a hybrid quantum–classical variational algorithm designed to estimate the ground-state energy of a quantum system, such as a molecule. It employs a parameterized quantum circuit, often referred to as an ansatz, whose quantum gates depend on tunable parameters (e.g., rotation angles in single-qubit gates). Formally, a parameterized circuit $\Phi(x, \theta)$ prepares a variational quantum state

$$|\psi(x, \theta)\rangle = \Phi(x, \theta) |0 \ldots 0\rangle ,$$

where x represents problem-specific input data and θ denotes a set of trainable parameters optimized via a classical feedback loop. The optimization objective is to minimize the expectation value of the Hamiltonian, thereby approximating the ground-state energy in accordance with the variational principle. As a demonstration, this work implements VQE for the H_2 molecule, using a two-qubit circuit with trainable layers composed of `U3` and `CU3` gates for state preparation.

4.1.5 QML

Quantum Machine Learning (QML) investigates the potential of leveraging quantum devices to perform machine learning (ML) tasks more efficiently or with enhanced capabilities compared to classical approaches. As a demonstration, this work implements a Quantum Neural Network (QNN) for the classification of MNIST handwritten digits. The encoding stage employs rotation gates to embed the pixel intensity values into the states of the qubits. Trainable layers are constructed using `U3` and `CU3` gates, enabling the QNN to learn task-specific transformations. Measurement is performed in the Z-basis, and the resulting expectation values are passed through a softmax function to obtain the probability distribution over digit classes. For the MNIST binary classification task (two digits), the circuit requires two qubits, while for the four-class classification task, four qubits are utilized.

4.2 Quantum Computer Virus Circuits

In the context of secure quantum computing, this work adopts the term *quantum computer virus circuit* to describe a quantum circuit that may not carry out any meaningful computation in isolation, yet can adversely affect the output fidelity or success probability of another (victim) circuit executed concurrently on the same device. Such circuits may also be designed to measure their own fidelity as a means of receiving leaked information from a victim's execution, or to probe the behavior of the underlying hardware in order to infer details of its design and architecture. The present study focuses exclusively on the first category, virus circuits that intentionally disrupt the operation of a victim program. Given that the output of a quantum program often represents the most valuable information, even minor disruptions can have practical consequences. For example, supplying an incorrect solution to a routing optimization problem could result in wasted resources in real-world scenarios such as package delivery.

For experimental purposes, the study defines ten preliminary virus circuit types, denoted V_1 through V_{10}. Their structural details are summarized in Fig. 5. A noteworthy design choice is the inclusion of delay gates with zero delay to prevent the Qiskit compiler from removing repeated gates during optimization. This ensures that the intended gate patterns critical for inducing crosstalk are preserved throughout the compilation process.

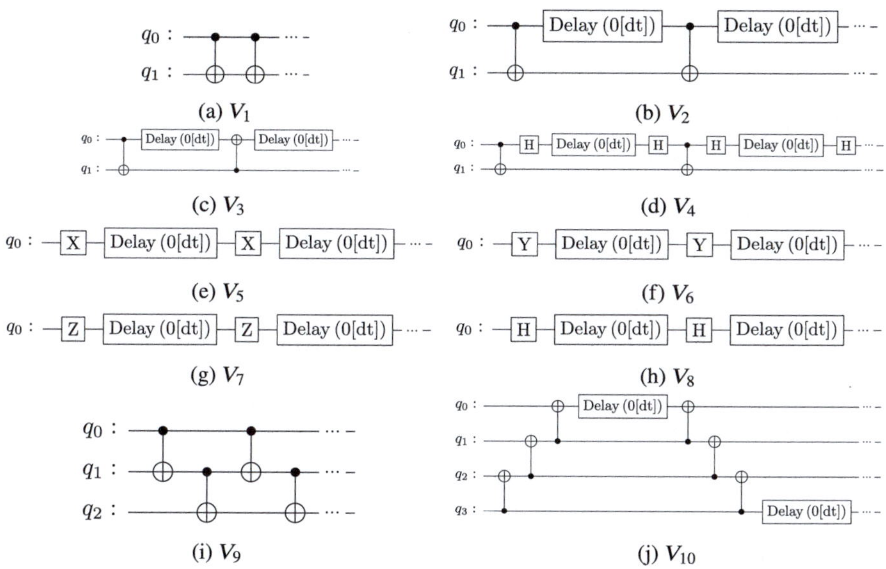

Fig. 5 Virus circuits, each subfigure is shown with 2 copies of the corresponding virus circuit type

4.3 Virus Attack Evaluation

For the experimental evaluation, this work focuses on 5-qubit and 7-qubit quantum processors that are freely accessible to researchers. To accommodate both the victim and virus circuits within these hardware constraints, the virus circuit size is restricted to a maximum of four qubits. Accordingly, the set of virus circuits comprises 1-qubit designs (V_5–V_8), 2-qubit designs (V_1–V_4), a 3-qubit design (V_9), and a 4-qubit design (V_{10}). Figure 6 illustrates an example experiment in which Grover's 2-qubit algorithm circuit (Fig. 4a) is executed concurrently with the V_2 virus circuit (Fig. 5b). This general setup is applied consistently across all benchmark circuits and virus circuit combinations. Non-parameterized circuits are executed with 4,000 shots, while parameterized circuits–Quantum Neural Networks (QNN) and Variational Quan-

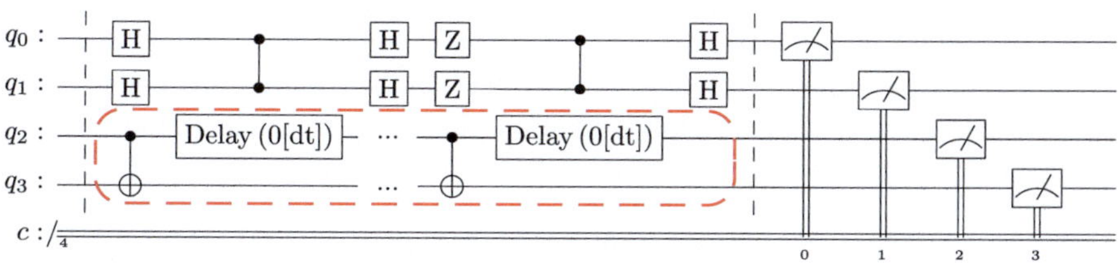

Fig. 6 Circuit diagram for 2-qubit Grover's Circuit alongside with our virus circuit V_2 (highlighted with dotted red line)

tum Eigensolver (VQE)–are executed with 8,192 shots. QNN and VQE training is implemented using the TorchQuantum library, following the methodology outlined in QuantumNAS [31].

Two evaluation procedures are employed, each tailored to the characteristics of the benchmark circuits:

1. **Probability-based crosstalk measurement:** For Grover's, Deutsch–Jozsa, and Bernstein–Vazirani benchmarks, crosstalk is quantified in terms of the output probability distribution. The experimental setup follows the configuration in Fig. 6, with the virus circuit duration parameter (D_M) varied by adjusting the number of repeated patterns in the virus design. The maximum D_M is capped according to the benchmark circuit duration (D_B).
2. **Accuracy-based crosstalk measurement:** For QML and VQE benchmarks, crosstalk is quantified in terms of classification or estimation accuracy. First, the benchmark circuits are executed in isolation (Figs. 4e and d) to establish baseline accuracy. Subsequently, these circuits are executed concurrently with a virus circuit using the maximum feasible D_M (based on D_B). The change in accuracy is then used to assess the impact of crosstalk.

4.3.1 Attack Evaluation Results

For the evaluation of Grover's, Deutsch–Jozsa, and Bernstein–Vazirani benchmarks, the results are grouped according to the size of the virus circuits: 1-qubit, 2-qubit, and 3- & 4-qubit designs. Figures 7a, 8a, and 9a present the performance of the 2-qubit virus circuits across different benchmark circuits. A consistent trend is observed whereby an increase in virus circuit duration, achieved by increasing the number of gates, results in a reduction in the output probability of all benchmark circuits. Figures 7b, 8b, and 9b depict the results for the 1-qubit virus circuits. Among these, only V_5 demonstrates a noticeable degradation in the output probability of the benchmark circuits, while the other designs show minimal or no measurable impact.

Figures 7c, 8c, and 9c present the results for the 3-qubit and 4-qubit virus circuits. Although these designs exhibit a measurable reduction in the output probability of the benchmark circuits, the impact is less pronounced compared to that of the 2-qubit virus circuits. This behavior can be attributed to the configuration of the `CX` gates: when connected step-wise, as in Fig. 5i and j, the overall circuit duration is longer than when the gates are connected in series, as in Fig. 5a and b. The extended duration limits the number of gates that can be included in the malicious pattern before reaching the benchmark circuit duration (D_B), thereby reducing the potential for interference. It is further observed that a higher number of `CX` gates within a virus circuit pattern correlates with a stronger crosstalk effect. Across all ten designs, the most effective virus circuits are identified as V_1 through V_5.

Figures 10 and 11 present the results of VQE circuits with 2 qubits executed on the `ibmq_lima` (V_1–V_8) and `ibmq_casablanca` (V_9–V_{10}) quantum processors, using victim circuits of approximately 15,000 dt (short) and 45,000 dt (long) duration, respectively. The ground-truth expectation value for the evaluated problem instance is -1.84. For the short-duration circuits, nine out of ten virus designs reduce performance, as evidenced by measured expectation values deviating further from the ground truth. For the long-duration circuits, eight out of ten virus designs exhibit a similar negative effect. It is further observed that virus circuits composed solely of single-qubit gates (V_5–V_8) have a comparatively smaller adverse impact on the performance of the victim circuits.

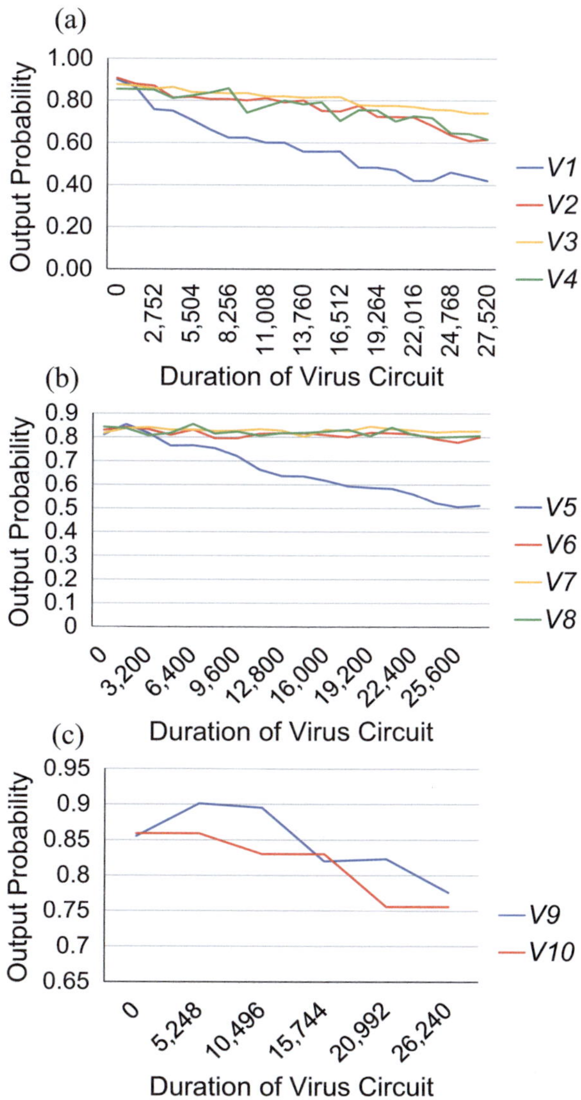

Fig. 7 Output probability for 2-qubit Grover's circuit vs. different duration of virus circuits. **a** Results for 2-qubit virus circuits (V_1, V_2, V_3 and, V_4), ran on `ibmq_lima` machine. **b** Results for 1-qubit virus circuits (V_5, V_6, V_7 and, V_8), ran on `ibmq_lima` machine. **c** Results for V_9 and V_{10}, ran on `ibmq_casablanca` machine

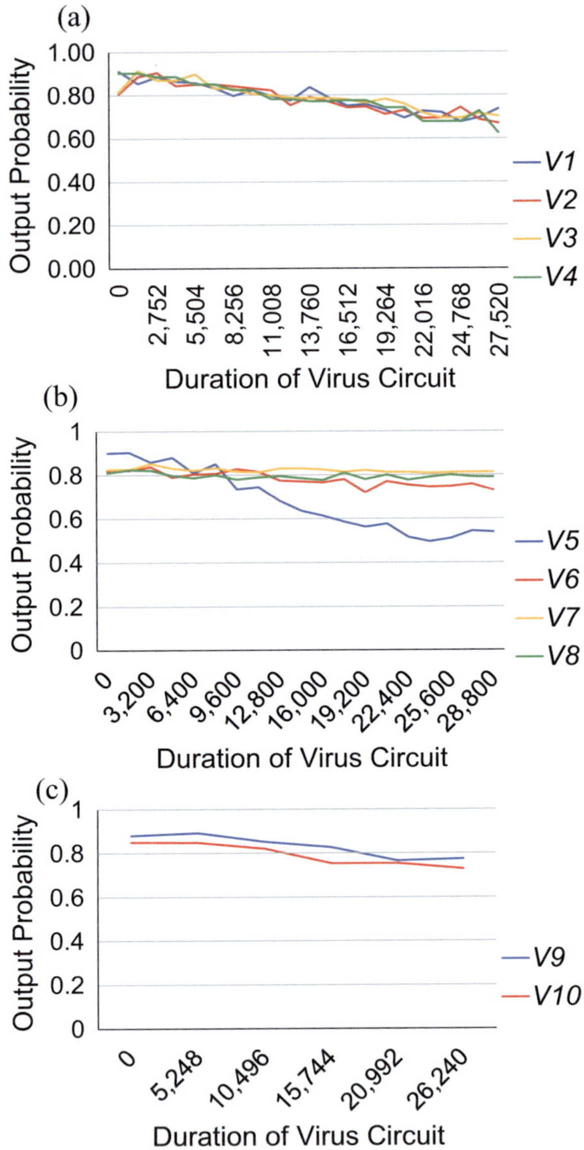

Fig. 8 Output probability for 2-qubit Deutsch–Jozsa circuit versus different duration of virus circuits. **a** Results for 2-qubit virus circuits (V_1, V_2, V_3 and, V_4), ran on `ibmq_lima` machine. **b** Results for 1-qubit virus circuits (V_5, V_6, V_7 and, V_8), ran on `ibmq_lima` machine. **c** Results for V_9 and V_{10}, ran on `ibmq_casablanca` machine

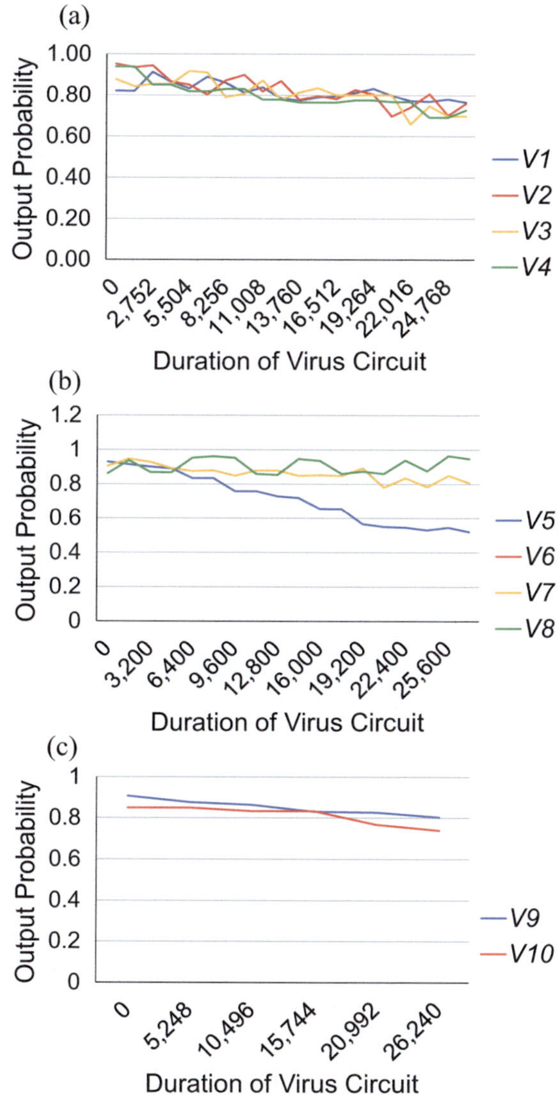

Fig. 9 Output probability for 2-qubit Bernstein–Vazirani circuit versus different duration of virus circuits. **a** Results for 2-qubit virus circuits (V_1, V_2, V_3 and, V_4), ran on `ibmq_lima` machine. **b** Results for 1-qubit virus circuits (V_5, V_6, V_7 and, V_8), ran on `ibmq_lima` machine. **c** Results for V_9 and V_{10}, ran on `ibmq_casablanca` machine

Figures 12 and 13 present the results for QNN circuits with 4 qubits executed on the `ibmq_casablanca` machine, using victim circuits of approximately 30,000 dt (medium) and 45,000 dt (long) duration, respectively. The V_{10} virus design is excluded from these experiments, as it requires 8 qubits, exceeding the maximum available (7 qubits). For the medium-duration circuits, eight out of nine virus designs reduce QNN classification accuracy, with an average accuracy drop of 6.4%. For the long-duration circuits, the overall accuracy is lower compared to the medium case; this is attributed to the similar noise-free simulation accuracy for both configura-

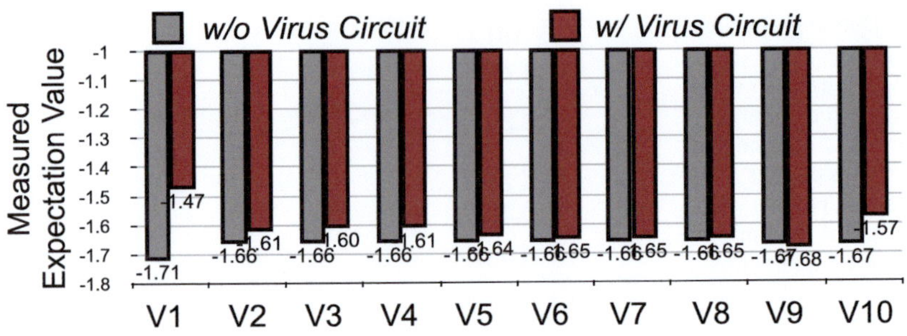

Fig. 10 VQE results for circuit of 15,000 dt duration, with different virus circuits. Lower value is more accurate

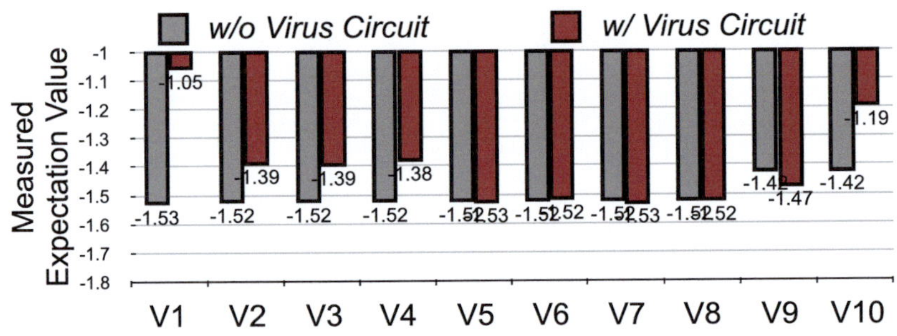

Fig. 11 VQE results for circuit of 45,000 dt duration, with different virus circuits. Lower value is more accurate

tions, combined with the additional noise introduced by the larger number of gates in the longer circuits. In this case, all virus designs negatively impact classification accuracy, with an average reduction of 3.3%. In addition to the four-class classification task, experiments are also performed on the two-class classification task. The accuracy degradation in the latter case is less pronounced, as two-class classification typically achieves accuracy above 90%, implying a larger decision boundary between classes and greater resilience to errors.

4.3.2 Attack Sensitivity to Qubit Mapping

The defense strategy proposed in [4] for mitigating crosstalk attacks is to introduce an idle qubit between the victim and the virus circuits. According to [4], this approach can reduce nearest-neighbor crosstalk effects. To evaluate this recommendation, experiments were conducted using a Grover's 2-qubit circuit (Sect. 4.1) as the victim circuit and the V_2 virus circuit (Sect. 4.2) with duration D_B. The experiments were performed on the 7-qubit `ibmq_casablanca` processor, with the number of shots set to $8,192$.

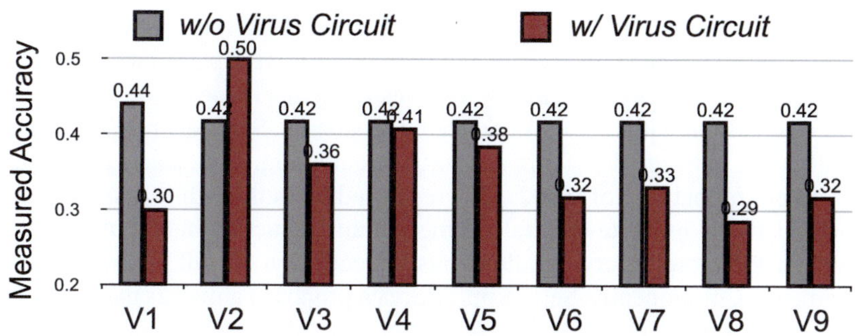

Fig. 12 QML results using circuit of around 30,000 dt duration, measured on `ibmq_casablanca`

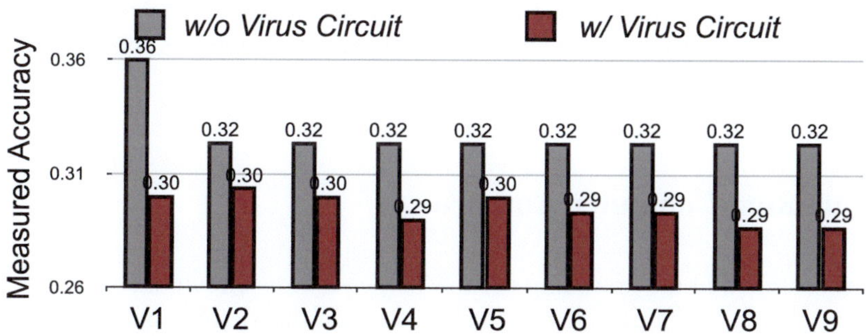

Fig. 13 QML results using circuit of around 45,000 dt duration, measured on `ibmq_casablanca`

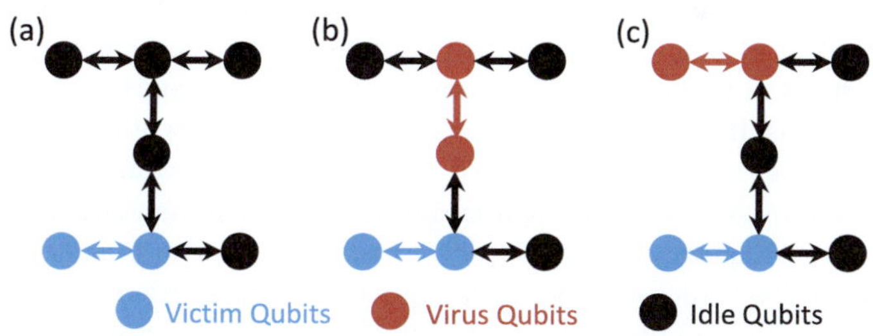

Fig. 14 **a** Coupling map for the experiment running *only* the benchmark victim circuit. **b** Coupling map for the experiment with virus circuit and victim circuit running alongside *without* an idle qubit between. **c** Coupling map for the experiment with benchmark circuit and victim circuits running alongside *with* an idle qubit between. Given topology is for `ibmq_casablanca`

First, the victim circuit was executed in isolation (Fig. 14a), yielding an output probability of 0.891. Next, the victim and virus circuits were placed without any idle qubits between them, as shown in Fig. 14b, resulting in a reduced output probability of 0.669, consistent with crosstalk-induced degradation. Subsequently, an idle qubit was introduced between the victim and virus circuits (Fig. 14c), leading to a slightly higher output probability of 0.692. This result indicates that while inserting an idle qubit can partially mitigate crosstalk, a measurable reduction in output fidelity still persists. Furthermore, reserving idle qubits increases the overall hardware resource requirements. Consequently, Sect. 5 introduces a proposed antivirus mechanism that aims to prevent the execution of malicious circuits at the scheduling stage, rather than relying solely on physical circuit isolation.

It is also noted that for NISQ-era quantum devices, short decoherence times can further impact results, particularly for longer circuits. Both decoherence and crosstalk effects may interfere with outcomes, especially when executions occur across different calibration periods, during which device parameters such as coherence times, gate errors, and other physical properties can vary significantly.

5 Quantum Computer Antivirus

As demonstrated in the preceding sections, when a virus circuit is co-located with a victim circuit, it can exploit crosstalk to degrade the victim's output fidelity. A variety of potentially malicious patterns have been evaluated, and given their ability to significantly impair victim circuit performance, these patterns are treated as *viruses* that should be detected and prevented.

The detection of such virus circuits can be performed at compile time using the proposed *antivirus* framework, described in this section. The fundamental concept of the antivirus is to analyze input quantum programs for the presence of predefined malicious patterns. If a circuit contains a considerable number of these patterns, it is classified as malicious. The frequency of malicious pattern occurrences serves as a strong indicator of the potential threat level posed by a quantum circuit.

In addition to frequency, the *continuity* of malicious patterns is another important factor. Consecutive malicious patterns tend to have a more pronounced detrimental effect compared to patterns that are spatially or temporally dispersed within the quantum circuit.

5.1 Architecture of the Quantum Computer Antivirus

This section outlines the proposed approach for a quantum computer antivirus. The method begins by identifying patterns considered to be malicious, such as those described in Sect. 4.2, and subsequently applies a pattern-matching algorithm to detect such patterns in quantum circuits. The antivirus represents each quantum circuit as a directed acyclic graph (DAG), thereby reducing the detection task to a *subgraph isomorphism problem*, in which the goal is to identify subgraphs of a given circuit that are isomorphic to the DAG representation of a malicious pattern. The algorithm locates all occurrences of the specified patterns and extracts relevant information such as the involved qubits and classical bits. Additionally, a bar graph function is used to analyze the *continuity* of the detected patterns, providing further insight into potential threats.

The DAG formalism has been widely used in prior work on quantum circuit optimization [3, 6, 17]. For instance, subsets of gates may be substituted or reordered based on commutation relations in order to reduce depth or gate count. However, such optimizations can alter the crosstalk characteristics of a circuit and, in turn, affect its maliciousness. To mitigate this issue, the proposed antivirus only treats gates acting on *disjoint* qubits and classical bits as commutable, ignoring commutation relations between gates sharing any resources. Accordingly, the DAG representation from [17] and the canonical form of quantum circuits from [24] are adapted to incorporate this restricted notion of commutativity.

In the resulting *DAG with non-commutativity* (DAGNC) representation, a quantum circuit is modeled as a multigraph in which vertices correspond to gates, and edges represent order between the gates. An edge from node i to node j indicates that the two gates share at least one qubit or classical bit, and that the gate at i executes before the gate at j. For example, in Figure 15(b), the five gates in the quantum circuit are represented by five vertices, each associated with an index in the DAGNC multigraph. Edges from vertex 0 and vertex 2–vertex 4 are created because the Hadamard gate and the Pauli `X` gate must execute before the `CX` gate. Similarly, an edge from vertex 1 to vertex 3 exists because the corresponding gates share qubits and execution order. There is no edge between the vertices in the upper group and those in the lower group, as no execution order is enforced between them. In contrast, in the canonical DAG representation, there is no edge from vertex 2 to vertex 4, since a Pauli `X` gate acting on the target qubit of a `CX` gate commutes with that `CX` gate.

Once the DAGNC representation is obtained, the task of malicious pattern detection becomes equivalent to finding subgraphs within the DAGNC of a given quantum circuit that are isomorphic to the DAGNC of a pattern. This problem is solved using the VF2 algorithm [7] as implemented in the `networkx` Python package [15]. The algorithm outputs all candidate mappings, which are then filtered to retain only *exact mappings*. A mapping is exact if there exists an assignment of qubits and classical bits that fully reconstructs the original pattern. For instance, if the pattern is "`H`(0); `CX`(0,1)" but the detected sequence is "`H`(1); `CX`(0,1)," no reordering of qubits can transform the latter into the former, and thus the mapping is discarded. Exactness is verified by traversing the vertices of each mapping and checking for consistency in qubit and classical bit assignments.

The antivirus produces, for each malicious pattern, the total number of occurrences in the target circuit. As shown in earlier evaluations, higher counts of malicious patterns (and correspondingly longer durations of the virus circuit) are correlated with increased crosstalk errors, thereby posing greater risks to co-located victim circuits. Furthermore, the pattern bar graph can be used to analyze pattern continuity, where

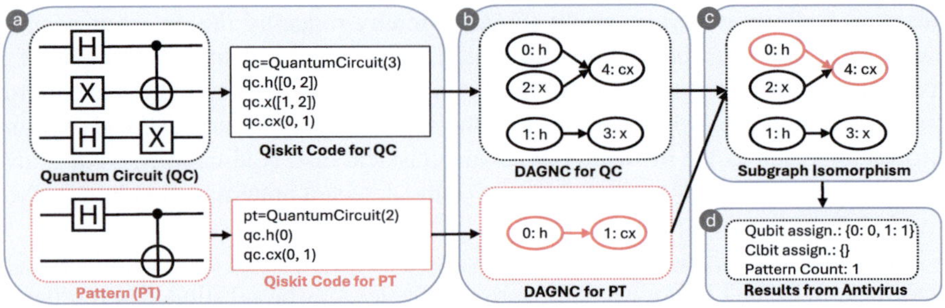

Fig. 15 Overview and workflow of the antivirus program. **a** Specify the quantum circuit (QC) and the pattern (PT) in Qiskit. **b** Convert the *QuantumCircuit* object to the directed acyclic graph with non-commutativity (DAGNC) representation. Vertices in DAGNC are gates in quantum circuits, and edges specify orders between gates. **c** Find patterns using subgraph isomorphism algorithm and check whether the found mappings are exact. **d** Analysis of the results. Qubit and clbit assignment is represented as a Python *Dict* object, whose keys (values) are qubit and clbit indices in QC (PT). Pattern count is the number of occurrences of found PTs in QC

continuity is defined as the number of identical patterns occurring in series without intervening gates and with the same operand mapping. This metric of continuity acts as an additional indicator of malicious activity, with higher values generally associated with greater performance degradation.

5.2 Antivirus Evaluation Results

To evaluate the proposed antivirus, this study involved testing it on the malicious patterns previously discussed. Additionally, it was applied to benign victim circuits to determine if any would be incorrectly classified as malicious.

Table 1 presents the results of scanning ten virus circuits using our antivirus. In all instances, the antivirus accurately identified both the malicious pattern and the number of copies of the pattern present in the circuit.

Table 2 presents the results of scanning the victim circuits using the proposed antivirus. The findings indicate that malicious patterns are generally absent in these circuits. In the case of the Bernstein-Vazirani algorithm, the antivirus identified a single instance of V_1 because V_1 consists of only one CX gate, and the algorithm itself contains one CX gate. However, as shown in the earlier evaluation, a single copy of V_1 is insufficient to induce noticeable crosstalk. It is worth noting that V_{10} can be malicious even with a single occurrence due to its long duration of 10,000 dt, whereas one copy of V_1 lasts only about 1,250 dt. Based on these observations, no false positives were recorded for the tested circuits.

Table 1 Malicious pattern counts found by antivirus program in the virus circuits. The duration of the virus circuits was set to about 10, 000dt, the same as used for attack evaluation in prior sections, and the # Copies lists how many repetitions of each malicious pattern fit in that time

Virus circuits	No. of copies	No. of detected	Detection accuracy (%)
V_1	8	8	100
V_2	8	8	100
V_3	4	4	100
V_4	6	6	100
V_5	65	65	100
V_6	65	65	100
V_7	65	65	100
V_8	65	65	100
V_9	2	2	100
V_{10}	1	1	100

Table 2 Malicious pattern counts found by antivirus program in the victim circuits

Victim circuits	No. of copies detected	Detected pattern	Labeled as virus?
Grover	0	–	No
Deutsch–Jozsa	0	–	No
Bernstein–Vazirani	1	V_1	No
VQE	0	–	No
QML	0	–	No

5.3 *Securing Antivirus from Attacks*

While the proposed antivirus can successfully detect all identified malicious patterns, a determined adversary may still attempt to circumvent it to execute virus circuits on quantum computers. One mitigation strategy is to perform compilation and antivirus scanning at the quantum computing provider's site. Alternatively, the process can be executed on the client side within a Trusted Execution Environment (TEE), ensuring that the antivirus remains protected and enabling TEE-based attestation to the provider that the circuits were compiled and verified.

Currently, users typically run the compiler (e.g., Qiskit) locally and submit the compiled circuits to IBM or other providers for execution on quantum backends. This workflow offers no safeguards against malicious or modified compilers. Figure 16 illustrates three possible workflows:

(a) The current practice with no antivirus protection.
(b) Compilation and antivirus scanning performed on the cloud provider's infrastructure.

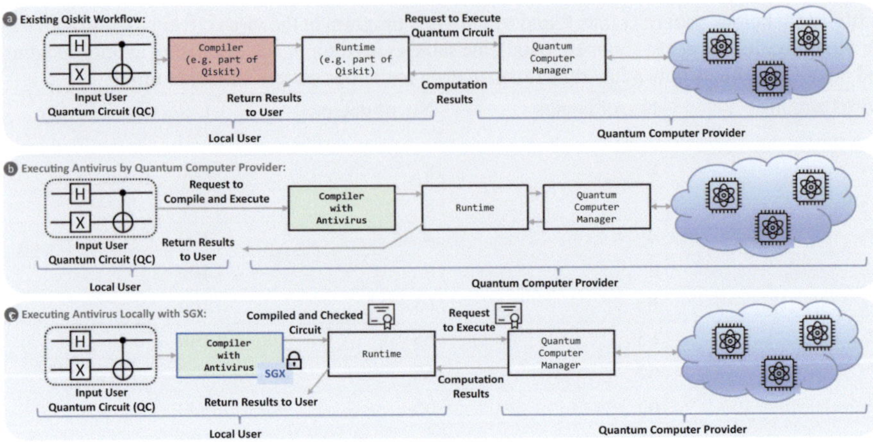

Fig. 16 Three quantum compiler workflows: **a** exiting workflow in Qiskit today, **b** proposed workflow with antivirus running on the quantum provider end, and **c** proposed workflow with antivirus running on the user end with SGX protections

(c) Compilation on the client side using SGX for secure execution and attestation.

5.3.1 Existing Qiskit Workflow

The current Qiskit workflow lacks any antivirus-like protection mechanisms. Moreover, because compilation is performed on the user's local environment, the quantum computing provider (e.g., IBM) has no assurance regarding the integrity or provenance of the compiled circuit beyond the trust placed in the submitting user. This process is depicted in Fig. 16a.

5.3.2 Executing Antivirus by Quantum Computer Provider

To incorporate antivirus protection, one potential approach is to execute the antivirus on the cloud provider's infrastructure. Given that toolchains such as Qiskit are already vendor-specific, the provider can readily perform compilation and malware scanning. This approach enables the provider to enforce antivirus checks and may reduce costs for users by offloading compilation and related computational tasks to the provider. However, it comes at the expense of reduced user control, as the provider gains access to the raw circuit code during compilation and verification. This workflow is illustrated in Fig. 16b.

5.3.3 Executing Antivirus Locally with SGX

An alternative approach, aimed at preserving user control over circuit code while still enabling antivirus protection, is to execute the compiler and antivirus locally within a Trusted Execution Environment (TEE). A TEE provides an isolated execution environment that offers strong integrity and confidentiality guarantees for applications. One such TEE is Intel's Software Guard Extensions (SGX) [2], which implements hardware-based memory encryption to isolate specific application code and data in protected memory regions, referred to as enclaves. These enclaves are designed to remain secure even against privileged system processes. In this context, Qiskit is ported to run inside SGX enclaves, ensuring that the antivirus cannot be maliciously modified or bypassed. Additionally, SGX supports remote attestation, enabling the quantum computing provider to verify that the compiled circuit and antivirus report originate from a genuine antivirus execution. This approach also reduces the exposure of potentially proprietary information, as both compilation and scanning occur locally. In this work, the Gramine framework[23] is used to execute Qiskit—implemented in Python—within SGX enclaves. The corresponding workflow is depicted in Fig. 16c.

5.4 *Antivirus Overhead Evaluation*

The performance overhead of executing the antivirus was evaluated on an Intel(R) Xeon E-2286G CPU operating at 4GHz. Execution time measurements were collected using the *perf* Linux profiler, with elapsed time extracted as the primary metric. The pattern-scanning phase itself is computationally lightweight, completing in approximately one second. The dominant runtime cost—both with and without SGX—stems from loading Qiskit and the antivirus code. When executed inside SGX, a substantial but nearly constant overhead of approximately 250s was observed, independent of the tested circuit duration. This constant overhead is primarily attributed to secure enclave initialization and the loading of Qiskit and its Python dependencies into the enclave. Additional time is required to generate SGX attestation reports and transmit the compiled circuit to the quantum computing provider. However, since IBM currently does not support receiving SGX attestation reports, this step was not included in the reported measurements.

Further analysis revealed that, for a fixed circuit execution duration, virus circuits V_5 through V_8 required more time to scan. These circuits employ fewer or simpler gates per copy, allowing a greater number of copies to fit within the same execution-time budget. As the complexity of the scanning algorithm is proportional to the circuit depth (measured by the total number of gates), circuits with a larger gate count despite having the same total execution time incur a longer scanning time.

6 Related Work

The body of literature addressing the security of quantum computers is currently limited but steadily expanding, with several surveys providing broad overviews of the emerging threat landscape [25, 26].

For superconducting quantum computers, prior work [4] has demonstrated that crosstalk errors can be exploited for fault injection attacks. The study further illustrated how an adversary could launch a denial-of-service (DoS) attack on a victim circuit through deliberate induction of crosstalk, a threat model closely related to the evaluation conducted in the present work. Another attack vector, qubit sensing, leverages differences in eigenstates, enabling malicious circuits to infer qubit states of victim circuits based on pre-existing statistical information [27]. Additional research has shown that quantum devices can be fingerprinted by characterizing device-specific error patterns, enabling identification of individual qubits or hardware instances [19, 22].

In trapped-ion quantum computers, repeated shuttle operations have been observed to increase the motional energy of the ion chain, thereby degrading the fidelity of co-located victim circuits [28, 29].

Motivated by such attack vectors, recent research has begun exploring preliminary defense mechanisms. Detection of crosstalk errors through execution-time analysis [30] has been followed by mitigation strategies such as connectivity reduction, qubit frequency tuning, and coupler tuning [11], as well as instruction scheduling modifications [20]. A common observation across these methods is that while crosstalk effects can be reduced, they cannot be fully eliminated. As demonstrated in this chapter, even defensive strategies such as idle-qubit placement (Figure 4.3.2) fail to fully prevent crosstalk propagation.

To address these limitations, the antivirus approach proposed herein targets the problem at compile time by detecting potentially malicious circuits before they are executed on quantum hardware and have the opportunity to induce crosstalk errors. Unlike many existing defenses, this approach does not rely on executing the circuit for maliciousness analysis, thereby eliminating runtime exposure to the attack.

7 Conclusion

This chapter introduced what is, to the best of current knowledge, the first antivirus framework for quantum computers. The study evaluated a range of virus circuits employing distinct gate patterns to induce crosstalk errors on adjacent qubits. Experimental results demonstrated that such circuits can adversely affect co-located victim circuits, altering their output probabilities through induced crosstalk. It was further shown that existing defenses, such as reserving idle qubits, do not provide complete protection, underscoring the need for proactive solutions. The attack scenarios were validated on publicly accessible, cloud-based IBM quantum computers, confirming that the threats are practical and observable on real hardware. In response, an antivirus mechanism was developed to identify malicious circuits at the source-code level prior to execution. The proposed approach achieved 100% detection accuracy with no false positives across the tested virus and victim circuits. Integration within Qiskit was demonstrated under two operational settings: without and with the use of SGX enclaves.

Acknowledgements This work was supported in part by NSF grants 1901901 and MIT–IBM Watson AI Lab. We would like to thank Dmitrii Kuvaiskii from Intel for help with Intel SGX and Gramine tools.

References

1. Intel sgx. https://www.intel.com/content/www/us/en/developer/tools/software-guard-extensions/overview.html.
2. Intel software guard extensions.
3. Abdessaied, Nabila, Mathias Soeken, Robert Wille, and Rolf Drechsler. 2013. Exact template matching using Boolean satisfiability. In *2013 IEEE 43rd international symposium on multiple-valued logic*, 328–333.
4. Ash-Saki, Abdullah, Mahabubul Alam, and Swaroop Ghosh. 2020. Analysis of crosstalk in nisq devices and security implications in multi-programming regime. In *Proceedings of the ACM/IEEE international symposium on low power electronics and design*, ISLPED '20, 25–30, New York, NY, USA. Association for Computing Machinery.
5. Bernstein, Ethan, and Umesh Vazirani. 1997. Quantum complexity theory. *SIAM Journal of Computing* 26 (5): 1411–1473.
6. Chen, Mingyu, Yu Zhang, and Yongshang Li. 2021. A quantum circuit optimization framework based on pattern matching. *SPIN* 11 (03): 2140008.
7. Cordella, L. P., P. Foggia, C. Sansone, and M. Vento. 2001. An improved algorithm for matching large graphs. In *In: 3rd IAPR-TC15 workshop on graph-based representations in pattern recognition, Cuen*, 149–159.
8. Das, Poulami, Swamit S. Tannu, Prashant J. Nair, and Moinuddin Qureshi. 2019. A case for multi-programming quantum computers. In *Proceedings of the 52nd annual IEEE/ACM international symposium on microarchitecture*, MICRO '52, 291–303, New York, NY, USA. Association for Computing Machinery.
9. de Lima Marquezino, Franklin, Renato Portugal, and Carlile Lavor. 2019. *Shor's Algorithm for Integer Factorization*, 57–77. Cham: Springer International Publishing.
10. Deutsch, David, and Richard Jozsa. 1992. Rapid solution of problems by quantum computation. *Proceedings of the Royal Society of London. Series A: Mathematical and Physical Sciences* 439: 553–558.
11. Ding, Yongshan and Frederic T. Chong. 2020. Quantum computer systems: Research for noisy intermediate-scale quantum computers. *Synthesis Lectures on Computer Architecture*.
12. Farhi, Edward and Hartmut Neven. 2018. Classification with quantum neural networks on near term processors. arXiv preprint arXiv:1802.06002.
13. Greene, James. Trusted execution environment.
14. Grover, Lov K. 1996. A fast quantum mechanical algorithm for database search. In *Proceedings of the twenty-eighth annual ACM symposium on theory of computing*, STOC '96, 212–219, New York, NY, USA. Association for Computing Machinery.
15. Hagberg, Aric A., Daniel A. Schult, and Pieter J. Swart. 2008. Exploring network structure, dynamics, and function using networkx. In *Proceedings of the 7th python in science conference*, eds. Gaël Varoquaux, Travis Vaught, and Jarrod Millman, 11–15, Pasadena, CA USA.

16. IBM. 2025. Quantum computing technology and roadmap. https://www.ibm.com/quantum/technology. Accessed 14 Aug 2025.
17. Iten, Raban, Romain Moyard, Tony Metger, David Sutter, and Stefan Woerner. 2022. Exact and practical pattern matching for quantum circuit optimization. *ACM Transactions on Quantum Computing*, 3(1).
18. Kandala, Abhinav, Antonio Mezzacapo, Kristan Temme, Maika Takita, Markus Brink, Jerry M Chow, and Jay M Gambetta. 2017. Hardware-efficient variational quantum eigensolver for small molecules and quantum magnets. *Nature* 549 (7671): 242–246.
19. Mi, Allen, Shuwen Deng, and Jakub Szefer. 2021. Short paper: Device- and locality-specific fingerprinting of shared nisq quantum computers. In *Proceedings of the workshop on hardware and architectural support for security and privacy*, HASP.
20. Murali, Prakash, David C. Mckay, Margaret Martonosi, and Ali Javadi-Abhari. 2020. Software mitigation of crosstalk on noisy intermediate-scale quantum computers. *Proceedings of the twenty-fifth international conference on architectural support for programming languages and operating systems*.
21. Peruzzo, Alberto, Jarrod McClean, Peter Shadbolt, Man-Hong Yung, Xiao-Qi Zhou, Peter J. Love, Alán Aspuru-Guzik, and Jeremy L. O'Brien. 2014. A variational eigenvalue solver on a photonic quantum processor. *Nature Communications* 5: 4213.
22. Phalak, Koustubh, Abdullah Ash Saki, Mahabubul Alam, Rasit Onur Topaloglu, and Swaroop Ghosh. 2021. Quantum puf for security and trust in quantum computing. *IEEE Journal on Emerging and Selected Topics in Circuits and Systems* 11 (2): 333–342.
23. Porzyck, Wojtek, and Michal kowalzyck. 2021. Gramineproject/gramine: A library OS for Linux multi-process applications, with intel sgx support.
24. Rahman, Md. Mazder, Gerhard W. Dueck, and Joseph D. Horton. 2015. An algorithm for quantum template matching. *Journal on Emerging Technologies in Computing Systems* 11 (3).
25. Saki, Abdullah Ash, Mahabubul Alam, and Swaroop Ghosh. 2021. Impact of noise on the resilience and the security of quantum computing. In *22nd International symposium on quality electronic design (ISQED)*, 186–191.
26. Saki, Abdullah Ash, Mahabubul Alam, Koustubh Phalak, Aakarshitha Suresh, Rasit Onur Topaloglu, and Swaroop Ghosh. 2021. A survey and tutorial on security and resilience of quantum computing.
27. Saki, Abdullah Ash, and Swaroop Ghosh. 2021. Qubit sensing: A new attack model for multi-programming quantum computing.
28. Saki, Abdullah Ash, Rasit Onur Topaloglu, and Swaroop Ghosh. 2021. Muzzle the shuttle: Efficient compilation for multi-trap trapped-ion quantum computers.
29. Saki, Abdullah Ash, Rasit Onur Topaloglu, and Swaroop Ghosh. 2022. Shuttle-exploiting attacks and their defenses in trapped-ion quantum computers. *IEEE Access*, 10: 2686–2699.
30. Sarovar, Mohan, Timothy Proctor, Kenneth Rudinger, Kevin Young, Erik Nielsen, and Robin Blume-Kohout. 2020. Detecting crosstalk errors in quantum information processors. *Quantum* 4: 321.
31. Wang, Hanrui, Yongshan Ding, Jiaqi Gu, Yujun Lin, David Z. Pan, Frederic T. Chong, and Song Han. 2021. Quantumnas: Noise-adaptive search for robust quantum circuits. arXiv preprint arXiv:2107.10845.

Trusted Execution Environments for Quantum Computers

Theodoros Trochatos, Chuanqi Xu, Sanjay Deshpande, Yao Lu, Yongshan Ding, and Jakub Szefer

Abstract Cloud-based quantum computing introduces significant security and privacy risks to users' intellectual property, including both program code and input data. In the absence of dedicated protections, quantum circuits submitted to remote services may be accessible to the cloud provider or compromised by malicious insiders. Furthermore, data embedded in these circuits remains susceptible to leakage during execution. This work proposes architectural and hardware enhancements for contemporary superconducting-qubit quantum computers to mitigate such threats. Drawing inspiration from classical Trusted Execution Environments (TEEs), we define Quantum Trusted Execution Environments (QTEEs): trusted hardware mechanisms designed to conceal or obfuscate quantum programs in cloud settings. We present two concrete instantiations, SoteriaQ and CASQUE, detailing their hardware support and required system software.

T. Trochatos (✉) · C. Xu · S. Deshpande · Y. Lu · Y. Ding · J. Szefer
Yale University, New Haven, CT, USA
e-mail: theodoros.trochatos@yale.edu

C. Xu
e-mail: chuanqi.xu@yale.edu

S. Deshpande
e-mail: sanjay.deshpande1@northwestern.edu

Y. Lu
e-mail: physics.lu@yale.edu

Y. Ding
e-mail: yongshan.ding@yale.edu

J. Szefer
e-mail: jakub.szefer@northwestern.edu

P. Mishra (ed.), *Design Automation for Quantum Computing*,
https://doi.org/10.1007/978-3-032-09303-5_13

1 Introduction

Noisy Intermediate-Scale Quantum (NISQ) computers are advancing rapidly, with systems exceeding 100 qubits already available [16], and industry roadmaps projecting devices with 4,000 or more qubits by the end of the decade [5]. Among the various quantum computing paradigms, superconducting qubit platforms are accessible today to researchers and the public via cloud-based services. Such systems are developed by multiple vendors, including IBM [4], Rigetti [7], and Quantum Circuits, Inc. [6]. Superconducting qubit processors operate using superconducting electronic circuits cooled to approximately 20mK by housing the quantum chips in dilution refrigerators.

Cloud platforms such as IBM Quantum [4], Amazon Braket [1], and Azure Quantum [25] provide remote access to superconducting and other NISQ quantum systems. Given the high capital and operational costs of quantum hardware, cloud-based access is likely to remain the predominant delivery model. In this setting, the cloud provider retains complete control over the quantum computers, including access to user-submitted circuits and the corresponding control pulses. Because control pulses can be reverse-engineered to recover the original circuits, providers can gain full visibility into users' workloads. While the cloud model offers on-demand scalability and broad accessibility, it also exposes users' intellectual property—namely, proprietary algorithms and circuits—to potential compromise by untrusted providers or malicious insiders. These risks necessitate new security mechanisms to safeguard user code and data in cloud-hosted quantum computing environments.

To address the risks posed by untrusted quantum cloud providers or malicious insiders, prior research has explored various forms of Blind Quantum Computation (BQC) [8, 10, 15, 19, 21, 24, 33, 36–42, 49, 50]. These schemes aim to conceal user workloads from the computing provider; however, most remain theoretical because they require a local trusted quantum processor and quantum networking to connect the local and remote systems. Another avenue of defense is Quantum Homomorphic Encryption (QHE) [13, 22, 32, 34, 48, 51, 63, 64], which enables computation on encrypted quantum data. Yet, as established by the "no-go theorem" [20], achieving fully secure QHE entails exponential computational overhead. This overhead is particularly prohibitive for near-term devices, where high error rates and limited qubit fidelity make it infeasible to tolerate the substantial noise accumulation inherent in such protocols.

As an alternative to BQC and QHE, this work proposes new hardware architectures to protect quantum circuits from untrusted cloud providers and malicious insiders. We demonstrate that incorporating trusted hardware directly into quantum processors can safeguard users' circuits and data during execution. Specifically, we introduce multiple architectures for Quantum Trusted Execution Environments (QTEEs), each tailored to distinct threat models.

1.1 *Contributions*

This chapter combines and extends our existing research on QTEEs. It incorporates results and designs from our work on QC-TEE [58] and CASQUE [54], pre-print work on SoteriaQ [56], and poster at QuantumOS workshop on the topic of software support for QTEEs [55]. It further introduces designs for system software and quantum operating system support needed for QTEEs. Combined, this chapter contributes:

1. design and evaluation of architecture for obfuscation of quantum circuit control pulses based on decoy control pulses, from QC-TEE and SoteriaQ work,
2. design and evaluation of architecture for obfuscation of quantum circuit control pulses based on switching of pulses among different qubit and control channels, from CASQUE work,
3. design for system software and quantum operating system support for QTEE, and
4. categorization of possible threat models and QTEE architectures (current and future) to address the different threat models.

2 Background

This section reviews the fundamentals of quantum computing. The basic unit of computation is the quantum bit (qubit), which encodes quantum states used in computation. Qubits can be measured, with outcomes determined probabilistically. Their states are manipulated through quantum gates, and, together with measurement operations, these gates define a quantum circuit, which is an ordered sequence specifying how qubits are transformed and measured.

2.1 *Quantum Bits*

Quantum states are represented by quantum bits, or qubits. Qubit is the fundamental unit of quantum computing, analogous to the classical bit used in conventional computing. Like a bit, a qubit has two basis states, typically denoted in bra-ket notation as $|0\rangle$ and $|1\rangle$. One widely used representation is the matrix-vector representation, where the quantum states are represented by vectors. For example, the two basis states $|0\rangle$ and $|1\rangle$ are usually represented as

$$|0\rangle = \begin{bmatrix} 1 \\ 0 \end{bmatrix}, |1\rangle = \begin{bmatrix} 0 \\ 1 \end{bmatrix} \tag{1}$$

Unlike a classical bit, which can assume only the values 0 or 1, a qubit can exist in an arbitrary superposition of the basis states, subject to the normalization constraint that the total probability equals one. This is often called *superposition*. Specifically, a qubit $|\psi\rangle$ can be expressed as

$$|\psi\rangle = \alpha\,|0\rangle + \beta\,|1\rangle = \begin{bmatrix} \alpha \\ \beta \end{bmatrix} \tag{2}$$

where α and β are complex numbers that satisfy the normalization condition $|\alpha|^2 + |\beta|^2 = 1$.

In general, an n-qubit system has 2^n basis states. The basis states span a 2^n-dimensional space, ranging from $|00\ldots0\rangle$ to $|11\ldots1\rangle$. An arbitrary n-qubit state $|\phi\rangle$ can be described as

$$|\psi\rangle = \sum_{i=0}^{2^n-1} a_i\,|i\rangle \tag{3}$$

where the coefficients a_i are complex numbers, and their magnitudes satisfy the normalization condition $\sum_{i=0}^{2^n-1} |a_i|^2 = 1$. i is the decimal number of a n-binary-digit number.

The above is also often called the *pure state*, and it cannot represent all quantum states. Quantum systems can also exist in *mixed states*, which represent a probabilistic combination of different quantum states. A mixed state occurs when, with probability p_i, a system is in the quantum state $|\psi_i\rangle$. Such a probabilistic mixture is denoted as $\{(p_i, |\psi_i\rangle)\}$ and is described mathematically using the density matrix formalism. The density matrix ρ is defined as

$$\rho = \sum_i p_i\,|\psi_i\rangle\,\langle\psi_i| \tag{4}$$

where $\langle\psi_i| := |\psi_i\rangle^\dagger$, indicating that ρ is a matrix. This representation is useful for describing systems that are not in a definite quantum state but rather in a statistical mixture of possible states.

2.2 Measurement

Measurement in quantum computing is a crucial process that connects the quantum world to classical outcomes. Unlike classical computing, where the measurement result is definite, the measurement result of a quantum state is often probabilistic. According to Born's rule, for a state described as in Eq. 3, the probability of measuring or observing $|i\rangle$ is given by

$$P(|i\rangle) = |a_i|^2 \tag{5}$$

For the density matrix representation in Eq. 4, the probability of measuring or observing $|i\rangle$ is given by

$$P(|i\rangle) = \langle i| \rho |i\rangle = \sum_j p_j \langle j| |\psi_j\rangle\langle\psi_j| |j\rangle = \sum_j p_j |a_{ji}|^2 \tag{6}$$

where $|a_{ji}|$ is the component of $|\psi_j\rangle$ on $|i\rangle$.

Another difference from classical computing is that in classical computing, the state of a bit can be directly observed without affecting the system. However, measuring a qubit fundamentally alters its state. Prior to measurement, a qubit can exist in a superposition of multiple states, but once a measurement is made, the qubit *collapses* to one of its basis states, typically $|0\rangle$ or $|1\rangle$, with probabilities determined by the above equations.

One concrete and interesting example is *quantum entanglement*. For a 2-qubit Bell state in the form of

$$|\Phi\rangle^+ = \frac{1}{\sqrt{2}}|00\rangle + \frac{1}{\sqrt{2}}|11\rangle \tag{7}$$

according to Eq. 5, the probabilities of individually measuring the first or the second qubit to get $|0\rangle$ or $|1\rangle$ are all $\frac{1}{2}$. However, if both two qubits are measured, the probabilities of measuring $|00\rangle$ and $|11\rangle$ are both $\frac{1}{2}$, and there will not be results of $|01\rangle$ and $|10\rangle$. This cannot be seen in classical computing. Consider two bits which can be $\frac{1}{2}$ probability in 0 and $\frac{1}{2}$ in 1. The measurement of them is independent and thus the result of one bit does not influence the result of the other bit, so there can be cases of 01 or 10.

Measurement in a quantum system not only extracts information but also induces an irreversible change in the state, making it inherently non-deterministic compared to classical measurement. The interaction between superposition, probabilistic outcomes, and state collapse is fundamental to the operation of quantum algorithms and the processing of quantum information.

2.3 Quantum Gates

In quantum computing, the fundamental operations are quantum gates, analogous to logic gates in classical computing. Quantum gates are unitary operators that transform qubit states, and quantum algorithms are realized by applying sequences of these gates to manipulate qubits into target states.

A quantum gate U must satisfy the property of unitarity, meaning $UU^\dagger = U^\dagger U = I$, where $U^\dagger$ is the conjugate transpose (or Hermitian adjoint) of U, and I is the identity matrix. When a quantum gate U operates on a qubit state $|\psi\rangle$, the transformation can be expressed as $|\psi\rangle \rightarrow U\,|\psi\rangle$. In matrix form, n-qubit quantum gates are represented by $2^n \times 2^n$ matrices that act on the vector representation of the qubit states.

For example, the Pauli-X gate, a single-qubit gate, performs a bit-flip operation, transforming $|0\rangle$ to $|1\rangle$ and $|1\rangle$ to $|0\rangle$. This gate is analogous to the classical NOT gate. Another crucial gate is the Controlled-NOT (CNOT) gate, also called the `CX` gate. The CNOT gate is a two-qubit gate where, if the control qubit is in the state $|1\rangle$, it applies a Pauli-X operation (bit-flip) to the target qubit; otherwise, it leaves the target qubit unchanged.

There are other quantum gates that do not have direct classical analogs. For instance, the `RZ` gate introduces a relative phase shift between the states $|0\rangle$ and $|1\rangle$ without changing their amplitudes. The `SX` gate performs a "half" of the Pauli-X gate's operation, and its function is unique to quantum computing without a clear classical counterpart.

The matrix representations of these gates, along with those of other quantum gates, are presented below. It should be noted that Qiskit's [60] qubit ordering convention is followed, where the leftmost qubit is the most significant, and the rightmost qubit is the least significant. As a result, the matrix representation of the `CX` gate may differ in other literature that uses different qubit ordering conventions:

$$\mathtt{ID} = \begin{bmatrix} 1 & 0 \\ 0 & 1 \end{bmatrix}, \quad \mathtt{X} = \begin{bmatrix} 0 & 1 \\ 1 & 0 \end{bmatrix}, \quad \mathtt{CX} = \begin{bmatrix} 1 & 0 & 0 & 0 \\ 0 & 0 & 0 & 1 \\ 0 & 0 & 1 & 0 \\ 0 & 1 & 0 & 0 \end{bmatrix} \tag{8}$$

$$\mathtt{RZ}(\theta) = \begin{bmatrix} e^{-i\frac{\theta}{2}} & 0 \\ 0 & e^{i\frac{\theta}{2}} \end{bmatrix}, \quad \mathtt{SX} = \frac{1}{2}\begin{bmatrix} 1+i & 1-i \\ 1-i & 1+i \end{bmatrix} \tag{9}$$

2.4 Gate-Level Quantum Circuit

Gate-level quantum circuits are the foundational model for executing quantum computations. Analogous to classical circuits that manipulate bits using logic gates like AND, OR, and NOT, quantum circuits use quantum gates to manipulate qubits. Quantum circuit transforms the quantum state of qubits in a way that leverages principles such as superposition and entanglement, both of which have no counterpart in classical computing.

A gate-level quantum circuit is composed of a series of quantum gates applied sequentially or in parallel to one or more qubits. Each qubit is initialized in a well-defined quantum state, typically $|0\rangle$, and then undergoes transformations as the gates act on it. The result of these transformations defines the quantum algorithm being implemented. The final step in most quantum circuits is a measurement, where the state of the qubits is collapsed to classical bits, yielding the result of the computation.

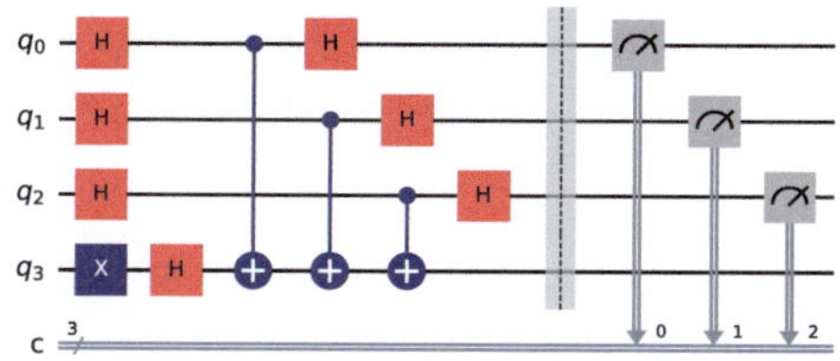

Fig. 1 An example of a quantum gate-level circuit, describing the Bernstein-Vazirani algorithm. This circuit uses 4 qubits, q_0, q_1, q_2 and q_3. A measurement operation would typically be applied at the end of the circuit on each qubit

Quantum circuits are typically represented diagrammatically, where time flows from left to right, and each line in the circuit diagram represents a qubit, with gates applied along the line to indicate operations on those qubits. An example of a gate-level circuit is shown in Fig. 1.

2.5 Control Pulses

Gate-level quantum circuits provide the structural basis for running quantum algorithms. They are easy to understand and are the abstract description of quantum algorithms and thus they are independent of the low-level hardware. Moving downward, the actual realization of qubits and quantum gates depends on the technology used. In superconducting qubit architectures, qubits are realized using Josephson junctions, while gates are realized using sets of analog RF control pulses which actuate the qubits or couplings between them to realize one-qubit and two-qubit gate operations, respectively.

A control pulse is generally characterized by five main parameters: the envelope, frequency, phase, duration, and amplitude. In the context of superconducting qubit control, the envelope determines the shape of the signal, which is generated using an arbitrary waveform generator (AWG). Envelopes are typically discretized into a series of time steps, where each element represents the amplitude at a particular time step. An alternative, more efficient approach involves the use of parameterized pulses. These pulses are defined by predefined shapes, requiring only a few parameters for storage. The frequency and phase define a periodic signal that modulates the envelope signal. In addition, duration specifies the length of the pulse, while the amplitude indicates the relative strength of the pulse. The combination of these components forms the control pulse that is sent to the qubit to perform the desired operation. There are separate control pulses for each qubit, as well as, control pulses used on “control” channels used to realize two-qubit operations. This precise manipulation of qubits using control pulses enables the implementation of quantum gates on superconducting qubit-based quantum computers. At the same time, the pulses totally define the operation of the quantum circuits being executed, thus knowing some or all of the pulse parameters can be used to reveal what circuit is being executed.

3 Cloud-Based Quantum Computers

Noisy Intermediate-Scale Quantum (NISQ) computers are currently accessible through cloud-based services such as IBM Quantum, with devices ranging from 5 to 127 qubits. In November 2022, IBM announced a 433-qubit system and has projected the availability of 4000-qubit or larger machines by the end of the decade [5]. These systems are accessible to both researchers and the public. In addition to free research access, a pay-per-use model is offered, charging 1.60 per runtime second via credit card or IBM Cloud credits [4]. As a practical example, executing a 2-qubit Grover's search algorithm [29] requires approximately 2μs per shot, with a usual 4096 shots per experiment the cost would be 0.013 USD.

Most quantum algorithms require thousands of shots for execution, where each shot corresponds to a single run of the algorithm. The measurement outcomes from all shots are aggregated to estimate output probabilities, which can either represent the final result of the algorithm, as in Grover's search, or guide optimization and subsequent iterations, as in the Variational Quantum Eigensolver (VQE) [46] and quantum machine learning[14]. Variations in the physical environment or operational errors of the NISQ device can directly affect these output probabilities, influencing the reliability and accuracy of algorithmic results.

4 Threat Models for QTEEs

The primary objective of this work is to safeguard users' quantum circuits from untrusted cloud-based quantum computing providers or malicious insiders. Potential adversaries may target different components of the quantum computing system to extract information about users' circuits. Accordingly, we define two threat models to guide the design of Quantum Trusted Execution Environment (QTEE) architectures. While other threat models may exist, they are left for future investigation.

4.1 Entities in the Threat Models

In all threat models considered in this work, we identify three entities: the users, the cloud provider (including potential malicious insiders), and the quantum computer manufacturer. Users possess sensitive data and computations that they wish to execute on a quantum computer. The cloud provider is responsible for managing the quantum computing infrastructure. The quantum computer manufacturer produces the physical quantum devices. Even when the cloud provider and manufacturer share the same brand, e.g., IBM, we treat them as distinct entities to differentiate between run-time security threats—arising from the cloud provider, and supply-chain or manufacturing security threats, arising from the manufacturer. This distinction is relevant in practice, as many cloud providers, such as Amazon Braket [1] and Microsoft Azure Quantum [25], do not manufacture the quantum hardware they provide.

4.2 Assumptions in the Threat Models

In this work, we assume an honest-but-curious cloud provider, meaning that the provider and any insiders faithfully execute requested operations while potentially attempting to exfiltrate sensitive information. Consequently, we consider side-channel attacks but do not address fault-injection attacks. Our focus is restricted to run-time attacks, and attacks on the supply chain or hardware modifications are deferred to future work.

4.3 Goals of the Attackers

We consider two primary adversarial objectives. First, the attackers may attempt to access the data processed by the quantum program. Second, they may seek to learn the quantum algorithm itself. The algorithm may constitute intellectual property that must remain confidential, or the owner may wish to avoid revealing implementation details.

4.4 Means for Attackers to Collect Information

We assume that attackers may use multiple methods to collect information. First, an attacker can directly access the programs or code submitted to the cloud provider. Currently, there are no technological mechanisms in place at IBM, Amazon, or Microsoft to prevent the cloud provider from viewing the code that users submit. This scenario corresponds to the untrusted cloud provider threat. Second, an attacker may collect traces of power, electromagnetic emissions, or other side-channel information. Even without direct access to the code, side-channel data could be analyzed to reverse engineer the program's control and data flow, potentially revealing both the algorithms and the data. This scenario corresponds to the malicious insider threat, in which an individual may lack logical access to user-submitted programs but can exploit physical access to perform side-channel attacks.

4.5 Threat Model A: Trusted QPU, Untrusted Controller

In this threat model, we assume that the user and the user's compiler are trusted, and that the user can generate transpiled circuits to be sent to the cloud provider for execution. We further assume that the user receives correct information about the properties of the quantum computer, enabling accurate transpilation for the target device. The quantum cloud provider is considered untrusted in the sense that it does not manipulate the transpiled circuits, the resulting control pulses, or any classical information sent for execution. However, the provider can observe this information in an attempt to infer the user's circuits or data.

We assume that the quantum processing unit (QPU) is trusted. In the case of superconducting qubit machines, the QPU corresponds to the dilution refrigerator that houses all qubits. This is analogous to the processor chip serving as the trust boundary in classical trusted execution environments, such as Intel SGX [18]. The dilution refrigerator maintains extremely low temperatures and target pressures, and any physical access or opening is easily detected. Any intrusion would disturb these conditions and effectively destroy the qubit states. Therefore, we assume that the cloud provider cannot access the dilution refrigerator without disrupting the quantum computation. All hardware inside the dilution refrigerator, including our added components and the existing hardware, is assumed to be correct, verified, and free of bugs.

We assume that the remaining components, including the controller and cloud servers, are untrusted. These components represent potential points of attack. For example, the cloud provider may directly read the code submitted by the user on the server, and a malicious insider may collect power traces from the controller to recover the executed control pulses. From these pulses, the corresponding quantum gates can be inferred, allowing the attacker to reverse engineer the quantum circuit. Details of this threat model are shown in Fig. 2a.

4.6 Threat Model B: Untrusted QPU, Trusted Controller

In this threat model, we assume that the user and the user's compiler are trusted, and that the user can generate transpiled circuits to be sent to the cloud provider for execution. We further assume that the user has correct information about the properties of the quantum computer, allowing proper transpilation for the target device. The quantum cloud provider is assumed not to manipulate the transpiled circuits, the resulting control pulses, or any classical information sent for execution. However, the provider may observe this information in an attempt to infer the user's circuits or data.

We assume that the quantum computer controller is trusted. Since these controllers are implemented using classical CPUs and FPGAs, there exist well-studied solutions, such as Intel SGX [18], that can protect these components. For example, any mod-

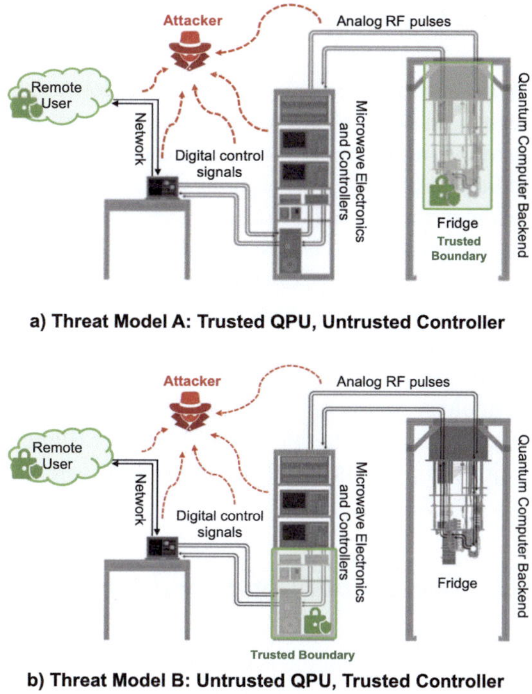

Fig. 2 Schematic of a typical superconducting quantum computer, showing an honest-but-curious cloud provider attempting to spy on the control pulses and equipment outside the trusted controller. The green boundaries are trusted components

ification to data or code provided to an SGX-enabled CPU can be detected using digital signatures, thereby preventing attacks at this level. Similar trusted execution environments for FPGAs have also been explored [44, 61].

We assume that the remaining components, including the cloud servers and the QPU, are untrusted. These components represent potential points of attack. For example, the cloud provider may directly read the code submitted by the user on the server, or the QPU could be manipulated to collect information about the executed quantum gates. A diagram illustrating this threat model is shown in Fig. 2b.

5 `SoteriaQ` and `CASQUE` Architectures

In the below sections, this book chapter presents two architecture for protection of quantum circuits: `SoteriaQ` and CASQUE. They adopt fundamentally different strategies for securing quantum computations, each with distinct trade-offs in performance and protection discussed below. The two architectures follow Threat Model A of untrusted controller, but trusted QPU. Our ongoing and future work explores Threat Model B, which is not discussed further in this chapter.

6 Hardware Architecture of SoteriaQ

In this section, we consider Threat Model A and leverage a trusted QPU to protect against an honest-but-curious and untrusted quantum cloud provider or malicious insiders. In this threat model, the trust placed in the QPU enables the integration of additional hardware protections within the quantum processor. The first architecture designed for this scenario is SoteriaQ, which targets superconducting qubit platforms. This section is based on our QC-TEE [59] and SoteriaQ [57] papers. The SoteriaQ work supersedes QC-TEE and the two are together simply referred to as the SoteriaQ architecture.

To realize SoteriaQ, several hardware components are integrated into the interior of the dilution refrigerator. All of these components are available with current technology and consume minimal power while occupying negligible area relative to existing quantum computing hardware. Figure 3 presents a block diagram of the major components and their connections, which are listed below.

6.1 Decryption Engine and Input Bitmap Memory

The Decryption Engine is responsible for decrypting the encrypted input bitmap and storing the resulting data in the Input Bitmap Memory. The engine first establishes a shared symmetric key using a post-quantum secure public-key cryptography protocol. It then applies post-quantum secure symmetric-key cryptography to decrypt the input bitmap. Both the public-key and symmetric-key algorithms are required to resist attacks from quantum-capable adversaries. All decrypted secrets are assumed

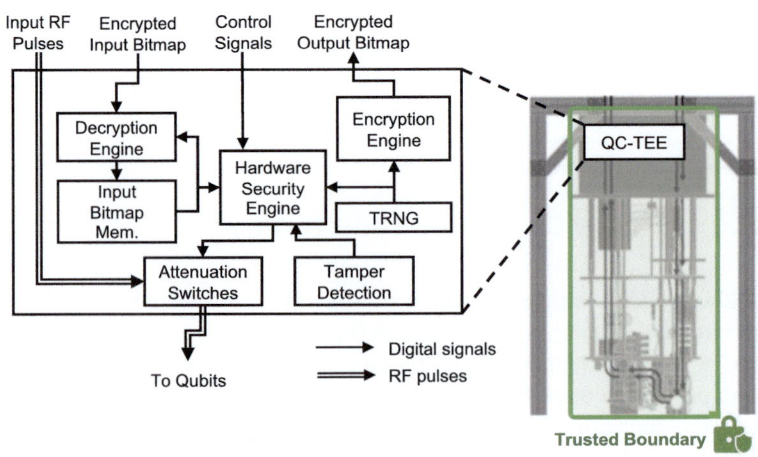

Fig. 3 Major hardware components of SoteriaQ design added to the internals of the dilution refrigerator holding the QPU

to remain secure from probing while inside the quantum computer's dilution refrigerator, consistent with the threat model described in Sect. 4.5.

6.2 Hardware Security Engine

The Hardware Security Manager transmits the bits of the decrypted input bitmap to the attenuation switches. Each drive and control channel is paired with a dedicated attenuation switch. During each 160dt time period, every drive and control channel for all qubits receives a single bit, where a value of 0 indicates no attenuation and a value of 1 indicates attenuation.

The Hardware Security Engine is implemented as a state machine that manages the attenuation switches described in Sect. 6.3. Once the Decryption Engine completes decryption of the ciphertext and stores the input bitmap in the Input Bitmap Memory (as shown in Fig. 3), the Hardware Security Engine is initialized and the circuit execution begins. As the RF control pulses are received, the Hardware Security Engine retrieves the corresponding bits from the Input Bitmap Memory and applies attenuation to the randomized pulses.

Once circuit execution begins, the Hardware Security Engine outputs one control bit per drive or control channel every 160dt. These bits are retrieved from the Input Bitmap Memory and instruct the switches to either attenuate the pulses (if the bit is 1) or pass them without attenuation (if the bit is 0) for the corresponding time period. Two-qubit gates are padded with delays to ensure their duration is an integer multiple of 160dt. As a result, multiple input bitmap bits are consumed for a single two-qubit gate, with one bit used for each 160dt interval.

In the randomize-output variant of the mixer algorithm (Algorithm 1), during execution of the final circuit layer, which consists of `X` gates, the Hardware Security Engine retrieves random bits from the TRNG. The number of bits matches the number of qubits. For each qubit, if the generated TRNG bit is 0, the corresponding signal is attenuated; if the bit is 1, the signal is passed without attenuation. This procedure is the inverse of the input bitmap specification. Under this convention, an output bitmap bit of 1 indicates that an `X` gate was applied and the qubit state was flipped. The value 1 can then be sent to the user, who applies an `xor` operation between the measured bit c_i and 1 to invert it. If no `X` gate is applied, the output bitmap bit is 0, and `xor` with 0 leaves the measured value unchanged.

6.3 Attenuation Switches

An advantage of superconducting qubit quantum computers, as leveraged in this work, is the presence of a dilution refrigerator that maintains qubits at cryogenic temperatures. The refrigerator establishes a physical boundary that is difficult to probe, thereby defining the region outside the refrigerator as the untrusted environment. This untrusted environment includes the cloud provider, who controls the pulse generators and classical control equipment responsible for delivering RF pulses to manipulate the superconducting qubits. However, the cooling power limitations of the refrigerator make it infeasible to introduce arbitrary additional equipment inside. As a result, naive TEE designs, such as placing a signal generator within the refrigerator, are currently impractical.

Attenuation switches are employed to attenuate decoy control pulses, which are introduced to mislead potential attackers and are not used for computation. Each RF switch requires a one-bit control signal to specify whether the incoming RF signal should be attenuated during the given time period.

Rather than introducing excessive and impractical equipment inside the dilution refrigerator, the proposed design incorporates simple RF switches to attenuate decoy control pulses sent to the quantum processor. The transpiled quantum program is augmented with additional gates that are not intended for execution but serve to obfuscate the computation from an honest-but-curious quantum cloud provider monitoring the RF pulse activity. The design operates as follows: (1) the user locally transpiles the quantum program and inserts dummy control and drive pulses, (2) a classical pulse mask indicating which pulses should be executed and which should be attenuated is encrypted, (3) the encrypted pulse mask is transmitted to the quantum computer, and (4) within the trusted boundary of the refrigerator, the pulse mask is decrypted and used to configure the RF switches to attenuate control pulses as specified.

The RF switches are passive components and do not generate control pulses themselves. Consider a single-pole double-throw (SPDT) RF switch, where the single pole is connected to the input line of the dilution refrigerator to receive the incoming RF signal. On the output side, one of the two throws is connected to the drive line leading to the qubit, while the other throw is terminated with a matched impedance. This way, the SPDT RF switch can either pass or isolate the drive pulses they use to operate the quantum processor, with its on-off state controlled by a DC gate voltage. In addition, they choose GaAs-based RF switches (such as CMD196C3 [2]) that can function at cryogenic temperatures, with a wide frequency range (DC-18GHz), low insertion loss (1.5 dB), high isolation (46 dB), and fast switching (2.5 ns). Existing work has already tested RF switches capable of routing microwave signals at cryogenic temperatures [45].

The control voltage of the switches is generated, achieved by the hardware security engine and the input bitmap. This "pulse mask" is encrypted by using quantum-safe encryption and it is only decrypted inside the refrigerator. They propose to mount the switches (power consumption of only 1 μW), the encryption and decryption engine (power consumption of only 20 mW), as well as the simple control state machine, all on the high-temperature stage (such as the 4 K stage) of the fridge, which provides ample cooling power (nearly 1 W) to suppress the heating effect of the switch operations. The control logic and switches need to operate at a rate between 1 and 200 MHz, which can be easily achieved. This is due to the fact that the single qubit gates in IBM machines each currently take 160dt, where 1 dt $= 0.222$ ns, and thus each gate has to be attenuated (or not to be attenuated) at a rate of 35.5 ns ($= 160 * 0.222$ ns) or equivalently at a frequency of 28.5 MHz. Due to the non-idealities of the on-state of the switch (such as the insertion loss or impedance mismatch), recalibrations of the phase and amplitudes of the single- and two-qubit gate pulses are required in order to optimize the gate infidelities.

6.4 Encryption Engine and TRNG

`SoteriaQ` provides an option to randomize circuit outputs. This is achieved by conditionally applying `X` gates immediately prior to the measurement operations at the end of the circuit. A true random number generator (TRNG) determines which `X` gates are applied, and this information is stored in the encrypted output bitmap sent to the user. Although numerous TRNG designs exist [35], the design of the TRNG is not the focus of this work.

The encryption engine is used when qubit flipping is enabled. It encrypts the output bitmap so that the user can correctly interpret the qubit measurement results. The engine employs symmetric-key encryption to secure the swapping bits using a randomly generated key and public-key encryption to encrypt the symmetric key with the user's public key. Post-quantum secure algorithms are required for the public-key encryption, while the symmetric-key encryption is implemented in hardware using AES-GCM [28].

The encryption engine can be implemented in hardware using AES-GCM [28]. It uses the shared secret established by the decryption engine (Sect. 6.1) as the symmetric key to encrypt the TRNG output, i.e., the output bitmap, when the *randomize-output* option is enabled. The client can later decrypt the encrypted output bitmap and conditionally apply logical NOT operations to any qubits indicated as flipped before measurement.

7 `SoteriaQ` Obfuscation Algorithm and Software

This section describes the algorithms for obfuscating control pulses through the insertion of decoy pulses on the user side and their subsequent removal in hardware within the trusted refrigerator of the quantum computer. The software algorithm accepts any transpiled quantum circuit as input, including circuits with custom gates, and performs gate-level operations to produce an obfuscated output circuit. Algorithm 1 provides a detailed description of the steps used to insert the decoy control pulses.

Algorithm 1 Decoy Pulse Insertion Algorithm. $Slot_{CX}$ is a slot that consists of at least one CNOT gate. $Slot_{SQ}$ is a slot that consists of only single-qubit gates.

```
Input:   Transpiled Quantum Circuit (QC_IN), Backend
         information (B_INFO), Obfuscation Level (OL)
Output:  Randomized Transpiled Quantum Circuit (QC_OUT),
         Input Bitmap (IB[m][n]) is used as input to the
         quantum computer backend

Step 1:  Split Input Circuit into Slots (Sec. 5.1.1)
         1a. Scan B_INFO and identify possible CNOT couplings
             and identify the longest CNOT duration
             (MAX_CNOT_DUR) and single qubit gate delay
             (SQ_DUR)
         1b. Round MAX_CNOT_DUR up to the next even multiple
         of single-qubit gate delay, CEIL_MAX_CNOT_DUR
         1c. Scan QC_IN and place a barrier before and after
             each CNOT gate to separate single-qubit gates
             and CNOT gates into separate slots
         1d. Insert delays in each slot:
             if (Slot_CX)
                 Insert delay to pad duration of the slot to
                 CEIL_MAX_CNOT_DUR
             else if (Slot_SQ) then
                 if (Obfuscation Level == one-sixteenth-delay)
                     Insert delays to pad duration of the
                     slot to CEIL_MAX_CNOT_DUR/16
                 else if (Obfuscation Level = quarter-delay)
                     Insert delays to pad duration of the
                     slot to CEIL_MAX_CNOT_DUR/4
                 else if (Obfuscation Level = max-delay)
                     Insert delays to pad duration of the
                     slot to CEIL_MAX_CNOT_DUR

Step 2:  Insert Decoy Gates into Sub-Slots (Sec. 5.1.2)
         2a. Replace delays with decoy gates:
             if (Slot_CX)
                 for each coupling without a CX gate randomly
                 insert CX gate, or a random mix of X and
                 SX gates totaling CEIL_MAX_CNOT_DUR; mark the
                 position of decoy gates in Input Bitmap (IB)
             if (Slot_SQ)
                 replace each delay with a random mix of X
                 and SX gates; mark the position of the decoy
                 gates in Input Bitmap

Step 3:  Generate Input Bitmap (Sec. 5.1.3)
         3a. Capture information from Steps 1 and Step 2
             for all drive and control channels C[1] to C[m]
             in IB[1][1] to IB[m][n] (where m is the number
             of drive and contorl channels, n is the number
             of sub-slots on each qubit) and generate the IB:

             for (i from C[1] to C[m])
                 for (j from IB[i][1] to IB[i][n])
                     if (gate == decoy gate)
                         B[i][j] = 1
                     else
                         B[i][j] = 0

Step 4:  Add X Gate for Randomized Output (Sec. 5.1.4)
         4a. if (randomize-output == True) then
             Append one X gate on each qubit drive channel
             at the end of the circuit
```

7.1 Circuit Preparation

The algorithm operates on an input circuit that has already been transpiled for the target backend and partitions it into discrete slots. During circuit processing, barriers are inserted before and after each CNOT gate to separate regions of single-qubit gates from those containing CNOT gates. Each barrier thereby marks the start of a slot. These barriers are removed after preparation and are used solely for organizing the circuit during the obfuscation process.

$Slot_{CX}$ is a slot that consists of at least one CNOT gate and optionally any single qubit gates on the other qubits where the CNOT gate is not connected. The duration of $Slot_{CX}$ slots is determined by choosing the longest CNOT gate duration out of all possible couplings on the backend, and then rounding it up to be an even integer multiple of the single-gate duration. Having set the $Slot_{CX}$ duration this way, all $Slot_{CX}$ slots are checked and delays are added so that each $Slot_{CX}$ slot has the same length.

$Slot_{SQ}$ is a slot that consists of only single qubit gates and optionally any delays if required. The duration of these slots is determined at compile time and chosen based on the target security level. Currently, three duration periods are suggested: *one-sixteenth-delay*, *quarter-delay* and *max-delay*. One-sixteenth-delay means the duration of $Slot_{SQ}$ is one sixteenth of the duration of $Slot_{CX}$. Quarter-delay means the duration of $Slot_{SQ}$ is a quarter of the duration of $Slot_{CX}$, recall $Slot_{CX}$ slots are set to be of duration this is even multiple of single-qubit gate duration. Max-delay means the duration of $Slot_{SQ}$ is equal to the duration of $Slot_{CX}$.

The output of this circuit preparation step is a circuit that consists of multiple slots. The number of slots is determined by the depth of the circuit in terms of CNOT gates: whichever qubit has the most CNOT gates on it, which determines how many slots the circuit will be divided into. In our experience it is unlikely, but possible, that the number of single-qubit gates between any two CNOT gates can be more than what can fit in a duration of $Slot_{CX}$. In such corner case, there will be two (or more) $Slot_{SQ}$ between consecutive $Slot_{CX}$. In the usual case, however, $Slot_{CX}$ and $Slot_{SQ}$ occur in alternating order. Extra $Slot_{CX}$ and $Slot_{SQ}$ can also be added to increase confusion for the attacker, as well as to make it harder to guess the structure of the user's circuit.

7.2 Decoy Gate Insertion

Each slot is further subdivided into equal-length sub-slots, with the length of a sub-slot corresponding to the duration of a single-qubit gate, which is 160dt on IBM quantum computers. Figure 4a illustrates the input circuit with its slots and sub-slots. Any sub-slots not occupied by an actual gate can be filled with decoy gates.

For each $Slot_{CX}$ in the circuit, on the unused qubits in that slot we either add a CNOT gate if CNOT coupling of the backend allows, or we add a random mix of X

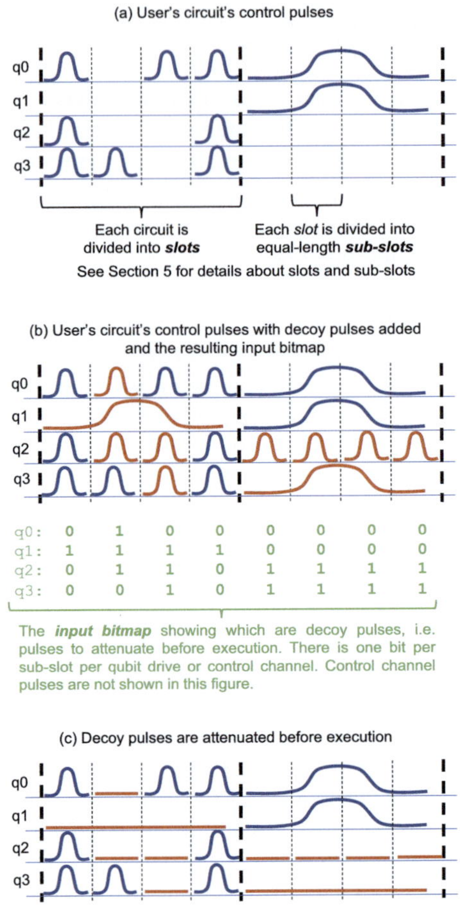

Fig. 4 Example illustration of how control pulses are obfuscated, the pulses shown are not real control pulses but their simplified graphical representation. **a** Control pulses correspond to the input circuit of the user. **b** Control pulses with decoy pulses added, along with the input bitmap, specifying which are real control pulses and which are decoy pulses to be attenuated. **c** Original control pulses are executed on the quantum computer, and after the attenuation of decoy pulses, attenuation is done based on the information from the input bitmap. The time is quantized into *sub-slots* of 160 dt. Each pulse takes an integer multiple of 160 dt, allowing for 1 a bit in the input bitmap per qubit per *sub-slot* period to specify if the pulse (in that time period on the corresponding qubit) should be attenuated or not. The input bitmap is encrypted before being sent to the cloud provider

and SX gates; addition of other and custom gates is possible as well. Figure 4b shows in red color the addition of a decoy CNOT gate next to user's CNOT gate in a $Slot_{CX}$. Note that the decoy CNOT gates are later attenuated, and the user's circuit's depth in terms of CNOT gates per qubit does not increase with our architecture.

For each $Slot_{SQ}$, in addition to the existing single-qubit gates from the original circuit, a random mix of X and SX gates is added in each empty sub-slot; addition of other and custom gates is possible as well. Figure 4b shows in red color the addition of decoy single-qubit gates in a $Slot_{SQ}$. The Rz gates from the original circuit are added to the slots but are not counted toward a sub-slot, because the Rz gates are virtual.

Our design supports dynamical decoupling as well as the optimizations provided by the target platform. We implement a decoy-to-identity gate conversion scheme. The output of the obfuscation process is a transpiled circuit that includes a number of decoy pulses. During decoy pulse insertion, sequences of two consecutive X decoy gates or four consecutive SX decoy gates may be introduced. Each such sequence implements an identity operation. These identity sequences can be executed by configuring the corresponding bits in the input bitmap to indicate that attenuation should not be applied. Additionally, any dynamical decoupling inserted in software prior to decoy gate insertion is preserved.

7.3 Input Bitmap Generation

While constructing the quantum circuit with added decoy gates, we keep track of each random single-qubit gate that is inserted. Each slot is divided into sub-slots of equal duration, corresponding to the duration of a single-qubit gate. One single-qubit gate fits in one sub-slot, and for each qubit in each sub-slot, we generate a binary bit to indicate whether the gate is an actual control pulse or a decoy pulse. A bit value of 1 indicates that a decoy gate has been inserted for obfuscation and must be attenuated before reaching the quantum computer's qubits, so it does not perform any operation. A bit value of 0 indicates that the gate is part of the original input circuit and should be executed without attenuation. These bits are stored in the input bitmap. An example of the input bitmap is shown in Fig. 4b.

7.4 Output Randomization

An additional layer of X gates can be added at the end of the circuit to randomize the output (shown in Step 4 of Algorithm 1). The randomizing of the output happens inside the dilution refrigerator on the server end. This process is described in Sect. 8.

7.5 Circuit Post Processing

At the end, the barriers are removed from the quantum circuit while the inserted gates remain. The resulting circuit is a valid circuit that can be executed on the target backend. However, unless the decoy gates are removed, i.e., attenuated, the result of the computation will be random. The purpose of `SoteriaQ` hardware, described later, is to attenuate the decoy gates, i.e., pulses, based on the (encrypted) *input bitmap* information, before these pulses actually reach qubits.

7.6 Circuit Execution

Once the algorithm completes, the input bitmap is encrypted and signed using quantum-safe cryptographic algorithms prior to transmission to the cloud provider. The encrypted input bitmap, along with the plaintext circuit containing decoy gates, is sent to the provider, which is considered untrusted and honest-but-curious. The cloud provider schedules and executes the circuit on the target quantum computer, while the hardware inside the dilution refrigerator is trusted. Specifically, the `SoteriaQ` hardware within the trusted boundary decrypts and validates the input bitmap and determines which control pulses to attenuate. All computationally intensive signal generation components remain outside the refrigerator.

8 `SoteriaQ` Deobfuscation and Hardware

The `SoteriaQ` hardware and state machines effectively implement the inverse of the obfuscation. When the circuit is about to execute, the ciphertext (the encrypted *input bitmap*) is sent to the decryption engine (described in Sect. 6.1) inside the fridge. The AES-GCM module which is part of the decryption engine decrypts the ciphertext and re-generates the input bitmap containing information about which control pulses in which sub-slots should be attenuated. As shown in Fig. 3, this input bitmap is stored in the input bitmap memory. After this, the hardware security manager and the signal/pulse generator are notified to start generating the RF signals involved in the quantum circuit (including the decoy pulses). The attenuation switches inside the fridge (shown in Fig. 3) filter the unwanted random gates based on the control signals generated by the hardware security manager. We propose to achieve the synchronization between the hardware security manager and signal/pulse generator using handshake signals. Since the original quantum circuit is only deobfuscated inside the dilution refrigerator (which we consider a trusted boundary), we achieve the required secrecy between the client and server.

To help hide the output of the circuit, `X` gates are inserted at the end of the circuit, and during execution, the control logic generates corresponding `X` control pulses.

However, when these X control pulses reach the quantum computer, SoteriaQ will randomly attenuate them, thus the cloud provide does not actually know which of the pulses were executed, i.e., which output was flipped. In software, only X gates are inserted. The decision to attenuate them is performed randomly at runtime on SoteriaQ hardware by using the TRNG. The random bits used to specify whether to attenuate or not the final X gates are also encrypted and sent back to the user so he or she can know how to interpret the outputs.

9 SoteriaQ Evaluation Setup

To evaluate the fidelity overhead of SoteriaQ's hardware additions, we use the 7-qubit IBM machine (backend) ibm_perth to test small-scale benchmarks, and 27-qubit ibm_algiers to test medium-scale benchmarks. We also use the Aer_Simulator with imported noise model from other different machines.

9.1 Benchmarks Used

QASMBench Benchmark Suite version 1.4 [30] was used in our work to analyze the impact of the SoteriaQ design on different circuits. We transpile each benchmark for the target backend. The transpiled code is then used as input to the obfuscation algorithm, which is described in Sect. 7. We test small-scale circuits on the free 7 qubit backend, as well as, we use the pay-as-you-27 qubits backend for medium-scale circuits. Overall, we test our design on 22 benchmarks. We keep the number of shots constant at 8192. For the simulator, we use the same set of benchmarks.

9.2 Obfuscation Levels Evaluated

We evaluated different configurations: *baseline*, *one-sixteenth-delay*, *quarter-delay*, *max-delay*; further for all the levels, except baseline, we analyze results with and without the *randomize-output* option. The *baseline* is simply the benchmark transpiled for the target backend without any of our modifications. The *one-sixteenth-delay* means the duration of $Slot_{SQ}$ is one sixteenth of the duration of $Slot_{CX}$. The *quarter-delay* means the duration of $Slot_{SQ}$ is a quarter of the duration of $Slot_{CX}$, recall $Slot_{CX}$ slots are set to be of duration, which is even multiple of single-qubit gate duration. The *max-delay* means the duration of $Slot_{SQ}$ is equal to the duration of $Slot_{CX}$. The *randomize-output* includes an additional layer of X gates before measurement to protect the output, it can be applied to any of the obfuscation levels.

9.3 Variational Distance

We measure the impact of our modifications on the circuits by using variational distance (VD). Informally, the variational distance of two output probability distributions is the measure of how one probability distribution is different from the other. In general, the total variation distance between P and Q is defined below. We compute variational distance between the circuit outputs generated without our modifications, and the outputs generated for different obfuscation levels.

$$VD(P, Q) = \frac{1}{2} \sum |P - Q|$$

9.4 Circuit Fidelity

We also measure fidelity (F) by performing state fidelity computations. The state fidelity computations use the density matrix of each state. Formally, the state fidelity F for density input states ρ_1 and ρ_2 is given by

$$F(\rho_1, \rho_2) = \left[\mathrm{tr} \left(\sqrt{\sqrt{\rho_1} \rho_2 \sqrt{\rho_1}} \right) \right]^2$$

In order to show the worst possible effects due to `SoteriaQ`, we compute the fidelity between the best and worst possible configuration, i.e., between baseline (the unmodified benchmark) versus the max-delay. This evaluation is discussed in greater detail in Sect. 10.4.

10 Evaluation of `SoteriaQ`

In this section, we present the evaluation of our design. Since we are not able to modify IBM Quantum computers to insert the RF switches and the `SoteriaQ`'s logic, we focus on the evaluation of the expected impact of the changes on the fidelity of the circuits.

10.1 Impact of Increased Circuit Duration

Our obfuscation technique introduces decoy pulses, which are subsequently attenuated prior to execution. Once attenuated, the decoy pulses effectively act as delays. With ideal RF switches, the executed circuit is identical to the user-provided circuit,

Table 1 Worst case variational distance (VD) for the QASMBench benchmarks considering: perfect RF switches, emulating imperfect RF switches which attenuate the decoy pulses to 0.01% amplitude, and considering our identity gate conversion scheme

Obfuscation level	Max-delay	
	w/o rand.-out.	w/ rand.-out.
Avg. VD (perfect RF switches)	0.2099	0.2453
Avg. VD (imperfect RF switches)	0.2652	0.2691
Avg. VD (imperfect RF switches w/ identity gate conversion)	0.1998	0.2188

except for additional delays corresponding to the decoy pulses. When RF switches are imperfect, the locations intended for full attenuation contain residual pulses, introducing noise and reducing the circuit fidelity.

Manufacturer datasheets indicate that high-isolation RF switches can achieve extremely low leakage rates. For instance, Pasternack specifies that their cryogenic RF switches can achieve isolation levels of up to 90 dB at cryogenic temperatures. This isolation level translates to a leakage of approximately 10^{-9} of the original signal amplitude. We model imperfect switches as switches that attenuate the decoy pulses to 0.01% of the original amplitude, so they do not fully remove them as ideal switches would. Regardless of the switch attenuation effectiveness, by using our identity gate conversion technique, some of the lost fidelity can be recovered. The variational distance results for the worst case scenario, which is the *max-delay* obfuscation level, are shown in Table 1. The assumption of 0.01% amplitude transmission is conservative and ensures that our security analysis remains robust. Even if the actual leakage is lower, our security guarantees would still hold, and the fidelity impact would be less significant than anticipated.

10.2 Sensitivity to Quantum Volume and Noise

As quantum computers improve, we expect the impact of the noise to decrease. Indeed, we explored if and how quantum volume (QV) and noise model may impact the variational distance. Table 2 reports the average VD over the set of the QASMBench benchmarks, and we observe a lower average VD for higher quantum volume machines.

Table 2 Worst case variational distance (VD) for the QASMBench benchmarks for machines with different quantum volume (QV). The VD is averaged over all the benchmarks and is reported for the *max-delay* obfuscation level, augmented with the *randomize-output* option. The data is for ideal RF switches

Obfuscation level	Max-delay	
	w/o rand.-out.	w/ rand.-out.
Real IBM Perth ($QV = 32$)	0.2099	0.2453
Sim. IBM Perth ($QV = 32$)	0.1461	0.1481
Sim. IBM Mumbai ($QV = 64$)	0.1653	0.1705
Sim. IBM Cairo ($QV = 128$)	0.1286	0.1299

10.3 Evaluation of Circuit Correctness

We evaluated the impact of the inserted decoy gates in the correctness of circuit outputs. In Table 3, we examined and show in detail the correctness of the output for 20 tested small-scale benchmarks and 2 medium-scale benchmarks for the three different configurations, with and without the *randomize-output* option considering imperfect switches that attenuate decoy gates to 0.01% of the original amplitude. We observe 100% correctness when we apply the *one-sixteenth-delay* configuration and nearly 90% correctness on average with *max-delay*. To emphasize, we observe correct output even for benchmarks with high depths, such as *dnn* and *qaoa*, which implies that our design can be scalable for circuits with more qubits and higher depths. Please note that for the evaluation of correctness, we compared the output of the baseline benchmark circuits without the addition of the decoy gates, with the output of modified benchmark circuits after inserting the decoy gates. If we are able to observe the true dominant state in both cases, we put a checkmark for this configuration. Although added decoy gates degrade the fidelity of the circuits, in all tested algorithms, we are able to observe correct outputs if the variational distance is less than 0.25.

10.4 Evaluation of Circuit Fidelity

We also evaluated the state fidelity between the quantum states of each modified with the inserted delays benchmark, comparatively with the quantum states of the baseline unmodified benchmark, as shown in the last column of Table 3. We only performed experiments for the worst-case scenario (max-delay). We observe that, except one benchmark, the fidelity remains high. We observe *basis_trotter* to have the worst behavior regarding the fidelity loss. This is understandable, as the circuit uses 4 qubits but is quite deep, challenging the decoherence times of a NISQ device, even if we run it without applying our `SoteriaQ` scheme.

Table 3 Evaluation of correctness and fidelity for 22 QASMBench benchmarks [30], assuming imperfect RF switches as simulated by gates representing decoy gates attenuated to 0.01% of the original amplitude of the gate on IBM *ibm_perth* quantum computer; number of Gates, CX, and Depth refer to pre-transpilation numbers. Checkmarks indicate that we are to observe the true, i.e., correct, dominant state. The fidelity experiments are performed on Aer_Simulator with an imported noise model from the 127-qubit IBM Brisbane quantum computer. The evaluation is done for the worst case of max-delay configuration

Benchmark	Qubits	Gates	CX	Depth	Correctness						Fidelity
					One-sixteenth-delay		Quarter-delay		Max-delay		Max-delay
					w/o rand.-out.	w/ rand.-out.	w/o rand.-out.	w/ rand.-out.	w/o rand.-out.	w/ rand.-out.	w/ rand.-out.
deutsch	2	5	1	5	✓	✓	✓	✓	✓	✓	0.978
iswap	2	9	2	8	✓	✓	✓	✓	✓	✓	0.984
quantumwalks	2	11	3	8	✓	✓	✓	✓	✓	✓	0.982
grover	2	16	2	12	✓	✓	✓	✓	✓	✓	0.972
dnn	2	226	42	155	✓	✓	✓	✓	✓	✓	0.998
teleportation	3	8	2	7	✓	✓	✓	✓	✓	✓	0.995
qaoa	3	15	6	12	✓	✓	✓	✓	✓	✓	0.994
toffoli	3	18	6	13	✓	✓	✓	✓	✓	✓	0.991
linearsolver	3	19	4	12	✓	✓	✓	✓	✓	✓	0.931
fredkin	3	19	8	12	✓	✓	✓	✓	✓	✓	0.906
basis_change	3	53	10	22	✓	✓					0.889
adder	4	23	10	12	✓	✓	✓	✓	✓	✓	0.825
bell	4	33	7	14	✓	✓	✓	✓	✓	✓	0.987
qft	4	36	12	9	✓	✓	✓	✓	✓	✓	0.994
variational	4	54	16	34	✓	✓					0.828
vqe	4	89	9	28	✓	✓	✓	✓	✓	✓	0.995
basis_trotter	4	1626	582	815	✓	✓					0.401
qec_en	5	25	10	18	✓	✓	✓	✓	✓	✓	0.862
error_correctiond 3	5	114	49	78	✓	✓	✓	✓	✓	✓	0.764
qaoa	6	270	54	110	✓	✓	✓	✓	✓	✓	0.945
bv	19	56	18	22	✓	✓	✓	✓	✓	✓	0.887
wstate	27	157	52	55	✓	✓	✓	✓	✓	✓	0.835

10.5 Evaluation of Power Requirements

Since we propose to add additional (minimal) hardware to the fridge, we consider the power consumption of these changes, since the hardware running our algorithms would be inside the dilution refrigerator that has limited cooling and power capabilities. For the public key cryptographic algorithm, we note that the existing work [53] shows that for CRYSTALS-Kyber with the highest security level, the Key Generation, Encapsulation, and Decapsulation consume 157 mW, 160 mW, and 162 mW, respectively on a Xilinx Artix 7 FPGA. And for symmetric key cryptography, AES-GCM [28] consumes 19 mW for encryption/decryption on Altera Cyclone V FPGA. We note that while we were writing this work, no prior work on running cryptographic algorithms in cryogenic temperatures was documented in the literature to the best of our knowledge. However, we note that other applications, such as [17], have been successfully tested in cryogenic temperatures ($4K$). Since all cryptographic operations do not need to run at the same time, we estimate a needed budget of 180 mW for our added hardware. Today, 1W cooling power at the $4K$ stage of the fridge should be easily achieved [26], thus our additions require only about 20% of the cooling budget, and this will become less with newer and better fridges.

10.6 Evaluation of Area Overhead

The main overhead of the scheme comes in terms of power, evaluated in above in Sect. 10.5, as well as physical overhead of the `SoteriaQ`'s logic and RF switches. Considering XLD Blue Fors dilution refrigerator [3], the volume of the refrigerator with height of about 1481 mm and radius of about 460 mm is $\pi \times 460^2 \times 1481 = 984,011,944\text{mm}^3$. Bigger refrigerators will be needed for larger quantum computers, but taking this as a conservative estimate, our analysis shows the percentage of volume is taken by the added hardware for quantum computers with different qubit sizes. We observed that the overhead in all cases is 0.002% or less.

10.7 Scalability Considerations

The proposed `SoteriaQ` can scale to handle large number of qubits. We note that publicly announced designs for large quantum computers are modular. For example, the forthcoming Flamingo processor [23] is projected to support 1,386 qubits, which will be located in three 462-qubit modules or chip. One `SoteriaQ` can be deployed per module or chip. Our analysis shows that in all cases the expected power consumption of the `SoteriaQ` logic for different size quantum computers, from 27 to 1386 qubits [23], will be nearly 180 mW. Note that larger quantum processors, such as Kookaburra, will be built from these smaller modules and would simply require a number of `SoteriaQ` hardware proportional to the number of the quantum computers.

11 Security Analysis of `SoteriaQ`

To evaluate the security of our proposed scheme, we compute the total number of possible circuits that the obfuscated circuit could represent. This is the attack complexity from the perspective of the honest-but-curious cloud provider or insider attacker. In the computation, we take into account the following aspects: number of qubits (n_{qubits}), number of `CNOT` gate slots ($n_{Slot_{CX}}$), number of single qubit gate slots ($n_{Slot_{SQ}}$), number of gates which fit in a slot $Slot_{SQ}$ on each qubit ($n_{SubSlots}$), number of `CNOT` gates on different qubits in a $Slot_{CX}$ ($n_{SubCXInSlotCX}$), number of single-qubit gates which fit on qubits not consisting `CNOT` gates in $Slot_{CX}$ ($n_{SubSlotsInSlotCX}$). The number of combinations ($Comb$) that the malicious cloud provider must try to find the correct circuit executed inside the fridge is

$$Comb = ((2^{n_{SubSlots}})^{n_{Slot_{SQ}}})^{n_{qubits}} \times ((2)^{n_{Slot_{CX}}})^{n_{SubCXInSlotCX}} \times ((2^{n_{SubSlotsInSlotCX}})^{n_{Slot_{CX}}})^{(n_{qubits} - 2 \times n_{SubCXInSlotCX})}$$

Figure 5 shows how many circuit combinations an attacker (without any knowledge of the algorithm) needs to guess the circuit, versus the circuit duration increase for different obfuscation levels. In general, we observe the number of possible combinations an attacker would need to guess is significantly beyond 2^{256} security level, making it nearly impossible for attackers to deduce the true circuit. Based on Fig. 5, we also observe that *one-sixteenth-delay*, the lowest obfuscation level, is sufficient to confuse the quantum cloud provider or insider attacker to be unable to apply brute force search in order to extract the circuit for most of the benchmarks. For example, for QAOA benchmark, which uses 6 qubits, 270 gates, 84 `CNOT` gates (after transpilation), the attack complexity for *max-delay* is 2^{12264}, here we assume $n_{SubSlots} = 18$, $n_{Slot_{SQ}} = 84$ (between each `CNOT` gate slot there is one non-`CNOT` gate slot), and $n_{qubits} = 6$; this is representative of the `ibm_perth` backend.

Our method can be also applied to hide the information of the victim circuits, even if attackers have some knowledge of the victim circuits. For some algorithms, the overall structure of the circuit, i.e., how gates are connected, may be fixed, while the specific configuration parameters of the circuit are encoded into parameters of the gates, e.g., rotational angles for VQE and QAOA. Both the algorithm and the specific configuration parameters need to be known to make them useful for attackers. Knowledge of the algorithm helps to decrease the attack complexity somewhat, but attackers still need to try to find the specific configuration parameters.

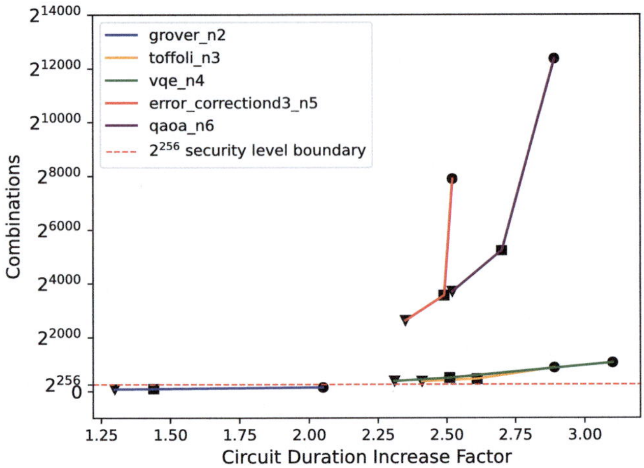

Fig. 5 Combinations required by an attacker to guess the circuit vs circuit duration increase factor for each of the three obfuscation levels. We assume the attacker does not have any knowledge of the circuit. ▼: *one-sixteenth-delay*, ■: *quarter-delay*, •: *max-delay*

As an example, for QAOA, there is typically one rotational gate between each CNOT gate pair. Therefore, if the attacker knows the circuit to be a QAOA circuit, then he or she knows that only one rotational gate should be between each CNOT gate pair, and he or she can use that knowledge to eliminate some circuit combinations from their guesses. Since we break down the circuit into slots, this means for a single-qubit gate slot between two CNOT gate slots, the approximated lower bound of the number of choices for the attacker to guess the correct QAOA configuration is about ${n_{SubSlots}}^{n_{cp}}$, where n_{cp} is the number of CNOT gate pairs in QAOA and $n_{SubSlots}$ is a number of gates which fit in a slot $Slot_{SQ}$ on each qubit.

We note that how the circuit is transpiled also affects the attack. Since not all qubits are connected on most quantum computers, swap operations may be inserted at transpliation time; swap operations are realized also using CNOT gates. The number of added swap operations is highly dependent on the circuit and the hardware, but an upper bound can be approximated. Suppose that there are $n_{qubitsInBackend}$ qubits in the quantum processor, then $n_{qubitsInBackend} - 1$ swap operations need to be added in the worst case to perform the CNOT gate between the two most far-apart qubits. Then, for n_{cp} CNOT gate pairs, in the worst case in the transpiled circuit there will be $n_{c\_switch} = n_{cp} \times 3 \times (n_{qubitsInBackend} - 1)$ CNOT gates since one switch gate consists of 3 CNOT gates. The approximated upper bound of the number of choices for an attacker to guess the correct QAOA configuration will be ${n_{SubSlots}}^{n_{cp}+n_{c\_switch}}$.

For instance, the same example of QAOA benchmark, which uses 6 qubits, 270 gates, 84 CNOT gates (after transpilation), if attacker knows this is QAOA benchmark, the attack complexity for *max-delay* is about $18^{798} \approx 2^{3330}$, here we assume

$n_{c\_switch} = 42 \times 3 \times (7 - 1) = 756$ for 7-qubit `ibm_perth` backend for example and $n_{SubSlots} = 18$. Even with some knowledge, the attack complexity for the attacker is significantly beyond 2^{256} security level.

11.1 Additional Possible Attacks

If the cloud provider has more knowledge, like a list of possible circuits and their duration in time, they can leverage this information plus the real execution timing to perform a circuit identification attack. In order to protect from such timing side-channel attacks, the next step is to make the number of slots to be variable and not directly depend on the circuits. For example, a random number of extra $Slot_{CX}$ and $Slot_{SQ}$ can be added into the circuits, where each such slot is full of decoy control pulses. Users can add such random gates in software independently of our architecture.

There can also be power side-channel attacks on quantum computers, which have been proposed in concurrent work in [62]. However, the authors only assume the power can be measured from the drive equipment. If the on and off states of RF switches have an influence on the power and can be measured by malicious providers, then the providers may be able to recover the real circuits. Our work assumes the fridge forms a trust boundary and power attacks on equipment inside the fridge are out of scope.

12 `SoteriaQ` Summary

`SoteriaQ` is a hardware-software framework designed to protect quantum programs from an honest-but-curious cloud provider. On the software side, it obfuscates quantum circuits by inserting decoy control pulses into a transpiled program. These decoy pulses do not perform computation and are tracked using an input bitmap, which indicates which pulses are real and which are decoys. The input bitmap is encrypted with quantum-safe cryptography and sent to the cloud along with the obfuscated circuit.

On the hardware side, trusted components inside the dilution refrigerator, including RF switches and a hardware security engine, use the decrypted bitmap to attenuate decoy pulses before they reach the qubits. Optional randomization of outputs is supported via conditional `X` gates controlled by a TRNG, with the results encrypted and returned to the user. All expensive RF generation remains outside the fridge, while the trusted hardware ensures that the cloud provider cannot distinguish real computation pulses from decoys, preserving privacy and security of the quantum program.

13 CASQUE Architecture

In this section, we again consider Threat Model A, and leverage the trusted QPU to provide protections from the honest-but-curious untrusted quantum cloud provider or insiders. In this threat model, since the QPU is trusted, additional hardware can again be introduced into the QPU to protect from the cloud providers. This second architecture for protection under this threat model is CASQUE. CASQUE targets superconducting qubit architectures. This section is based on our CASQUE [54] paper.

13.1 Pulse Switching

The key idea of CASQUE is pulse switching. Today, a cloud provider can directly see which pulses execute on which qubit. However, if new hardware is added to allow to switch any control pulse to any channel, then for each time period there are $\binom{n}{k}$, i.e., *n choose k*, possibilities that *k* pulses can execute on *n* qubits. Within limitations discussed later, after a circuit is transpiled, the control pulses can be re-arranged in $\binom{n}{k}$ ways in each time period and the re-arrangement information can be saved so that the pulses can be switched back during execution. The re-arrangement in software is a simple modification of digital data representing the circuit. In the hardware, the re-arrangement can be achieved by use of a Beneš Networks.

13.1.1 Single-Qubit Gates

Single qubit gates such as X and SX gates can be switched between any qubit channel. The amplitude of the pulses is different for different qubits, thus if there is any switching, the amplitude has to be adjusted. To simplify the CASQUE design, we assume all single qubit pulses on all channels will be initially sent at maximum amplitude, and then the trusted hardware attenuates them according to the target qubit. Since every incoming pulse has the same initial amplitude regardless of the channel, the attenuation hardware only needs to know the target channel and does not require knowledge of the initial amplitude or the channel of origin.

13.1.2 Two-Qubit Gates

Two qubit gates such as CNOT gate cannot be switched between channels, at least in current IBM Quantum computer designs. The reason is that for different qubits and couplings, the exact pulses, not just their amplitudes, are different for different couplings. In base CASQUE, we assume that CNOT gates will not be switched. In CASQUE+ we discuss how to increase attack complexity by adding dummy qubits to the user's programs (as long as the quantum computer backend has sufficient physical qubits to accommodate the original qubits and the dummy qubits). Dummy qubits increase the number of possible locations for switching pulses, increasing n in the $\binom{n}{k}$ number of possible combinations. We further leverage the dummy qubits to enable the addition (and elimination) of dummy single-qubit gates, as well as two-qubit gates, such as `CNOT` gates. The dummy gates can be switched to the dummy qubits, effectively neutralizing them, while the cloud provider remains unaware, as they only observe the transpiled circuit specification, which includes both actual and dummy qubits and gates.

13.2 Pulse Protection Map (PPM)

We propose storing the re-arrangement information in a new data structure called the Pulse Protection Map (PPM). The PPM can be interpreted as a list specifying the switching configuration for each time interval. The transpiled circuit can be readily adapted to quantize time into fixed intervals. For example, single-qubit gates on IBM Quantum today execute in 160dt time, while two-qubit gates have variable timing, but can be easily padded so that each two-qubit gate takes a duration that is a fixed multiple of 160dt. As a result, the PPM can be viewed as a list of control bits specifying the switching of pulses at each time period. The PPM needs to be encrypted so that the cloud provider does not access it. Inside the trusted hardware, the PPM can be decrypted. The decrypted control bits can be used directly to control the Beneš Network, and no real-time computation is needed—simply the control bits from PPM can be sent to the RF switches.

13.3 Attenuation and Phase Map (APM)

While control pulses can be switched across different channels, the same pulse exhibits slight variations depending on the channel. For example, the control pulse specifying `X` gate executing on qubit 0 may have a different amplitude than the control pulse specifying `X` gate executing on qubit 1. The phase of the pulses may also have to be adjusted. Further, since we want to enable switching any pulse to any channel, the input pulse should have the maximum amplitude of all the channels. Thus, in the transpiled circuit, all the control pulses are at maximum amplitude. This means, regardless of the circuit, the input pulses going to the quantum computer will be at maximum amplitude, and the attenuation amount does not depend on the circuit. The attenuation amount and any phase change information can be stored in the Attenuation and Phase Map (APM). This is public information, since the properties of the quantum computer, such as shapes and amplitudes of control pulses, are known. The APM can be stored on trusted hardware without any protection. The APM is used by our trusted `CASQUE` logic for amplitude and phase controller (APC) which we introduce into the dilution refrigerator.

13.4 Measurement Obfuscation

After all the pulses are switched and the circuit finishes execution, there is a need to protect the measurements. To obfuscate the measurement results, the best approach is to randomly flip qubits before measurement by using `X` gates at the end of the circuit. However, gates cannot be simply "added" to the circuit in the dilution refrigerator. As an alternative, we assume the user transpiles his or her circuit with a layer of `X` gates at the end of the circuit, on half of the qubits used by the circuit qubits. During execution of the circuit, each `X` gate will be randomly switched to a different channel. Half of the qubits will not have `X` gate applied and their output is not flipped, while the other half will have the `X` gate applied and the output will be flipped. Assuming there are n qubits used, then there are $\binom{n}{\frac{n}{2}}$ possibilities for which qubits are flipped and which are not.

The switching of the final gates can be determined at run-time, allowing different qubits to have their outputs flipped in each shot. A true random number generator (TRNG) can be employed to generate the randomness that decides which qubits to flip. The information specifying which `X` gates were applied, that is, which qubits were flipped—must be encrypted and sent back to the user to enable correct recovery of the outputs. We note that, as a side benefit, the addition of `X` before measurements may also help to reduce measurement errors [52].

14 Software Overview of `CASQUE`

We summarize the software steps required by `CASQUE` below and we show them schematically in Fig. 6.

1. Pad each `CNOT` gate with delays such that all `CNOT` gates with their associated padding take the same amount of time, the duration including padding should be a fixed multiple of single-qubit gate duration.
2. Add padding between other gates if necessary to "line up" all the gates so that each gate starts at a time that is a multiple of single-qubit gate duration.
3. Increase the amplitude of each control pulse to the maximum amplitude needed by any channel.

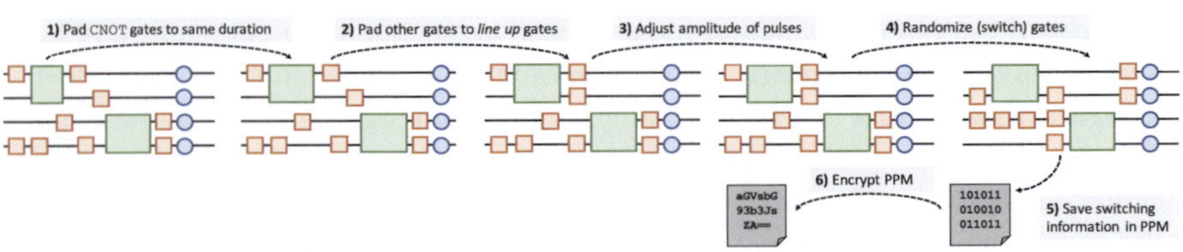

Fig. 6 Overview of the software steps for the update of user's circuits for `CASQUE` protection

4. For each time period, randomly switch the pulses between channels and add a layer of gates for flipping the output.
5. Save the switching information in the pulse protection map (PPM).
6. Encrypt PPM with the cryptographic key associated with the trusted `CASQUE` hardware inside the target.

The transpiled circuit (with all the modifications and with pulses switched between channels) is sent to the cloud provider, along with the encrypted PPM.

15 Hardware Architecture of `CASQUE`

On the hardware side, control pulses must be routed to the correct channels and, if necessary, have their amplitude or phase adjusted. We assume that the cloud provider correctly generates the control pulses based on the transpiled circuit provided by the user. An untrusted cloud provider, who may attempt passive observation, cannot infer details of the circuit because the mapping of pulses to qubits is hidden. The encrypted Pulse Protection Map (PPM) remains inaccessible to the cloud provider, as they do not possess the decryption keys. However, the encrypted PPM is sent to the hardware before the circuit executes so that the trusted `CASQUE` Logic hardware can decrypt the PPM and send the control bits to the switches and amplitude and phase control. Compared to an unmodified quantum computer, a number of operations are performed on the control pulses which are input to the dilution refrigerator.

1. For each time period, based on the control bits from the PPM, the incoming pulses need to be switched to the correct channels; at each time period, the control bits are loaded into the Beneš Network to re-configure the routing of the incoming signals.
2. After the pulses pass the Beneš Network, they pass through amplitude and phase control; the APM specifies the attenuation and any phase shifting needed.
3. Finally, the control pulses are now sent to the mixers so that the incoming control signals can be mixed with the carrier signal at the frequency matching the channel.
4. For the output protection, at runtime, TRNG generates random bits used to determine how the final `X` gates will be switched, this information is also encrypted and sent back to the user.

Switching control pulses requires more than just redirecting the control signals to different qubits. We need to consider other features of the control signals, such as the frequency or amplitude of the signals.

15.1 Pulse Switching with Beneš Network

Swapping of the control pulses between qubits and couplings can be realized by a Beneš network. A Beneš network can be used to switch signals between N inputs and N outputs, where any rearrangement of the inputs can be achieved without blocking. Beneš network of N inputs has $2 \times log_2(N) - 1$ stages, each containing $N/2$ two-by-two crossbar switches, and use a total of $N \times log_2(N) - N/2$ two-by-two crossbar switches. For specifying switching, rather than specify for each qubit or coupling channel its target channel, we can instead directly provide control bits for each switch in the Beneš network at each time period. Recall that all gates in transpiled circuit are padded to that they each start at a time that is multiple of single-qubit duration. In effect, the circuits are quantized into fixed time periods. In each time period, $N \times log_2(N) - N/2$ bits are needed (i.e., one bit per switch to determine if the switch should exchange its inputs, or let them pass unchanged).

The Beneš network's switches can be realized with standard RF switches. For example, the CMD272P3 [47] is a low loss broadband positive control double-pole, double-throw (DPDT) transfer switch. The CMD272P3 covers DC to 10 GHz and offers a low insertion loss of 1.6 dB and high isolation of 43 dB at 5 GHz, which is the target frequency for many superconducting qubit quantum computers. The CMD272P3 operates using complementary control voltage logic lines of 0/+5V that can be easily generated by the control logic by converting typical single-ended outputs to differential pair outputs.

15.2 Frequency Adjustment

In a superconducting qubit quantum computer, each qubit has a designated target frequency, and pulses generated for one qubit will not function correctly on another qubit with a different frequency. Figure 7 illustrates a typical superconducting quantum computer, highlighting the arbitrary waveform generators (AWGs) with local oscillators (LOs) and mixers used to upconvert the I and Q pulses to the target qubit or coupling frequency. The AWGs, LOs, and mixers are all located outside the dilution refrigerator.

In `CASQUE`, the Beneš network can switch pulses, but cannot adjust the frequency. Further, we do not want to introduce any high-power equipment, such as signal generators, into the dilution refrigerator. A solution we propose for `CASQUE` is to move the mixers into the dilution refrigerator. The I and Q pulses are at the same frequency for all qubits, and it is only the frequency of the LO that changes for each qubit. Thus, the I and Q pulses can be sent through the switching network to switch the gates. Once the pulses are switched, they can be mixed with the LO signal.

15.3 Amplitude and Phase Adjustment

One of the main differences between the control pulses, besides the frequency, is the amplitude of the pulse. As discussed before, in CASQUE all pulses are sent at the highest amplitude, and then attenuated. The amplitude adjustment can be achieved with a voltage-controlled attenuator, for example, F2258NLGK8 from Renesas [43]. The example attenuator works in range up to 6 GHz, with 1.4 dB insertion loss and the control voltage can be from 0 to 3.6 V. To generate the attenuation control voltage from digital information, a digital-to-analog converter is needed. Assuming 10 bits of resolution for the digital-to-analog, we can control the amplitude with resolution of over 0.001%. Detailed analysis of the type of the needed attenuation levels, and any residual impact of the attenuation on the control signals, is left for future work.

The phase of the control pulses also affects their operation. The phase information is stored in I and Q signals. If needed, the phase may have to be modified before the I and Q signals reach the mixers. Phase adjustment can be achieved with a phase shifter, for example, HMC649A from Analog Devices [12]. The example phase shifter works in 3 GHz to 6 GHz frequency range, with 8dB insertion loss, and has a resolution of 6 bits, corresponding to phase adjust resolution of 5.625 degrees.

15.4 Combined Hardware Modifications

Figure 7 illustrates the combined hardware modifications introduced by CASQUE. The trusted components added inside the dilution refrigerator include the CASQUE logic, switching network, APC, and mixers. The CASQUE logic consists of one or

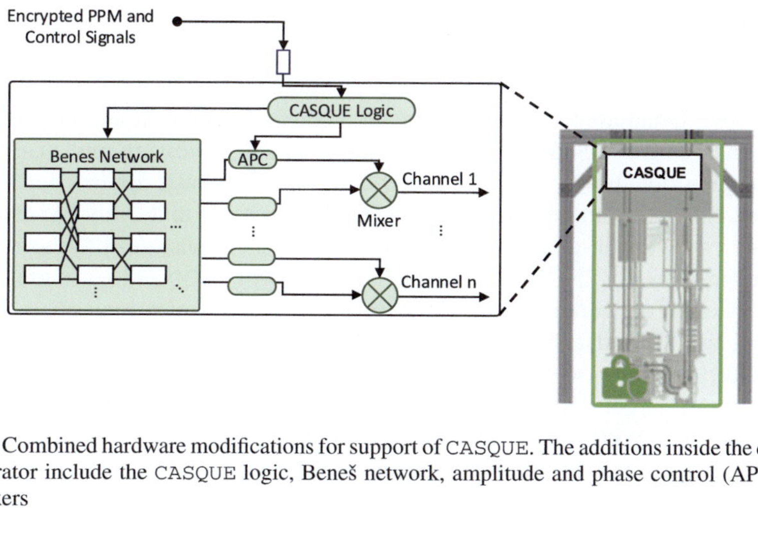

Fig. 7 Combined hardware modifications for support of CASQUE. The additions inside the dilution refrigerator include the CASQUE logic, Beneš network, amplitude and phase control (APC), and the mixers

more decryption tiles and engines, which are designed to support varying sizes of switching networks and the number of control bits they require. The only modification to the untrusted hardware outside the dilution refrigerator is the removal of the mixers, which are relocated inside the fridge.

15.5 PPM and APM Data Size and Bandwidth

The Beneš network requires $N \times log_2(N) - N/2$ control bits for each time period to allow for arbitrary switching of N channels. Further, $10 \times N$ control bits are needed for the amplitude control on N channels, and $6 \times N$ control bits may be needed for phase adjustment. The control bits for Beneš network are stored in PPM, and at each time period The Beneš network requires $N \times log_2(N) - N/2$ bits need to be provided from the PPM to switch the switches. Meanwhile, control data related to amplitude and phase is stored in the APM. These control bits from APM only need to be provided before the circuit starts to execute. At each time period, for each channel, it is only needed to specify the type of gate, so the appropriate APM data can be used to adjust the amplitudes, for example. In IBM Quantum there are `X`, `SX`, and `CX` basis gates that use real control pulses, thus $2 \times N = log_2(3) \times N$ control bits are needed for each time period, in addition to the switching bits. The time period corresponding to the duration of single-qubit gates is today 160 dt in IBM quantum computers, which is equivalent to 35.5 ns. This corresponds to a frequency of 28.5 MHz and which the switching network control bits need to be provided.

15.6 Security-Performance Tradeoff: Tiled Switching Network Design

According to recent IBM roadmaps, large quantum computer systems will be built from smaller quantum computers or quantum computer chips on the order of 1000 qubits, such as the $1,386$-qubit Kookaburra quantum processing unit that can be replicated multiple times to build a larger computer. In a basic approach, `CASQUE` hardware can be instantiated for each quantum processing unit. Multiple `CASQUE` hardware can work in parallel, each with its associated quantum processing unit. While approximately 75 GB/s bandwidth between each `CASQUE` logic and its associated switching network is high, it is within reach of today's electronics (e.g., DDR5 can provide 51 GB/s data bandwidth per module, or HBM can provide 256 GB/s data bandwidth). However, so many switches are not needed in practice. From a security perspective, if we consider even circuits with an order of 6 qubits and 100s of gates, our evaluation in Sect. 10 shows the complexity can be above 2^{256}. Thus, if we only switch a few channels, the attack complexity is huge and there is no need to switch all the channels each time and the 75GB/s bandwidth is unlikely to be actually needed, order or two smaller bandwidth could be sufficient when only smaller number of qubits are switches—while still having high security level.

15.6.1 Tiling of the Switching Network

$N \times N$ Beneš network is actually built from two $N/2 \times N/2$ networks with added layers. As a result, it is simple to trade off the number of possible permutations versus the number of control bits (and thus bandwidth) needed. Table 4 shows different sizes of networks and the associated number of permutations, needed control bits, and bandwidth. Based on the different sizes of the network, a different number of control bits is needed. We note that in our design of the decryption engine which decrypts the PPM, discussed next, we use the AES algorithm, which has a block size of 128 bits. Each decryption tile can provide 128 bits in a time period. Depending on the size of the switching network, multiple tiles can be used in parallel.

15.7 CASQUE *Logic and Decryption Engine Design*

In this work, we also implement a hardware design for the decryption engine, which is a key part of the CASQUE logic. The decryption engine helps decrypt the encrypted PPM inside the trusted hardware. For our evaluation, we consider AES-GCM algorithm to encrypt the PPM on the user's end. We use an existing AES-GCM module [27] which can perform both encryption and decryption and implement a wrapper consisting of a controller which helps in loading the encrypted PPM from the BRAM or DRAM, decrypt it, and write it back to the same memory location block by block. We name this module as `decrypt_tile` (shown in Fig. 8). The `decrypt_tile` has the capability of generating parameterized output width for these decrypted PPM bits. Users can choose the width arbitrarily based on

Table 4 Number of control bits, number of possible permutations, and corresponding bandwidth for different sizes of Beneš network. Bandwidth computation is based on the number of control bits that need to be provided in each 35.5ns time period

Size	Perm.	Ctrl. bits	Bandwidth (MB/s)	Num. dec. tiles
2×2	2^1	1	3.5	1
4×4	$\sim 2^{4.5}$	6	21.1	1
8×8	$\sim 2^{15.3}$	20	70.4	1
16×16	$\sim 2^{44.3}$	56	197.2	1
32×32	$\sim 2^{117.6}$	144	507	2
64×64	$\sim 2^{296.0}$	352	1,873	3

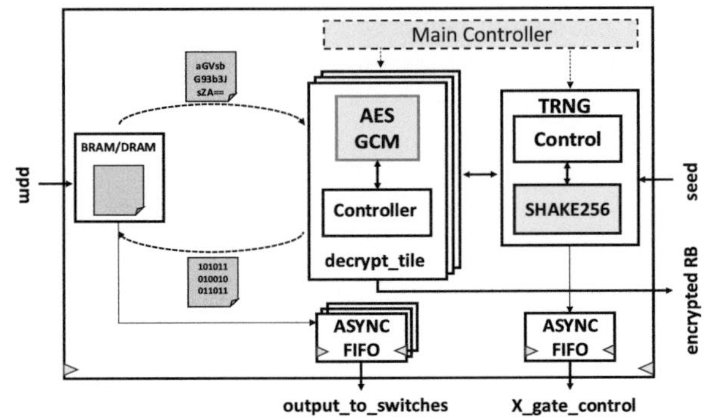

Fig. 8 Top-level design of CASQUE hardware module, which implements key parts of the CASQUE logic. Quantum-safe public-cryptographic modules (mceliece38864 decapsulation) for establishing the shared secret for use in the AES would be extra hardware

the bandwidth requirement. Furthermore, the decrypt_tile also supports a high-performance mode where multiple decrypt_tile could be stacked together to decrypt large PPMs efficiently. The key required by the AES-GCM module is established using a quantum-safe public key algorithm, e.g., Classic McEliece [11]. In our hardware design, for the public key algorithm, we use the mceliece38864 decapsulation module described in Sect. 5.4 of [9]. Other quantum-safe algorithms could be used as well.

In addition to the decryption engine, to protect the output generated from the quantum computer (discussed in Sect. 13.4), we also implement a True Random Number Generator (TRNG). For the TRNG module, we use an existing SHAKE256 module [9] and implement a wrapper around it to feed an arbitrary size seed (we assume that there is a random number generator (RNG) inside the trusted hardware boundary that provides uniformly distributed random bits to our module as an initial seed) and squeeze out the required number (N) of random bits (*RB*), where N is the number of qubits for the given quantum computer. As discussed in Sect. 13.4, based on these bits, an X gate is either applied or not on the output of the quantum circuit. The wrapper also facilitates the encryption of *RB* using the AES-GCM module from the decrypt_tile, which is sent to the user.

The TRNG module uses the SHAKE256 with the smallest performance parameter configuration, i.e., parallel_slices = 1 (described in [9]). The top_module combines all other modules as shown in Fig. 8. The hardware utilization of our CASQUE hardware module is as follows: 3,340 LUTs, 1,158 FFs, 10 BRAMs, and it operates at a frequency of 103 MHz when targeted to Xilinx Artix 7 xc7a100t. To handle a 1 MB PPM, our hardware module takes 7.7 ns. These results do not include the resources used for the public-key algorithm. We do not report the BRAM utilization required for the PPM storage. This is because the size of the PPM changes as per the target quantum computer and quantum circuit. For large quantum computers and large circuits, the size of the on-chip BRAM may not be enough for the PPM storage, in which case we could use off-chip storage units such as DRAM. Our CASQUE hardware design supports the usage of either of them. We also note that we chose to target a lightweight hardware implementation for this evaluation. However, the parameterizable capability of our design allows switching to a high-speed parallel implementation easily.

15.7.1 Serializer-Deserializer for Outputs

Although the operating frequency of the CASQUE hardware module is higher (103 MHz), the frequency at which the pulse switching happens is much slower (i.e., 28.5 MHz, as described in Sect. 15.5). One possible solution to tackle this is to run the complete design at the (slower) switching frequency, but the bandwidth requirement of the Beneš network may not be met. Consequently, we run our design at the fast frequency and use an asynchronous FIFO (ASYNC_FIFO) to handle the crossing clock domains. Figure 8 shows usage of two sets of ASYNC_FIFOs, one for switching network and the other to control the application of X gate on the output. From the (ASYNC_FIFO), the decrypted PPM output is then loaded into the Serializer-Deserializer to arrange bits as per required bandwidth. We note that we successfully conducted practical experiments by running our hardware design on Xilinx Artix 7 xc7a100t FPGA and interfacing them with the RF switches (described in Sect. 15.1). We simulate the quantum computer control pulses, going into the RF switches, using a lab signal generator, and an oscilloscope is used to validate that the switches attenuate the pulses when needed.

16 Evaluation of CASQUE

In this section we present evaluation of CASQUE.

16.1 CASQUE *Evaluation Setup*

Our evaluation focuses on fidelity, measured using Variational Distance (VD), and on the computational complexity for an attacker to reconstruct the circuit based on the obfuscation introduced by pulse switching. We use selected benchmarks from QASMBench [31] for this evaluation. Fidelity measurements are performed on the 7-qubit IBM Perth quantum processor.

16.2 *Fidelity Evaluation of* CASQUE

Assuming ideal operation of the switches and other added components, we focus on the impact of added delays due to "lining up" of gates done as part of CASQUE software steps. When CNOT gates are padded with delays to have fixed duration, we observe that with more delays, the variational distance increases, i.e., the fidelity decreases. For this reason, we measure the Variational Distance (VD) between each of the eight selected benchmarks and the unmodified benchmark from the QASM-Bench benchmarks suite. Informally, the variational distance of two output probability distributions is the measure of how one probability distribution is different from the other. In general, the total variation distance between P and Q is defined as $\delta(P, Q) = \frac{1}{2}\sum |P - Q|$. The impact of the extended duration of the selected benchmarks is presented in Table 5.

16.3 *Security Analysis of* CASQUE

In the computation of the security level and the security analysis, we take into account the following aspects: number of qubits (n_{qubits}), total number of single-qubit slots (p), number of single-qubit gate pulses in each slot (ms_i), total number of CNOT slots (q), number of single-qubit slots within duration of a CNOT slot (r), number of CNOT gate pulses we have in each CNOT slot (c_j), number of single-qubit gate pulses in each slot within a CNOT slot ($mc_{j,k}$):

$$C = \prod_{i=0}^{p-1} \binom{n_{qubits}}{ms_i} \times \prod_{j=0}^{q-1} \left(\prod_{k=0}^{r-1} \left(\binom{(n_{qubits} - (2 \times c_j))}{mc_{j,k}} \right) \right) \tag{10}$$

Table 5 Variational Distance (VD) metric for selected QASMBench benchmarks [31]. The Qubits, Gates (single-qubit gates), and CNOT numbers are pre-transpilation

Benchmark	Qubits	Gates	CNOT	VD
wstate	3	30	9	0.126
basis_change	3	53	10	0.116
variational	4	54	16	0.080
vqe	4	89	9	0.195
qec_en	5	25	10	0.642
error_correctiond3	5	114	49	0.184
simon	6	44	14	0.195
qaoa	6	270	54	0.203

Table 6 Approximate attack complexity on selected QASMBench benchmarks [31]. Complexity calculated for zero, two, four, and eight number of added dummy qubits. The Qubits, Gates (single-qubit gates), and CNOT gates numbers are pre-transpilation

Benchmark	Qubits	Gates	CNOT	Complexity			
				0 dummy qubits (CASQUE)	w/ 2 dummy qubits (CASQUE+)	w/ 4 dummy qubits (CASQUE+)	w/ 8 dummy qubits (CASQUE+)
wstate	3	30	9	2^{8}	2^{15}	2^{19}	2^{25}
basis_change	3	53	10	2^{43}	2^{66}	2^{80}	2^{100}
variational	4	54	16	2^{20}	2^{26}	2^{30}	2^{36}
vqe	4	89	9	2^{45}	2^{64}	2^{77}	2^{95}
qec_en	5	25	10	2^{17}	2^{22}	2^{26}	2^{32}
error_correctiond3	5	114	49	2^{103}	2^{132}	2^{155}	2^{191}
simon	6	44	14	2^{20}	2^{25}	2^{28}	2^{33}
qaoa	6	270	54	2^{268}	2^{324}	2^{365}	2^{428}

In the above equation, we use $\binom{n}{k}$ notation to represent $\binom{n}{k}$ computation. In our implementation, each CNOT gate slot is padded with delays such that all CNOT gates within the circuit take the same duration of time (equal to the duration of the longest CNOT gate on any of the coupling in the target quantum computer). The CNOT gate duration of time including padding is set to be a multiple of the single-qubit gate slot duration, currently 160dt. Since all CNOT gate slots are of the same duration, regardless which CNOT slot j is considered, the number of single-qubit gate slots that fit within the CNOT slot is the same and equal to r. As result r depends on the quantum computer backend which determines the duration of CNOT gates. Meanwhile, p, q, ms_i, $mc_{j,k}$, and c_j depend on the user's circuit, its structure, and the gates used. n_{qubits} depends on the number of qubits used by the user's circuit, but clearly cannot be larger than the number of qubits available on the target quantum computer. The approximate attack complexity on selected bencharks is shown in Table 6. The table also shows the complexity for CASQUE+ extended architecture, discussed next.

17 CASQUE Summary

CASQUE is a hardware-based security framework for superconducting quantum computers that protects quantum circuits from an untrusted cloud provider. It operates by dynamically switching control pulses to different qubit channels and optionally adjusting their amplitude or phase, according to an encrypted Pulse Protection Map (PPM). This ensures that even if the cloud provider observes the RF pulses, they cannot determine which qubits are actually being operated on. Trusted components inside the dilution refrigerator, including switching networks, decryption engines, and mixers, execute the mapping while keeping the cloud provider oblivious to the actual pulse routing.

18 CASQUE+ Architecture

Our proposed CASQUE provides very good protection at a very high obfuscation level. However, we observed that a novel application of the switches becomes possible if additional dummy qubits are incorporated into the user's circuit. Specifically, when qubits not used by the original circuit are added, control pulses from the functional qubits can be redirected to these dummy qubits. Pulses routed to dummy qubits do not affect the computation and are effectively neutralized. Moreover, additional pulses can be applied to dummy qubits, and pulses can be switched among dummy qubits, without impacting the execution of the original circuit.

18.1 *Increasing Obfuscation: Dummy Qubits and Dummy Gates*

Adding dummy qubits and dummy gates increases the number of possibilities for switching the gates—increasing the complexity for the attacker. Considering our prior Eq. 1, adding dummy qubits increases n_{qubits}. Adding dummy single-qubit gates increases ms_i and $mc_{j,k}$. Adding two-qubit gates increases c_j. However, the complexity increases further as in Eq. 1 and baseline CASQUE design we do not alter the two-qubit CNOT gates. Now, with dummy qubits, we can add dummy two-qubit gates on original qubits and then switch them to the dummy qubits to eliminate them. As a result, the attacker (cloud provider) no longer is certain that a CNOT gate will execute (as it did in CASQUE). Rather, each CNOT could be a real gate that executes or could be a gate that is eliminated (by switching it to a dummy qubit) before execution. The updated complexity for the number of possible circuits that the untrusted cloud provider would have to guess from is:

$$C' = \prod_{i=0}^{p-1} \binom{n_{qubits}}{ms_i} \times \prod_{j=0}^{q-1} \left(\left(\prod_{k=0}^{r-1} \binom{(n_{qubits} - (2 \times c_j))}{mc_{j,k}} \right) \times 2^{c_j} \right) \tag{11}$$

18.2 *Preventing Dummy Qubit Detection*

Although the attacker cannot directly identify which qubits are dummy qubits, they could attempt to infer them from the circuit structure. First, if the circuit can be partitioned into two disjoint subcircuits with no connecting two-qubit gate, the attacker may assume that one subcircuit consists entirely of dummy qubits. To prevent this, we ensure that at least one dummy two-qubit gate connects an original qubit with a dummy qubit. Second, qubits without measurement gates could be identified as ancillary or dummy qubits. To mitigate this, we place a measurement gate on all qubits, making every qubit appear actively involved in the computation, even if some measurement results are discarded by the user.

18.3 Switching of Multi-qubit Gates

The current design of CASQUE does not switch multi-qubit gates due to the complexity of the waveform of the multi-qubit gates. However, this can also be done with enough knowledge of the waveform of the multi-qubit gates. For example, if the CNOT gates on different qubit pairs have the same waveform, then this can be easily done with the same scheme as in CASQUE. Nevertheless, the current design of the CNOT gate on the IBM cloud introduces different duration and pulse patterns, which may require additional hardware, such as the hardware to change the duration of pulses. The study of switching multi-qubit gates is left for future work, but we acknowledge that switching CNOT gates would further increase attack complexity.

18.4 Operation of CASQUE+

Figure 9 shows schematically the steps of extending the user's circuit with CASQUE+ protections. We assume that the input is the CASQUE protected circuit. The steps are done on top of, or in addition to, the protection applied by CASQUE. The steps to add dummy qubits and dummy gates are

1. add dummy qubits to the circuit,
2. randomly add two-qubit gates in CNOT gate slots, ensure that at least in one slot one of the added gates uses both real and dummy qubits,
3. randomly add single qubit gates,
4. add measurement gates on all qubits, and
5. update pulse protection map.

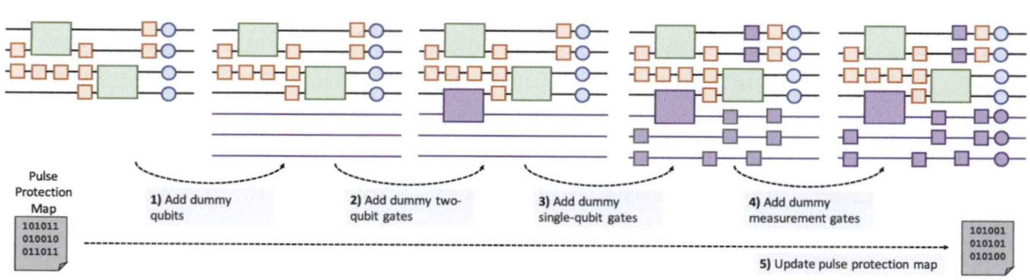

Fig. 9 Software steps in the extended CASQUE+ architecture

18.5 Analysis of Increased Complexity of CASQUE+

Based on Table 6, the circuit complexity can be increased by adding additional dummy qubits. We evaluated this approach using two, four, and eight extra dummy qubits per application. The highest complexity was observed for the 6-qubit benchmark *qaoa*, which contains the largest number of total gates and CNOT gates among all tested benchmarks. We note that in the CASQUE+ architecture, some additional qubits are reserved by the machine and cannot be used for other applications running in parallel. However, this limitation is expected to have minimal impact in larger-scale quantum systems.

19 CASQUE+ Summary

The proposed CASQUE+ architecture is an extension of base CASQUE architecture. It leverages ideas and mechanisms of CASQUE, but adds the dummy qubit and dummy gates. This can further help increase the obfuscation level.

20 Comparison of SoteriaQ and CASQUE

SoteriaQ and CASQUE adopt fundamentally different strategies for securing quantum computations, each with distinct trade-offs in performance and protection. In general, CASQUE introduces lower latency compared to SoteriaQ. This advantage stems from its avoidance of extensive decoy operations; instead, CASQUE achieves obfuscation by effectively rearranging quantum channels through the permutation of logical operations onto physical qubits. By not requiring uniform-duration slots or time-padding, CASQUE maintains shallower circuits and faster execution, making it particularly suitable for platforms with tight coherence budgets or scenarios where minimizing latency is critical. SoteriaQ, in contrast, emphasizes stronger security guarantees. It employs a larger number of decoy gates and slot equalization to conceal timing information, mitigating an adversary's ability to infer circuit structure via side-channel analysis. While this approach incurs additional latency, potential fidelity loss, and higher resource usage, it provides enhanced protection against adversaries capable of monitoring fine-grained execution characteristics. Both architectures are application-agnostic. Collectively, they represent different points in the design space: CASQUE prioritizes efficiency and performance, whereas SoteriaQ accepts increased overhead to achieve stronger obfuscation, making it more suitable for threat models anticipating sophisticated attacks.

21 Conclusion

This chapter presents a comprehensive framework for establishing Trusted Execution Environments (TEEs) in quantum computers, enabling secure execution even in potentially untrusted quantum cloud environments. The proposed architectures, designed for superconducting quantum machines and exemplified by CASQUE and SoteriaQ, provide robust mechanisms for obfuscating quantum control pulses and measurements, significantly reducing the risk of side-channel attacks and insider threats. Through layered hardware and software protections—including decoy pulse insertion, randomized output gates, and encrypted Pulse Protection Maps (PPM)—the system effectively conceals sensitive quantum circuits and user data from unauthorized access. Our results demonstrate the feasibility and scalability of these solutions in quantum cloud infrastructure, achieving strong security guarantees with minimal performance overhead. Leveraging trusted controllers further enables the extension of this framework to other quantum architectures beyond superconducting qubits. Future work will focus on adapting these security measures to additional quantum computing paradigms and integrating emerging advances in TEE technology to support the evolving quantum ecosystem.

Acknowledgements This work was supported in part by NSF grants 2312754 and 2245344.

References

1. Amazon braket. https://aws.amazon.com/braket/.
2. Cmd196c3. https://www.qorvo.com/products/d/da007444.
3. Cryogen-free XLDsl dilution refrigerator measurement system—Bluefors—bluefors.com. https://bluefors.com/products/xldsl-dilution-refrigerator/. Accessed 8 May 2023.
4. Ibm quantum. https://quantum-computing.ibm.com/.
5. Ibm's target: A 4000-qubit processor by 2025. https://spectrum.ieee.org/ibm-quantum-computer.
6. Quantum circuits. https://quantumcircuits.com/.
7. Rigetti computing. https://www.rigetti.com/.
8. Sep 2005. https://arxiv.org/pdf/quant-ph/0111046.pdf.
9. Complete and improved FPGA implementation of classic McEliece. 2022: 71–113.
10. Aharonov, Dorit, M. Ben-Or, and E. Eban. 2008. Interactive proofs for quantum computations. https://arxiv.org/pdf/0810.5375.pdf.
11. Albrecht, Martin R., Daniel J. Bernstein, Tung Chou, Carlos Cid, Jan Gilcher, Tanja Lange, Varun Maram, Ingo von Maurich, Rafael Misoczki, Ruben Niederhagen, Kenneth G. Paterson, Edoardo Persichetti, Christiane Peters, Peter Schwabe, Nicolas Sendrier, Jakub Szefer, Cen Jung Tjhai, Martin Tomlinson, and Wen Wang. 2020. Classic McEliece. Technical report, National Institute of Standards and Technology. Available at https://csrc.nist.gov/projects/post-quantum-cryptography/round-3-submissions.
12. Analog. Hmc649a documentation.
13. Armknecht, Frederik, Colin Boyd, Christopher Carr, Kristian Gjøsteen, Angela Jäschke, Christian A Reuter, and Martin Strand. 2015. A guide to fully homomorphic encryption. Cryptology ePrint Archive.

14. Biamonte, Jacob, Peter Wittek, Nicola Pancotti, Patrick Rebentrost, Nathan Wiebe, and Seth Lloyd. 2017. Quantum machine learning. *Nature* 549 (7671): 195–202.
15. Broadbent, Anne, Joseph Fitzsimons, and Elham Kashefi. 2009. Universal blind quantum computation. In *50th Annual IEEE symposium on foundations of computer science*, 517–526. https://ieeexplore.ieee.org/stamp/stamp.jsp?tp=&arnumber=5438603.
16. Chow, Jerry, Oliver Dial, and Jay Gambetta. 2021. Ibm quantum breaks the 100-qubit processor barrier. *IBM Research Blog* 2.
17. Conway Lamb, I. D., J. I. Colless, J. M. Hornibrook, S. J. Pauka, S. J. Waddy, M. K. Frechtling, and D. J. Reilly. 2016. An FPGA-based instrumentation platform for use at deep cryogenic temperatures. *Review of Scientific Instruments*, 87 (1): 014701.
18. Costan, Victor, and Srinivas Devadas. 2016. Intel sgx explained. Cryptology ePrint Archive.
19. Dunjko, Vedran, Elham Kashefi, and Anthony Leverrier. 2012. Blind quantum computing with weak coherent pulses. *Physical Review Letters* 108 (20). https://arxiv.org/pdf/1108.5571.pdf.
20. Fang, Kun, and Zi-Wen. Liu. 2020. No-go theorems for quantum resource purification. *Physical Review Letters* 125 (6): 060405.
21. Fitzsimons, Joseph F., and Elham Kashefi. 2017. Unconditionally verifiable blind quantum computation. *Physical Review A* 96 (1). https://arxiv.org/pdf/1203.5217.pdf.
22. Fontaine, Caroline, and Fabien Galand. 2007. A survey of homomorphic encryption for non-specialists. *EURASIP Journal on Information Security* 1–10: 2007.
23. Gambetta, Jay. 2020. Ibm's roadmap for scaling quantum technology. *IBM Research Blog*.
24. Giovannetti, Vittorio, Lorenzo Maccone, Tomoyuki Morimae, and Terry G. Rudolph. Efficient universal blind quantum computation. *Physical Review Letters*, 111 (23). https://arxiv.org/pdf/1306.2724.pdf.
25. Hooyberghs, Johnny, and Johnny Hooyberghs. 2022. Azure quantum. In *Introducing microsoft quantum computing for developers: Using the quantum development kit and Q#*, 307–339.
26. Oxford Instruments. Principles of dilution refrigeration.
27. Käsper, Emilia, and Peter Schwabe. 2009. Faster and timing-attack resistant AES-GCM. In *International workshop on cryptographic hardware and embedded systems*, 1–17. Springer.
28. Koteshwara, Sandhya, Amitabh Das, and Keshab K. Parhi. 2017. FPGA implementation and comparison of AES-GCM and Deoxys authenticated encryption schemes. In *IEEE international symposium on circuits and systems (ISCAS)*, 1–4.
29. Lavor, Carlile, LRU Manssur, and Renato Portugal. 2003. Grover's algorithm: Quantum database search. arXiv preprint quant-ph/0301079
30. Li, Ang, Samuel Stein, Sriram Krishnamoorthy, and James Ang. 2020. Qasmbench: A low-level qasm benchmark suite for nisq evaluation and simulation. arXiv preprint arXiv:2005.13018.
31. Li, Ang, Samuel Stein, Sriram Krishnamoorthy, and James Ang. 2022. Qasmbench: A low-level quantum benchmark suite for nisq evaluation and simulation. *ACM Transactions on Quantum Computing*.
32. Liu, Jiang, Qin Li, Junyu Quan, Can Wang, Jinjing Shi, and Haozhen Situ. 2022. Efficient quantum homomorphic encryption scheme with flexible evaluators and its simulation. *Designs, Codes and Cryptography* 90 (3): 577–591.
33. Liu, Wen-Jie, Zhen-Yu Chen, Jin-Suo Liu, Zhao-Feng Su, and Lian-Hua Chi. 2020. Full-blind delegating private quantum computation. arXiv preprint arXiv:2002.00464.
34. Mahadev, Urmila. 2020. Classical homomorphic encryption for quantum circuits. *SIAM Journal on Computing* : FOCS18–189.
35. Majzoobi, Mehrdad, Farinaz Koushanfar, and Srinivas Devadas. 2011. FPGA-based true random number generation using circuit metastability with adaptive feedback control. In *Workshop on cryptographic hardware and embedded systems*.
36. Mantri, Atul, Carlos A. Pé rez-Delgado, and Joseph F. Fitzsimons. 2013. Optimal blind quantum computation. *Physical Review Letters* 111 (23). https://arxiv.org/pdf/1306.3677.pdf.
37. Morimae, Tomoyuki. 2012. Continuous-variable blind quantum computation. *Physical Review Letters* 109 (23). https://arxiv.org/pdf/1208.0442.pdf.
38. Morimae, Tomoyuki. 2014. Verification for measurement-only blind quantum computing. https://arxiv.org/pdf/1208.1495.pdf.

39. Morimae, Tomoyuki, Vedran Dunjko, and Elham Kashefi. 2011. Ground state blind quantum computation on AKLT state. https://arxiv.org/pdf/1009.3486.pdf.
40. Morimae, Tomoyuki, and Keisuke Fujii. 2012. Blind topological measurement-based quantum computation. *Nature Communications* 3 (1). https://arxiv.org/pdf/1110.5460.pdf.
41. Morimae, Tomoyuki, and Keisuke Fujii. 2013. Blind quantum computation protocol in which alice only makes measurements. *Physical Review A* 87 (5). https://arxiv.org/pdf/1201.3966.pdf.
42. Morimae, Tomoyuki, and Takeshi Koshiba. 2013. Composable security of measuring-alice blind quantum computation. https://arxiv.org/pdf/1306.2113.pdf.
43. Newark. F2258NLGK8 RF attenuator documentation.
44. Hyunyoung, Oh., Kevin Nam, Seongil Jeon, Yeongpil Cho, and Yunheung Paek. 2021. Meetgo: A trusted execution environment for remote applications on FPGA. *IEEE Access* 9: 51313–51324.
45. Pechal, M., J.C. Besse, Mintu Mondal, M. Oppliger, S. Gasparinetti, and A. Wallraff. 2016. Superconducting switch for fast on-chip routing of quantum microwave fields. *Physical Review Applied* 6(2): 024009.
46. Peruzzo, Alberto, Jarrod McClean, Peter Shadbolt, Man-Hong Yung, Xiao-Qi Zhou, Peter J. Love, Alán Aspuru-Guzik, and Jeremy L. O'Brien. 2014. Variational eigenvalue solver on a photonic quantum processor. Nature Communications 5 (1).
47. Qorvo. Cmd272p3 dpdt transfer switch documentation.
48. Rivest, Ronald L., Len Adleman, Michael L. Dertouzos, et al. 1978. On data banks and privacy homomorphisms. *Foundations of Secure Computation* 4 (11): 169–180.
49. Shan, Rui-Ting., Xiubo Chen, and Kai-Guo. Yuan. 2021. Multi-party blind quantum computation protocol with mutual authentication in network. *Science China Information Sciences* 64 (6): 162302.
50. Sueki, Takahiro, Takeshi Koshiba, and Tomoyuki Morimae. 2013. Ancilla-driven universal blind quantum computation. *Physical Review A* 87: 060301.
51. Tan, Si-Hui, Joshua A. Kettlewell, Yingkai Ouyang, Lin Chen, and Joseph F. Fitzsimons. 2016. A quantum approach to homomorphic encryption. *Scientific reports* 6 (1): 33467.
52. Tannu, Swamit S., and Moinuddin K. Qureshi. 2019. Mitigating measurement errors in quantum computers by exploiting state-dependent bias. In *Proceedings of the 52nd annual IEEE/ACM international symposium on microarchitecture*, 279–290.
53. Tasopoulos, George, Charis Dimopoulos, Apostolos P. Fournaris, Raymond K. Zhao, Amin Sakzad, and Ron Steinfeld. 2023. Energy consumption evaluation of post-quantum TLS 1.3 for resource-constrained embedded devices. Cryptology ePrint Archive, Paper 2023/506. https://eprint.iacr.org/2023/506.
54. Trochatos, Theodoros, Sanjay Deshpande, Chuanqi Xu, Yao Lu, Yongshan Ding, and Jakub Szefer. 2024. Dynamic pulse switching for protection of quantum computation on untrusted clouds. In *IEEE international symposium on hardware oriented security and trust (HOST)*, 404–414.
55. Trochatos, Theodoros, and Jakub Szefer. 2024. Quantum operating system support for quantum trusted execution environments.
56. Trochatos, Theodoros, Chuanqi Xu, Sanjay Deshpande, Yao Lu, Yongshan Ding, and Jakub Szefer. 2023. Hardware architecture for a quantum computer trusted execution environment. arXiv preprint arXiv:2308.03897.
57. Trochatos, Theodoros, Chuanqi Xu, Sanjay Deshpande, Yao Lu, Yongshan Ding, and Jakub Szefer. 2023. Hardware architecture for a quantum computer trusted execution environment.
58. Trochatos, Theodoros, Chuanqi Xu, Sanjay Deshpande, Yao Lu, Yongshan Ding, and Jakub Szefer. 2023. A quantum computer trusted execution environment. *IEEE Computer Architecture Letters*.
59. Trochatos, Theodoros, Xu. Chuanqi, Sanjay Deshpande, Lu. Yao, Yongshan Ding, and Jakub Szefer. 2023. A quantum computer trusted execution environment. *IEEE Computer Architecture Letters* 22 (2): 177–180.

60. Wille, Robert, Rod Van Meter, and Yehuda Naveh. 2019. Ibm's qiskit tool chain: Working with and developing for real quantum computers. In *Design, automation & test in Europe conference & exhibition (DATE)*, 1234–1240. IEEE.
61. Xia, Ke, Yukui Luo, Xiaolin Xu, and Sheng Wei. 2021. SGX-FPGA: Trusted execution environment for CPU-FPGA heterogeneous architecture. In *2021 58th ACM/IEEE design automation conference (DAC)*, 301–306. IEEE.
62. Xu, Chuanqi, Ferhat Erata, and Jakub Szefer. 2023. Exploration of power side-channel vulnerabilities in quantum computer controllers. In *Proceedings of the 2023 ACM SIGSAC conference on computer and communications security*, CCS '23, 579–593, New York, NY, USA. Association for Computing Machinery.
63. Zeuner, Jonas, Ioannis Pitsios, Si-Hui Tan, Aditya N. Sharma, Joseph F. Fitzsimons, Roberto Osellame, and Philip Walther. Experimental quantum homomorphic encryption. *npj Quantum Information* 7 (1): 25.
64. Zhang, Yuanjing, Tao Shang, and Jianwei Liu. 2021. A multi-valued quantum fully homomorphic encryption scheme. *Quantum Information Processing* 20: 1–25.

High-Performance Post-Quantum Cryptographic Engineering in Hardware

Bardia Taghavi, Merve Karabulut, Reza Azarderakhsh, and Mehran Mozaffari Kermani

Abstract Advances in quantum computing pose a serious threat to classical public-key cryptographic schemes, particularly those based on the hardness of factoring and discrete logarithms. Post-Quantum Cryptography (PQC) aims to develop algorithms that remain secure even in the presence of quantum-capable adversaries. This chapter focuses on lattice-based cryptography, with particular emphasis on the Module-Lattice-based Key-Encapsulation Mechanism (ML-KEM) and the Module-Lattice-based Digital Signature Algorithm (ML-DSA)—two schemes recently standardized by the U.S. National Institute of Standards and Technology (NIST). We examine the mathematical foundations of these constructions, their role in the standardization process, and key considerations related to their performance and implementation.

1 Introduction

Quantum computing is advancing rapidly, transitioning from theoretical models to practical hardware. Recent developments, particularly in neutral atom-based platforms, have demonstrated improved qubit fidelity, connectivity, and scalability [55]. These advances bring renewed urgency to the field of cryptography, as quantum algorithms threaten the foundations of widely used public-key systems. Shor's algorithm [50] can efficiently solve the underlying problems behind RSA [48], Diffie–Hellman [14], and elliptic curve cryptography [25] in polynomial time of $O((log_2 n)^3)$

B. Taghavi · M. Karabulut · R. Azarderakhsh (✉)
Florida Atlantic University, Boca Raton, FL, USA
e-mail: razarderakhsh@fau.edu

B. Taghavi
e-mail: staghavi2024@fau.edu

M. Karabulut
e-mail: mkarabulut2024@fau.edu

M. M. Kermani
University of South Florida, Tampa, FL, USA
e-mail: mehran2@usf.edu

P. Mishra (ed.), *Design Automation for Quantum Computing*,
https://doi.org/10.1007/978-3-032-09303-5_14

when n is the input size, rendering these schemes insecure once scalable quantum computers become available. In contrast, symmetric schemes such as AES [36] are less vulnerable; Grover's algorithm [19] reduces their adequate security level quadratically, which can typically be addressed by increasing key sizes. As a result, AES-256 on a quantum computer is as secure as AES-128 on a classic computer, which is considered an acceptable level of security. As a result, asymmetric cryptography requires entirely new designs.

Figure 1 presents the current quantum computing capability landscape, plotting physical qubit count against gate error rate. The chart identifies distinct operational regions: current devices reside in the lower-left, where high error rates and limited qubit counts prevent practical computation. The purple region marks where surface codes become effective, enabling the construction of fault-tolerant logical qubits. The red curves indicate the estimated thresholds for breaking RSA using Shor's algorithm; the curve for RSA-2048 reflects recent reductions in resource estimates. Green-striped areas represent problems—such as quantum chemistry simulations—that may be solvable without full error correction. Although current systems are far from reaching cryptographically significant thresholds, improvements in qubit fidelity, architecture, and error correction are steadily reducing this gap [24].

In response to this threat, the NIST launched a standardization initiative in 2016 to identify cryptographic algorithms resistant to quantum attacks [34]. Figure 2 shows how PQC fits into NIST's broader framework, which includes public-key and private-key cryptography, utility primitives, and implementation guidance. PQC specifically addresses the public-key components, replacing mechanisms for key establishment and digital signatures that are vulnerable to Shor's algorithm. In 2024, NIST announced five selected algorithms: ML-KEM and Hamming Quasi-Cyclic (HQC) for key establishment, and ML-DSA, SLH-DSA, and FALCON for digital signatures [15, 31, 37–39]. Among them, ML-KEM, ML-DSA, and SLH-DSA have been published as finalized Federal Information Processing Standards: FIPS 203, 204, and 205. FALCON and HQC remain in draft status pending final publication.

This chapter begins with an overview of the NIST standardization timeline, highlighting the selection of ML-KEM and ML-DSA and the advantages they offer in security, efficiency, and broad applicability. It then introduces the core modules that underpin both algorithms. Next, we present the detailed processes of ML-KEM and ML-DSA, walking through each step of key generation, encapsulation, decapsulation, signing, and verification. Finally, we survey a range of high-performance hardware implementations designed for real-world deployment, comparing their throughput, resource usage, and architectural trade-offs.

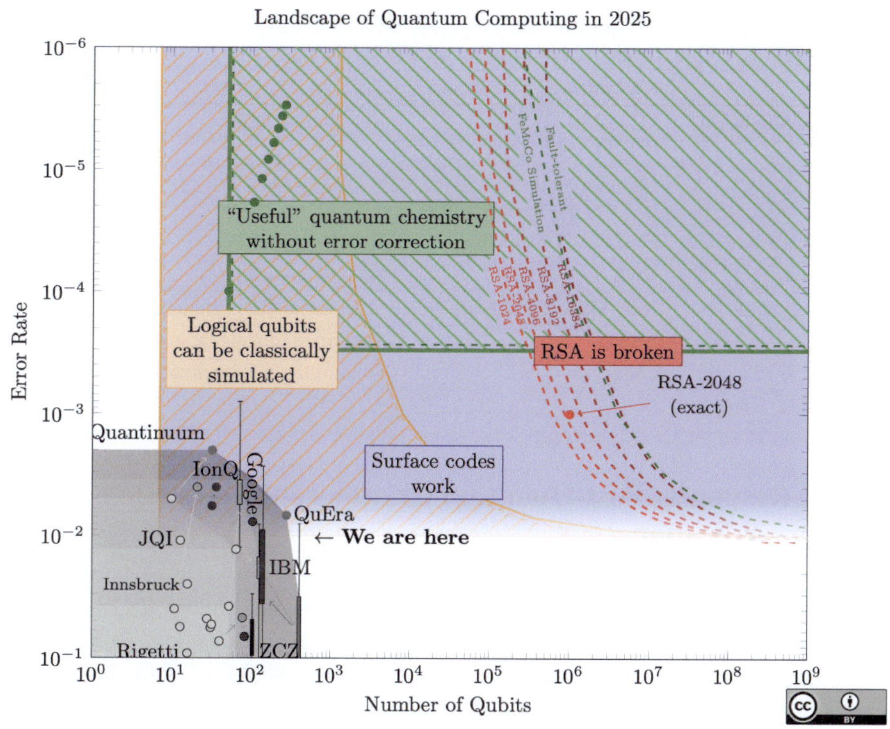

Fig. 1 Quantum computing capability space showing the number of physical qubits versus gate error rate. Red curves indicate updated thresholds for breaking RSA-2048 using Shor's algorithm [24]

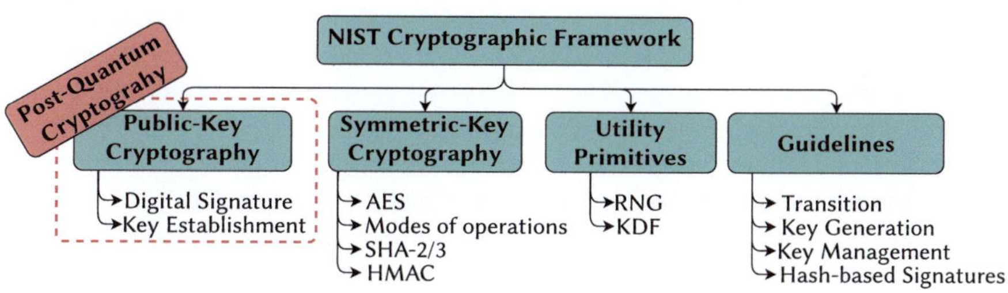

Fig. 2 NIST cryptographic framework highlighting the role of post-quantum cryptography within public-key cryptography

Table 1 NIST-approved post-quantum cryptographic algorithms categorized by underlying hardness assumptions, cryptographic functionality, and FIPS publication

Category	Algorithm	Type	FIPS #
Module-lattice-based	ML-KEM (CRYSTALS-Kyber)	Key encapsulation	203
	ML-DSA (CRYSTALS-Dilithium)	Digital signature	204
NTRU-lattice-based	FN-DSA (FALCON)	Digital signature	206
Code-based	HQC	Key encapsulation	207
Hash-based	SLH-DSA (SPHINCS+)	Stateless digital signature	205
	XMSS	Stateful digital signature	–
	LMS	Stateful digital signature	–

2 The Response: Post-Quantum Cryptography and NIST Standardization

To coordinate a global transition to quantum-resistant cryptography, NIST launched its PQC project in 2016 [34]. The process began with 82 submissions in Round 1, narrowed down to 26 candidates in Round 2, and concluded with 7 finalists and 8 alternates in Round 3 [35]. After extensive public review and cryptanalysis, NIST announced its first selections in 2022 and published finalized standards in 2024. These include ML-KEM, formerly CRYSTALS-Kyber, for key establishment, and ML-DSA, formerly CRYSTALS-Dilithium, along with SPHINCS+ and FALCON for digital signatures [15, 37–39]. In 2025, HQC was added as an additional KEM [31]. Separately, NIST has also standardized stateful hash-based signature schemes—XMSS and LMS—under NIST SP 800-208 [11], providing options for constrained or short-lived signature use cases. The full list of selected and documented algorithms, categorized by underlying hardness assumptions, is summarized in Table 1.

With these selections, NIST aims to finalize a portfolio of post-quantum cryptographic algorithms that balance security, performance, and implementation maturity. Among the standardized schemes, lattice-based cryptography has gained significant attention due to its well-understood mathematical foundations, high efficiency across software and hardware platforms, and suitability for resource-constrained environments [40]. Compared to HQC, the other standardized KEM, ML-KEM, offers significantly smaller key and ciphertext sizes, as summarized in Table 2, making it more practical for real-world deployment.

For digital signatures, NIST recommends ML-DSA as the general-purpose solution. ML-DSA offers a strong balance between efficiency and security. In contrast, SLH-DSA, based on the stateless hash-based SPHINCS+ design [39], provides a highly conservative option with small keys but significantly larger signatures and

Table 2 Key and ciphertext sizes (in bytes) for selected NIST-approved KEM algorithms [31, 37]

Scheme	Encapsulation key	Decapsulation key	Ciphertext size	Shared secret	Security level
ML-KEM-512	800	1,632	768	32	1
ML-KEM-768	1,184	2,400	1,088	32	3
ML-KEM-1024	1,568	3,168	1,568	32	5
HQC-128	2,249	2,289	4,497	64	1
HQC-192	4,522	4,562	9,042	64	3
HQC-256	7,245	7,285	14,485	64	5

Table 3 Key and signature sizes (in bytes) for selected NIST-approved digital signature schemes [4, 38, 39]

Scheme	Signing key	Verification key	Signature size	Security level
ML-DSA-44	2,560	1,312	2,420	2
ML-DSA-65	4,032	1,952	3,309	3
ML-DSA-87	4,896	2,592	4,627	5
SLH-DSA-128s	64	32	7,856	1
SLH-DSA-192s	96	48	16,224	3
SLH-DSA-256s	128	64	29,792	5
FN-DSA-512	1,281	897	666	1
FN-DSA-1024	2,305	1,793	1,280	5
XMSS-SHA2-20-256	2,573	64	2,820	3
LMS-SHA256-H10-W4	60	64	2,512	3

slower performance. SLH-DSA has different variants, where some of them use SHA2 as the hashing function, and some of them use SHAKE. Each of them has two versions: "f" for fast computation, and "s" for small signature size. The lattice-based FALCON scheme [15] achieves much smaller signatures but requires complex floating-point arithmetic, complicating constant-time and side-channel-safe implementations. FALCON, or FN-DSA, comes with two standards: 512 for the lowest security and 1024 for the highest level of security. Table 3 also includes representative parameter sets for stateful hash-based schemes. In particular, XMSS-SHA2_20_256 refers to the use of SHA-2 with 256-bit output and a Merkle tree of height 20, while LMS_SHA256_H10_W4 denotes SHA-256 with a Merkle tree of height 10 and a Winternitz parameter of 4. Stateful hash-based schemes like LMS and XMSS [11] offer small keys and efficient signing but are limited by the need for state management, making them more suitable for short-lived credentials or constrained environments.

The remainder of this paper focuses on module-lattice-based cryptography. We begin by introducing the underlying mathematical structures—polynomial rings and

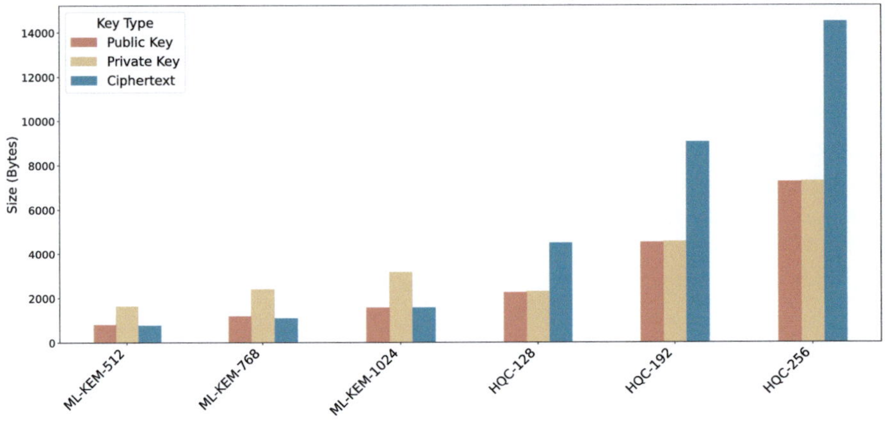

(a)

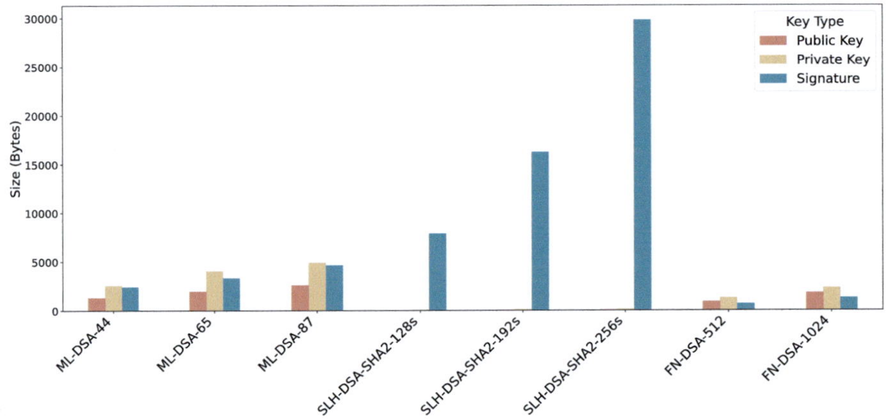

(b)

Fig. 3 Comparison of sizes of NIST accepted **a** KEM schemes and **b** DSA schemes

lattices—followed by a detailed discussion of the core algorithms standardized by NIST and their implementation considerations.

Figure 3 illustrates the public key, private key, and cipher/signature sizes for the mentioned PQC algorithms that have won the NIST competition. This picture demonstrates the superiority of ML-KEM and ML-DSA, as they can make the perfect balance between security and efficiency.

3 Polynomial Arithmetic Computation in Lattice-Based Cryptography

In this section, we provide the mathematical background necessary to understand the core operations used in module-lattice-based cryptographic schemes, including ML-DSA and ML-KEM. These algorithms rely on structured polynomial rings and efficient arithmetic techniques to achieve both security and performance. We begin with the algebraic structure of polynomial rings and their role in defining the lattice framework. We then introduce the Number Theoretic Transform (NTT), which enables efficient polynomial multiplication essential for practical implementations.

3.1 Lattices, Modules, and Module-Lattices

A lattice is a discrete subgroup of a vector space (e.g., $\mathbb{R}^n$ or $\mathbb{Z}_q^n$) formed by all integer linear combinations of a set of basis vectors. Lattices are grid-like structures with no additional algebraic operations beyond addition. Formally, given basis vectors $\mathbf{b}_1, \mathbf{b}_2, \ldots, \mathbf{b}_k \in \mathbb{R}^n$, the lattice $\mathcal{L}$ is:

$$\mathcal{L} = \left\{ \sum_{i=1}^{k} a_i \mathbf{b}_i \mid a_i \in \mathbb{Z} \right\}.$$

On the other hand, a module is a generalization of a vector space. While a vector space is defined over a field, where every non-zero element has a multiplicative inverse, a module is defined over a ring, which does not have this requirement for all elements.

A module-lattice combines these two concepts. It is a lattice whose points are vectors of elements from a specific ring. This structure allows for the creation of cryptographic schemes with strong security guarantees and efficient implementations. As a result, the basis vectors $\mathbf{b}_i$ are selected from a specific ring instead of $\mathbb{R}^n$. In the NIST standardized module-lattice-based schemes, this ring is the polynomial ring. This structure provides the lattice with additional algebraic properties inherited from the ring, enabling more complex cryptographic constructions.

3.2 Polynomial Rings in Module-Lattice-Based Cryptography

At the core of many module-lattice-based cryptographic schemes lies a structured ring of polynomials, commonly defined as

$$\mathcal{R}_q = \mathbb{Z}_q[x]/(x^n + 1),$$

where $\mathbb{Z}_q$ is the set of integers modulo a prime q, and the irreducible polynomial $x^n + 1$ defines the ring structure. Each element in $\mathcal{R}_q$, i.e., a scalar in the ring, is a polynomial of degree less than n with coefficients in $\mathbb{Z}_q$. For instance, the polynomial $\mathbf{a}(x) \in \mathcal{R}_q$ is:

$$\mathbf{a}(x) = a_0 + a_1x + a_2x^2 + a_3x^3 + \cdots + a_{n-1}x^{n-1}, \quad a_i \in \mathbb{Z}_q.$$

The polynomial $\mathbf{a}(x) \in \mathcal{R}_q$ often shows with its coefficients:

$$\mathbf{a} = [a_0, a_1, a_2, \ldots, a_{n-1}].$$

This construction enables compact data representations and efficient computations, which are essential for practical cryptographic systems.

As explained in the previous part, the secret key and other variables are selected from a module-lattice structure, where the module is defined over the polynomial ring $\mathcal{R}_q$. So, each point on these lattices is a k-dimension vector or $(k \times \ell)$-dimension matrix, where the elements of the vector are from a polynomial ring. For instance, the secret key $\mathbf{S} \in \mathcal{R}_q^k$ is from a module-lattice structure:

$$\mathbf{S} = \begin{bmatrix} \mathbf{s}_1 \\ \mathbf{s}_2 \\ \vdots \\ \mathbf{s}_k \end{bmatrix}, \quad \mathbf{s}_i \in \mathcal{R}_q.$$

3.3 Hardness of Module-Lattice Based Schemes

ML-KEM and ML-DSA use the polynomial ring to define lattices based on the hardness of Module-LWE (MLWE) and Selftarget-MSIS (or Inhomogeneous MSIS or IMSIS) problems [37, 38]. FALCON, although based on a different security assumption known as NTRU, also uses the same polynomial ring to define its lattice structure [15]. MLWE problem states that given $\mathbf{A}$ and $\mathbf{t}$ in $\mathbf{t} = \mathbf{A} \cdot \mathbf{s} + \mathbf{e}$, finding s in the average case where the points are from a lattice selected randomly, is at least as hard as solving hard lattice problems like Closest Vector Problems (CVP) and Shortest Vector Problems (SVP), which has been proven to be NP-hard problems [46]. IMSIS is a problem to find a short solution for $\mathbf{s}$ in $\mathbf{t} = \mathbf{A} \cdot \mathbf{s}$. IMSIS has been proven to be slightly harder than MLWE.

3.4 Arithmetic Operations Over the Ring $\mathcal{R}_q$

In all three schemes, the ring enables arithmetic operations that are both structured and efficient, supporting the performance and compactness of the algorithms. The ring supports two main operations: addition and multiplication. Addition is performed by adding coefficients component-wise modulo q. As a result, addition (and also subtraction) can be done in two steps: (1) component-wise addition and (2) reduction of the result modulo q, which is straightforward and studied deeply in the previous work. Note that for adding two polynomials $\mathbf{a}$ and $\mathbf{b}$, each coefficient is in $\mathbb{Z}_q$, so after the addition, the result will be in $\mathbb{Z}_{2q-1}$. As a result, only one subtraction would be enough to convert the elements in the result to the ring of polynomials $\mathcal{R}_q$.

However, the multiplication of two polynomials is harder than the addition and involves more steps since the multiplication will change both coefficients and the degree of the polynomial. Multiplication involves three steps: (1) multiplication of two polynomials, (2) reduction modulo q, and (3) reduction modulo $x^n + 1$. For the reduction over modulo q, several algorithms have been proposed, like Montgomery [33], Barrett [43], Plantard [45], K^2-RED [8], Solinas [18], and other modified versions of these algorithms, such as Word-Level Montgomery [32].

The reduction modulo $x^n + 1$ introduces a negacyclic structure, meaning that terms exceeding degree $n - 1$ wrap around to the beginning with inverted signs. In other words, $x^n \equiv -1$, and $x^{n+m} \equiv -x^m, \forall m \geq 0$. This property is essential for the security and correctness of many module-lattice-based constructions [49].

As can be seen, the multiplication of two polynomials is an expensive operation, and it will be the bottleneck in the PQC schemes' computations. There are several methods to compute the multiplication of two polynomials $\mathbf{a}(x) = a_0 + a_1x + \cdots + a_{n-1}x^{n-1}$ and $\mathbf{b}(x) = b_0 + b_1x + \cdots + b_{n-1}x^{n-1}$ and reduce modulo $x^n + 1$. First, it can be computed by schoolbooks multiplication to obtain $\mathbf{c} = c_0 + c_1x + \cdots + c_{2n-2}x^{2n-2}$, and then using the equation $x^n \equiv -1 \pmod{x^n + 1}$, substituting all x^{n+m} by $-x^m, \forall m \geq 0$ in $\mathbf{c}$. This naive method is obviously slow since the time complexity for this method is $O(n^2)$.

Another algorithm utilizes negacyclic convolution, or negative-wrapped convolution (NWC), which can compute the multiplication and the reduction over the irreducible polynomial at the same time, with the following equation:

$$c_j = \sum_{i=0}^{k} a_i \cdot b_{j-i} - \sum_{i=k+1}^{n-1} a_i \cdot b_{j-i+n} \pmod{q}. \tag{1}$$

It is worth mentioning that if the ring is defined modulo $x^n - 1$, cyclic convolution, or positive-wrapped convolution (PWC) will be used.

However, these algorithms are still slow, since the time complexity of NWC is still $O(n^2)$. As one can use the Discrete Fourier Transform (DFT) to speed up the computation of convolution over real or complex numbers, a generalization of DFT over the rings (or fields), called Number Theoretic Transform (NTT), will be used to compute NWC faster [28].

3.5 Number Theoretic Transform

Efficient polynomial multiplication in $\mathcal{R}_q = \mathbb{Z}_q[x]/(x^n + 1)$ is essential for lattice-based cryptographic primitives. The Number Theoretic Transform (NTT) speeds up this operation by converting polynomials into a domain where multiplication becomes pointwise (PWM) and much faster to compute.

NTT is the finite field or ring version of the Discrete Fourier Transform. In DFT, the n-th root of unity $\omega = e^{-i2\pi/n}$ is defined such that $\omega^n = e^{-i2\pi} = \cos(-2\pi) + i\sin(-2\pi) = 1$, and $\omega^{n/2} = -1$. For computation of NTT for the PWC, it requires a primitive n-th root of unity $\omega \in \mathbb{Z}_q$, meaning $\omega^n \equiv 1 \pmod q$, and $\omega^{n/2} \equiv -1 \pmod q$ and no smaller positive power of ω equals 1. To compute the NTT for NWC (which is the case in all PQC module-lattice-based schemes), $2n$-th root of unity ζ is required, i.e., $\zeta^{2n} \equiv 1 \pmod q$ and $\zeta^n \equiv -1 \pmod q$.

The formal definition of NTT of a polynomial $\mathbf{a}(x)$ is the polynomial evaluation of $\mathbf{a}(x)$ when $x = \zeta^{2\mathrm{brv}_l(i)+1}$:

$$\mathrm{NTT}(\mathbf{a}) := \left(\mathbf{a}(\zeta^{2\mathrm{brv}_l(0)+1}), \mathbf{a}(\zeta^{2\mathrm{brv}_l(1)+1}), \ldots, \mathbf{a}(\zeta^{2\mathrm{brv}_l(255)+1})\right) \in \mathcal{T}_q,$$

where brv_l is the bit-reverse function when the total number of bits is $l = \log_2 n$. The NTT domain $\mathcal{T}_q$ is defined as $\mathcal{T}_q := \prod_{i=0}^{n-1} \mathbb{Z}_q$. In other words, in the NTT domain, there are n elements, each in $\mathbb{Z}_q$. So, at least in appearance, an element in $\mathcal{R}_q$ and an element in $\mathcal{T}_q$ look the same. The NTT is a function that maps elements from $\mathcal{R}_q$ to $\mathcal{T}_q$, and iNTT is a function to map from $\mathcal{T}_q$ to $\mathcal{R}_q$. However, the time complexity of polynomial evaluation is $O(n^2)$, which is considered a bad time complexity.

Another way of defining the forward NTT of a polynomial $\mathbf{a} = [a_0, \ldots, a_{n-1}]$ is as follows:

$$\hat{a}_j = \sum_{i=0}^{n-1} \zeta^{2ij+i} \cdot a_i \pmod q,$$

where $\mathrm{NTT}(\mathbf{a}) = \hat{\mathbf{a}} = [\hat{a}_0, \ldots, \hat{a}_{n-1}]$. Once transformed, multiplication becomes pointwise:

$$\hat{c}_j = \hat{a}_j \cdot \hat{b}_j \pmod q.$$

After this step, an inverse NTT is applied to return the result to the coefficient domain:

$$c_i = n^{-1} \sum_{j=0}^{n-1} \zeta^{-2ij-i} \cdot \hat{a}_j \pmod q.$$

The computation time of this method is still unacceptable, since each NTT conversion needs n^2 operations, and PWM requires n operations, so the asymptotic time complexity for this method will still be $O(n^2)$. However, as FFT is the faster method of computation of DFT, one can use FFT-like algorithms to compute NTT faster. To achieve $O(n \log_2 n)$ time complexity, NTT implementations use recursive divide-

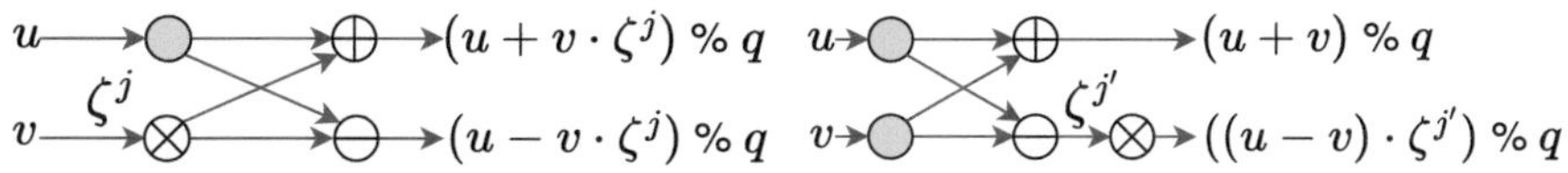

Fig. 4 Butterfly structures for fast NTT computation: **a** Cooley–Tukey unit for forward NTT, **b** Gentleman–Sande unit for inverse NTT

and-conquer methods. Two commonly used approaches are the Cooley–Tukey (CT) [10] for forward NTT and Gentleman–Sande (GS) [17] for inverse NTT (iNTT) algorithms. Both use butterfly operations, which combine modular addition or subtraction with multiplication by a power of ζ, known as a twiddle factor (TF).

Figure 4 shows the basic butterfly units (BFU) for both algorithms. In part (a), the CT butterfly performs a modular multiplication by the twiddle factor, then computes a modular addition and a modular subtraction, which is the building block of the forward NTT converter. In part (b), the GS butterfly computes an inverse NTT stage by summing and subtracting inputs, then a multiplication of the result of the subtraction by twiddle factors. These small building blocks are reused throughout the NTT multiplier.

Figure 5 illustrates the full structure of an 8-point NTT and iNTT. The transform proceeds in $\log_2 n$ stages, where each stage consists of butterfly units operating on pairs of coefficients. In the forward transform, the distance between input pairs gets halved in each stage, which is essential in data locality and in-place computation of NTT. Forward NTT unit converts a polynomial to the NTT domain in $\log_2 n$ stages, and in each stage, n values will be computed. The distance between the pairs for the first stage is $n/2$, and the algorithm is continued by halving the distance for each stage, until the distance between the pairs becomes 1. The twiddle factors will be precomputed and stored in memory to be used during the conversion. To achieve in-place computation, the twiddle factors have to be stored in the bit-reverse order, i.e., $\mathrm{TF}[i] = \zeta^{\mathrm{brv}_l(i)}$. The first twiddle factor, i.e., TF[0], is never used, and in the first stage, TF [1] will be used for all $n/2$ butterfly units. When the pair distance is d, after each d butterfly computation, the next twiddle factor will be used. For instance, in the first stage, the pair distance is $n/2$, so TF [1] will be used for $n/2$ butterfly units. In the second stage, where the spacing between the pairs is $n/4$, TF [2] will be used at first, then after $n/4$ butterfly units, TF[3] will be used.

In the inverse transform, the spacing between pairs is doubled in each stage, starting from 1. The computation of inverse NTT uses the GS butterfly unit, which has the same computation units as the CT butterfly unit, but with a different order. There is an extra step at the end of the iNTT computation, which is multiplying each value by $n^{-1} \pmod q$. However, to save time and resources, in iNTT BFU, after each addition and multiplication, before outputting the result, the result will be multiplied by $2^{-1} \pmod q$, which can be done by a shift to the right for the even numbers, and adding to q and then a shift to the right for the odd numbers, and it will be done for all $\log_2 n$ stages. Both methods achieve the same result, since

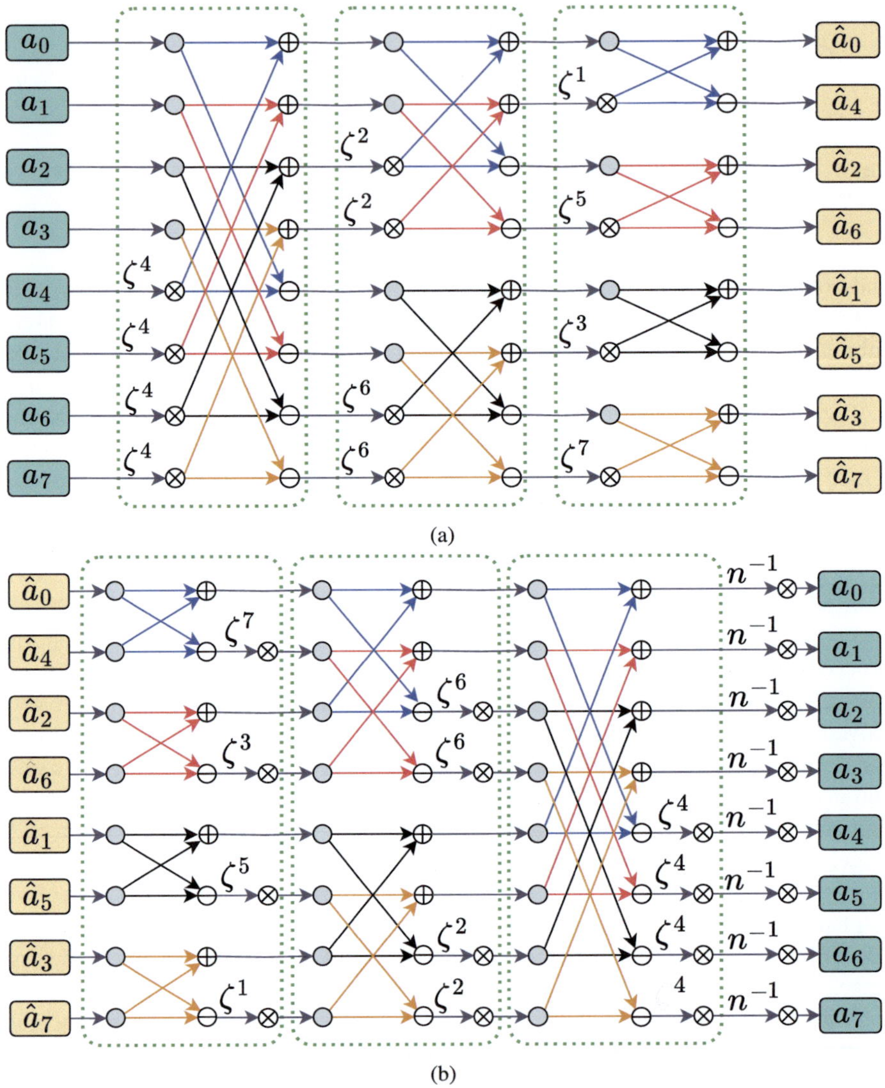

Fig. 5 Structure of **a** an 8-point NTT and **b** iNTT using recursive butterfly units across multiple stages. Colored lines show data flow and twiddle factor multiplication. The blue boxes represent coefficients in the normal domain, and the yellow boxes represent them in the NTT domain

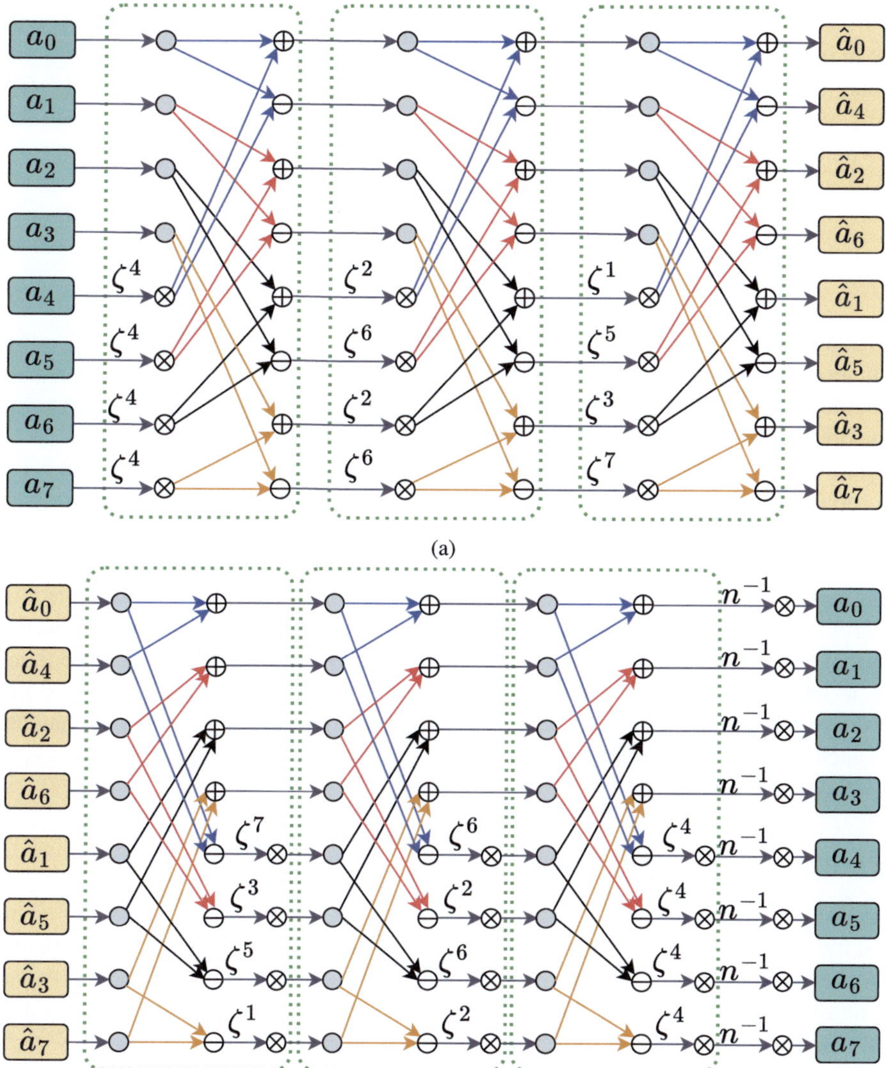

(a)

(b)

Fig. 6 Diagram of constant-geometry **a** NTT and **b** iNTT converter

$n^{-1} = (2^{-1})^{\log_2 n}$. Using the mentioned steps, the output of NTT using CT BFU will be in bit-reversed order, i.e., $\hat{a} = [\hat{a}_{\text{brv}_l(0)}, \hat{a}_{\text{brv}_l(1)}, \hat{a}_{\text{brv}_l(2)}, \ldots, \hat{a}_{\text{brv}_l(n-1)}]$. It does not make any trouble computing the polynomial multiplication since fortunately, GS BFU gets the coefficients in bit-reversed order and generates the coefficients in the normal order (Fig. 6).

3.6 Constant-Geometry NTT Computation

In addition to the explained NTT computation, which is called in-place NTT computation, another approach has also been explored, called out-of-place, constant-geometry (CG), or Pease NTT computation, which is proposed in [44] by Pease for the first time for FFT. In recent years, some interest has been shown in CG NTT computation [16, 51, 52]. In CG NTT computation, BFUs are exactly like in-place NTT computation, but the spacing between the inputs is always the same for all stages. For NTT conversion, the spacing between inputs of each BFU is always $n/2$ for all stages, whereas the spacing is 1 for the outputs of the BFU. It means each BFU gets the pair of $[a_i, a_{i+n/2}]$, and after the butterfly operation, generates two outputs and stores them in $[\hat{a}_{2i}, \hat{a}_{2i+1}]$. For iNTT, input spacing is 1 for all stages and $n/2$ for the output spacing. This feature enables us to compute NTT more efficiently since there is no need to change the distance between pairs in different stages. However, the order of feeding the twiddle factors is a challenge in this approach, since it changes in each BFU in stages of 2 to the end.

3.7 Incomplete NTT Trick

All the explained algorithms are for when the $2n$-th root of unity exists, which means there exists a ζ where $\zeta^{2n} \equiv 1 \pmod 1$ and $\zeta^n \equiv -1 \pmod q$. The rule is ζ exists if and only if $(q-1) \mid 2n$. This is the case for ML-DSA, so for this scheme, one can use the mentioned algorithms to multiply polynomials in ML-DSA. However, in ML-KEM, because of efficiency, modulo q has been selected in a way that there is no ζ exists that satisfies the requirement. In other words, ML-KEM parameters are not NWC NTT-friendly. However, for ML-KEM, the n-th root of unity exists, i.e., $(q-1) \mid n$. Note that the irreducible polynomial is still $x^n + 1$, which has been proven to be more secure than $x^n - 1$ [30]. As a result, negacyclic NTT should be used for ML-KEM. In case the n-th root of unity exists and the $2n$-th root of unity does not exist, to compute the negacyclic NTT, the incomplete NTT trick can be used. In FIPS 203, the n-th root of unity for ML-KEM is denoted with ζ, the same as $2n$-th root of unity for ML-DSA. Incomplete NTT includes two differences from the full NTT: (1) the number of steps, and (2) PWM. Both differences stem from the fact that incomplete NTT converts the polynomial of degree $n-1$ to $n/2$ pairs of degree-1 polynomials of form $\hat{a}_j + \hat{a}_{j+1}x$, instead of n degree-0 polynomials of form $\hat{a}_j$ in the full NTT conversion.

In the NTT computation, the number of stages is $\log_2 n - 1$. For the forward NTT, the last stage, and for the inverse NTT, the first stage, will not be done for ML-KEM and will be ignored. So the number of required twiddle factors is $n/2 - 1$, instead of $n - 1$ in full NTT.

In PWM, instead of computing $\hat{a}_j \cdot \hat{b}_j$, two degree-1 polynomials of $\hat{a}$ and $\hat{b}$ should be multiplied as follows:

$$\hat{c}_j + \hat{c}_{j+1}x = (\hat{a}_j + \hat{a}_{j+1}x) \cdot (\hat{b}_j + \hat{b}_{j+1}x) = \hat{a}_j \cdot \hat{b}_j + (\hat{a}_{j+1} \cdot \hat{b}_j + \hat{a}_j \cdot \hat{b}_{j+1})x + \hat{a}_{j+1} \cdot \hat{b}_{j+1}x^2.$$

However, this result should be reduced to a degree-1 polynomial, i.e., reduced modulo $x^2 - \zeta^{2\mathrm{brv}_l(j/2)+1}$, where $ln = \log_2 n - 1$. Therefore, x^2 can be substituted by $\zeta^{2\mathrm{brv}_l(j/2)+1}$. As a result, the PWM can be done by this for all $j \in \{0, 2, 4, \ldots, n-1\}$:

$$\hat{c}_j = \hat{a}_j \cdot \hat{b}_j + \hat{a}_{j+1} \cdot \hat{b}_{j+1} \cdot \zeta^{2\mathrm{brv}_l(j/2)+1},$$
$$\hat{c}_{j+1} = \hat{a}_{j+1} \cdot \hat{b}_j + \hat{a}_j \cdot \hat{b}_{j+1}.$$

Finally, the formal definition of incomplete NTT is as follows:

$$\mathrm{NTT}(\mathbf{a}) := \big(\mathbf{a} \mod (x^2 - \zeta^{2\mathrm{brv}_l(0)+1}), \ldots, \mathbf{a} \mod (x^2 - \zeta^{2\mathrm{brv}_l(n/2-1)+1})\big) \in \mathcal{T}_q.$$

The NTT domain, when the incomplete NTT has been used, is defined as follows:

$$\mathcal{T}_q := \bigoplus_{i=0}^{n/2-1} \mathbb{Z}_q[x]/(x^2 - \zeta^{2\mathrm{brv}_l(i)+1}),$$

where $\bigoplus$ is the direct sum. However, like the full NTT, this definition is not used in practice because of the large time complexity.

4 Structural View of Lattice-Based Cryptography

In this section, we examine the algorithmic structure and core operations of the two module-lattice-based schemes selected by NIST for standardization: ML-KEM and ML-DSA. We describe the key generation, encapsulation, and decapsulation procedures of ML-KEM, followed by an overview of the key generation, signing, and verification processes in ML-DSA (Fig. 7).

4.1 Guide to Read the Diagrams

Throughout this paper, all the internal functions of ML-KEM and ML-DSA have been illustrated in flow diagrams. The yellow boxes represent the random data, selected randomly from a distribution, denoted by $\in_R$. Blue boxes are function blocks for internal cryptographic operations. Green boxes indicate outputs. Mauve-colored boxes denote the received data from the other party. Lavender-colored boxes indicate the stored data, which was generated by this party. Teal diamonds show a decision-making process with an if statement. Red boxes denote a big internal function like K-PKE.Encrypt. Arrows indicate the flow of data or control between different functional blocks.

4.2 Notation

Throughout the paper, all scalars in a module-lattice structure, i.e., a single polynomial in $\mathcal{R}_q$, which has n elements of $b = \log_2 q$ size, are shown in normal face. Vectors of polynomials in $\mathcal{R}_q^k$ are shown in bold-faced and small letters, and matrices of polynomials, i.e., elements in $\mathcal{R}_r^{k\times k}$ are shown in boldface and capital letters. The infinity norm $||\mathbf{x}||_\infty$ shows the maximum value in the matrix or vector of x, which is the maximum value of coefficients of polynomials inside the vectors or matrices. $\mathbb{B}$ is used to denote a byte, or 8 bits. So, for example, $\mathbb{B}^{32}$ shows 32 bytes, or 256 bits.

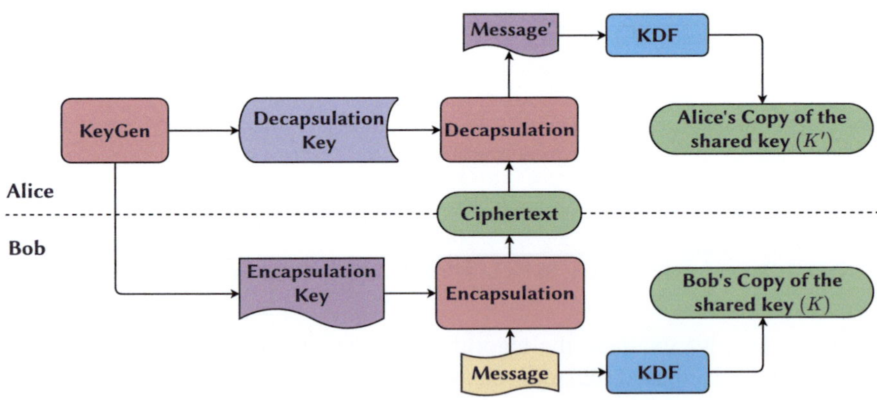

Fig. 7 Flow of the ML-KEM key establishment protocol. Alice generates a key pair and shares her public key. Bob encapsulates a random message into a ciphertext and sends it to Alice. Alice decapsulates it, so both parties can generate the shared key using the random message by passing it through a Key Derivation Function (KDF)

$\mathcal{S}_\eta$ shows the ring of small polynomials, where the coefficients are much smaller than q.

4.3 ML-KEM

ML-KEM is a post-quantum key establishment scheme standardized by NIST FIPS 203 to replace vulnerable classical methods. It is derived from the CRYSTALS–Kyber [9] construction and is based on the hardness of the decisional MLWE (DMLWE) problem. ML-KEM achieves IND-CCA2 security by applying the Fujisaki–Okamoto (FO) transform to a CPA-secure base public key encryption scheme, known as K-PKE. At its core, ML-KEM consists of three main algorithms: Key Generation, Encapsulation, and Decapsulation. These correspond to the generation of a public/private key pair, the secure encapsulation of a shared secret, and the decapsulation process to recover that secret, respectively. The Encapsulation and Decapsulation algorithms internally rely on the primitives K-PKE.Encrypt and K-PKE.Decrypt, while the key generation step is based on K-PKE.KeyGen [37] (Fig. 8).

Figure 7 illustrates the overall key establishment workflow between two parties, Alice and Bob. Alice, as the receiver, first runs Key Generation to generate a public/private key pair. She keeps the decapsulation key and shares the encapsulation key with Bob. Bob, as the sender, uses this public encapsulation key to run Encapsulation on a random message m, generating a ciphertext c and a shared secret $K = \text{KDF}(m)$, where KDF is a Key Derivation Function. The ciphertext is sent to Alice, who then uses her decapsulation key to execute Decapsulation to get m', and then recovers the same shared secret $K' = \text{KDF}(m')$. If the ciphertext has not been tampered with, the derived keys K and K' match, enabling both parties to securely establish a symmetric key. To detect any tampering in the ciphertext, the decapsulated message will be encapsulated again by Alice, and a ciphertext c' will be obtained by her. Then, Alice makes a comparison between c and c', and if they are equal, the derived secret key K' will be considered as a valid shared key. This key can be used in a symmetric encryption scheme such as AES-256. In other words, ML-KEM will be used only to exchange the symmetric key between two parties, and the encryption of the connection will be continued using the encrypted key transmitted with ML-KEM. ML-KEM comes with three approved parameter sets: ML-KEM-512, ML-KEM-768, and ML-KEM-1024. These correspond to NIST's intended security levels of AES-128 (Cat 1), AES-192 (Cat 3), and AES-256 (Cat 5) classical security, respectively. Each parameter set is defined by five variable parameters–k, η_1, η_2, d_u, d_v–and three fixed constants $n = 256$ and $q = 3329 = 2^{12} - 2^9 - 2^8 + 1$, and the n-th root of unity ζ,

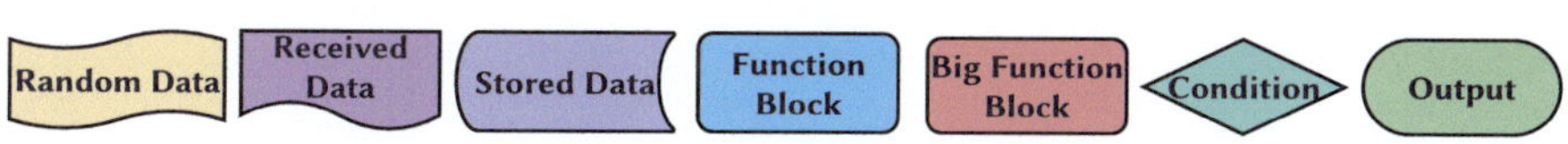

Fig. 8 The guide to understand the flow diagrams

which will be used in the NTT computation, is 17 for all three sets. The parameter k determines the matrix and vector dimensions in the key generation and encryption algorithms, i.e., each matrix is a matrix of $k \times k$, and vectors are of length k. Each element of vectors or matrices is a polynomial of degree n, i.e., a vector of length n, where the coefficients are in $[0, q)$, which is defined by the module-lattice. The noise parameters η_1 and η_2 govern the sampling distributions used in generating secret and error polynomials. The compression parameters d_u and d_v affect the size of the ciphertext through the compression and decompression routines. The naming of each parameter set stems from n and k, i.e., each standard of ML-KEM is ML-KEM-$(n \cdot k)$, representing different security levels of ML-KEM. Table 4 summarizes these parameter sets and their values. Among them, ML-KEM-1024 offers the strongest security level [37].

4.3.1 Key Generation

The general idea of the key generation part is to generate a length k vector of polynomials $\mathbf{s}$ as the secret key, and compute $\mathbf{t} = \mathbf{A} \cdot \mathbf{s} + \mathbf{e}$, and send the pair of $(\mathbf{A}, \mathbf{t})$ as the public key. MLWE, which protects the secret in key generation, states that the attacker cannot find $\mathbf{s}$ given the public key, and decisional MLWE states that the attacker cannot even distinguish a randomly generated tuple of $(\mathbf{A}', \mathbf{t}')$ and a true public key generated by a valid $\mathbf{s}$. This assumption proved to be true, and this problem is at least as hard as some NP-hard problems of lattices. As a result, this scheme will be secure, even on a quantum computer.

In the whole process, all multiplications have to be done using the NTT. Therefore, when a matrix or a vector is needed afterward, or a variable has to be generated from a certain distribution, it would be generated in the normal domain, and then it will be converted to NTT to simplify the polynomial multiplications. For example, the secret key $\mathbf{s}$ is generated from a Centered Binomial Distribution (CBD), where the coefficients of $\mathbf{s}$ are in $[-\eta, \eta]$; consequently, $\mathbf{s}$ will be generated in the normal domain $\mathcal{R}_q$ and then will be transformed into the NTT domain for the fast multiplication.

Table 4 Parameter sets of ML-KEM [37]

Parameter	ML-KEM-512	ML-KEM-768	ML-KEM-1024
n	256	256	256
q	3329	3329	3329
ζ	17	17	17
k	2	3	4
η_1	3	2	2
η_2	2	2	2
d_u	10	10	11
d_v	4	4	5

In contrast, when there are no constraints in generating a variable, like **A**, or the variable itself will not be needed in the future, the generated matrix or vector of polynomials will be considered in the NTT domain $\mathcal{T}_q$.

The ML-KEM key generation procedure, shown in Fig. 9, is based on the K-PKE.KeyGen algorithm and deterministically derives a public/private key pair from two 32-byte random seeds d and z. The first seed d is expanded using SHA3-512 into two values: $\rho \in \mathbb{B}^{32}$, used as the seed to generate the public matrix $\hat{\mathbf{A}}$, and $\sigma \in \mathbb{B}^{32}$, used to sample the secret vector of polynomials $\mathbf{s} \in \mathcal{S}_{\eta_1}^k$ and a noise vector of polynomials $\mathbf{e} \in \mathcal{S}_{\eta_1}^k$. It means that $||\mathbf{s}||_\infty \leq \eta_1$ and $||\mathbf{e}||_\infty \leq \eta_1$. The public matrix $\hat{\mathbf{A}} \in \mathcal{T}_q^{k \times k}$ is generated pseudorandomly using an XOF (SHAKE128) with ρ. The reason for this is efficiency; instead of sending a $k \times k$ matrix of polynomials of size $b \times n$, where $b = \log_2 q$, a random seed of 32 bytes will be sent to the other parties as the public key component, and using SHA3-512, a deterministic expanding function, all parties can re-generate the public key **A**. This is highly efficient because it dramatically reduces the public key's transmission size. Moreover, as **A** is a random matrix of polynomials, there are no constraints in generating it. As a result, it will be considered in the NTT domain, so no NTT conversion will be required for **A**.

The secret vector **s** and the noise vector **e**, both in $\mathcal{S}_{\eta_1}^k$, are sampled from centered binomial distributions using SHAKE256-based pseudorandom functions seeded by σ. The CBD distribution forces the polynomial coefficient to be in the range $[-\eta_1, \eta_1]$. Since both **s** and **e** have a constraint in the distribution, they will be generated in the normal domain first, and then using an NTT converter, they will be

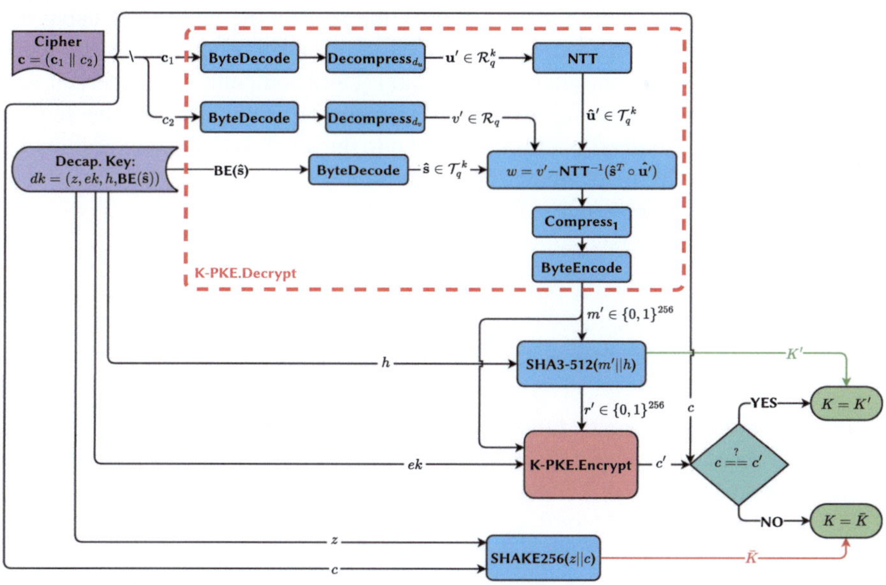

Fig. 9 Key generation process of ML-KEM

transformed to the NTT domain $\mathcal{T}_q^k$ to produce $\hat{\mathbf{s}}$ and $\hat{\mathbf{e}}$. The value $\hat{\mathbf{t}} = \hat{\mathbf{A}} \circ \hat{\mathbf{s}} + \hat{\mathbf{e}}$ is then computed in the NTT domain, where $\circ$ is the pointwise multiplication. Then, the target vector **t** is encoded using a byte encoding function.

The encapsulation key ek consists of the encoded vector $\hat{\mathbf{t}}$ and the matrix seed ρ, enabling regeneration of $\hat{\mathbf{A}}$ during encapsulation. The decapsulation key dk includes the encoded $\hat{\mathbf{s}}$, the encapsulation key, the hash of the encapsulation key with SHA3-256, and the second random seed z, which supports implicit rejection during decapsulation. The encapsulation key is sent to the other parties for encapsulating their messages, and the decapsulation key is stored by this party to open the encapsulated message. We can continue this procedure without the need to compute inverse NTT now, and we will apply only one inverse NTT at the end. In the meantime, all computations are done in the NTT domain. This key generation process ensures compact and secure key material, grounded in the hardness of the DMLWE problem [37].

4.3.2 Encapsulation

Given a public encapsulation key, ML-KEM encapsulation produces a ciphertext with two components $\mathbf{c}_1$ and c_2 and a shared secret K, which can be used as the key for a symmetric encryption scheme like AES. As shown in Fig. 10, the algorithm first samples a 32-byte random message m, which is hashed together with the hash of the public key to derive both a shared secret K and encryption randomness r. This randomness is then used to encrypt the message m via the K-PKE.Encrypt primitive, yielding ciphertext $\mathbf{c} = (\mathbf{c}_1, c_2)$. Internal of the K-PKE.Encrypt, the encryption process regenerates the public matrix $\hat{\mathbf{A}}$ from the seed in the public key using the deterministic expansion of seed with SHAKE128, samples ephemeral small vector of polynomials $\mathbf{y}$, $\mathbf{e}_1$, and the small polynomial e_2, and computes the ciphertext as a pair of polynomials $(\mathbf{u}, v)$ using NTT-based polynomial arithmetic. The message m is encoded into a polynomial μ and embedded into the second component v, allowing it to be later recovered by the holder of the secret key.

The compress and decompress functions are defined as follows:

$$\text{Compress}_d(x) = \lceil (\frac{2^d}{q}) \cdot x \rfloor \pmod{2^d},$$
$$\text{Decompress}_d(y) = \lceil (\frac{q}{2^d}) \cdot y \rfloor.$$

Compress_d and decompress_d functions are for reducing the size of variables and recovering them. They also have another use case in ML-KEM. Using the $\text{decompress}_1(\text{m}) = \lceil (\frac{q}{2^1}) \cdot m \rfloor$ function in the encryption process, the message range will be converted from [0,1] to $[0, q/2]$, so by using the compress_1 in the decryption process, message can be seen through the noises added in MLWE instance in the key generation part. **u** and v are essentially computed with the following equations:

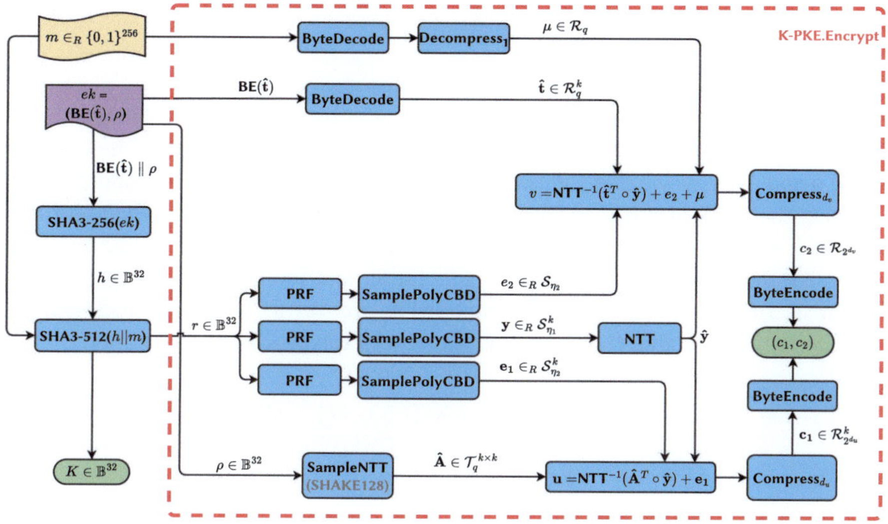

Fig. 10 Encapsulation process of ML-KEM. The red dashed region highlights the K-PKE.Encrypt core, which encrypts m into ciphertext using randomness r

$$\mathbf{u} = \mathbf{A}^T \cdot \mathbf{y} + \mathbf{e}_1,$$
$$v = \mathbf{t}^T \cdot \mathbf{y} + e_2 + \lceil (\frac{q}{2}) \cdot m \rfloor.$$

The pair $(\mathbf{u}, v)$ is then compressed and encoded into byte form as the final ciphertext. This procedure ensures that both the sender and the receiver can independently derive the same 32-byte shared secret K, assuming the ciphertext remains unaltered [37].

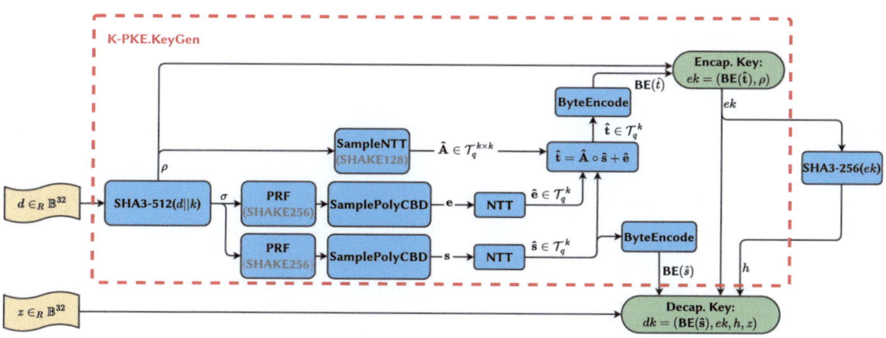

Fig. 11 Decapsulation process of ML-KEM. The red dashed region highlights the K-PKE.Decrypt core, which recovers the message m' from the ciphertext. Re-encryption and comparison ensure correctness through implicit rejection

4.3.3 Decapsulation

The decapsulation algorithm securely derives the shared secret key from a received ciphertext using the decapsulation key. As shown in Fig. 11, the ciphertext $\mathbf{c} = (\mathbf{c}_1 \| c_2)$ is parsed and decoded into polynomials $\mathbf{u}' \in \mathcal{R}_q^k$ and $v' \in \mathcal{R}_q$. They are not called $\mathbf{u}$ and v yet, as we don't know if the ciphertext is changed in the middle or not.

Using the NTT of the secret vector $\hat{\mathbf{s}}$ from the decapsulation key, the algorithm computes $w = v' - \text{NTT}^{-1}(\hat{\mathbf{s}}^T \circ \hat{\mathbf{u}}')$, which is converted from $[0, q/2]$ to $[0, 1]$ to remove the added noise in the key generation and the encryption stages, using compress$_1$ function. Then it will be encoded to recover the candidate message m'. Considering the fact that $\mathbf{t} = \mathbf{A} \cdot \mathbf{s} + \mathbf{e}$, the decryption process is as follows:

$$\begin{aligned} w = v' - \mathbf{s}^T \cdot \mathbf{u}' &= \mathbf{t}^T \cdot \mathbf{y} + e_2 + \lceil (\frac{q}{2}) \cdot m \rfloor - \mathbf{s}^T \cdot \mathbf{A}^T \cdot \mathbf{y} + \mathbf{e}_1 \\ &= (\mathbf{A} \cdot \mathbf{s} + \mathbf{e})^T \cdot \mathbf{y} + e_2 + \lceil (\frac{q}{2}) \cdot m \rfloor - \mathbf{s}^T \cdot \mathbf{A}^T \cdot \mathbf{y} + \mathbf{e}_1 \\ &= \mathbf{s}^T \cdot \mathbf{A}^T \cdot \mathbf{y} + \mathbf{e}^T \cdot \mathbf{y} + e_2 + \lceil (\frac{q}{2}) \cdot m \rfloor - \mathbf{s}^T \cdot \mathbf{A}^T \cdot \mathbf{y} + \mathbf{e}_1 \\ &= \mathbf{e}^T \cdot \mathbf{y} + e_2 + \lceil (\frac{q}{2}) \cdot m \rfloor + \mathbf{e}_1. \end{aligned}$$

By using the compress$_1$ function, the small polynomials $\mathbf{e}^T \cdot \mathbf{y} + e_2 + \mathbf{e}_1$, which are added to the message, most of the time will be removed efficiently. As a result, compressing the whole expression will remove the noise from the message, and the message will be recovered:

$$\begin{aligned} m' &= \text{Compress}_1(\mathbf{e}^T \cdot \mathbf{y} + e_2 + \lceil (\frac{q}{2}) \cdot m \rfloor + \mathbf{e}_1) \\ &= \lceil (\frac{2}{q}) \cdot (\mathbf{e}^T \cdot \mathbf{y} + e_2 + \mathbf{e}_1 + \lceil (\frac{q}{2}) \cdot m \rfloor) \rfloor \pmod 2 \\ &\approx m, \end{aligned}$$

since $||\mathbf{e}||_\infty, ||\mathbf{e}_1||_\infty, ||\mathbf{y}||_\infty, ||e_2||_\infty \ll q$. However, there is a very small chance that the noise remained in the message and was not removed after the compression. The decapsulation failure rate for ML-KEM-512 is $2^{-138.8}$, for ML-KEM-768 is $2^{-164.8}$, and for ML-KEM-1024 is $2^{-174.8}$. Although this number is very small, there are some discussions that the existence of failure can make this scheme vulnerable [13].

Now that the message candidate m' is recovered, the decrypter has to verify that the recovered message m' is the same message that has been sent, i.e., no attacker has been involved throughout the process. For this, first the recovered message is encrypted again. The stored hash of the public key $h = \text{SHA3-256}(ek)$ will be concatenated with m' and hashed to derive $(r', K') = \text{SHA3-512}(m' \| h)$. A new ciphertext c' is computed by re-encrypting m' using the encapsulation key ek and the randomness r'. If the received ciphertext c matches c', the derived shared key $K = K'$ is returned as the valid symmetric encryption key. It means that no alter-

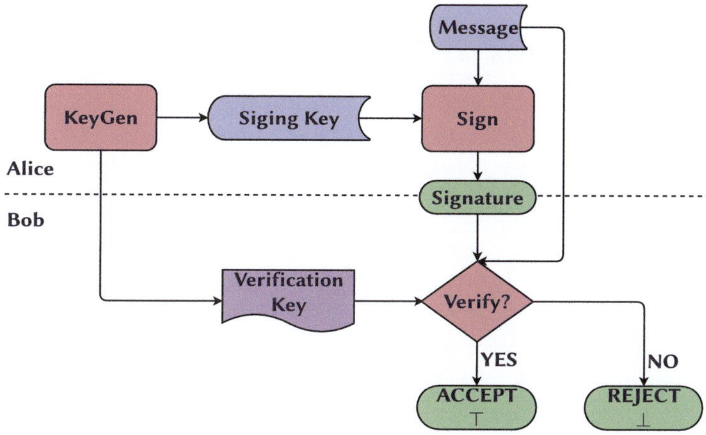

Fig. 12 Overview of the ML-DSA digital signature protocol. Alice signs a message with the signing key and sends the message and the signature to Bob. Bob verifies the signature using the verification key, received from Alice

ation happened in the process. Otherwise, the algorithm applies implicit rejection and outputs $\bar{K} = \text{SHAKE256}(z\|c)$, where z is a random secret value embedded in the decapsulation key. As a result, the attacker won't get any new information by changing the ciphertext since it will generate a seemingly valid key $\bar{K}$, which is, in fact, a completely invalid key that is independent of the message. This design ensures robustness against chosen-ciphertext attacks by preventing any leakage about decryption success.

4.4 ML-DSA

ML-DSA is a quantum-resistant digital signature scheme standardized by NIST in FIPS 204, based on CRYSTALS-Dilithium [3]. It is based on the hardness of the Module Learning With Errors (Module-LWE) and Module Short Integer Solution (Module-SIS) problems [27], which are believed to be intractable even for quantum computers. ML-DSA was selected as the primary post-quantum signature standard due to its strong security guarantees, efficiency, and flexible parameterization.

Similar to classical digital signature schemes, ML-DSA includes three main algorithms: Key Generation, Sign, and Verify. In this scheme, the signer Alice generates a key pair through the Key Generation algorithm and uses the signing key to create a signature for a message. The verifier, Bob, checks the validity of the signature using the public verification key and the original message. Figure 12 illustrates the overall interaction between the signer and verifier. A message is signed using the secret key and sent together with the signature to the verifier, who decides to accept or reject it based on a series of correctness checks.

Table 5 Parameter sets of ML-DSA [38]

Parameter	ML-DSA-44	ML-DSA-65	ML-DSA-87
n	256	256	256
q	8380417	8380417	8380417
d	13	13	13
ζ	1753	1753	1753
λ	128	192	256
γ_1	2^{17}	2^{19}	2^{19}
γ_2	$\frac{q-1}{88}$	$\frac{q-1}{32}$	$\frac{q-1}{32}$
(k, ℓ)	(4, 4)	(6, 5)	(8, 7)
τ	39	49	60
η	2	4	2
$\beta = \tau \cdot \eta$	78	196	120
ω	80	55	75

ML-DSA includes three approved parameter sets—ML-DSA-44 (Cat 2), ML-DSA-65 (Cat 3), and ML-DSA-87 (Cat 5)—designed to achieve NIST's security SHA-256, AES-192, and AES-256, respectively. These sets define parameters that control the security and efficiency of the digital signature operations. The number element in each scalar $n = 256$, the modulus $q = 8380417 = 2^{23} - 2^{13} + 1$, dropped bits $d = 13$, and 2n-th root of unity $\zeta = 1753$ are fixed across all sets, while other parameters such as the matrix dimensions (k, ℓ), sparsity of the challenge polynomial τ, and noise level η vary. Specifically, τ defines the number of non-zero coefficients in the challenge polynomial, and γ_1, γ_2 govern coefficient bounds during signature compression and rejection sampling. 2λ defines the bit length of the signature component $\tilde{c}$. The hint bound ω, derived error bound $\beta = \tau \cdot \eta$, and the rejection probability (which can be computed by the parameters) indirectly influence signing performance. k and ℓ appear in the naming of different parameter sets, and each standard is in the form of ML-DSA-$k\ell$. Among the three, ML-DSA-87 offers the highest level of post-quantum security, and NSA predicted that ML-DSA-87 will be used in most cases [1]. Table 5 summarizes these parameters for all variants.

4.4.1 Key Generation

As shown in Fig. 13, the procedure begins with a 32-byte seed ξ that is expanded using SHA3 into values $\rho \in \{0, 1\}^{256}$, $\rho' \in \{0, 1\}^{512}$, and $K \in \{0, 1\}^{256}$, totally 1024 bits. The matrix $\hat{\mathbf{A}} \in R_q^{k\times\ell}$ is deterministically generated from ρ in NTT domain, and secret polynomials $\mathbf{s}_1 \in \mathcal{S}_q^{\ell}$ and $\mathbf{s}_2 \in \mathcal{S}_q^{k}$ are sampled from a CBD using ρ'. CBD sampling is done by a rejection sampling; i.e., it samples uniformly from $\mathcal{R}_q$, and then checks that each coefficient is in the bound $[-\eta, \eta]$. If the generated number is valid, the sampler outputs the numbers; otherwise, the sampler discards the number and generates another number instead of the invalid number. The sampler continues this procedure until all $k \times n$ and $\ell \times n$ valid numbers are generated. The rejection sampler works as follows: (1) Using SHAKE256 and ρ' as the seed, 1 byte will be generated randomly called z. (2) Two numbers will be produced by z: $z_1 = z \pmod{16}$ and $z_2 = \lfloor z/16 \rfloor$. (3) When $\eta = 2$, if $z_i < 15$, $2 - (z_i \pmod 5)$ will be the value of coefficient, when $i = \{1, 2\}$; otherwise, it will discard z_i and generate another pseudorandom number by squeezing SHAKE256 one more time. When $\eta = 4$, if $z_i < 9$, $4 - z_i$ will be the valid value for the coefficients; otherwise, it will discard it and generate another random number. This procedure will be continued until all required coefficients are generated.

The public vector $\mathbf{t} = \mathbf{A} \cdot \mathbf{s}_1 + \mathbf{s}_2$ is split into high-order $\mathbf{t}_1$ and low-order $\mathbf{t}_0$ via Power2Round. In ML-DSA, $\mathbf{t}_1$ consists of the higher 13 bits of coefficients in $\mathbf{t}$, and the rest of the coefficients are stored in $\mathbf{t}_0$. The reason for splitting $\mathbf{t}$ is the efficiency; only 13 bits of vector $\mathbf{t}$ are sent to the verifiers, and later, the signer hints to the verifiers about how they can do the computations on $\mathbf{t}$ without losing the precision. The value $\mathbf{t}_1$, alongside with the seed ρ that can generate $\hat{\mathbf{A}}$, forms the public key.

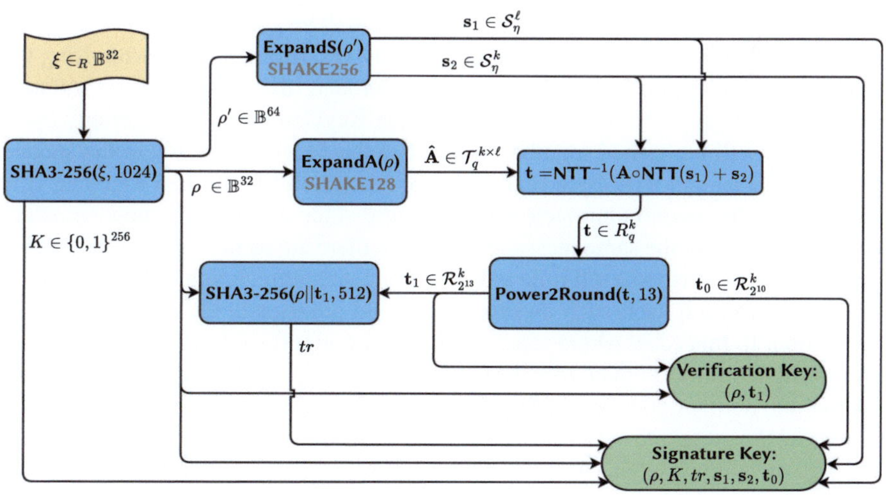

Fig. 13 The key generation process of ML-DSA

For the private key, the hash of $\rho||\mathbf{t}_1$ using SHA3-256 will be computed, and 512 bits will be generated by this unit, called tr. Finally, the private key consists of $(\rho, K, tr, \mathbf{s}_1, \mathbf{s}_2, \mathbf{t}_0)$.

4.4.2 Signature Generation

The signature generation procedure, illustrated in Fig. 14, begins by regenerating matrix $\hat{\mathbf{A}}$ using the ExpandA function and the seed ρ. In the signing process, as with many other signature algorithms, the hash of the message will be signed, instead of signing the whole message. The most secure way of hashing the message is to produce the hash using both the message and a component from the public key. Although another algorithm, called HashML-DSA, has been proposed in FIPS 204, which first hashes the message before feeding it to the algorithm, and then the hash of the message is hashed with a component of the public key, the NSA prohibits the use of HashML-DSA. The reason for this is inefficiency since the message has to be hashed anyway inside the signature block, and hashing the message without the public key is considered an insecure way of signing. In addition, HashML-DSA is not foolproof, and it may easily cause problems in the implementation [1]. As a result, the message should not be hashed before feeding it into the signature block, and the concatenation of the message M and the hash of the public key tr will be hashed by SHA3-256, to generate a 512-bit message representative μ.

The next step is to generate the private randomness seed ρ'' using the message and seed K. Note that ρ'' is the root of all randomness during the signature part. As a result, the same ρ'' will produce the same signature for the same message and the same public key. In this step, two methods have been proposed by NIST in FIPS 204: (1) *Deterministic*, when with the same public key and the same message, a unique seed ρ'', and therefore, the same signature will be generated, and (2) *Hedged (randomized)*, when the generated signature is always different, even with using the same message and public key. For the first approach, the message representative and randomness K, which was generated in the KeyGen part, and 32 empty bytes (256 zeros), will be hashed together to produce the 512-bit seed ρ''. In the second approach, K, μ, and a 32-byte random number rnd will be hashed together. It is obvious that the second approach is more secure since each time, a new signature will be produced for the same message. However, there are some applications that the user needs to be verified each time with the same signature, and a unique signature is dedicated to each user, such as smart cards and other methods of authentication verification. In this case, where repeatability is important, we have to use the deterministic signature. Moreover, if in the implementation, no fully secure cryptographic true random number generator (TRNG) has been implemented, a less secure TRNG can expose the system to even more dangers. As a result, in this case, using deterministic signing is more secure. Finally, adding a TRNG to the system for signing may reduce efficiency, and when efficiency is more important than security, deterministic signing is recommended. Note that the hedged version can be more secure than the deterministic one, but the latter is also secure enough. Ultimately, the choice between

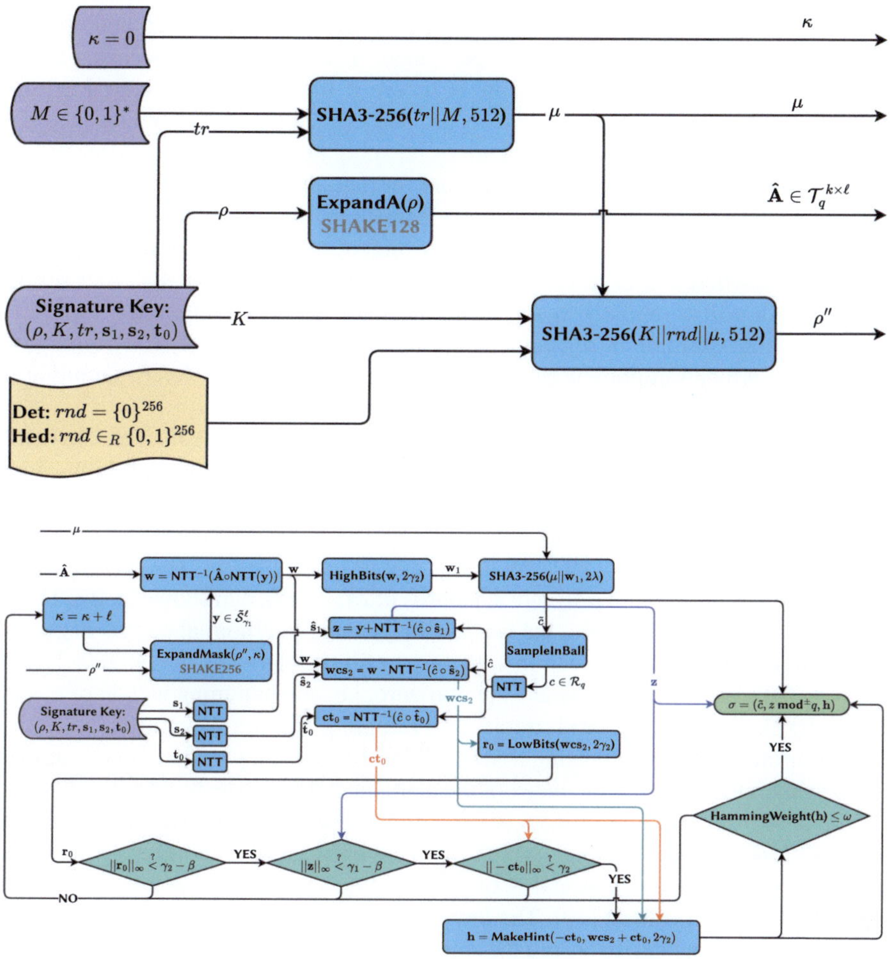

Fig. 14 ML-DSA signing process. The first part is the initialization and computing one-time values (up), and the second part is the main loop to generate the signature by rejection sampling (down)

deterministic and hedged signing depends on the specific security requirements and operational environment of the application.

After the initialization, the rejection sampling loop is started. In this loop, first, a signature will be generated, using the randomness $\mathbf{y} \in \tilde{S}_{\gamma_1}^{\ell}$. Note that γ_1 is a large number, but it is still much smaller than q, so it will be considered as a small polynomial. So, the ring of *relatively* small polynomials is shown by $\tilde{S}_{\gamma}$. This vector of length ℓ of relatively small polynomials is generated by the ExpandMask function, which uses SHAKE256 when the inputs are ρ'' and a counter κ, which starts counting from 0. A signature will be generated using $\mathbf{y}$, and at the end, four conditions will be checked. If all of them were true for this signature, it is a valid signature, and this block outputs the signature; otherwise, it is an invalid signature, and another signature for this message has to be generated. The reason for this is that it has been proven that this message scheme is secure if the signature is from a certain distribution. The signature should be secure and efficient, and if it cannot pass the final examinations, it is either not secure or it cannot be verified later. To generate a new signature, a new vector $\mathbf{y}$ is needed, and for that, the inputs of the ExpandMask function should be changed. In the ExpandMask function, ρ'' is constant, so κ has to be changed. Inside the function, there are ℓ iterations, and in each iteration $\kappa + i$ is consumed, where $i \in \{1, 2, \ldots, \ell - 1\}$, so the next unused κ is $\kappa + \ell$. As a result, ExpandMask will generate $\mathbf{y}$ again, using the same ρ'', but a different $\kappa = \kappa + \ell$. This is a prime example of rejection sampling: sampling uniformly (or from a distribution for which sampling is easy enough), checking for validation, resampling if invalid, and outputting when valid.

After a vector $\mathbf{y}$ is generated, it will be multiplied by the public matrix $\hat{\mathbf{A}}$ to compute the *signer's commitment* $\mathbf{w} = \mathrm{NTT}^{-1}(\hat{\mathbf{A}} \circ \mathrm{NTT}(\mathbf{y})) = \mathbf{A} \cdot \mathbf{y}$. The high-order bits of $\mathbf{w}$, denoted $\mathbf{w_1}$, are extracted and used along with a message hash to compute the *commitment hash* $\tilde{c} = \text{SHA3-256}(\mu || \mathbf{w}_1, 2\lambda)$. Splitting $\mathbf{w}$ is done by writing it in a canonical way as $\mathbf{w} = \mathbf{w}_1 \cdot 2\gamma_2 + \mathbf{w}_0$, where $||\mathbf{w}_0||_\infty \leq \gamma_2$. The length of the polynomial $\tilde{c}$ is different in different standards of ML-DSA, and shown by 2λ. Then, using a SampleInBall function, the *verifier's challenge* polynomial $c \in B_\tau$ is generated. SampleInBall uses a rejection sampling to fill the elements in c with 0, 1, and -1, such that exactly τ elements in c are non-zero, i.e., the hamming weight of c is τ. To fill the elements with 1 or -1, a rejection sampling loop has been used inside the SampleInBall function, and using a SHAKE256 function, a pseudorandom number will be generated, which is the index of the element that has to be filled with $\{1, -1\}$.

The core of the signing process is the computation of the *signer's response* $\mathbf{z} = \mathbf{y} + \mathrm{NTT}^{-1}(\hat{c} \circ \hat{\mathbf{s}}_1) = \mathbf{y} + c \cdot \mathbf{s}_1$, which links the ephemeral randomness $\mathbf{y}$ with the secret key $\mathbf{s}_1$, and hides the structure of $\mathbf{s}_1$ by adding a random value $\mathbf{y}$ to it to obtain the publicly known value $\mathbf{z}$ [29]. The low-order bits of $\mathbf{w} - \mathrm{NTT}^{-1}(\hat{c} \circ \hat{\mathbf{s}}_2)$, denoted as $\mathbf{r_0}$, are then extracted. To ensure the signature's validity, $\mathbf{z}$, $\mathbf{r}_0$, and an intermediate value involving $\mathbf{t}_0$ must satisfy specific bounds. If these checks fail, a new $\mathbf{y}$ is sampled and the process restarts. Once the bounds are satisfied, a hint vector $\mathbf{h}$ is generated to compress information needed for verification. The purpose of the hint vector is to recover the dropped bit of $\mathbf{t}$, since the verifier has only access to $\mathbf{t}_1$. Using the hint vector $\mathbf{h}$, the verifier can find what was the original $\mathbf{t}$, using only $\mathbf{t}_1$, and the signer can send the compressed $\mathbf{t}$, leading to increasing efficiency by reducing the size of public key. To make sure that $\mathbf{h}$ does not add extra bytes, the Hamming weight of $\mathbf{h}$ has to be less than a bound ω.

The final signature output is $\sigma = (\tilde{c}, \mathbf{z}, \mathbf{h})$, binding the message to the private key through structured lattice-based operations. The first bound $||\mathrm{LowBits}(\mathbf{w} - c \cdot \mathbf{s}_2)||_\infty < \gamma_2 - \beta$ ensures that $\mathbf{w}_1 = \mathrm{HighBits}(\mathbf{w}, 2\gamma_2)$ can be recovered by the verifier. The second bound $||\mathbf{y} + c \cdot \mathbf{s}_1||_\infty < \gamma_1 - \beta$ is for the security reasons. We know that $||\mathbf{y}||_\infty \leq \gamma_1$, $c \in B_\tau$, and $\mathbf{s}_1 \in \mathcal{S}_\eta^\ell$, therefore, $||c \cdot \mathbf{s}_1||_\infty \leq \beta$, where $\beta = \tau\eta$. Hence, $||\mathbf{y} + c \cdot \mathbf{s}_1||_\infty \leq \gamma_1 + \beta$. If $||\mathbf{y} + c \cdot \mathbf{s}_1||_\infty = \gamma_1 + \beta$, it will be clear for the attacker that all $\mathbf{y}$, c, and $\mathbf{s}_1$ are at their own maximum value, which is γ_1, τ, and η,

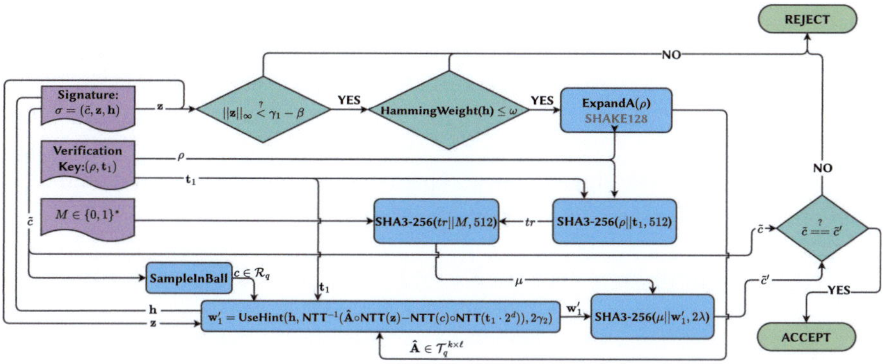

Fig. 15 Verification process of ML-DSA. The decision is based on norm and weight checks on $\mathbf{z}$ and $\mathbf{h}$, and whether the recomputed commitment hash $\tilde{c}'$ matches $\tilde{c}$

respectively. With the same reasoning, it will reveal the information about the values of these secret parameters when $||\mathbf{y} + c \cdot \mathbf{s}_1||_\infty \in [\gamma_1 - \beta, \gamma_1 + \beta]$. So, if the second condition is not satisfied, the attacker can easily find the values of secret parameters, and it can forge a new signature for the message.

The third condition states that $|| - c \cdot \mathbf{t}_0||_\infty < \gamma_2$. If this condition is not satisfied, adding $-c \cdot \mathbf{t}_0$ to $\mathbf{w} - c \cdot \mathbf{s}_2 + c \cdot \mathbf{t}_0$ to obtain $\mathbf{w} - c \cdot \mathbf{s}_2$ in the verification (indirectly), will change the value of HighBits of it by more than 1 or less than -1, so the verifier cannot verify the signature. If the condition is met, using the hint bits $\mathbf{h}$, the signer will notify the verifier to change the values of $\mathbf{w} - c \cdot \mathbf{s}_2 + c \cdot \mathbf{t}_0$ by ± 1.

The last condition is that the Hamming weight of the hint vector $\mathbf{h}$ has to be less than ω. The only reason for this condition is efficiency and reducing the signature size. If any of these conditions are not satisfied, another signature will be generated. The expected number of iterations for generating a valid signature can be found by $e^{n\beta(\ell/\gamma_1 + k/\gamma_2)}$, which is 4.25 for ML-DSA-44, 5.1 for ML-DSA-65, and 3.85 times for ML-DSA-87. The signature is comprised of $\sigma = (\tilde{c}, \mathbf{z} \bmod^{\pm} q, \mathbf{h})$, which will be sent to the verifiers to verify the message and/or the authenticity of the sender (Fig. 15).

4.4.3 Signature Verification

The verification process, shown in Fig. 17, ensures the integrity, which states that the message has not been altered, and authenticity, which states that the message has been signed by the authentic sender, of the received signature $\sigma = (\tilde{c}, \mathbf{z}, \mathbf{h})$ with respect to the message M and the public key $(\rho, \mathbf{t}_1)$. The verifier first checks that $\mathbf{z}$ and $\mathbf{h}$ meet the norm and the Hamming weight bounds. If the requirement is not satisfied, it means that something changed in the middle, or the signature is not valid, so it wouldn't bother with the entire process of checking the signature, and the verifier will reject the signature at the first step. If it passes the bounds checks, it then reconstructs the matrix $\hat{\mathbf{A}}$ using ρ in the verification key, computes the hash of the message with the hash of the public key to obtain μ, and recomputes the verifier's challenge c by using SampleInBall($\tilde{c}$). Then, using public matrix $\hat{A}$, the signer's response $\mathbf{z}$, the hint bits $\mathbf{h}$, the higher bits of $\mathbf{t}$ which is $\mathbf{t}_1$, and the challenge polynomial c, it regenerates the higher bits of commitment $\mathbf{w}_1'$, and computes a new commitment hash $\tilde{c}'$. The signature is accepted only if $\tilde{c}'$ matches the received $\tilde{c}$. To illustrate the correctness of the verification, we start by

$$\mathbf{w}_1' = \text{UseHint}(\mathbf{h}, \mathbf{A} \cdot \mathbf{z} - c \cdot \mathbf{t}_1 \cdot 2^d, 2\gamma_2) = \text{HighBits}(\mathbf{A} \cdot \mathbf{z} - c \cdot \mathbf{t}, 2\gamma_2),$$

since $||c \cdot \mathbf{t}_0|| < \gamma_2$. The expression inside the parentheses can be computed as follows:

$$\begin{aligned}
\mathbf{A} \cdot \mathbf{z} - c \cdot \mathbf{t} &= \mathbf{A} \cdot (\mathbf{y} + c \cdot \mathbf{s}_1) - c \cdot \mathbf{t} \\
&= \mathbf{A} \cdot \mathbf{y} + \mathbf{A} \cdot c \cdot \mathbf{s}_1 - c \cdot (\mathbf{A} \cdot \mathbf{s}_1 + \mathbf{s}_2) \\
&= \mathbf{A} \cdot \mathbf{y} - c \cdot \mathbf{s}_2 \\
&= \mathbf{w} - c \cdot \mathbf{s}_2.
\end{aligned}$$

So:

$$\mathbf{w}_1' = \text{HighBits}(\mathbf{w} - c \cdot \mathbf{s}_2, 2\gamma_2).$$

Since $||\text{LowBits}(\mathbf{w} - c \cdot \mathbf{s}_2, 2\gamma_2)|| < \gamma_2 - \beta$, and $||c \cdot \mathbf{s}_2|| \leq \beta$:

$$\text{HighBits}(\mathbf{w} - c \cdot \mathbf{s}_2, 2\gamma_2) = \text{HighBits}(\mathbf{w}, 2\gamma_2) = \mathbf{w}_1.$$

As a result, if the message has been signed by the authentic sender to provide the valid $\mathbf{A}$ and $\mathbf{t}$, the verifier can generate the exact same $\mathbf{w}_1$. If the message has been intact, the hash of the message with the hash of the public key and $\mathbf{w}_1$ will generate the same $\tilde{c}$ as the one received by the signer.

4.5 On the Hardness of ML-DSA

The hardness of ML-DSA is based on two hard mathematical problems on module-lattices: MLWE and Selftarget-MSIS (lMSIS). MLWE protects the secret keys $\mathbf{s}_1$ and $\mathbf{s}_2$, so the adversary doesn't have any information on secret keys, and thus, cannot impersonate as the original signer. MLWE in ML-DSA states that given $\mathbf{A}$ and $\mathbf{t}$, finding $\mathbf{s}_1$ and $\mathbf{s}_2$ is infeasible in the equation $\mathbf{t} = \mathbf{A} \cdot \mathbf{s}_1 + \mathbf{s}_2 \pmod q$.

MSIS (Module-Short Integer Solution), states that given $\mathbf{A}$ in $\mathbf{A} \cdot \mathbf{v} = 0 \pmod q$, finding a solution for $\mathbf{v}$ such that $0 < \mathbf{v} \leq \beta$ is infeasible, where $\mathbf{A}$ is a $k \times \ell$ matrix of polynomials, and $\mathbf{v}$ is a vector of length ℓ [29]. The lMSIS is the modified version of the MSIS problem, where the problem is to find $\mathbf{v}$ in $\mathbf{A} \cdot \mathbf{v} = \mathbf{t} \pmod q$, where $\mathbf{v}$ is a small non-zero solution for the equation. If $\mathbf{v}$ is not small, i.e., not bounded by β,

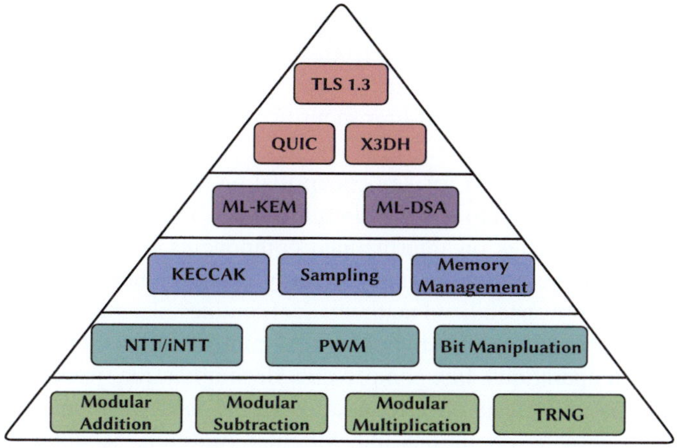

Fig. 16 Computation hierarchy in ML-KEM and ML-DSA

the solution would be trivial. However, considering the bound β, finding the solution for this equation is extremely hard. The IMSIS problem in ML-DSA requires a little explanation. As explained in the previous part, we have

$$\mathbf{w}_1 = \text{HighBits}(\mathbf{w}, 2\gamma_2) = \text{HighBits}(\mathbf{A} \cdot \mathbf{z} - c \cdot \mathbf{t}, 2\gamma_2) \Rightarrow \mathbf{A} \cdot \mathbf{z} - c \cdot \mathbf{t} = \mathbf{w} = 2\gamma_2 \mathbf{w}_1 + \mathbf{w}_0,$$

where $||\mathbf{w}_0||_\infty \leq \gamma_2$. As a result, we have

$$\mathbf{A} \cdot \mathbf{z} - \mathbf{w}_0 = 2\gamma_2 \mathbf{w}_1 + c \cdot \mathbf{t} \Rightarrow [\mathbf{A}|\mathbf{I}_k][\mathbf{z}| - \mathbf{w}_0]^T = 2\gamma_2 \mathbf{w}_1 + c \cdot \mathbf{t}.$$

The Selftarget-MSIS problem in ML-DSA is to find a short vector $[\mathbf{z}| - \mathbf{w_0}]^T$, such that $||\mathbf{z}||_\infty < \gamma_1 - \beta$ and $||\mathbf{w}_0||_\infty \leq \gamma_2$, given the public key $\mathbf{A}$ and publicly known values c, $\mathbf{t}$, γ_2, and $\mathbf{w}_1$ in the equation $\mathbf{A} \cdot \mathbf{z} - \mathbf{w}_0 = c \cdot \mathbf{t} + 2\gamma_2 \mathbf{w}_1$. If no one can find a solution to $\mathbf{z}$ and $\mathbf{w}_0$, it means that no one can forge a signature. ML-DSA is proven to be a strong existentially unforgeable digital signature scheme against a chosen-message attack.

5 High-Performance Implementations of Module-Lattice-Based Cryptography

This section surveys and compares FPGA-based implementationsLattice-based cryptography and hardware-software co-designs of module-lattice-based cryptography. Specifically, we focus on ML-DSA and ML-KEM and evaluate their performance, resource utilization, and architectural trade-offs.

To ground our discussion, we refer to Fig. 16, a hierarchical cryptographic stack, showcasing the layered dependencies from high-level security protocols down to fundamental hardware operations. At the top, application-level protocols like TLS 1.3, QUIC, and X3DH provide end-user security, which can execute a post-quantum algorithm for safe communication. These are built upon post-quantum cryptographic algorithms, such as ML-KEM for key encapsulation and ML-DSA for digital signatures, which are designed to resist quantum attacks. The implementation of these advanced algorithms relies on intermediate components like the KECCAK hash function, cryptographic Sampling, and Memory Management. These, in turn, are constructed from more specialized mathematical routines like the Number Theoretic Transform (NTT/iNTT) for efficient polynomial multiplication, alongside PWM and Bit Manipulation operations, like compression/decompression, packing/unpacking, etc. The entire pyramid rests on a foundation of the most basic and critical components: core arithmetic operations like Modular Addition, Modular Subtraction, and Modular Multiplication, and a True Random Number Generator (TRNG) that provides the essential element of randomness for the entire security system.

5.1 The Role of Hardware in Cryptographic Systems

The implementation of cryptographic algorithms spans a spectrum from pure software libraries to dedicated hardware, each presenting distinct trade-offs between performance, security, and cryptographic-agility–the ability to adapt or replace cryptographic primitives in response to new threats or standards. Hardware acceleration is critical for achieving the high throughput required in modern applications, but it introduces significant challenges related to cryptographic-agility.

At the lowest level, hardware support can manifest as specialized instructions within a general-purpose CPU's instruction set architecture (ISA), such as the Intel SHA Extensions designed to accelerate hash computations. More commonly, cryptographic functions are offloaded to dedicated hardware accelerators. These can be functional blocks within a larger System-on-Chip (SoC) that implement specific sub-operations, or they can be entire application-specific integrated circuits (ASICs) designed for a single algorithm. In these architectures, firmware is essential for managing memory and sequencing the low-level hardware functions in the correct order.

A key characteristic of these hardware implementations is their immutability; once fabricated, the integrated circuits cannot be altered. This physical permanence provides a strong defense against malicious modification but creates a significant barrier to cryptographic transitions. If a design flaw is discovered later or an algorithm is deprecated, the chip itself must be replaced. This is particularly critical for systems with long deployment lifecycles and limited field-update capabilities, such as the secure boot mechanisms in embedded devices. In such cases, the public key and verification code are burned into boot code that cannot be updated, demanding that the chosen signature scheme remains secure for the entire lifetime of the device.

Table 6 Hardware design priorities for different PQC implementation targets. The priority indicates the importance of optimizing for that criterion in the specified environment (e.g., a "High" priority for silicon area in a constrained device means minimizing resource usage is a primary goal)

Design criterion	High-end server	Mobile/IoT device	Constrained embedded
Speed (throughput/latency)	High	High	Medium
Silicon area (resource usage)	Low	Medium	High
Power consumption	Low	High	High
Flexibility (crypto-agility)	High	Medium	Low

Several classes of devices are built around these principles:

- **Hardware Security Modules (HSMs)** are robust, often rack-mounted or portable appliances that provide high-assurance cryptographic services and secure key storage. They typically integrate a microprocessor with multiple accelerator chips and feature tamper-detection mechanisms to protect private keying material, which never leaves the HSM's security boundary.
- **Trusted Platform Modules (TPMs)** and **Subscriber Identity Module (SIM)** cards are smaller-scale cryptographic co-processors integrated into larger systems like laptops and mobile phones. They perform a limited set of cryptographic operations and ensure that the private keys they store are never exposed to the host system. Upgrading their algorithmic capabilities requires replacing the chip.
- Personal cryptographic tokens, such as **Personal Identity Verification (PIV)** cards and USB tokens, function as miniature, portable HSMs for individual users, storing private keys and performing cryptographic operations on behalf of the user's computer.

While hardware provides undeniable benefits in performance and the physical protection of secret keys, its inherent lack of agility complicates the transition to post-quantum cryptography. A cryptographic API must be designed to accommodate this entire spectrum, from software-only libraries to systems where data must be moved to a specific hardware device for processing. The design of PQC hardware, therefore, requires a forward-looking approach, selecting state-of-the-art and conservative algorithm variants to maximize the secure operational lifetime of the device [5]. Table 6 highlights the fundamental trade-offs between performance, resource utilization, power consumption, and algorithmic flexibility that dictate architectural design choices for different PQC hardware implementations.

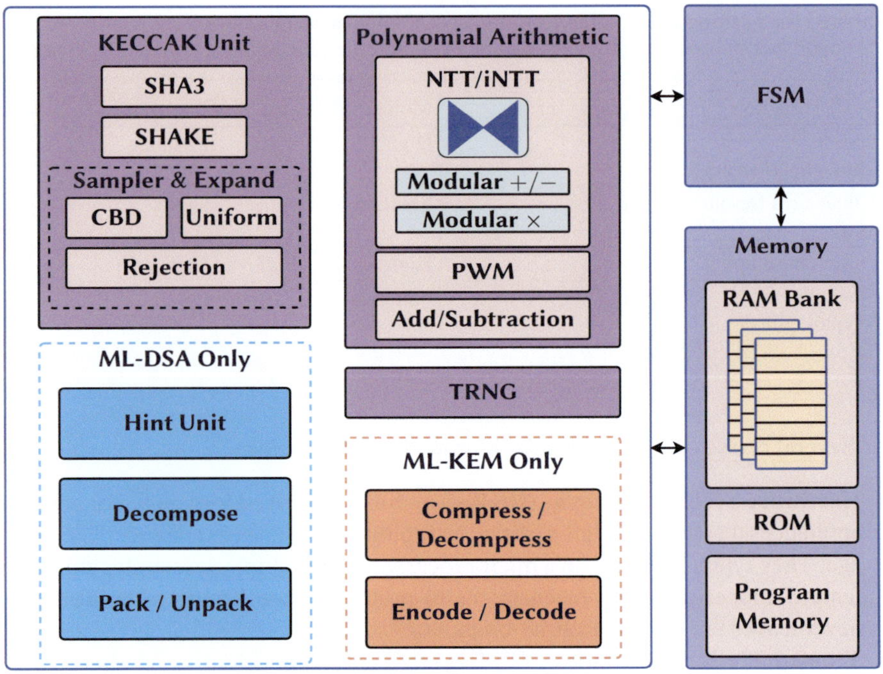

Fig. 17 A high-level overview of a unified design for ML-KEM and ML-DSA

5.2 High Level Hardware Architecture of Module-Lattice-Based PQC

The general architecture of module-lattice-based cryptography in hardware is shown in Fig. 17. It presents a high-level, conceptual overview of the key hardware components required to support both ML-KEM and ML-DSA. The blocks that are designed specifically for ML-KEM are shown in orange, the blocks that are only for ML-DSA are shown in blue, and the blocks that will be used in both algorithms are shown in purple. It can be seen that the fundamental operations of these module-lattice-based schemes are the same, and a unified design would benefit from this, since most power and area are consumed by the shared blocks.

On the left, the KECCAK Unit provides SHA-3 and SHAKE functions, which are used for hashing, seed expansion, and message digest operations. This unit feeds into the Sampler and Expand block, which includes uniform sampling, CBD, and rejection sampling—mechanisms essential for generating secret keys and noise vectors, such as in SampleNTT, SampleA, and SampleMask. At the center of the architecture is the Polynomial Arithmetic unit, which performs the core computations of module-lattice-based schemes. It includes an NTT/iNTT engine for domain transformation, as well as modular adders, subtractors, and multipliers. The pointwise multiplication and coefficient-wise add/subtraction units enable the required mathematical operation inside ML-KEM and ML-DSA. A TRNG module provides fresh entropy to support secure sampling and randomness needs. A secure implementation of TRNG is very important in the security of the whole system. All PRNGs (Pseudo-random number generators) are implemented in the KECCAK unit, using an XOF (eXtendable Output Function), and for a true random number, a TRNG will be needed. Below the shared blocks, the architecture branches into two paths with scheme-specific components: the ML-DSA block contains logic for hint generation (UseHint and MakeHint), coefficient decomposition, which can be used for Power2Round, HighBits, and LowBits, and packing/unpacking. The ML-KEM block includes modules for compression/decompression and encoding/decoding during key encapsulation/decapsulation. An FSM (Finite State Machine) coordinates control flow across all modules and manages communication with the memory subsystem. The memory unit contains RAM for intermediate storage, ROM for storing constants like twiddle factors, and a dedicated program memory segment.

Overall, this architecture represents a generalized view of how the fundamental components of module-lattice-based cryptography can be organized in hardware. It highlights the importance of resource sharing, modularity, and the separation between scheme-agnostic and scheme-specific operations in designing scalable and efficient PQC hardware.

5.3 High-Performance Implementations of ML-KEM

The first pure hardware implementation of the CRYSTALS-Kyber algorithm was presented by Huang et al. [23], targeting the Round 2 specification. The design implements encapsulation and decapsulation operations, excluding key generation, and reports a 129× speedup compared to a software implementation.

Xing et al. [56] presented a hardware implementation of CRYSTALS-Kyber on the Xilinx Artix-7 FPGA, scheduling NTT operations via a predefined control order table and using unified butterfly units for NTT, inverse NTT, and pointwise multiplication. Memory usage was minimized through on-the-fly access and FIFO-based data transmission. Similarly, Bisheh-Niasar et al. [7] proposed a Kyber coprocessor that optimized the NTT core by merging computation layers and introducing a configurable butterfly unit for modular reduction.

Dang et al. [12] and Ni et al. [42] both proposed hardware architectures for CRYSTALS-Kyber that were evaluated on Artix-7 and Zynq-UltraScale+ platforms, demonstrating platform scalability. The design by Dang et al. follows a traditional RTL-based methodology focused on latency and throughput, while Ni et al. introduced a pipelined architecture with a shared Radix-2 Multipath Delay Commutator NTT unit. By eliminating BRAM and using FIFO-based buffering, Ni et al. design improved area-time efficiency, albeit with increased LUT and FF usage. Both implementations achieved consistent performance improvements across all three Kyber security levels, as summarized in Table 7.

Guo et al. introduced two Kyber hardware accelerators incorporating optimized NTT architectures [20, 21]. The first design [20] utilized a mixed radix-2/4 NTT combined with the "lazy-last-layer" technique to improve average memory bandwidth during NTT computation. The second design [21] explored a split-radix-based NTT and integrated the divide-by-2 operation into the modular adder/subtractor unit in INTT, reducing the pipeline latency. Both architectures demonstrated improvements in hardware efficiency over previous designs.

Ni et al. [41] proposed a resource-efficient hardware architecture for ML-KEM, structured into three computational layers to enable pipelined and parallel execution of cryptographic operations. Memory usage is optimized by eliminating BRAM and reducing FIFO size requirements. The architecture achieves notable reductions in computation time and area compared to existing designs, targeting all three NIST-recommended security levels.

Overall, FPGA-based implementations of ML-KEM reflect a wide spectrum of design strategies, targeting various trade-offs in latency, area, and throughput. As summarized in Table 7, results show continued improvements across all three NIST security levels. This section focuses on functional and architectural efficiency; side-channel countermeasure techniques are not addressed.

5.4 High-Performance Implementations of ML-DSA

Ricci et al. [47] presented the first manually designed hardware implementation of the CRYSTALS-Dilithium signature scheme, targeting the Round 2 version of the algorithm. Their architecture used parallelization to achieve high signing throughput but relied on substantial hardware resources. Area and frequency results were provided only for security level 2. As the Dilithium specification changed significantly from Round 2 to Round 3—particularly with the addition of lower security levels—

Table 7 Comparison of FPGA-based implementations for different security levels of ML-KEM

Sec. lvl	Design	Device	Freq. (MHz)	LUT	FF	DSP	BRAM	Keygen (μs)	Enc (μs)	Decap (μs)
I	[56]	Artix-7	167	7412	4644	2	3	23.4	30.5	41.3
	[7]	Artix-7	200	10502	9859	8	13	9.4	12.0	18.8
	[12]	Artix-7	220	9347	8186	4	6	9.7	14.8	20.3
	[12]	ZU+	450	9435	8605	4	6	4.8	7.2	9.9
	[42]	Artix-7	208	16147	13868	2	0	5.3	7.2	10.1
	[42]	ZU+	435	16004	13890	2	0	2.5	3.4	4.8
	[20]	Artix-7	200	7300	3200	4	5	12.1	15.0	21.2
	[21]	Artix-7	200	6910	3270	3	2	14.3	21.8	30.4
	[41]	Artix-7	270	10076	8227	4	0	6.7	9.9	13.7
III	[56]	Artix-7	167	7412	4644	2	3	39.2	47.6	62.3
	[7]	Artix-7	200	11783	10424	12	14	13.3	16.3	24.0
	[12]	Artix-7	220	10434	9473	6	8.5	12.3	17.7	22.9
	[12]	ZU+	450	9347	8186	4	6	6.0	8.6	11.2
	[42]	Artix-7	208	16834	13722	2	0	8.2	11.5	14.4
	[42]	ZU+	435	16797	13705	2	0	3.9	5.5	6.9
	[20]	Artix-7	200	7300	3200	4	5	21.2	25.2	34.9
	[21]	Artix-7	200	6910	3270	3	2	25.4	30.8	38.5
	[41]	Artix-7	270	10781	8299	4	0	12.3	15.5	20.0
V	[56]	Artix-7	167	7412	4644	2	3	58.2	67.9	86.2
	[7]	Artix-7	185	13347	11639	16	16	17.3	20.6	31.3
	[12]	Artix-7	220	11527	11767	8	10.5	16.3	21.8	27.1
	[12]	ZU+	450	9347	8186	4	6	8.0	10.6	13.2
	[42]	Artix-7	208	17827	13979	2	0	13.0	16.3	19.7
	[42]	ZU+	435	17812	14002	2	0	6.2	7.8	9.4
	[20]	Artix-7	200	7300	3200	4	5	34.1	40.9	51.1
	[21]	Artix-7	200	6910	3270	3	2	40.1	44.4	58.0
	[41]	Artix-7	263	11639	8893	4	0	19.3	22.6	28.7

the results from this work are not directly comparable to Round 3 implementations and are therefore excluded from Table 8.

One of the earliest Round 3 implementations came from Land et al. [26], who targeted the Xilinx Artix-7 platform. The design includes an NTT module optimized for minimizing DSP usage. Separate modules are used for NTT, INTT, and multiplication, though integration could potentially reduce resource overhead.

Table 8 includes results from Beckwith et al. [6], who presented a high-performance FPGA implementation of CRYSTALS-Dilithium targeting all three NIST security levels. The two-stage pipelining of the rejection sampling loop significantly improves average-case signing latency. The implementation achieves

Table 8 Comparison of FPGA-based implementations for different security levels of ML-DSA signature scheme

Sec. lvl	Design	Device	Freq. (MHz)	LUT	FF	DSP	BRAM	Keygen (μs)	Enc (μs)	Decap (μs)
II	[2]	ZU+	270	23277	9798	4	24	54.0	117/–	57.0
	[26]	Artix-7	163	27433	10681	45	15	115.0	119/470	121.0
	[54]	Zynq-7k	159	18558	7342	10	17	49.0	–/327	48.0
	[57]	Artix-7	96.9	29998	10366	10	11	43.0	86.3/326.1	45.6
	[6]	VUs+	256	53907	28435	16	29	19.0	43/117	26.0
	[53]	VUs+	300	49753	23477	16	27.5	11.1	36/97.4	13.8
III	[2]	ZU+	270	23277	9798	4	24	87.0	171/–	97.0
	[26]	Artix-7	145	30900	11372	45	21	229.0	186/852	222.0
	[54]	Zynq-7k	159	19614	8466	10	21	82.0	–/561	71.0
	[57]	Artix-7	96.9	29998	10366	10	11	60.4	117.6/510.8	63.8
	[6]	VUs+	256	53907	28435	16	29	32.0	63/193	39.0
	[53]	VUs+	300	49753	23477	16	27.5	17.8	49.5/154.2	19.5
V	[2]	ZU+	270	23277	9798	4	24	147.0	260/–	173.0
	[22]	ZU+	391	13975	6845	4	35	162.0	291/–	174.0
	[26]	Artix-7	140	44653	13814	45	31	364.0	261/1042	377.0
	[54]	Zynq-7k	159	20973	9677	10	28	127.0	–/589	100.0
	[57]	Artix-7	96.9	29998	10366	10	11	90.5	163/570.9	93.3
	[6]	VUs+	256	53907	28435	16	29	55.0	95/215	57.0
	[53]	VUs+	300	49753	23477	16	27.5	27.4	70.7/169.5	26.9

* Sign time varies due to rejection sampling; best/avg cases shown

one of the lowest reported latencies among Dilithium designs on an FPGA, with resource usage traded off for higher parallelism.

Aikataet al. [2] proposed KaLi, a unified hardware implementation of the CRYSTALS-Dilithium and CRYSTALS-Kyber algorithms. The main contribution is a shared polynomial arithmetic unit with a 24-bit datapath, designed to support both Dilithium's 23-bit and Kyber's 12-bit moduli within a single architecture. A unified Keccak-based SHAKE and sampler module was used to reduce redundant computation across both schemes. KaLi was deployed on a Zynq UltraScale+ (ZU+).

Wang et al. [54] implemented a software/hardware codesign 3 Dilithium on a Xilinx Zynq-7000 SoC platform. The main computation is handled in hardware, including an NTT architecture based on a radix-4 algorithm with conflict-free memory mapping and a dedicated modular multiplication unit. A multipurpose SHAKE-based hashing unit with a dedicated sampling mode is also included. The design uses three independent hardware cores for different security levels, while bit packing and unpacking are performed by the processor.

Table 8 includes results from Zhao [57] and Gupta [22], both of whom focus on compact hardware architectures for CRYSTALS-Dilithium. Zhao's design adopts a segmented pipelined processing strategy and performs on-the-fly matrix generation while omitting packing and unpacking stages, thereby reducing storage requirements at the cost of flexibility. Similarly, Gupta targets embedded applications with a lightweight accelerator that achieves the smallest reported resource usage for security level 5 on Zynq UltraScale+. While both designs prioritize area efficiency, Gupta's implementation is limited to a single security level and reports best-case signing latencies without average-case data.

Truong et al. [53] proposed a low-latency hardware implementation of ML-DSA. Their architecture emphasizes optimized submodules, including a fully pipelined NTT engine and independent SHAKE modules. Unlike several previous designs, their implementation includes packing/unpacking modules. They report the best latency among FPGA-based implementations. The design includes several specialized modules, which increase architectural complexity compared to simpler implementations.

Overall, FPGA-based implementations of ML-DSA exhibit a variety of design strategies tailored to different performance and resource constraints. A key challenge in high-performance ML-DSA implementations is the efficient handling of the rejection sampling loop, which can lead to variable signing times. Some architectures prioritize low latency through pipelined submodules and parallel execution, while others focus on minimizing area for use in embedded environments. Combined implementations support multiple schemes but may introduce additional complexity. As summarized in Table 8, these designs reflect a range of compromises and provide insight into practical deployments of ML-DSA in reconfigurable hardware. This section focuses solely on performance and resource utilization; implementations with side-channel countermeasures are beyond the scope of this discussion.

References

1. Agency, N.S. 2024. The commercial national security algorithm suite 2.0 and quantum computing FAQ. National Security Agency. https://media.defense.gov/2022/Sep/07/2003071836/-1/-1/0/CSI_CNSA_2.0_FAQ_.PDF.
2. Aikata, A., A.C. Mert., M. Imran, S. Pagliarini, and S.S. Roy. 2022. KaLi: A crystal for post-quantum security using kyber and dilithium. Cryptology ePrint Archive, Paper 2022/1086. https://doi.org/10.1109/TCSI.2022.3219555, https://eprint.iacr.org/2022/1086
3. Bai, S., L. Ducas, E. Kiltz, T. Lepoint, V. Lyubashevsky, P. Schwabe, G. Seiler, and D. Stehlé. 2021. Crystals-Dilithium. Algorithm specifications and supporting documentation. https://pq-crystals.org/dilithium/data/dilithium-specification-round3-20210208.pdf.
4. Banerjee, A., T. Reddy, D. Schoinianakis, and T. Hollebeek. 2024. Post-quantum cryptography for engineers. https://www.ietf.org/archive/id/draft-ietf-pquip-pqc-engineers-04.html.
5. Barker, E., L. Chen, D. Cooper, D. Moody, A. Regenscheid, M. Souppaya, W. Newhouse, R. Housley, S. Turner, W. Barker, and K. Scarfone. 2025. Cswp 39, considerations for achieving cryptographic agility: Strategies and practices. https://csrc.nist.gov/pubs/cswp/39/considerations-for-achieving-cryptographic-agility/2pd
6. Beckwith, L., D.T. Nguyen, and K. Gaj. 2021. High-performance hardware implementation of crystals-Dilithium. Cryptology ePrint Archive, Paper 2021/1451. https://eprint.iacr.org/2021/1451

7. Bisheh-Niasar, M., R. Azarderakhsh, and M. Mozaffari-Kermani. 2021. High-speed NTT-based polynomial multiplication accelerator for Crystals-Kyber post-quantum cryptography. Cryptology ePrint Archive, Paper 2021/563. https://eprint.iacr.org/2021/563
8. Bisheh-Niasar, M., Azarderakhsh, R., Mozaffari-Kermani, M.: High-speed NTT-based polynomial multiplication accelerator for post-quantum cryptography. In: 2021 IEEE 28th Symposium on Computer Arithmetic (ARITH). pp. 94–101 (2021). https://doi.org/10.1109/ARITH51176.2021.00028
9. Bos, J., L. Ducas, E. Kiltz, T. Lepoint, V. Lyubashevsky, J.M. Schanck, P. Schwabe, G. Seiler, and D. Stehlé. 2018. Crystals-Kyber: A CCA-secure module-lattice-based KEM. In *2018 IEEE European symposium on security and privacy (EuroS&P)*, 353–367. IEEE.
10. Cooley, J. W., and J. W. Tukey. 1965. An algorithm for the machine calculation of complex Fourier series. *Mathematics of Computation* 19 (90): 297–301.
11. Cooper, D., D. Apon, Q. Dang, M. Davidson, M. Dworkin, and C. Miller. 2020. Recommendation for stateful hash-based signature schemes. Technical Report NIST SP 800-208, National Institute of Standards and Technology. https://doi.org/10.6028/NIST.SP.800-208
12. Dang, V.B., K. Mohajerani, and K. Gaj. 2021. High-speed hardware architectures and FPGA benchmarking of CRYSTALS-Kyber, NTRU, and Saber. Cryptology ePrint Archive, Paper 2021/1508. https://eprint.iacr.org/2021/1508
13. D'Anvers, J., Q. Guo, T. Johansson, A. Nilsson, F. Vercauteren, and I. Verbauwhede. 2019. Decryption failure attacks on IND-CCA secure lattice-based schemes. In *Public-key cryptography—PKC 2019—22nd IACR international conference on practice and theory of public-key cryptography*, eds. D. Lin, and K. Sako, Beijing, China, Apr 14–17, Proceedings, Part II. Lecture Notes in Computer Science, vol. 11443, 565–598. Springer. https://doi.org/10.1007/978-3-030-17259-6_19
14. Diffie, W., and M. Hellman. 1976. New directions in cryptography. *IEEE Transactions on Information Theory* 22 (6): 644–654. https://doi.org/10.1109/TIT.1976.1055638.
15. Fouque, P., J. Hoffstein, P. Kirchner, V. Lyubashevsky, T. Pornin, T. Prest, T. Ricosset, G. Seiler, W. Whyte, and Z. Zhang. 2020. FALCON: Fast-Fourier lattice-based compact signatures over NTRU, specification v1.2. Technical Report v1.2, École normale supérieure, Paris.
16. Geng, Y., X. Hu, M. Li, and Z. Wang. 2023. Rethinking parallel memory access pattern in number theoretic transform design. *IEEE Transactions on Circuits and Systems II: Express Briefs* 70 (5): 1689–1693. https://doi.org/10.1109/TCSII.2023.3260811.
17. Gentleman, W.M., and G. Sande. 1966. Fast Fourier transforms: For fun and profit. In *Proceedings of the fall joint computer conference*, 7–10 Nov. AFIPS Conference Proceedings, vol. 29, 563–578. https://doi.org/10.1145/1464291.1464352.
18. Greuet, A., S. Montoya, and C. Vermeersch. 2022. Quotient approximation modular reduction. In *29th IEEE symposium on computer arithmetic, ARITH 2022*, Lyon, France, Sept 12–14, 103–110. IEEE. https://doi.org/10.1109/ARITH54963.2022.00028
19. Grover, L.K. 1996. A fast quantum mechanical algorithm for database search. https://arxiv.org/abs/quant-ph/9605043
20. Guo, W., and S. Li. 2023. Highly-efficient hardware architecture for crystals-Kyber with a novel conflict-free memory access pattern. *IEEE Transactions on Circuits and Systems I: Regular Papers* 70 (11): 4505–4515. https://doi.org/10.1109/TCSI.2023.3306347.
21. Guo, W., and S. Li. 2024. Split-radix based compact hardware architecture for crystals-Kyber. *IEEE Transactions on Computers* 73 (1): 97–108. https://doi.org/10.1109/TC.2023.3320040.
22. Gupta, N., A. Jati, A. Chattopadhyay, and G. Jha. 2022. Lightweight hardware accelerator for post-quantum digital signature crystals-Dilithium. Cryptology ePrint Archive, Paper 2022/496. https://eprint.iacr.org/2022/496
23. Huang, Y., M. Huang, Z. Lei, and J. Wu. 2020. A pure hardware implementation of CRYSTALS-KYBER PQC algorithm through resource reuse. *IEICE Electronics Express* 17: 20200234. https://doi.org/10.1587/elex.17.20200234
24. Jaques, S. 2024. Quantum computing landscape. https://sam-jaques.appspot.com/quantum_landscape. Accessed 8 June 2025

25. Koblitz, N. 1987. Elliptic curve cryptosystems. *Mathematics of Computation* 48 (177): 203–209. http://www.jstor.org/stable/2007884
26. Land, G., P. Sasdrich, and T. Güneysu. 2021. A hard crystal—implementing dilithium on reconfigurable hardware. Cryptology ePrint Archive, Paper 2021/355. https://eprint.iacr.org/2021/355
27. Langlois, A., and D. Stehlé. 2012. Worst-case to average-case reductions for module lattices. Cryptology ePrint Archive, Paper 2012/090. https://eprint.iacr.org/2012/090. Accessed 4 Dec 2024
28. Liang, Z., and Y. Zhao. 2022. Number theoretic transform and its applications in lattice-based cryptosystems: A survey. arXiv preprint arXiv:2211.13546
29. Lyubashevsky, V. 2012. Lattice signatures without trapdoors. In *Advances in Cryptology–EUROCRYPT 2012—31st annual international conference on the theory and applications of cryptographic techniques*, eds. D. Pointcheval and T. Johansson, Cambridge, UK, Apr 15–19. Proceedings. Lecture Notes in Computer Science, vol. 7237, 738–755. https://doi.org/10.1007/978-3-642-29011-4_43.
30. Lyubashevsky, V., C. Peikert, and O. Regev. 2013. On ideal lattices and learning with errors over rings. *Journal of ACM* 60 (6): 43:1–43:35. https://doi.org/10.1145/2535925
31. Melchor, C.A., N. Aragon, S. Bettaieb, L. Bidoux, O. Blazy, J. Bos, J.C. Deneuville, A. Dion, P. Gaborit, J. Lacan, E. Persichetti, J.M. Robert, P. Véron, and G. Zémor. 2025. Hamming quasi-cyclic (hqc): Fourth round specification. Technical report, NIST Post-Quantum Cryptography Standardization Project. https://pqc-hqc.org/doc/hqc-specification_2025-02-19.pdf, fourth Round Version, updated on 19 Feb 2025
32. Mert, A.C., E. Öztürk, and E. Savas. 2019. Design and implementation of a fast and scalable ntt-based polynomial multiplier architecture. IACR Cryptol. ePrint Arch, p 109. https://eprint.iacr.org/2019/109
33. Montgomery, P.L. 1985. Modular multiplication without trial division. *Mathematics of Computation* 44: 519–521. https://api.semanticscholar.org/CorpusID:119574413
34. Moody, D. 2016. Post-quantum cryptography standardization: Announcement and outline of NIST's call for submissions. In *International conference on post-quantum cryptography (PQCrypto)*. https://csrc.nist.gov/presentations/2016/announcement-and-outline-of-nist-s-call-for-submis
35. Moody, D., G. Alagic, D. Apon, et al. 2020. Status report on the second round of the NIST post-quantum cryptography standardization process. NIST Interagency/Internal Report (NISTIR) 8309, National Institute of Standards and Technology, Gaithersburg, MD. https://doi.org/10.6028/NIST.IR.8309.
36. National Institute of Standards and Technology. 2023. Advanced encryption standard (AES). Technical report, National Institute of Standards and Technology (U.S.). https://nvlpubs.nist.gov/nistpubs/FIPS/NIST.FIPS.197.pdf.
37. National Institute of Standards and Technology. 2024. FIPS 203: Module-lattice-based key-encapsulation mechanism standard. Technical report, U.S. Department of Commerce. https://doi.org/10.6028/NIST.FIPS.203. Accessed 8 June 2025.
38. National Institute of Standards and Technology. 2024. FIPS 204: Module-lattice-based digital signature standard. Technical report, U.S. Department of Commerce. https://doi.org/10.6028/NIST.FIPS.204. Accessed 8 June 2025
39. National Institute of Standards and Technology. 2024. FIPS 205: Stateless hash-based digital signature standard. Technical report, U.S. Department of Commerce. https://doi.org/10.6028/NIST.FIPS.205, Accessed 8 June 2025.
40. Nejatollahi, H., N. Dutt, S. Ray, F. Regazzoni, I. Banerjee, and R. Cammarota. 2017. Software and hardware implementation of lattice-based cryptography schemes. Technical report CECS Technical Report 17-04, Center for Embedded and Cyber-Physical Systems, University of California, Irvine. https://www.cecs.uci.edu/files/2018/06/2017-tr-1.pdf.
41. Ni, Z., A. Khalid, W. Liu, and M. O'Neill. 2025. A highly hardware efficient ML-KEM accelerator with optimised architectural layers. *ACM Transactions on Embedded Computing Systems* 24 (2). https://doi.org/10.1145/3708469

42. Ni, Z., A. Khalid, D. e Shahwar Kundi, M. O'Neill, and W. Liu. 2022. HPKA: A high-performance CRYSTALS-Kyber accelerator exploring efficient pipelining. Cryptology ePrint Archive, Paper 2022/1093. https://doi.org/10.1109/TC.2023.3296899, https://eprint.iacr.org/2022/1093.
43. Özerk, Ö., C. Elgezen, A. C. Mert, E. Öztürk, and E. Savaş. 2022. Efficient number theoretic transform implementation on GPU for homomorphic encryption. *The Journal of Supercomputing* 78 (2): 2840–2872. https://doi.org/10.1007/s11227-021-03980-5.
44. Pease, M. C. 1968. An adaptation of the fast Fourier transform for parallel processing. *Journal of ACM* 15 (2): 252–264. https://doi.org/10.1145/321450.321457.
45. Plantard, T. 2021. Efficient word size modular arithmetic. *IEEE Transactions on Emerging Topics in Computing* 9 (3): 1506–1518.
46. Regev, O. 2009. On lattices, learning with errors, random linear codes, and cryptography. *Journal of the ACM (JACM)* 56 (6): 1–40.
47. Ricci, S., L. Malina, P. Jedlicka, D. Smekal, J. Hajny, P. Cibik, and P. Dobias. 2021. Implementing CRYSTALS-Dilithium signature scheme on FPGAs. Cryptology ePrint Archive, Paper 2021/108. https://eprint.iacr.org/2021/108.
48. Rivest, R. L., A. Shamir, and L. Adleman. 1978. A method for obtaining digital signatures and public-key cryptosystems. *Communications of the ACM* 21 (2): 120–126.
49. Satriawan, A., R. Mareta, and H. Lee. 2024. A complete beginner guide to the number theoretic transform (NTT). Cryptology ePrint Archive, Paper 2024/585. https://doi.org/10.1109/ACCESS.2023.3294446, https://eprint.iacr.org/2024/585.
50. Shor, P.W. 1994. Algorithms for quantum computation: Discrete logarithms and factoring. In *Proceedings of the 35th annual symposium on foundations of computer science (FOCS)*, 124–134. https://doi.org/10.1109/SFCS.1994.365700.
51. Taghavi, B., R. Azarderakhsh, and M. Mozaffari Kermani. 2025. LightNTT: A high-efficiency NTT/iNTT core for ML-DSA featuring a constant-geometry pipelined design. In: *LightSEC*. Springer.
52. Taghavi, B., R. Azarderakhsh, and M. Mozaffari Kermani. 2025. ParallelNTT: Maximizing performance of forward and inverse NTT on FPGA for ML-DSA and ML-KEM. In *Proceedings of the great lakes symposium on VLSI 2025*, 372–378. GLSVLSI '25. https://doi.org/10.1145/3716368.3735214.
53. Truong, Q. D., P. N. Duong, and H. Lee. 2024. Efficient low-latency hardware architecture for module-lattice-based digital signature standard. *IEEE Access* 12: 32395–32407.
54. Wang, T., C. Zhang, P. Cao, and D. Gu. 2022. Efficient implementation of dilithium signature scheme on fpga soc platform. *IEEE Transactions on Very Large Scale Integration (VLSI) Systems* 30 (9): 1158–1171.
55. Wintersperger, K., F. Dommert, T. Ehmer, A. Hoursanov, J. Klepsch, W. Mauerer, G. Reuber, T. Strohm, M. Yin, and S. Luber. 2023. Neutral atom quantum computing hardware: Performance and end-user perspective. *EPJ Quantum Technology* 10 (1). https://doi.org/10.1140/epjqt/s40507-023-00190-1.
56. Xing, Y., and S. Li. 2021. A compact hardware implementation of CCA-secure key exchange mechanism CRYSTALS-KYBER on fpga. *IACR Transactions on Cryptographic Hardware and Embedded Systems* (2): 328–356. https://doi.org/10.46586/tches.v2021.i2.328-356, https://tches.iacr.org/index.php/TCHES/article/view/8797.
57. Zhao, C., N. Zhang, H. Wang, B. Yang, W. Zhu, Z. Li, M. Zhu, S. Yin, S. Wei, and L. Liu. 2022. A compact and high-performance hardware architecture for CRYSTALS-Dilithium. *IACR Transactions on Cryptographic Hardware and Embedded Systems* (1): 270–295. https://doi.org/10.46586/tches.v2022.i1.270-295, https://tches.iacr.org/index.php/TCHES/article/view/9297.

Conclusion

The Future of Fault-Tolerant Quantum Computing

Prabhat Mishra

Abstract Quantum computing has the potential to significantly outperform classical computing for solving specific classes of simulation and optimization problems. In order to utilize the potential of quantum computing, scientists and engineers need to solve various design automation challenges starting from specification to implementation and analysis. This book presented a wide variety of tools and techniques for design automation of quantum computing systems. This chapter concludes the book with a summary of ideas presented in the previous chapters, and outlines the roadmap of future design automation challenges and opportunities.

1 Summary

Quantum computers offer the ability to solve specific classes of simulation and optimization problems significantly faster than the classical computers. In order to develop a robust quantum computing ecosystem, we need to address challenges in different stages of the design cycle, including development of quantum algorithms to solve specific optimization problems, compilation of quantum algorithms to quantum circuits, and mapping of quantum circuits to physical quantum computers. Specifically, we need to develop automated tools and techniques for various design automation phases, such as quantum compilation, quantum error correction, quantum state preparation, quantum measurement, quantum security, and verification of quantum systems. This book provided a comprehensive overview of quantum computing and design automation challenges from the perspective of computer scientists and engineers. The topics covered in this book can be broadly divided into the following five categories.

P. Mishra (✉)
University of Florida, Gainesville, FL, USA
e-mail: prabhat@ufl.edu

P. Mishra (ed.), *Design Automation for Quantum Computing*,
https://doi.org/10.1007/978-3-032-09303-5_15

1.1 Introduction

The first chapter introduced the readers to the fundamentals of quantum computing as well as associated design automation challenges. Specifically, Chapter "Design Automation Challenges in Quantum Computing" covered the following topics.

- It introduced the basic building blocks for designing quantum computing systems, such as superposition and entanglement of qubits, quantum gates and circuits, quantum algorithms [24], and quantum architectures.
- It provided an overview of various design automation challenges in quantum state preparation [29], quantum compilation [25, 27], quantum security [32], quantum error correction [26], quantum simulation, quantum verification [30, 34], quantum benchmarking [28], and quantum measurement [22, 23].

1.2 Quantum Algorithms and Architectures

The next three chapters focused on describing diverse quantum computing architectures as well as quantum algorithms.

- The second chapter explored different quantum computing architectures based on various quantum technologies. Specifically, it presented the architectures of single-core quantum processing units from various providers, including IBM, Google, and Rigetti. It also described both computation and communication in modular multi-core configurations.
- The third chapter described an efficient state preparation framework for variational quantum algorithms [21, 26, 29, 31]. Specifically, it provided an overview of variational quantum algorithms and state preparation methods. It also demonstrated that measurement-induced steering can be utilized for efficient state preparation for variational quantum algorithms.
- The fourth chapter provided an overview of quantum AI algorithms. Specifically, it explored how the AI algorithms deal with different data sources (classical versus quantum) and machine learning models (classical versus quantum), including classical model processing classical data, classical model processing quantum data, quantum model processing classical data, and quantum model processing quantum data.

1.3 Quantum Compilation and Error Correction

The next two chapters deal with quantum compilation and error correction methods.

- The fifth chapter described quantum compilation for different quantum computing platforms. Specifically, it presented an overview of hardware-aware quantum

circuit compilation tailored to each of the three major hardware platforms: superconducting qubits, neutral atoms, and trapped ions. It focused on key compilation stages (mapping, routing, and scheduling) emphasizing how platform-specific features influence algorithm design.
- The sixth chapter provided an overview of design automation for quantum error correction. Specifically, it described theoretical aspects as well as practical considerations in implementing near-term quantum error correction methods and their integration into scalable hardware platforms.

1.4 Quantum Verification and Debug

The next four chapters described different quantum verification and debug methods.

- The seventh chapter explored various quantum verification paradigms to ensure fidelity and functionality in quantum systems. Specifically, it presented an overview of quantum verification using formal equivalence checking, simulation-based validation, quantum fuzzing, randomized benchmarking, and quantum characterization.
- The eighth chapter presented a framework for quantum circuit simulation. Specifically, it described how to ensure the correctness of quantum circuits by simulating them on top of classical computers. It also outlined the utilization of symbolic simulation for robust verification of quantum circuits.
- The ninth chapter provided an overview of various quantum testing and validation techniques. Specifically, it presented a wide range of testing techniques, including mutation testing, fuzz testing, coverage testing, combinatorial testing, and search-based testing.
- The tenth chapter provided a comprehensive survey of automated debugging techniques developed to support quantum program correctness, with a particular focus on assertion-based debugging approaches.

1.5 Quantum Security and Post-quantum Cryptography

The next four chapters presented various methods for addressing the security implications in quantum computing systems.

- The eleventh chapter described security attacks abusing pulse-level quantum circuits. Specifically, it presented various attacks by manipulating the custom gate specification, including qubit plunder, qubit block, qubit reorder, timing mismatch, frequency mismatch, phase mismatch, and waveform mismatch.
- The twelfth chapter presented the design of quantum computer antivirus. Specifically, it developed a quantum computer antivirus to protect the quantum computer hardware from quantum computer viruses. The antivirus can analyze quantum

circuits to detect possibly malicious ones before they are executed on quantum computer hardware.
- The thirteenth chapter presented the design of a trusted execution environment (TEE) for quantum computers to protect from cloud providers who may try to steal or learn what quantum program the user is executing. The quantum circuits are obfuscated with decoy control pulses during circuit transpilation by the user. The decoy pulses are removed by the TEE hardware before they reach the qubits.
- The fourteenth chapter explored different post-quantum cryptographic algorithms. Specifically, it examined the mathematical foundations of lattice-based cryptography, their role in the standardization process, and key considerations related to their performance and implementation.

2 Future Directions

In spite of extensive research efforts in developing design automation over the years, there are many remaining challenges to design robust quantum computing systems. In this section, we discuss a few promising avenues to extend the proposed approaches to deal with future quantum computing systems.

2.1 Quantum Compilation for Emerging Quantum Technologies

While superconducting qubits, trapped ions, and neutral atoms currently dominate experimental efforts, a range of alternative quantum hardware platforms is rapidly emerging. Systems such as photonic and measurement-based quantum computing, quantum dots with spin shuttling, and silicon-based qubits introduce fundamentally different operational principles and constraints. Extending the existing compilation techniques offers a promising avenue for developing compilation tools tailored to these next-generation platforms. Given that noise is a major concern in today's noisy intermediate scale quantum computers, there is a critical need to develop noise-resilient compilation strategies for reliable quantum computing.

2.2 Quantum Error Correction Challenges and Opportunities

Quantum error correction is vital today to protect the information stored in qubits from errors and decoherence caused by noise in today's noisy intermediate scale quantum computers. There are major challenges in error correction due to qubit reliability concerns as well as scalability considerations. For example, to construct a fault-tolerant (logical) qubit, we need multiple physical qubits and associated control electronics, which can introduce unacceptable overhead when solving real-world applications. Future quantum error correction (QEC) should focus on development of efficient error correcting code as well as improving existing techniques to reduce the runtime for current optimization ideas in QEC implementation and modifying the near-term quantum computing architecture to accommodate multiple qubit technologies. There are promising avenues to enable fault-tolerant quantum computing, including modifications in the modern QEC codes for accounting for defects introduced due to radiation and calibration errors in quantum computers and accelerating the decoding algorithms using Application-Specific Integrated Circuits (ASICs) and Field-Programmable Gate Arrays (FPGAs).

2.3 Quantum Verification and Debug Complexity

This book described a multifaceted verification stack for quantum computing, spanning algebraic equivalence checking, statistical benchmarking, tomographic reconstruction, inline assertions, and hardware-in-the-loop fuzzing. While assertion-based validation is widely used in classical systems [33], it is still in its infancy for validation and debugging of quantum systems [34]. For example, efficient assertion strategies and slicing techniques can significantly reduce the number of program executions required during debugging, thereby minimizing resource overhead and improving its applicability on emerging quantum systems. Similarly, there is a need to develop standardized fault models, benchmark suites, and efficient quantum testing framework for reliable quantum computing. Similar to verification of classical computing systems [6, 8, 9, 12, 13, 16, 35], there is a need for developing hybrid approaches for verifying quantum systems using an effective combination of simulation-based validation and formal methods.

2.4 Quantum Computation and Communication Security

This book explored various security perspectives related to quantum computing. It is necessary to develop automated vulnerability analysis and mitigation techniques for future quantum computing systems. For example, similar to security solutions for classical computing systems [5, 7, 10, 11, 14, 17–20], there is a need for developing hybrid approaches for verifying quantum security using an effective combination of simulation-based validation, formal methods, and side-channel analysis. Like classical network-on-chip security [2–4, 15], there is a need for investigating on-chip communication security vulnerabilities and suitable countermeasures for modular multicore quantum architectures.

2.5 Fault-Tolerant Quantum Computers

IBM has provided a roadmap to design fault-tolerant quantum computers by 2029 [1]. Specifically, IBM plans to develop a modular, error-corrected quantum-centric supercomputer with 200 logical qubits capable of running quantum circuits with 100 million gates. This is expected to enable the development of more sophisticated circuit libraries. Development of fault-tolerant quantum computers by IBM and other provides will create opportunity for creating design automation ecosystem, starting from development of high-level specification languages to efficient error correction and compilation methods. For example, the designers need to develop fast and automated tools and techniques to compile 100 million-gate circuits to these fault-tolerant quantum computers. Similarly, there would be design automation challenges to scale electronics and infrastructure while minimizing design cost and energy requirements.

This book considered design automation challenges for today's noisy intermediate scale quantum computers. In the future, we have to deal with fault-tolerant quantum computers with diverse quantum technologies. There will be a need for developing automated tools and techniques for design automation of emerging quantum computing systems.

References

1. IBM quantum roadmap. 2025. https://www.ibm.com/quantum.
2. Charles, Subodha, Yangdi Lyu, and Prabhat Mishra. 2020. Real-time detection and localization of distributed DoS attacks in NoC-based SoCs. *IEEE Transactions on Computer-Aided Design of Integrated Circuits and Systems* 39 (12): 4510–4523.
3. Charles, Subodha, and Prabhat Mishra. 2020. Reconfigurable network-on-chip security architecture. *ACM Transactions on Design Automation of Electronic Systems (TODAES)* 25 (6): 1–25.
4. Charles, Subodha, and Prabhat Mishra. 2021. A survey of network-on-chip security attacks and countermeasures. *ACM Computing Surveys (CSUR)* 54 (5): 1–36.
5. Farahmandi, Farimah, Yuanwen Huang, and Prabhat Mishra. 2020. *System-on-chip security: Validation and verification*. Springer.
6. Farahmandi, Farimah, and Prabhat Mishra. 2018. Automated test generation for debugging multiple bugs in arithmetic circuits. *IEEE Transactions on Computers* 68 (2): 182–197.
7. Jayasena, Aruna, Richard Bachmann, and Prabhat Mishra. 2025. Ciseleaks: Information leakage assessment of cryptographic instruction set extension prototypes. *IEEE Transactions on Information Forensics and Security (TIFS)* 20: 1551–1565.
8. Jayasena, Aruna, and Prabhat Mishra. 2024. Directed test generation for hardware validation: A survey. *ACM Computing Surveys* 56 (5): 1–36.
9. Jayasena, Aruna, and Prabhat Mishra. 2024. Hive: Scalable hardware-firmware co-verification using scenario-based decomposition and automated hint extraction. *IEEE Transactions on Computer-Aided Design of Integrated Circuits and Systems* 43 (10): 3278–3291.
10. Jayasena, Aruna and Prabhat Mishra. 2025. Firmwall: Directed symbolic execution of firmware binaries for defending against unauthorized system calls. *IEEE Transactions on Information Forensics and Security (TIFS)*.
11. Jayasena, Aruna, Sai Suprabhanu Nallapaneni, and Prabhat Mishra. 2025. Fuss: Coverage-directed hardware fuzzing with selective symbolic execution. *ACM Transactions on Embedded Computing Systems (TECS)*.

12. Lyu, Yangdi, and Prabhat Mishra. 2020. Scalable concolic testing of RTL models. *IEEE Transactions on Computers* 70 (7): 979–991.
13. Lyu, Yangdi, Xiaoke Qin, Mingsong Chen, and Prabhat Mishra. 2018. Directed test generation for validation of cache coherence protocols. *IEEE Transactions on Computer-Aided Design of Integrated Circuits and Systems* 38 (1): 163–176.
14. Mishra, Prabhat, Swarup Bhunia, and Mark Tehranipoor. 2017. *Hardware IP security and trust*. Springer.
15. Mishra, Prabhat and Subodha Charles. 2021. *Network-on-chip security and privacy*. Springer.
16. Mishra, Prabhat and Farimah Farahmandi. 2019. *Post-silicon validation and debug*. Springer.
17. Pan, Zhixin, and Prabhat Mishra. 2022. A survey on hardware vulnerability analysis using machine learning. *IEEE Access* 10: 49508–49527.
18. Pan, Zhixin and Prabhat Mishra. 2023. *Explainable AI for Cybersecurity*. Springer.
19. Pan, Zhixin, Jennifer Sheldon, and Prabhat Mishra. 2022. Hardware-assisted malware detection and localization using explainable machine learning. *IEEE Transactions on Computers* 71 (12): 3308–3321.
20. Sanjaya, Sahan, Aruna Jayasena, and Prabhat Mishra. 2025. Information leakage through physical layer supply voltage coupling vulnerability. *IEEE Transactions on Very Large Scale Integration (VLSI) Systems* 33 (6): 1715–1728.
21. Sanjaya, Sahan, Daniel Volya, and Prabhat Mishra. 2024. Variational quantum algorithms via measurement-induced passive steering. In *2024 IEEE international conference on quantum computing and engineering (QCE)*, vol. 1, 481–487. IEEE.
22. Utt, Zachery, Daniel Volya, and Prabhat Mishra. 2023. Quantum measurement discrimination using cumulative distribution functions. In *Design, automation & test in Europe conference & exhibition (DATE)*, 1–6. IEEE.
23. Utt, Zachery, Daniel Volya, and Prabhat Mishra. 2024. Quantum measurement classification using statistical learning. *ACM Transactions on Quantum Computing (TQC)* 5 (2): 1–16.
24. Volya, Daniel and Prabhat Mishra. 2021. Quantum spectral clustering of mixed graphs. In *2021 58th ACM/IEEE design automation conference (DAC)*, 463–468. IEEE.
25. Volya, Daniel and Prabhat Mishra. 2023. Quantum data compression for efficient generation of control pulses. In *Asia and south pacific design automation conference (ASPDAC)*, 216–221.
26. Daniel Volya and Prabhat Mishra. Quantum steering of surface error correcting codes. In *2023 IEEE International Conference on Quantum Computing and Engineering (QCE)*, volume 1, pages 1394–1400. IEEE, 2023.
27. Volya, Daniel and Prabhat Mishra. 2023. Qudcom: Towards quantum compilation for qudit systems. arXiv preprint arXiv:2311.07015.
28. Volya, Daniel and Prabhat Mishra. 2024. Quantum benchmarking via random dynamical quantum maps. In *2024 IEEE international conference on quantum computing and engineering (QCE)*, vol. 1, 1277–1283. IEEE.
29. Volya, Daniel, and Prabhat Mishra. 2024. State preparation on quantum computers via quantum steering. *IEEE Transactions on Quantum Engineering (TQE)* 5: 1–14.
30. Volya, Daniel, Andrey Nikitin, and Prabhat Mishra. 2024. Fast quantum characterization via riemannian optimization. In *IEEE international conference on quantum computing and engineering (QCE)*, vol. 1, 1270–1276. IEEE.
31. Volya, Daniel, Zhixin Pan, and Prabhat Mishra. 2023. Feedback-based steering for quantum state preparation. In *IEEE international conference on quantum computing and engineering (QCE)*, vol. 1, 1308–1318. IEEE.
32. Volya, Daniel, Tao Zhang, Nashmin Alam, Mark Tehranipoor, and Prabhat Mishra. 2023. Towards secure classical-quantum systems. In *2023 IEEE international symposium on hardware oriented security and trust (HOST)*, 283–292. IEEE.
33. Witharana, Hasini, Yangdi Lyu, Subodha Charles, and Prabhat Mishra. 2022. A survey on assertion-based hardware verification. *ACM Computing Surveys (CSUR)* 54 (11s): 1–33.
34. Witharana, Hasini, Daniel Volya, and Prabhat Mishra. 2024. Qcassert: Quantum device testing with concurrent assertions. In *Asia and south pacific design automation conference (ASP-DAC)*, 491–496. IEEE.

35. Witharana, Hasini, Hansika Weerasena, and Prabhat Mishra. 2024. Formal verification of virtualization-based trusted execution environments. *IEEE Transactions on Computer-Aided Design of Integrated Circuits and Systems* 43 (11): 4262–4273.

Index

P. Mishra (ed.), *Design Automation for Quantum Computing*,
https://doi.org/10.1007/978-3-032-09303-5

D

E

F

G

H

I

K

MIX
Papier aus verantwortungsvollen Quellen
Paper from responsible sources
FSC® C105338

If you have any concerns about our products,
you can contact us on
ProductSafety@springernature.com

In case Publisher is established outside the EU,
the EU authorized representative is:
Springer Nature Customer Service Center GmbH
Europaplatz 3, 69115 Heidelberg, Germany

Printed by Libri Plureos GmbH
in Hamburg, Germany